Global Perspectives and Practices for Reform-Based Mathematics Teaching

Ozgul Kartal
University of Wisconsin, Whitewater, USA

Gorjana Popovic
Illinois Institute of Technology, USA

Susie Morrissey
Mercer University, USA

IGI Global
PUBLISHER of TIMELY KNOWLEDGE

A volume in the Advances in Educational Technologies and Instructional Design (AETID) Book Series

Published in the United States of America by
IGI Global
Information Science Reference (an imprint of IGI Global)
701 E. Chocolate Avenue
Hershey PA, USA 17033
Tel: 717-533-8845
Fax: 717-533-8661
E-mail: cust@igi-global.com
Web site: http://www.igi-global.com

Library of Congress Cataloging-in-Publication Data

Names: Kartal, Ozgul, 1979- editor. | Popovic, Gorjana, 1967- editor. | Morrissey, Susie (Glenda Susan), 1962- editor.
Title: Global perspectives and practices for reform-based mathematics teaching / Ozgul Kartal, Gorjana Popovic, Susie Morrissey, editor.
Description: Hershey PA : Information Science Reference, [2022] | Includes bibliographical references and index. | Summary: "This book explores international perspectives on diverse reform-based practices in teaching and learning of mathematics, describes challenges and issues for teachers and teacher educators, promotes reflection and academic discussion at various levels and in various educational systems, and raises questions for the field of mathematics education"-- Provided by publisher.
Identifiers: LCCN 2021052946 (print) | LCCN 2021052947 (ebook) | ISBN 9781799894223 (hardcover) | ISBN 9781799894230 (paperback) | ISBN 9781799894247 (ebook)
Subjects: LCSH: Mathematics--Study and teaching. | Teaching--Methodology. | Student-centered learning.
Classification: LCC QA11.2 .G56 2022 (print) | LCC QA11.2 (ebook) | DDC 510.71--dc23/eng/20211208
LC record available at https://lccn.loc.gov/2021052946
LC ebook record available at https://lccn.loc.gov/2021052947

This book is published in the IGI Global book series Advances in Educational Technologies and Instructional Design (AE-TID) (ISSN: 2326-8905; eISSN: 2326-8913)

British Cataloguing in Publication Data
A Cataloguing in Publication record for this book is available from the British Library.

Advances in Educational Technologies and Instructional Design (AETID) Book Series

Lawrence A. Tomei
Robert Morris University, USA

ISSN:2326-8905
EISSN:2326-8913

Mission

Education has undergone, and continues to undergo, immense changes in the way it is enacted and distributed to both child and adult learners. In modern education, the traditional classroom learning experience has evolved to include technological resources and to provide online classroom opportunities to students of all ages regardless of their geographical locations. From distance education, Massive-Open-Online-Courses (MOOCs), and electronic tablets in the classroom, technology is now an integral part of learning and is also affecting the way educators communicate information to students.

The **Advances in Educational Technologies & Instructional Design (AETID) Book Series** explores new research and theories for facilitating learning and improving educational performance utilizing technological processes and resources. The series examines technologies that can be integrated into K-12 classrooms to improve skills and learning abilities in all subjects including STEM education and language learning. Additionally, it studies the emergence of fully online classrooms for young and adult learners alike, and the communication and accountability challenges that can arise. Trending topics that are covered include adaptive learning, game-based learning, virtual school environments, and social media effects. School administrators, educators, academicians, researchers, and students will find this series to be an excellent resource for the effective design and implementation of learning technologies in their classes.

Coverage

- Adaptive Learning
- Instructional Design
- Collaboration Tools
- Bring-Your-Own-Device
- Digital Divide in Education
- Web 2.0 and Education
- Hybrid Learning
- Higher Education Technologies
- Game-Based Learning
- Classroom Response Systems

IGI Global is currently accepting manuscripts for publication within this series. To submit a proposal for a volume in this series, please contact our Acquisition Editors at Acquisitions@igi-global.com or visit: http://www.igi-global.com/publish/.

The Advances in Educational Technologies and Instructional Design (AETID) Book Series (ISSN 2326-8905) is published by IGI Global, 701 E. Chocolate Avenue, Hershey, PA 17033-1240, USA, www.igi-global.com. This series is composed of titles available for purchase individually; each title is edited to be contextually exclusive from any other title within the series. For pricing and ordering information please visit http://www.igi-global.com/book-series/advances-educational-technologies-instructional-design/73678. Postmaster: Send all address changes to above address.

Titles in this Series

For a list of additional titles in this series, please visit: http://www.igi-global.com/book-series/advances-educational-technologies-instructional-design/73678

Technology-Supported Interventions for Students With Special Needs in the 21st Century
Xiongyi Liu (Cleveland State University, USA) and Patrick Wachira (Cleveland State University, USA)
Information Science Reference • © 2022 • 289pp • H/C (ISBN: 9781799888741) • US $195.00

Cases on Practical Applications for Remote, Hybrid, and Hyflex Teaching
Valerie Harlow Shinas (Lesley University, USA) Chu N. Ly (Framingham State University, USA) and Sule Yilmaz Ozden (Sakarya Üniversitesi, Turkey)
Information Science Reference • © 2022 • 355pp • H/C (ISBN: 9781799891680) • US $195.00

Affordances of Film for Literacy Instruction
Jason D. DeHart (Appalachian State University, USA)
Information Science Reference • © 2022 • 277pp • H/C (ISBN: 9781799891369) • US $195.00

Applications of Machine Learning and Artificial Intelligence in Education
Seda Khadimally (North American University, USA)
Information Science Reference • © 2022 • 271pp • H/C (ISBN: 9781799877769) • US $195.00

Handbook of Research on Promoting Global Citizenship Education
Jared Keengwe (University of North Dakota, USA)
Information Science Reference • © 2022 • 371pp • H/C (ISBN: 9781799895428) • US $265.00

Assistive Technologies for Differently Abled Students
Sangeeta Dhamdhere-Rao (Modern College of Arts, Science and Commerce, India) and Frederic Andres (National Insitute of Informatics, Japan)
Information Science Reference • © 2022 • 305pp • H/C (ISBN: 9781799847366) • US $195.00

Impact of Digital Transformation in Teacher Training Models
Ana Afonso (LE@D, Universidade Aberta, Portugal & CISUC, University of Coimbra, Portugal) Lina Morgado (Universidade Aberta, Portugal) and Licínio Roque (University of Coimbra, Portugal)
Information Science Reference • © 2022 • 313pp • H/C (ISBN: 9781799895381) • US $195.00

Enhancing Teaching and Learning With Socratic Educational Strategies Emerging Research and Opportunities
Frank G. Giuseffi (William Woods University, USA)
Information Science Reference • © 2022 • 172pp • H/C (ISBN: 9781799871729) • US $175.00

IGI Global
PUBLISHER of TIMELY KNOWLEDGE

701 East Chocolate Avenue, Hershey, PA 17033, USA
Tel: 717-533-8845 x100 • Fax: 717-533-8661
E-Mail: cust@igi-global.com • www.igi-global.com

Editorial Advisory Board

Table of Contents

Detailed Table of Contents

Section 3
Teacher Experiences With Reform-Based Mathematics Teaching

The chapters in this section present studies that explore teachers' experiences with and opportunities to learn reform-based mathematics teaching in various contexts. There are implications for teacher development and practice.

The relationship between enactment of reform math (namely Common Core State Standards – Mathematics [CCSSM]), student misconceptions, and teacher variables was investigated. After providing professional development on enactment for the Standards of Mathematical Practice (SMP) of the CCSSM, observations were conducted to determine the degree of enactment of SMP in 22 classrooms of nine teachers in an urban charter school network consisting of three high schools. Students were all boys, 98% African American, and predominantly of low socio-economic status. Data included curricular documents, observation notes, quarterly assessments, and a teacher survey. Results indicated experienced teachers with high efficacy who expected students to reason and discuss their work were related to higher levels of SMP enactment and fewer misconceptions. Implications for teacher education and teacher practice are discussed.

The focus of this chapter is to describe efforts to implement reform-based mathematics practices in Kenya preprimary classrooms. As much of the research behind reform-based practices comes from high-income contexts, the authors explore how practices can be adapted to middle income contexts through a case study with seven teachers. Using observations, surveys, and interviews, they present data around two key strategies common in reform-based math: questioning strategies and small group, hands-on work. They show that teachers struggled to use open ended questions, which may be related to their pedagogical knowledge. Teachers were successful with implementing small group work, especially after they were comfortable with the content. They discuss what types of research are needed to better understand how reform-based math practices can be adapted to be used across countries with varying levels of resources.

Improving opportunities for under-served (and all) students to learn mathematics requires shifts in what teachers know and can do. This chapter explores teachers' opportunities to learn: How might collaborative inquiry support them to make such shifts? In particular, the chapter explores co-teaching

Foreword

Reform mathematics was initially referenced in the National Council of Teachers of Mathematics (NCTM) Standards (1989), emphasising conceptual thinking and problem solving. Other terms have been used to describe reform-based mathematics perspectives and practices but to summarise, the focus is on the learner and the ways students engage with mathematics in classrooms. While there are different views about the best approaches to teaching and learning in school mathematics, many argue a need for greater student autonomy with more discussion and reflection on learning than might be observed in a 'traditional' mathematics classroom (e.g., Boaler, 2002). Traditionally, mathematics has been taught using didactical approaches with teachers delivering content, explaining methods and procedures, and students practicing independently. While such approaches appear to have been successful for a mathematics elite (Schoenfeld, 2004), they have been less successful for many students who find mathematics challenging, repetitive, and uninteresting. Rather than either reform or traditional approaches, the *Standards* and many authors of publications since argue for a balanced approach with a combination of "appropriate project work; group and individual assignments, discussion between teacher and students and among students; practice on mathematical methods; and exposition by the teacher" (NCTM, 1989, p. 10).

Reform-based approaches frequently include students conjecturing, arguing, and justifying, with teachers guiding, challenging, and scaffolding learning. These approaches do not suggest teachers are not planning and directing the learning, nor that students are discovering knowledge without guidance. Rather, teachers have a key role in designing learning experiences which challenge understanding, and engage students with powerful mathematical ideas and yet cater for diversity (Sullivan, 2015). While these reform-based approaches have been promoted in the international literature since the Eighties, there is evidence they have not been embraced and are not evident in many mathematics classrooms (Desimone, 2013). While some have proposed reasons for the lack of response to the reform agenda (e.g., Boaler, 2002), mathematics education researchers, school leaders and teachers have embraced a range of reform approaches with research publications, books and teacher articles providing evidence of successful outcomes. This volume adds to that body of work in several ways.

1. THROUGH DIVERSITY IN PERSPECTIVES OF KEY STAKEHOLDERS

Several chapters considered the perspective of the practicing teacher, others examined reform approaches in preservice or prospective teacher education programs, and finally there is some focus specifically on the needs of students in reform classrooms at various levels of schooling. Sitabham et al. (Chapter 12) explored the use of, and challenges associated with implementing reform practices by seven elementary teachers in Kenya while Eden (13) investigates co-teaching as an approach to addressing the complexity of implementing reform mathematics pedagogy.

Given the importance of exposing prospective teachers to reform practices they may not have experienced in their own schooling, several authors analysed and reported on successful programs and practices used in preservice education programs. A range of practices have been successful such as scaffolded reflection (Krejci et al. – 6), building a community of practice (Baker – 9), employing the *Five Practices for Orchestrating Productive Mathematical Discussions* (Smith & Stein, 2018) (Rupe & Borowski – 7), and developing argumentation (Magiera et al. – 8).

Finally, Nab (14) used the *Five Practices* to analyse classroom examples for monitoring, selecting, sequencing and connecting student work to improve equitable learning for all students, London (4) examined the potential for using non-routine problems for students to see connections with significant problems connected to themselves, and Morrissey and Popovic (11) examined the impact of higher levels of reform practice and higher teacher expectations for students to reason and discuss their thinking leading to fewer student mathematical misconceptions.

2. THROUGH THEORETICAL AND PRACTICAL IMPLICATIONS

Several authors analysed recent peer-reviewed publications to determine the extent of promotion of reform-based mathematics teaching linking theory with practice. One approach was to consider research-informed publications for the last 10 years in the International Group for the Psychology of Mathematics Education (IGPME) annual conference proceedings. Morrisey et al. (1) identified the predominance of six reform practices including lesson planning, noticing, representations, problem solving and posing, discourse and assessment leading to a set of beneficial practices including lesson analysis, student work analysis, allowing creativity, discussion, reflection, and piloting assessments. Karatas et al's (2) reviewed articles from the last two years of *Mathematics Teacher: Learning and Teaching PK-12* (US) and revealed a more recent focus on mathematical discourse, conceptual understanding, task selection and real-life representations, with less focus on equity, mathematical modelling and using technology to solve problems. Both chapters revealed how the research community continues to seek solutions to promoting more student engagement through discourse and appropriate task selection, particularly in relation to problem solving.

With a focus on using other theoretical perspectives to identify practical implications, Gujarati (10) used the mathematics histories of 75 preservice teachers to analyse the primary influences on their identities noting connections with the teacher, care, the learning environment, understanding the processes of mathematics, the reasons for using manipulatives, as well as the role of gender, and the implications of timed assessment tasks. Saralar-Aras (3) developed a new model for teaching geometry based on the UK context and compared the resultant RETA model (Realistic, Exploratory, Technology-enhanced, Active) with the typical teaching approach in Turkey.

3. THROUGH DIVERSITY IN RESEARCH AND INQUIRY METHODS

A diversity of research methods was employed to develop the chapters in this volume. Such diversity enables new researchers to see how different methods can be used to seek answers to important questions about the implementation of reform-based mathematics teaching. Some of the methods employed in this volume include analysing the focus of articles in journals or proceedings (Karatas et al., Morrisey et al.); classroom observations (Eden, Sitabkhan et al.); analysing mathematics histories (Gujarati); comparing one model of practice with the approach used in another context (Saralar-Aras), case studies (Sitabkhan et al.), and interventions (Baker, Krejci et al., Rupe & Borowski).

4. THROUGH DIVERSITY IN CURRICULUM AND OR TEACHING APPROACHES

One of the reasons implementing reform-based mathematics teaching is so challenging is the diversity of approaches promoted in the literature – this volume is no different. Most chapters focus on a small number of approaches and investigate their potential to support teachers. Whether it is through the use of particular strategies to promote productive discussions or the use of particular types of tasks, all have potential to support the implementation of reform practices. However, the biggest challenge going forward is to identify the best ways to implement such strategies at scale to make a real and ongoing difference in classrooms. This challenge is yet to be realised.

Judy Anderson
The University of Sydney, Australia

Judy Anderson *is an Honorary Associate Professor in Mathematics Education in the Faculty of Arts and Social Sciences at the University of Sydney. She worked at the university for more than 18 years as a secondary mathematics education lecturer and researcher. A former secondary mathematics teacher and leader, she lectured in secondary mathematics curriculum across pre-service and postgraduate teacher education programs. She was Director of the STEM Teacher Enrichment Academy from 2016 to 2020, working with academics from the faculties of Science and Engineering to develop an innovative professional learning program for STEM teachers. The Academy reached over 1200 teachers from 220 primary and secondary school settings in NSW, Australia. Prior to her role at the University, Judy worked as a curriculum team leader, responsible for the development of the mathematics syllabuses for NSW schools. To support mathematics curriculum innovation in NSW, she has regularly conducted professional development for teachers and provided advice on the first national mathematics curriculum in Australia. Judy is an active researcher in STEM education particularly in the field of teachers' beliefs and practices, as well as in exploring the impact of professional development on classroom practice. She has presented papers at local, national and international research conferences and has published widely in research and teacher education journals. She uses both qualitative and quantitative research methodologies to explore teaching practices, particularly in relation to integrated STEM education, problem solving and inquiry-based learning. Community engagement and leadership are evident in Judy's commitment to serve in a range of capacities to support education research and schooling communities. She was Secretary of the International Committee for the International Group for the Psychology of Mathematics Education (IGPME) from 2018 to 2021, and was recently appointed Associate Editor of the new journal, Research in Integrated STEM Education.*

REFERENCES

Boaler, J. (2002). *Experiencing school mathematics: Traditional and reform approaches to teaching and their impact on student learning*. Taylor and Francis. doi:10.4324/9781410606365

Desimone, L. M. (2013). Teacher and administrator responses to standards-based reform. *Teachers College Record, 115*(8), 1–53. doi:10.1177/016146811311500805

National Council of Teachers of Mathematics. (1989). *Curriculum and evaluation standards for school mathematics*. NCTM.

Schoenfeld, A. (2004). The math wars. *Educational Policy, 18*(1), 253–286. doi:10.1177/0895904803260042

Smith, M. S., & Stein, M. K. (2018). *Five practices for orchestrating productive mathematical discussions*. NCTM.

Sullivan, P. (2015). Maximising opportunities in mathematics for all students: Addressing within-school and within-class differences. In A. Bishop, H. Tan, & T. N. Barkatsas (Eds.), *Diversity in mathematics education: Towards inclusive practices* (pp. 239–254). Springer. doi:10.1007/978-3-319-05978-5_14

Preface

We live in a world where mathematics is increasingly used to characterize societal problems and formulate proposed solutions (Larson, 2018). There is a common agreement that reform teaching is needed in order to help all students develop the mathematical thinking and knowledge required to actively participate in today's world and society (Conference Board of the Mathematical Sciences, 2001, 2019; National Council of Teachers of Mathematics [NCTM], 2000; Rand Mathematics Study Panel, 2003); and that there is a need for teachers who embrace a reform teaching style, and educate children by taking these societal needs into account.

Reform-based ***mathematics*** teaching is broadly described as using instructional practices that build on students' informal knowledge of mathematics, connect mathematics with other disciplines and "real world," provide opportunities for students to contribute and invent their own methods during problem solving, emphasize classroom discourse and analysis of multiple methods for solving a given problem, expect students to provide explanations for their answers, and emphasize instructional goals related to student engagement in and deep understanding of mathematical reasoning, concepts, and procedures (Gabriele & Joram, 2007; Hiebert, 2003; Krulik et al. 2003; NCTM, 2000, 2014).

Reform-based mathematics has been an important topic for the mathematics education research community for many years. Studies across the world have found that reform-based mathematics teaching led to significant and substantial achievement gains in schools (e.g., Balfanz et al., 2006; Borman et al., 2003; Boaler & Staples, 2008; Franco et al., 2007). Therefore, investigating teachers' adherence to reform ideas, understanding their practices and approaches to reform teaching, and investigating how they succeed in incorporating reform approaches has been in the focus of attention of researchers (e.g., Franco et al. 2007, Handal & Herrington, 2003; Superfine, 2009; Van Zoest & Bohl, 2005). In accordance, teacher educators across the world have constantly investigated the best practices, helpful tools, activities, instruments, and pedagogies to better prepare preservice teachers for reform-based teaching of mathematics (e.g., Althauser, 2018; Evans et al., 2013; Guillaume & Kirtman,2010; Jao, 2017; Kartal et al., 2020; Kartal & Tillett, 2021). Much work has been done; yet more work needs to be done to make reform-based mathematics teaching happen.

This book provides valuable information for teacher educators, teachers, practitioners, and researchers by exploring perspectives on diverse reform-based practices in teaching and learning of mathematics, describing challenges and issues for teachers and teacher educators, promoting reflection and academic discussion at various levels and in various educational systems, and raising questions for the field of mathematics education. The individual chapters of this book reflect a variety of research- and practice-based perspectives on reform-based teaching and learning from a range of U.S. and international authors.

The focus of Section 1 is on theoretical and practical perspectives and approaches for reform-based mathematics teaching. The five chapters in this section discuss practices from around the world and U.S. by providing systematic analysis and vivid examples. For example, Morrissey, Kartal, and Popovic, in their chapter "Reform-Based Mathematics Teaching Practices in Teacher Education Research Across the Globe," elaborate on a variety of views and defining features of reform-based mathematics teaching; summarize a decade of research on mathematics pre-service teacher education programs around the world, that focused on teaching practices; and identify six salient reform-based practices. The authors discuss each of these practices to describe how reform-based teaching has been approached in preservice teacher education across the world; and provide recommendations for broadening the research and practice on reform-based mathematics teaching.

Karatas and colleagues, in their chapter "Reform-Based Mathematics Teaching in the United States," provide a systematic analysis of the reform-based practices in the U.S. elementary and secondary school classrooms in recent years. The authors share a brief overview of reform movements in school mathematics in the U.S. over the last two decades; and describe the trending reform practices based on their analysis of a leading practitioner journal. They describe a variety of shared strategies, and discuss each strategy in relation to the recent reform movements in the U.S.

In "RETA Model for Teaching Mathematics: From the United Kingdom to Turkey," Saralar-Aras introduces a novel reform-based mathematics teaching model with realistic, exploratory, technology-enhanced, and active (RETA) principles. The author discusses principles of RETA model and its practical applications in middle schools in Turkey. The chapter provides sample lesson plans and activities along with a discussion of design decisions.

In "A Holistic Approach to Teaching the Process of Problem Solving: A Curriculum of Nonroutine Problems," London defines the characteristics of non-routine problems, and discusses the concept of curriculum of non-routine problems in relation to the most recent reform movements and practices in the U.S. Examples of non-routine problems are included in the chapter, along with a list of suggestions and guidelines for implementation.

Campo, in "When Multidisciplinarity Meets Mathematics Teaching," discusses the concept of learning without boundaries through a complex-fuzzy-multidisciplinary-humanistic model. The author provides sample episodes from Israeli schools along with a theoretical discussion of the concept of multidisciplinary.

The five chapters in Section 2 discuss practices implemented in mathematics methods and mathematics content courses in order to support teacher candidates in developing reform-based teaching skills. Implemented practices include reflection, five practices for orchestrating meaningful discussions, community of practice, integrated learning spaces, and writing autobiographies.

In "Fostering Commitment to Reform-Based Mathematics in Preservice Teachers: Shifting Sources of Mathematics Teaching Efficacy," Krejci and colleagues describe theoretically motivated activities designed to promote reform-based mathematics instruction. Specifically, the authors describe instructional tasks that required teachers to reflect on reform-based teaching from videos, reform-based lesson planning, and reform-based readings. The authors, also, make explicit connections between these activities and the shifts observed across preservice teacher's (PT) observations and reflections, and discuss implications for teacher educators.

Rupe and Borowski, in "Supporting Preservice Elementary Teachers in Planning for Mathematical Discussions," describe a series of learning experiences, including formal lesson planning, that were implemented with PTs in order to support them in developing knowledge and skills to plan for and facilitate mathematical discussions with their future students. The activities include asking PTs to solve rich tasks in multiple ways, reading articles in order to develop collective knowledge about Five Practices for Orchestrating Productive Discussions, facilitating explicit discussion around rich tasks that show a variety of solution approaches, and developing a lesson plan using the Five Practices.

Magiera and colleagues, in their chapter "Supporting Pre-service Teachers' Argumentation-Focused Visions of Mathematics Teaching and Learning in a Teacher Preparation Program," focus on creating Integrated Learning Spaces (ILS) in teacher education programs in order to support PTs' development of mathematics teaching and learning consistent with the reform-oriented pedagogies. The authors document that consistent focus on argumentation through ILS supports the PTs developing a wide range of positive views and dispositions about argumentation as a teaching/learning practice.

Baker's chapter, " 'Who Else Would I Even Talk To?' Supporting Prospective Teachers' Engagement With Student Thinking Through a Community of Practice," discusses the ways in which a Community of Practice facilitates PTs' engagement with student thinking in the area of fractions. Baker suggests two key components that should be included in supporting PTs during their student teaching: a community of practice structure in which PTs have the opportunity to practice their instructional ideas, take risks, have disagreements, and receive feedback, both through a content expert's facilitation and on their own; and focused and substantive time for PTs to invest in the analysis of student work.

In "Disrupting Their Histories: Preservice Teachers' Mathematics Autobiographies as Catalysts for Change," Gujarati shares results of the analysis of the PTs' autobiographies written in mathematics methods courses from 2010 to 2020 and across three different teacher education programs. The author details seven common themes identified across PTs' autobiographies and suggests that, despite the PTs' anxiety and negative feelings towards mathematics, PTs wanted to promote positive relationships with mathematics for their students. The chapter discusses implications for practice and research.

The four chapters in Section 3 present authors' experiences in working with teachers to enact reform-based practices. Asking students to explain their reasoning, allowing students to use manipulatives to explore mathematical concepts, noticing students' thinking, using the practices of goal setting, developing tasks, anticipating, monitoring, selecting, and sequencing student work are strategies that were recommended to teachers. Although levels of enactment varied, teachers in each study came to appreciate the ways reform-based practices could benefit student learning.

In "Implementing Reform-Based Mathematics Teaching Practices to Address Student Errors," Morrissey and Popovic investigate how teachers enacted student-centered teaching practices after participating in professional development, and how their enactment of the practices relates to student errors. Teachers themselves engaged in reform-based practices during the professional development, discussed how this could look in their own classrooms, and enacted the practices in their classrooms. A sample activity from the professional development is provided, and implications that are especially important to school administrators, who interact often with teachers and set the tone within a school, are discussed.

Sitabkhan and colleagues, in "Reform-Based Mathematics in Kenya: A Case Study," examine how questioning strategies and small group work is used by teachers in implementing reform-based practices. Further, the chapter looks at how teachers' knowledge and beliefs are changed through the experience. The chapter includes "teachable moments" lessons and mini lessons. Implications point to the importance of addressing barriers to enacting reform-based practices, while considering the culture of the classroom.

In "Co-Teaching to Develop Reform-Based Practices in Mathematics: Classrooms as Shared Spaces for Teacher Learning," Eden explores the importance of inquiry in two dimensions: teachers' inquiry into their own teaching, and engaging their students in mathematical inquiry. The author presents how teachers' own inquiry into their teaching, talking about their teaching practices, engaging in co-teaching practices, and noticing students' thinking led to engaging in reform-based practices in a way that made a difference in their classrooms. A co-teaching model is provided, with teachers' comments as they progressed through the model. Implications for teachers highlight the importance of a shared community as a safe space in which to discuss the challenges of enacting reform-based mathematics.

Nabb provides a practical, in-depth look at implementing one active-learning model in "Using the Five Practices to Promote Active and Equitable Learning Environments." The author focuses on the six practices of goal setting/developing tasks, anticipating student responses, monitoring student work, selecting student work to share, sequencing the work to be shared, and connecting the student work. The author provides examples of the practices in use in an equitable fashion in the classroom, including sample tasks, student work, and a chart for use in organizing the practices during class. Implications support the use of these practices as a means of ensuring equity in mathematics learning.

CONCLUSION

Chapters in this book serve to remind us that teaching mathematics effectively is a world-wide imperative, as authors from across the globe work to improve mathematical understanding in teachers and students. Mathematics provides the underpinnings for many careers (e.g., computer science, engineering, medicine) (Malone, 2018) that are necessary for advances in technology, research and health care. These careers should be accessible to everyone; teaching mathematics in a way that provides that access is essential. Reform-based mathematics practices provide that access (NCTM, 2014), and thus are important across the globe.

Throughout this book, we provide various viewpoints, trends, experiences, and recommendations for practices suitable for teachers and teacher educators. Perspectives of researchers, educators, and practitioners are represented from several countries. Several authors have taken the time to provide instances of reform practices as they were used in the classroom, through recommendations, examples and lesson plans. Successful and not-so-successful attempts at implementing reform-based practices have been included, to allow readers to learn from what works and what has not yet worked in our quest to reform how students learn mathematics.

Overall, we have attempted to draw together research and ideas from across the globe in hopes that recommendations found throughout will enable teachers and teacher educators to enact reform-based mathematics practices. These practices are essential for students, for their access to equitable learning, and for the future of our intertwined societies.

REFERENCES

Althauser, K. L. (2018). The emphasis of inquiry instructional strategies: Impact on preservice teachers' mathematics efficacy. *Journal of Education and Learning*, *7*(1), 53–70. doi:10.5539/jel.v7n1p53

Balfanz, R., Mac Iver, D. J., & Byrnes, V. (2006). The implementation and impact of evidence-based mathematics reforms in high-poverty middle schools: A multi-site, multi-year study. *Journal for Research in Mathematics Education*, *37*(1), 33–64.

Boaler, J., & Staples, M. (2008). Creating mathematical futures through an equitable teaching approach: The case of Railside School. *Teachers College Record, 110*(3), 608–645. doi:10.1177/016146810811000302

Borman, G. D., Hewes, G. M., Overman, L. T., & Brown, S. (2003). Comprehensive school reform and achievement: A Meta-analysis. *Review of Educational Research, 73*(2), 125–230. doi:10.3102/00346543073002125

Conference Board of the Mathematical Sciences (CBMS). (2001). *The mathematical education of teachers* (Vol. 11). Providence, RI: American Mathematical Society.

Conference Board of the Mathematical Sciences (CBMS). (2019). *Preparing students for future*. Reston, VA: American Mathematical Society.

Evans, B. R., Leonard, J., Krier, K., & Ryan, S. (2013). The influence of a reform-based mathematics methods course on preservice teachers' beliefs. *Journal of Educational Research and Practice, 3*(1), 79–92.

Franco, C., Sztajn, P., & Ortigão, M. I. R. (2007). Mathematics teachers, reform, and equity: Results from the Brazilian national assessment. *Journal for Research in Mathematics Education, 38*(4), 393–419.

Gabriele, A. J., & Joram, E. (2007). Teachers' reflections on their reform-based teaching in mathematics: Implications for the development of teacher self-efficacy. *Action in Teacher Education, 29*(3), 60–74. doi:10.1080/01626620.2007.10463461

Guillaume, A. M., & Kirtman, L. (2010). Mathematics stories: Preservice teachers' images and experiences as learners of mathematics. *Issues in Teacher Education, 19*(1), 121–143.

Handal, B., & Herrington, A. (2003). Mathematics teachers' beliefs and curriculum reform. *Mathematics Education Research Journal, 15*(1), 59–69. doi:10.1007/BF03217369

Hiebert, J. (2003). *Teaching mathematics in seven countries: Results from the TIMSS 1999 video study.* DIaNe Publishing.

Jao, L. (2017). Shifting pre-service teachers' beliefs about mathematics teaching: The contextual situation of a mathematics methods course. *International Journal of Science and Mathematics Education, 15*(5), 895–914. doi:10.100710763-016-9719-9

Kartal, O., Morrissey, S. G., & Popovic, G. (2020). Analysis of elementary preservice teachers' identification of mathematical problem-solving tasks and anticipated student solutions. *The Mathematics Educator, 29*(1), 85–111.

Kartal, O., & Tillett, W. (2021). Transforming preservice teachers' conceptions and lesson plan designs for teaching mathematics from a direct-teaching approach to a reform-based approach. *Investigations in Mathematics Learning, 13*(2), 107–124. doi:10.1080/19477503.2020.1847988

Krulik, S., Rudnik, J., & Milou, E. (2003). *Teaching mathematics in middle schools. A practical guide.* Pearson Education Inc.

Larson, M. (2018, February 21). *Why teach mathematics?* NCTM. https://my.nctm.org/blogs/matthew-larson/2018/02/21/why-teach-mathematics

Malone, M. (2018). List of careers involving math. *Chron*. https://work.chron.com/list-careers-involving-math-10640.html

National Council of Teachers of Mathematics (NCTM). (2000). *Principles and standards for school mathematics*. Author.

National Council of Teachers of Mathematics (NCTM). (2014). *Principles to actions: Ensuring mathematical success for all.* Author.

RAND Mathematics Study Panel. (2003). *Mathematical proficiency for all students: Toward a strategic research and development program in mathematics education*. Author.

Superfine, A. M. C. (2009). Planning for mathematics instruction: A model of experienced teachers' planning processes in the context of reform mathematics curriculum. *The Mathematics Educator*, *18*(2).

Van Zoest, L. R., & Bohl, J. V. (2005). Mathematics teacher identity: A framework for understanding secondary school mathematics teachers' learning through practice. *Teacher Development*, *9*(3), 315–345. doi:10.1080/13664530500200271

Acknowledgment

The publication of this book entailed the contributions of many people. First, we would like to express our appreciation to both those who submitted chapter proposals and the chapter authors for their thought-provoking manuscripts and their responsiveness to requests for revisions within tight timelines. Second, we would like to acknowledge Jason Celal Aydar at University at Buffalo, Dr. Omar Sillah at Mercer University, and Dr. Ayse Ozturk at Ohio State University for their assistance in reviewing manuscripts. Additionally, we would like to express our superb appreciation to Dr. Judy Anderson at the University of Sydney for writing a foreword within a tight timeline. Finally, we would like to acknowledge editorial advisory board members—Dr. Kathryn Rupe, Dr. Marta T. Magiera, Dr. Raewyn Eden, and Dr. Christy Peterson—for their assistance in disseminating the chapter call, reviewing chapter manuscripts, and authoring or co-authoring significant chapters (that also underwent the blind-review process).

Section 1

Reform-Based Mathematics Perspectives and Approaches

The chapters in this section present theoretical and practical perspectives and approaches by the international research and practitioner community by looking into trends in teacher education and K-12 classrooms; and by discussing implementation of models with sample tasks, lesson plans, or scenarios.

Chapter 1
Reform-Based Mathematics Teaching Practices in Teacher Education Research Across the Globe

Susie Morrissey
Mercer University, USA

Ozgul Kartal
University of Wisconsin, Whitewater, USA

Gorjana Popovic
Illinois Institute of Technology, USA

ABSTRACT

In accordance with views of reform-based mathematics teaching from across the globe, the authors reviewed a decade of research published in the International Group-Psychology of Mathematics Education (IG-PME) conference proceedings on reform teaching practices for mathematics pre-service teachers. Six salient practices were identified in the decade of research: lesson planning, noticing, representations, problem solving and posing, discourse and assessment. The authors found that the beneficial practices of lesson analysis, student work analysis, allowing creativity, discussion, reflection, and piloting assessments promoted reform-based practices. Further research and emphases are needed on preservice teachers' abilities to make decisions based on student thinking, understand the importance of linking multiple representations, and require explanations from students. Recommendations for broadening the research are included.

DOI: 10.4018/978-1-7998-9422-3.ch001

INTRODUCTION

Researchers over the years have agreed on some aspects of reform mathematics, such as that students should have a deep understanding of mathematics, which could be achieved through discourse around solving problems that have high cognitive demand. For example, in the United States (US), Ball (1993) saw reform mathematics as embodying classrooms where "students would conjecture, experiment, and make arguments; they would frame and solve problems; and they would read, write, and create things that mattered to them" (p. 374). Boaler (2002) likewise saw discourse as an important part of problem solving. While conducting research in the US, she described international ideas of reform-based mathematics as productive learning experiences, provided by "open-ended problems that encourage students to choose different methods, combine them, and discuss them with their peers" (p. 239). This is similar to the description of the reform curriculum in Greece described by Christou et al. (2002), as engaging students in "thought-provoking, original problems that involve challenging themes" (p. 257). Students were intended to be engaged in interdisciplinary, discovery activities. Chapman (2004), in her research in Canada, emphasized the importance of peer interactions as part of reform mathematics. She focused on how teachers implement discourse among students in the classroom. In Australia, reform mathematics had a similar view, with Anderson and Bobis (2005) identifying aspects of reform mathematics in Australian curricular documents of "questioning, applying strategies, communicating, reasoning and reflecting" (p. 65), which are intended to support problem solving and high order thinking. These views are also seen in the US, as Van den Heuvel-Panhuizen and Fosnot (2001) point out, a basic component of reform math is that it focuses more on processes and strategies than on answers, to provide insight into students' thinking.

Researchers also believe that solving problems in the context of real life can be an important part of reform mathematics, as seen by Carroll (1996), in the US, who focused on the belief that a reform curriculum would use problem solving in broader mathematical contexts. Ball (1993) also included authentic tasks in her view of reform mathematics, as did Vos and Bos (2001), who described reform mathematics in the Netherlands as based on Realistic Mathematics Education, which relied on the belief that mathematics is an important part of real life, and students should create mental images to aid in problem solving. The curriculum emphasized open-ended mathematics problems set in the context of real life.

Reform mathematics has also been seen as student-centered. Sakonidis et al. (2001) summarized reforms in Greece under the umbrella of constructivism, such that the teacher provides opportunities for students to make sense of new ideas. Meanwhile, in Israel, Chissick (2002) studied teachers' attempts to implement reform mathematics with the characteristics of open-ended tasks, use of technology, and pupil-centered teaching. Oh (2003) characterized reform teaching in South Korea by asserting that "reform-oriented teachers have been seen as 'facilitators' or 'stimulators' given that they assist students' mathematical learning as they create a learning environment that reflects students' ideas" (p. 406).

Anderson and Bobis (2005) included reasoning and reflecting in their reform-mathematics ideas. Reasoning was also seen in the earlier National Council of Teachers of Mathematics (NCTM, 2000) document, as part of the Process Standards of problem solving, reasoning and proof, connections, communication, and representation, as well as in Adding It Up: Helping Children Learn Mathematics (National Research Council [NRC], 2001), which identified five strands that defined mathematical proficiency: adaptive reasoning, strategic competence, conceptual understanding, procedural fluency, and productive disposition. Reflection, as well as productive disposition as seen in the NRC (2001) document, were included as part of Schoenfeld's (2006) characterization of reform mathematics as including "strategic competence, metacognitive ability (including monitoring and self-regulation), and productive beliefs and affect (or disposition)" (p. 25).

Reviews of reform mathematics literature have yielded synthesized definitions. For example, Ross et al. (2002), in their review of reform in mathematics education 1993-2000, defined reform mathematics based on their review of the literature and NCTM policy documents. Their definition includes multiple mathematics' strands, access for all students to complex mathematical ideas, open-ended problems in context with multiple solutions, students' discourse, teacher as co-learner rather than expert, use of manipulatives and tools, student-student interactions, authentic formative assessments, the belief that mathematics is a dynamic discipline, and the belief that student efficacy is important. In their chapter in the NCTM handbook, Franke et al. (2007) summarized the research on effective classroom practice, and noted that a research-based consensus supports moving towards a reform in mathematics education that includes opportunities for students to develop concepts, procedural skills, flexibility, and higher cognitive level skills. Franke et al. recommended research that would lead to norms of practice that teachers could engage in to enact reform-based teaching.

In the US, the Common Core State Standards - Mathematics (CCSSI, 2010) were developed under the leadership of the National Governor Association and the Council of Chief State School Officers. The NCTM Process Standards (NCTM, 2000) and the five strands that defined mathematical proficiency from NRC (2001) became the basis for the Standards for Mathematical Practice of the Common Core State Standards - Mathematics, which have eight standards that teachers are expected to develop in their students, such as persevere at solving more challenging problems, model situations using a mathematical model, defend their reasoning, and critique the reasoning of others.

Overall, reform mathematics has been described as a student-centered approach to teaching, such that students' thinking is revealed to teachers. Students are expected to engage in reasoning and discourse in order to solve contextual problems that have high cognitive demand. Students should reflect on their problem-solving strategies in order to build a strong conceptual understanding. Engaging in such activities is anticipated to lead to a positive disposition in mathematics.

AIM OF THE CHAPTER

Reform-based mathematics teaching requires teachers to have a deep understanding of mathematics they will teach, how that knowledge develops across grade levels, and to be able to teach in ways that support learning of all students (NCTM, 2014). However, most mathematics teachers have not experienced learning in the ways research suggests students learn best (Popovic, 2011). Moreover, the dominant cultural belief of teachers, as well as parents, is that students should be taught the way they were taught, by the rote memorization of the rules and procedures for solving a narrow set of problems and then practicing skills over and over again (Karp, 2010; NCTM, 2014). It seems obvious that, in order to ensure the implementation of reform-based mathematics teaching, teacher education programs should provide teacher candidates with the opportunities to experience learning in the way they will be required to teach.

In this chapter we summarize a decade of research on mathematics pre-service teacher education programs that focused on teaching practices. We chose to review research published in the International Group - Psychology of Mathematics Education (IG-PME) conference proceedings to gain a global perspective on methods utilized in mathematics teacher education programs. First, we describe the criteria used for the selection of research reports to be included in this review. Next, we present the review of research that focused on teaching practices that are student centered—reveal student thinking, engage students in reasoning and discourse while solving problems that have higher cognitive demands. We end the chapter with the discussion and conclusions, as well as recommendations for future research.

CRITERIA FOR SELECTION

We reviewed a decade of IG-PME conference proceedings from 2010 to 2021, and identified a total of 159 research reports from around the world (see Figure 1 for the geographic distribution of the reports) that studied Preservice Teacher (PT) education. We examined the reports, and created codes for the focus and content of each research paper.

Figure 1. Geographic distribution of the PME research reports included in the analysis

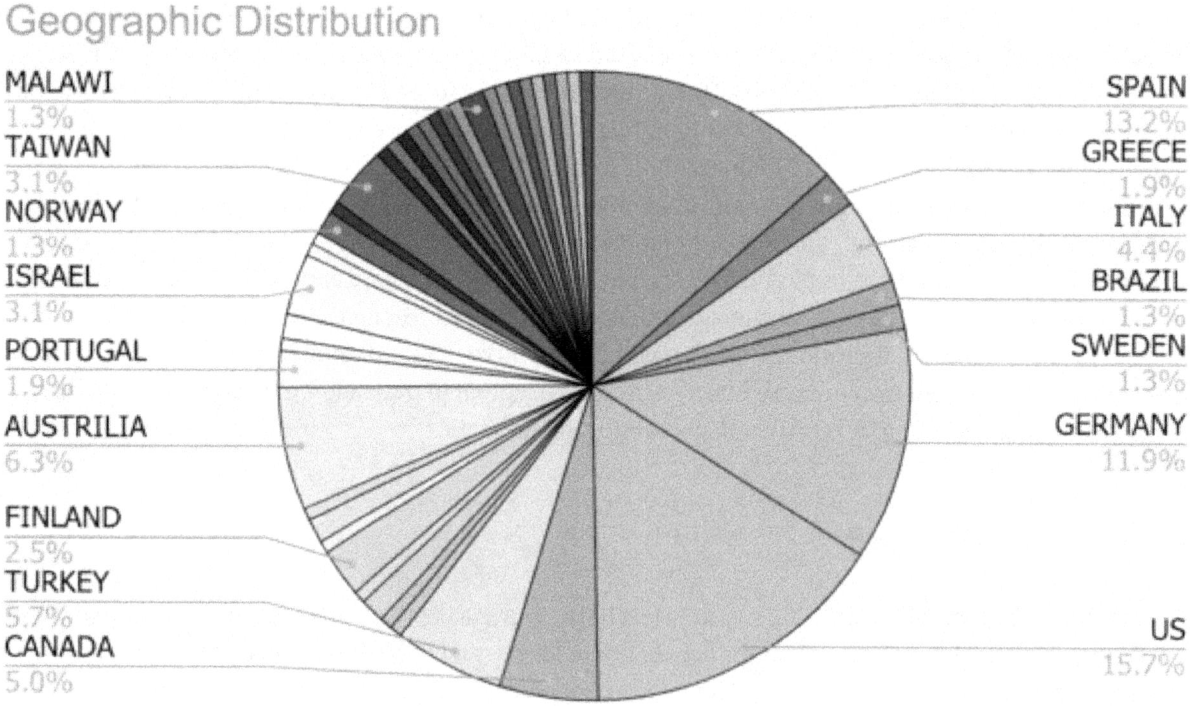

A total of 28 different categories resulted from the analysis of the focus and content of the 159 PME papers of the past decade (see Table 1). Eleven categories became apparent, where four of them were PTs' knowledge and beliefs, and not about teaching practices. Although there is an argument that knowledge, beliefs, and practices are intertwined, we only included the papers that solely focused on practices for the purpose of this chapter. We found six salient practices for structuring the chapter that were studied in the field of PT education during the period of 2010-2021 in the IG-PME proceedings: lesson planning, noticing, representations, problem solving/posing, discourse, and assessment. In the next sections, we discuss each of these practices to describe how reform-based teaching has been approached in PT education across the world, based on the perspectives taken in the reported research.

Table 1. Focus and content of the PME research papers included in the analysis

Code	Frequency Percent	Code	Frequency Percent
Common Content Knowledge (CCK)	22.0%	Lesson Planning	5.7%
Affect and Belief	11.9%	Problem Solving/Posing	5.0%
Noticing	10.7%	Specialized Content Knowledge (SCK)	4.4%
Assessment	7.5%	Discourse	4.4%
Pedagogical Content Knowledge (PCK)	7.5%	Task	2.5%
Representations	5.7%	Miscellaneous	<1.3%

REFORM-BASED PRACTICES

Lesson Planning

Research on lesson planning involved the analysis of lessons and curricular materials to look for reform practices. Overall, research showed a focus on collaborative planning to teach student-centered lessons, with an emphasis on student ideas and participation.

Bausch et al. (2011) asked PTs to analyze lesson plans, while both Pang (2011) and Baki et al. (2016) asked PTs to analyze videos of lessons being taught. In Bausch et al. and Pang, the analyses were open-ended, in which PTs selected what components to analyze. PTs in these studies evidenced common themes in their analyses of goals, teacher actions, tasks, and student participation. PTs in the Baki et al. study completed reflections focused on student difficulties, sources of misconceptions, and proposing solutions for the observed difficulties and misconceptions. Lesson study, the action of collaboratively planning, observing, commenting, discussing and reflecting on each other's lessons was studied by Elipane (2011, 2012), Livy et al. (2017), and Bussi and Funghi (2019). Elipane (2011) focused on habits of mind when observing a PT engaged in lesson study within a community of practice, while Bussi and Funghi focused on PTs' beliefs about planning and teaching. Livy et al. (2017) asked PTs to focus on sequencing and connecting student solutions during their lesson study.

Researchers also investigated PTs' abilities to modify given curricular materials in order to create reform-based mathematics materials (Amador & Earnest, 2018; Kim & Son, 2017). Kim and Son asked PTs to analyze videos of teaching and written lesson plans through a reform-based teaching lens, specifically mathematical explorations. PTs then modified existing direct-instruction lesson plans into reform-based lesson plans. Amador and Earnest tasked PTs with documenting their decision-making process as they adapted materials for a lesson on fractions, then used an animation tool to envision the lesson in a classroom.

Analysis of lessons was seen as a beneficial practice for PTs, as they were able to evaluate effective mathematics teaching, noticing student ideas, and teacher feedback in a more sophisticated and meaningful way (Baki et al., 2016; Bausch et al., 2011; Pang, 2011). Additionally, engaging in a community of practice in a school that supports reform-based teaching led to a positive transformation in the PT (Elipane, 2011, 2012). Elipane found four themes, which were: making sense of resources, using school contexts for inquiry, critical reflection, and collaboration. These reform actions were seen in the PTs' focus on conceptual understanding and linking to previous knowledge. In analyzing sequencing and connecting student solutions, PTs most often noticed that students learned from each other, how students learn, student dispositions, scaffolding, timing, learning how to teach, and strategies used by students (Livy et al., 2017). Bussi and Funghi (2019) found that, although PTs initially struggled with the idea that each portion of a lesson was timed, they came to see timing as a benefit. They also saw value in anticipating students' responses. In Kim and Son (2017), PTs agreed that providing a solution strategy to be followed limited students' opportunities to engage in problem solving. PTs modified the lessons to have a focus on sense-making, problem solving, justification, and productive struggle. Amador and Earnest (2018) found examples of changes PTs made to curricular materials included bringing in actual segmented candy bars in place of the brownie mentioned in the problem, moving all vocabulary terms to the beginning of the lesson, adding a discussion about what students could share, and adding a discussion about parts of a whole.

Lesson analysis, coupled with discussions of reform practices seen in the lessons, was seen as an essential component of reinforcing reform-based practices (Baki et al., 2016; Pang, 2011). Elipane (2011, 2012) stressed the importance of a community of practice, and noted that the common culture that existed between the university teacher education program and the university laboratory school where student teaching took place served to reinforce methods that were taught in teacher education courses. Bussi and Funghi (2019) found that PTs came to see timing as a benefit, and also saw value in anticipating students' responses. Livy et al. (2017) felt that PTs benefitted from seeing the specific strategies of sequencing and selecting employed by teachers in the summarize phase of the lesson. Anticipating, sequencing, and selecting are important elements of effective teaching, especially when coupled with monitoring, and making connections between student responses (Stein et al., 2008).

Kim and Son (2017) concluded that PTs gained critical thinking skills in analyzing lessons in the curriculum. However, Amador and Earnest (2018) found that PTs were most concerned about real life connections, and did not consider that some of their changes to curricular materials would decrease the cognitive demand of the lesson. Based on the conclusions reached by researchers over the past decade, teacher educators must continue the practices of analyzing lessons and curricular materials through a reform-based lens. PTs need practice attending to the five practices described in Stein et al. (2008). Additionally, the inclusion of real-life contexts must be seen as part of an overall effective teaching strategy rather than the sole focus of a lesson.

Noticing

Professional noticing is closely related to reform-based teaching as it is central to student-centered instruction (Stockero, 2013). A good deal of the research within the PT education found in the IG-PME proceedings during the period 2010–2021 studied professional noticing. In the reviewed research reports, noticing is commonly defined as interpreting students' thinking and reasoning and making instructional decisions. The reviewed papers (e.g., Fernández et al., 2011; Ivars et al., 2016, 2017, 2018, 2019) utilized Jacobs et al.'s (2010) conceptualization of professional noticing of children's mathematical thinking as a set of three interrelated skills (i) attending to children's strategies, (ii) interpreting children's mathematical understandings, and (iii) deciding how to respond on the basis of children's understandings. In this means, IG-PME papers investigated how PTs attended, interpreted, and responded to student thinking; and how PTs improved on these skills, i.e., noticing, through the course of methods classes and or student teaching.

The researchers studied the development and improvement of PTs' noticing by means of an online discussion (Fernandez et al., 2011), writing narratives (Ivars & Fernandez, 2016), analysis of student mathematics work (Buforn et al., 2017, 2018; Ivars et al., 2019), a hypothetical learning trajectory (Ivars et al., 2017, 2018, 2019; Sanchez-Matamoros et al., 2018), and analysis of classroom vignettes (Kuntze et al., 2021). The review of a decade of IG-PME studies revealed that teacher education programs around the world have increasingly incorporated activities in an effort to support PTs' professional noticing, and have found promising results. The development of noticing is perceived as a shift from general strategy descriptions to descriptions that included teachers' reasoning based on mathematically relevant details of students' mathematical thinking (Ivars, 2018, p. 148) which hypothetically leads to making effective instructional decisions. Although, PTs showed improvement in noticing students' thinking and reasoning, they had difficulty in making appropriate decisions to support student progression and propose specific instructional decisions (e.g., Sanchez-Matamoros et al., 2018; Buforn et al., 2018; Ivars & Fernandez, 2016). More specifically, PTs had difficulty in providing activities focused on the mathematical element of the problem (Buforn et al., 2018), instead they focused on correct answers, general evidence of confusion, and non-mathematical concerns (Stockero, 2013). Instead of proposing specific instructional decisions, they proposed general teaching actions such as re-teaching (Buforn et al., 2018). Thus, a decade of IG-PME papers focused on PTs' professional noticing revealed that making specific instructional decisions based on students' mathematical thinking remains a challenge for the PTs.

Representations

Research on the reform practice of linking multiple representations was evident in many countries, at various content levels. According to Ainsworth (1999), multiple representations lead to deeper understanding, complement each other and inhibit misunderstandings from forming. However, PTs were found to not realize the importance of multiple representations (Dreher et al., 2012) or to not have sufficient conceptual understanding of either the content or of multiple representations to make use of them in problem solving (Barmby & Milinkovic, 2011; Ho & Lai, 2012; Lim et al., 2017; Milinkovic, 2012; Sevimli & Delice, 2011). Researchers successfully used multiple representations to further student understanding in studies by Amato (2011), Gagatsis and Monoyiou (2011), and Ozmantar et al. (2011).

Several studies investigated PTs' views on multiple representations, and their choices in representations to use for solving problems. In most cases, PTs were provided with multiple representations, and asked to select the representation to use for solving problems (Lim et al., 2017; Sevimli & Delice, 2011) or for posing problems when teaching content (Barmby & Milinkovic, 2011; Milinkovic, 2012). In another case, PTs were asked to solve problems, and researchers investigated the representations they used (Ho & Lai, 2012). Three studies asked students to flexibly move between representations (Amato, 2011; Gagatsis & Monoyiou, 2011; Ozmantar et al., 2011).

Overall, studies concluded that PTs lacked awareness of reasons for the importance of using multiple representations, and lacked a full understanding of how multiple representations are important to mathematical thinking and learning (Dreher et al., 2012; Lim, et al., 2017; Sevimli & Delice, 2011). Sevimli and Delice additionally concluded the predominant use of algebraic representations and lack of use of numeric representations in the course influenced PTs. Barmby and Milinkovic (2011) indicated that subject matter knowledge was an issue; Ho and Lai (2012) found a similar deficit in conceptual understanding. Amato (2011) saw increases in PTs' understandings, but they continued to struggle with some content. In addition to confirming their hypotheses that problem solving ability and flexible use of function representations influenced conceptual understanding, Gagatsis and Monoyiou (2011) also concluded transformations were related to flexible use of representations and conceptual understanding of functions. Results in Ozmantar et al. (2011) showed that PTs increased their vocabulary, understandings, and ability to link multiple representations.

In summary, findings in the reviewed IG-PME studies support successful strategies such as the flexible use of varied representations, converting between representations, and explicitly linking multiple representations in order to develop deeper understandings of the concept (Amato, 2011; Gagatsis & Monoyiou, 2011; Ozmantar et al. 2011). Needs are also illuminated, such as the need to model multiple representations and how they are related, as well as making clear to PTs why multiple representations are an important part of mathematical thinking. This need is echoed in Milinkovic's (2012) conclusion that PTs needed more support in providing different representations for problem posing. To further this goal, Ho and Lai (2012) recommended the use of multiple representations in real life contexts to improve conceptual understanding, as was seen with some success in Amato (2011). Future research on specific strategies for allowing PTs to link representations and see the importance of multiple representation is recommended.

Problem Solving and Posing

Problem solving and posing are viewed as an important component of reform-based mathematics. The IG-PME studies in teacher education explored PTs' beliefs about, abilities in, and understanding of problem solving and posing through a variety of interventions.

Chapman's (2013) 1st intervention required PTs to solve three problems, describe their processes in narrative form, develop a model for the solution process, share and discuss in small groups, and develop a group model of the solution and represent it in a flow chart. The 2nd intervention was similar, but the PTs were required to describe their processes by recording every time they were stuck and every time they got an idea (an Aha! moment). The results showed that both intervention approaches were helpful for the PTs' learning of problem solving. In Approach 1, PTs included affective, while in Approach 2, PTs focused on the cognitive aspects of problem solving in their descriptions of the experience. Chapman suggested that a combination of the two approaches would be best for improving PTs' problem-solving experiences.

Piñero et al. (2018) used a questionnaire with closed, dichotomous questions regarding the concept of what constitutes a problem, and issues related to problem solving. They found that PTs' ideas about what constitutes a problem aligned with the conceptualization found in literature. However, they agreed that the tasks commonly found in textbooks, mostly routine tasks, were problems. While they understood that there are more solution paths to a problem, they did not have the knowledge needed to select problems to afford students such choices. Although PTs' responses regarding problem solving stages and characterization were consistent with a dynamic, cyclical, and genuine interpretation of the phases defined by Pólya (1945, as cited in Piñero et al., 2018), their further responses suggested that their classroom practice would remain linear. Lastly, PTs agreed that in problem solving, affective factors invoke intellect to find the solution. The researchers concluded that PTs' knowledge was complex with contradictions that may have a detrimental effect on students' learning.

Pyper and MacGregor (2018) also used a questionnaire and an interview to unpack aspects of PTs' ontological and epistemological beliefs about the practice of implementing problem solving. Four themes emerged from data analysis, two related to changing ontological beliefs—a contextual attempt and a sense of boundedness to problem solving, and two related to changing epistemological beliefs—how students work as learners and what needs to be taught. The PTs' contextual attempts changed from the initial use of 'real-world' examples from the curriculum that are not connected to the students in the classroom, towards using real-world problems with contexts that focus on students' interests and concerns. The sense of boundedness pertained to the PTs' beliefs about the level of control and ownership in their classroom practice, which changed from using problem solving confined to the curriculum resources, to problems that use some concepts to learn other concepts, and finally to contexts engage students in problem solving in a meaningful way. Lastly, the PTs initially believed that students must work individually and be presented with all the steps, strategies and thinking of the problem. By experiencing working collaboratively on problem solving, PTs began to understand that students may be more imaginative and creative in problem solving, and that the students may need less structure and control over their problem solving. Consistently, they changed from teaching everything they know about problem solving for every problem to teaching the critical thinking unique to specific problems.

The idea of creativity in problem solving was explored by Chapman (2011), Vale et al. (2012), and Kar et al. (2019). Vale et al. explored in what way a didactical experience through challenging figural pattern problems is a context for developing creativity in elementary PTs' solutions. In contrast, Chapman (2011) examined the ways in which PTs made sense of mathematical problem posing and how posing various problems impacted PTs' learning; similarly, Kar et al. examined how the scores achieved in different creativity dimensions (fluency, flexibility, and originality), measured using problem-posing, affected PTs' problem-solving achievement.

Vale et al. (2012) presented three types of tasks, which required PTs (1) to see the arrangement in different ways, connecting previous knowledge about numbers, relationships, and their connections with basic geometric concepts; (2) to look for a pattern in a figurative sequence, to describe it, and produce arguments to validate it using different representations; and (3) to formulate additional data according to the solution presented by each PT. Authors found that two components of creativity, fluency and flexibility, were largely identified mainly in the first two tasks. PTs' solutions for the third task showed less creativity. Vale et al. recommended that classroom teachers should provide an environment and problem solving that stimulates creativity that will enable the development of mathematics competence of all students.

Chapman (2011) engaged participants in comparing problems of similar and different structure and responding to tasks involving posing problems with various requirements. The researcher analyzed PT's written work for the problem posing experience and their reflective journals on their thinking. Results showed that PTs' sense-making and problem posing depended on their mathematical knowledge, imagination or creativity, and past experience with problem solving. Posing problems that were open-ended, related to a specific mathematics concept, and derived from a given picture of a mathematics concept were the most challenging for PTs. The initial ways of PTs' sense making of problem posing was limited due the lack of their exposure to problem posing and exposure to traditional ways of experiencing problems and problem solving.

In the Kar et al. (2019) study, PTs completed an instrument that engaged them in posing problems about the given patterns, and a corresponding problem-solving test. Next, selected PTs were interviewed in order to clarify the reasons behind their different fluency, flexibility, and originality scores when posing problems. Authors found that the flexibility and originality dimensions of creativity predicted PTs' success on the problem-solving test, while the fluency dimension did not have any predictive power. A higher level of flexibility and originality allowed PTs to carry out an in-depth analysis of the mathematical concepts involved in the problem solving tasks and thus be more successful in problem-solving. Kar et al. also pointed out that in their study, the flexibility dimension of creativity had the strongest effect on problem solving success, and not originality that is usually emphasized in the definition of creativity.

Abu-Elwan (2016), similar to Chapman (2011) and Kar et al. (2019), related problem posing to problem solving skills. Abu-Elwan explored the effectiveness of using problem-posing skills on PTs' problem-solving performance, and whether there were differences between PTs who used problem posing strategies and those who did not. The pre- and post-achievement tests on Mathematical Problem Posing-Solving showed that PTs, in the experimental group, performed significantly better on the Mathematical Problem Posing-Solving post-test than the PTs in the control group.

Iskenderoglu (2016) analyzed the fraction multiplication and division problems posed by PTs. The data, problems posed by PTs for two operations: one multiplication and one division, were analyzed using the qualitative descriptive analysis method. The results showed that PTs were more successful in posing problems regarding multiplication than division regardless of the year in their training. Overall, the study showed that, regardless of the year in the program, PTs demonstrated difficulties with multiplication and division with fractions, mostly reflecting lack of conceptual understanding. Such a situation is worrisome, since the PTs' conceptual weaknesses may have a significant effect on the education they provide to their future students. In summary, all reviewed studies recommended implementation of problem solving and posing into the PT education programs.

Discourse

Mathematical discussions are emphasized in reform mathematics, because they help students learn mathematical discourse practices, espouse student thinking, and allow students to evaluate their own, as well as each other's mathematical ideas (Stein et al., 2008). In a typical reform-oriented mathematics lesson, a whole class discussion has three phases. First, the teacher poses a problem that involves important mathematical ideas. Next, the students work on the problem, discuss it in small groups, and prepare the explanation of their solution approach. In the final phase, different student approaches to solving the problem are discussed. The reports included in the IG-PME Proceedings described the use of the discussions in PT education programs.

In their study, Sánchez and Garcia (2011) identified commognitive conflicts between socio-mathematical (SMN) and mathematical norms (MN) in the context of solving a task related to a specific topic. The researchers examined the verbal dialogues of students and the final written responses to the question. Using Sfard's (2008) commognitive framework, the researchers identified four commognitive conflicts between MN and SMN. The conflict between MN defining—criterion of minimality, and the SMN "the more you write, the better you do" lead students to incorporate descriptive features/aspects that are not necessary but might be relevant. The conflict between MN defining—criterion of minimality, and the SMN "everything you see in a figure that goes with the presentation of a task has to necessarily indicate something" leads students to incorporate descriptive features/aspects that are not necessary and not relevant. The conflict between MN that points out that a definition is required to discriminate between instances and non-instances of a concept and SMN "you have to answer every question; it does not matter if the answers are the same" leads students not to use discriminatory aspects of the mathematical objects. The last conflict between the MN of hierarchy that considers that in the meaning of a mathematical object the meanings of the objects that define it are included and the SMN "text/word in a mathematical task has only a linguistic meaning" leads to accepting that the word that expresses a property as 'one property' and do not assume that the already mentioned property involves other properties that give meaning to it.

Cheng et al. (2012) analyzed PTs' small group discussions while solving a high-level cognitive demand task. They documented that PTs who experience giving mathematical explanations and justifications, and reflecting on the mathematical connections would be better prepared to engage their own students in similar activities. After identifying the cognitive levels of PTs' discussions, authors concluded that shifting roles during small group interactions between PTs can maintain high levels of cognitive demand as well as help explain seemingly unstructured productive talk.

Xenofontos and Kyriakou (2016) engaged PTs in solving non-routine mathematical problems. The aim was to promote two types of classroom talk, as defined by Alexander (2008): discussion—the exchange of ideas in view of sharing information and solving problems; and dialogue—achieving common understanding through structured, cumulative questioning and discussion which guide and prompt, reduce choices, minimize risk and error, and expedite the 'handover' of concepts and principles (p. 4-396). Using discourse analysis (Brown & Yule, 1983), the researchers identified ten tentative qualities of talk: brainstorming, explanations, realization of errors, "Aha!" moments, complementary, decision making, confusion, moments of silence, disagreements, and external input.

Hoynes et al. (2018) used two analytical frameworks by Drageset to code the teacher actions (2014) and the pupil comments (2015) to investigate one dialogue conducted by a PT. The PT, together with the mentor and lecturer, planned for productive whole class dialogs where students could reason in mathematics and develop a deep understanding of mathematical concepts, and conducted dialogs with 3rd graders. Hoynes et al. found that the PT posed questions and asked follow-up questions in order for students to expand their answers. However, it was noticed that the students primarily talked to the teacher, and that the teacher assumed the responsibility to articulate the students' explanations. They found that, although the PTs read a lot about conducting dialogue, planned their dialogue with peers and mentors, and thus were prepared for the task, conducting a whole class discussion is a challenging task. Their findings clearly show the importance of incorporating conducting whole-class dialogue teaching practice into teacher education programs.

Broza and Lifshit (2021) studied the learning of PTs who employed a reflective model developed specifically for the PTs practicum course. The researchers analyzed transcripts of complete lessons, lesson plans, and reflections. The results showed an active process of PTs' changing views about teaching and learning processes demonstrated by awareness of the quality of discourse and their own roles as teachers in mediating teaching. The results also showed a development in PTs ability to critically observe the discourse they conducted in the lesson they taught and its effect on them as teachers.

Assessment

Reform-based mathematics teaching involves authentic formative assessment techniques (Ross et al., 2002). Assessing students' understanding is a main component of reform-based teaching, because the way teachers evaluate their students can either reinforce or undermine the learning process (Lavy & Shriki, 2012). Therefore, developing formative assessment competencies in PTs was the main focus of number of IG-PME research papers from around the world over the last decade (e.g., Buforn et al., 2015; Friesen & Kuntze, 2016; Lavy & Shriki, 2012; Lin & Tsai, 2013). Educational policies around the world have advocated a greater focus on teachers' use of formative assessment practices, in order to motivate students with task-specific feedback, and to adjust the teaching approaches during the learning process

(Andrade & Cicek, 2010; Black & Wiliam, 2009 as cited in Ayalon & Wilkie, 2018). Effective formative assessment requires the development of useful tasks that gauge deeper understanding and development of appropriate criteria for assessment, and then providing constructive feedback so that students could improve their learning (Danielson & Marquez, 2016 as cited in Ayalon & Wilkie, 2018). In this respect, teacher educators and researchers engaged PTs in developing assessment tasks and criteria, analyzing student work and actions, and providing constructive feedback. For example, PTs acted as both assessors and assessees, in Lavy and Shriki (2012), in a peer-assessment activity. The researchers reported that PTs initially had difficulties in generating a satisfactory set of criteria for evaluating their peers' work; yet they improved in this skill through engaging in multiple peer assessment activities. Similarly, in Avalon and Wilkie (2018), PTs developed initial assessment criteria, analyzed a selection of student responses, and revised their criteria. PTs found the activity helpful, and they showed noticeable improvement from the initial set of criteria to the revised criteria through analyzing student responses.

Developing a set of criteria has been considered crucial, as it allows teachers to provide justification and explanation for their scoring, beyond simply determining whether an answer is "right or wrong." In this means, Buforn et al. (2015) asked PTs to determine the learning objective for the number of math problems; analyze students' understanding in response to the problems, and make instructional decisions in order to support or extend their understanding. The results showed that some prospective teachers provided general comments on the correctness of the answers even though they had identified the key mathematical content. Beyond having PTs analyze students' math work for formative assessment, other IG-PME researchers engaged PTs in analyzing classroom situations (e.g., Jeschke et al., 2017; Friesen & Kuntze, 2016) or conducting one-on-one interviews (e.g., Reinhold, 2015). Friesen and Kuntze raised the need for empirically evaluated assessment instruments in formats such as text, comics, and videos. More specifically, they used text, comics, and video formats to assess PTs' competence in analyzing student understanding. The researchers found that the vignette formats text, comic and video all appeared to be comparably effective to tap PTs' competence of analyzing. Jeschke et al. engaged PTs in video-based instructional situations, and found that PTs lack the ability to give helpful feedback in instructional situations.

A few IG-PME papers in the field of PT education studied analysis of errors and misconceptions in the context of assessment. A view of misconceptions and errors as springboards for learning (NCTM, 2000), and a window into students' thinking align with the view of reform-based teaching. However, there has been limited attention on enhancing PTs' knowledge of errors and misconceptions in the literature as reported by Lin and Tsai (2013). In this respect, Lin and Tsai examined the effect of research-based cases on pre-service teachers' knowledge of students' errors. They provided cases of episodes or scenarios of how a teacher experienced a problem and what strategies were used; and engaged PTs in case discussions which initiated reflective peer conversations. Through case discussions, PTs became competent in identifying specific errors and suggesting possible pedagogical strategies. Royea et al. (2015) also studied PTs' interpretation of student errors through analysis of students' erroneous solutions. PTs in Royea et al.'s study described what the student should have done to get the correct answer, rather than analyzing the student's work for what the student actually did.

To sum up, a variety of activities have been integrated in the methods courses around the world to enhance PTs' enhancement of assessment skills, such as collaborative analysis of student work, developing assessment criteria, analysis of classroom situations through video, comics, text, cases of episodes and teaching scenarios, and case discussions. The results revealed that the teacher education courses need to work on moving PTs away from the "right/wrong" perspective in the assessment, in order to be effective in responding to their students.

DISCUSSION

Although our analyses by no means cover the entirety of reform-based mathematics teaching research with PTs, we have focused on research found in the proceedings of the IG-PME over the past decade on six reform-based practices. From these studies, we have found both successful teacher education practices and needs for future research.

According to NCTM (2014), effective reform-based mathematics lessons begin with clear goals that guide decisions about instructional strategies and content before and during the lesson. Lesson analysis, particularly when engaged in as part of a community of practice, has been successful in helping PTs incorporate reform-based strategies into their lessons (Baki et al., 2016; Elipane, 2011; Elipane, 2012; Pang, 2011). Elipane (2011, 2012) also stressed that a common culture between the university and the student teaching school placement is important to reinforce methods in practice. Observing and practicing specific strategies, such as the five practices of anticipating, monitoring, selecting, sequencing, and making connections between student responses (Stein et al., 2008) were also seen to improve lesson planning and enactment skills in PTs. Real life contexts have been seen as part of reform-based mathematics teaching, but the emphasis must remain on reasoning and problem solving, with real life contexts used as a way to make the content more accessible to students.

Student-centered lesson planning needs teacher actions that include noticing students' thinking and reasoning in order to make instructional decisions (Baki et al., 2016). Researchers found promising results in improving PTs' noticing skills through use of an online discussion (Fernandez et al., 2011), writing narratives (Ivars & Fernandez, 2016), analysis of student math work (Buforn et al., 2017, 2018; Ivars et al., 2019), a hypothetical learning trajectory (Ivars et al., 2017, 2018, 2019; Sanchez-Matamoros et al., 2018), and analysis of classroom vignettes (Kuntze et al., 2021). However, making specific instructional decisions based on students' mathematical thinking continues to be a challenge for the PTs, and remains an area of needed research, given the importance placed on engaging students in reasoning in reform-based mathematics teaching.

Additionally, lessons in reform-based teaching include the flexible use of multiple representations in solving problems. Researchers reported some successes with the strategies of flexibly using varied representations, converting between representations, and explicitly linking multiple representations in order to develop deeper understandings of the concept (Amato, 2011; Gagatsis & Monoyiou, 2011; Ozmantar et al. 2011). However, a disconnect exists between the use of multiple representations and the reasoning behind the use of multiple representations. PTs do not see the importance of linking multiple representations, or how different representations highlight different aspects of a mathematical idea, and thus limit misconceptions. Overall, the view of how multiple representations support mathematical thinking is missing, and therefore additional research is warranted.

Chapman (2011) found creativity was important in problem posing, and Vale et al. (2012) and Kar et al. (2019) both found evidence that creativity was an important part of success in problem solving, even more so than flexibility. Vale et al. recommended that teachers provide classrooms that stimulate creativity, and Kar additionally stated that teachers should provide students with less structure and control in order to allow for creative problem solving. Chapman (2013) found that the process of narrating and modeling problem solving pathways, including difficulties and discoveries, and discussing those with classmates improved their conceptions of problem solving. Likewise, fully exploring the problem solving process aided PTs in Abu-Elwan's (2016) study in improving their problem solving as well as problem posing skills. Problem posing skills were found to be hampered by lack of content knowledge in Iskenderoglu (2016) and Piñero et al. (2018). PTs provided incorrect problems to accompany given fraction expressions, or were unable to select problems that had multiple solution paths. Additional research into the relationship of content knowledge to problem posing would be insightful.

Discourse was successfully enacted in studies where PTs experienced mathematical discussions, engaged in reflection, and conducted whole-class discourse as part of their PT experience (Broza & Lifshit, 2021; Cheng et al., 2012; Hoynes et al., 2018). Socio-mathematical norms that conflict with problem solving (Sánchez & Garcia, 2011) need to be further addressed, both in teacher education and in research.

Developing formative assessment skills is an additional important component of reform-based mathematics teaching. Successful strategies included having PTs engage in multiple peer assessment activities, as well as having them develop initial assessment criteria, analyze a selection of student responses, and revise their criteria (Avalon & Wilkie, 2018; Lavy & Shriki, 2012). Analysis of student misconceptions also holds potential in developing formative assessment skills, if PTs focus on student thinking rather than simply noting mistakes (Lin & Tsai, 2013). Various other strategies have been incorporated in the research; however, the need is still strong to move PTs away from the "right/wrong" perspective and incorporate a set of criteria that provide justification and explanations for scoring (Buforn et al., 2015). This would allow PTs to make instructional decisions based on assessment results, leading back to effective lesson planning.

RECOMMENDATIONS

We provided an overview of the research presented in the IG-PME proceedings during the period 2010–2021, focusing on PT education pertaining to reform-based mathematics teaching; identified six salient reform-based practices that have been researched by the IG-PME teacher educator community; and discussed the needs for future research on those six practices. We now recommend ways for broadening the research and practice on reform-based mathematics teaching. Reviewing 159 IG-PME research reports over the past decade, it can be concluded that many aspects of reform-based mathematics teaching remain in need of further attention by the teacher educator community across the world. Issues and questions that need further attention, indeed, evolve along with the reform movements in different parts of the world. However, based on how reform-based mathematics teaching has been framed in a broader context by a variety of mathematics educators and researchers, it is noteworthy that mathematical modeling was not a focus in the IG-PME teacher education community. Not only that, with the release of the CCSSM, modeling situations using a mathematical model has become at the forefront of reform in the US, but also the increased global prevalence of models and modelling was highlighted by researchers (e.g., English et al., 2016). Thus, mathematical modeling warrants increased attention from the IG-PME teacher educator community. Furthermore, research on practices that PTs could engage in to enact reform-based teaching that includes opportunities for students to develop concepts is needed (Franke et al., 2007). Building conceptual understanding in students, as a teaching practice, was not a main focus of any of the IG-PME teacher education research. Last but not least, engaging students in activities that lead to a positive disposition in mathematics needs increased attention by the teacher education community. Affect and belief studies in the reviewed research focused mainly on PTs' affect and belief about mathematics, not on teaching practices that would lead to positive affect and belief outcomes in PTs' future students. In this chapter, we have highlighted only a few areas in enhancement of reform-based mathematics teaching that need attention from the international teacher education community; there remains many other questions in need of further attention.

REFERENCES

Abu-Elwan, R. (2016). Mathematics problem posing skills in supporting problem solving skills of prospective teachers. In C. Csikos, A. Rausch, & J. Szitanyi, (Eds.), *Proceedings of the 40th Conference of the International Group for the Psychology of Mathematics Education* (Vol. 2, pp 3-10). PME.

Ainsworth, S. (1999). The functions of multiple representations. *Computers & Education, 33*(2-3), 131–152. doi:10.1016/S0360-1315(99)00029-9

Amador, J., & Earnest, D. (2018). Preservice mathematics teachers' curriculum visualization. In E. Bergqvist, M. Osterholm, C. Granberg, & L. Sumpter, (Eds.), *Proceedings of the 42nd Conference of the International Group for the Psychology of Mathematics Education* (Vol 2, pp. 19-26). PME.

Amato, S. A. (2011). Ratio: A neglected division at schools. In B. Ubuz, (Ed.), *Proceedings of the 35th Conference of the International Group for the Psychology of Mathematics Education* (Vol. 2, pp 25-32). PME.

Anderson, J., & Bobis, J. (2005). Reform-oriented teaching practices: A survey of primary school teachers. In H. L. Chick, & J. L. Vincent, (Eds.), *Proceedings of the 29th Conference of the International Group for the Psychology of Mathematics Education* (Vol. 2, pp. 65–72). PME.

Ayalon, M., & Wilkie, K. (2018). Learning to assess: Exploring changes to pre-service teachers' criteria for a quadratics task. In E. Bergqvist, M. Österholm, C. Granberg, & L. Sumpter (Eds.), *Proceedings of the 42nd Conference of the International Group for the Psychology of Mathematics Education* (Vol. 2, pp. 91-98). PME.

Baki, M., Baki, A., Celik, D., Guler, M., & Sonmez, N. (2016). Improving prospective mathematics teachers' knowledge of student through lesson analysis. In C. Csikos, A. Rausch, & J. Szitanyi, (Eds.), *Proceedings of the 40th Conference of the International Group for the Psychology of Mathematics Education* (Vol. 2, pp 51-58). PME.

Ball, D. L. (1993). With an eye on the mathematical horizon: Dilemmas of teaching elementary school mathematics. *The Elementary School Journal, 93*(4), 373–397. doi:10.1086/461730

Barmby, P., & Milinkovic, J. (2011). Pre-service teachers' use of visual representations of multiplication. In B. Ubuz, (Ed.), *Proceedings of the 35th Conference of the International Group for the Psychology of Mathematics Education* (Vol. 2, pp 105-112). PME.

Bausch, I., Bruder, R., & Prescott, A. (2011). Personal constructs of planning mathematics lessons. In B. Ubuz, (Ed.), *Proceedings of the 35th Conference of the International Group for the Psychology of Mathematics Education* (Vol. 2, pp 113-120). PME.

Boaler, J. (2002). Learning from teaching: Exploring the relationship between reform curriculum and equity. *Journal for Research in Mathematics Education, 33*(4), 239–258. doi:10.2307/749740

Broza, O., & Lifshit, A. (2021). The influence of analytic model on critical reflective thought of pre-service mathematics teachers for elementary school. In M. Inprasitha, N. Changsri, & N. Boonsena (Eds.). *Proceedings of the 44th Conference of the International Group for the Psychology of Mathematics Education*, (Vol. 2, pp. 128-136). PME.

Buforn, A., Fernandez, C., Coles, A., & Brown, L. (2015). The meaning of ratio: Prospective mathematics teachers' knowledge about the teaching and learning of proportional reasoning. In K. Beswick, T. Muir, & J. Fielding-Wells (Eds.). *Proceedings of 39th Conference of the International Group for the Psychology of Mathematics Education*, (Vol. 2, pp. 129-136). PME.

Buforn, A., Fernandez, C., & Llinares, S. (2018). Decision-making in noticing students' proportional reasoning. In E. Bergqvist, M. Österholm, C. Granberg, & L. Sumpter (Eds.), *Proceedings of the 42nd Conference of the International Group for the Psychology of Mathematics Education* (Vol. 2, pp. 187-194). PME.

Buforn, A., Fernandez, C., Llinares, S., & Badillo, E. (2017). Pre-service primary teachers' profiles of noticing students' proportional reasoning. In B. Kaur, W.K. Ho, T.L. Toh, & B.H. Choy (Eds.), *Proceedings of the 41st Conference of the International Group for the Psychology of Mathematics Education* (Vol. 2, pp. 193-200). PME.

Bussi, M. G. B., & Funghi, S. (2019). Lesson study in primary pre-service teachers' education: Influences on beliefs about lesson planning and conduction. In M. Graven, H. Venkat, A. Eddien, & P. Vale, (Eds.), *Proceedings of the 43rd Conference of the International Group for the Psychology of Mathematics Education* (Vol 2, pp. 81-88). PME.

Carroll, W. M. (1996). Use of invented algorithms by second graders in a reform mathematics curriculum. *The Journal of Mathematical Behavior, 15*(2), 137–150. doi:10.1016/S0732-3123(96)90011-5

Chapman, O. (2004). Facilitating peer interactions in learning mathematics: Teachers' practical knowledge. *Proceedings of the 28th Conference of the International Group for the Psychology of Mathematics Education* (Vol. 2, pp. 191-198). PME.

Chapman, O. (2011). Prospective teachers' ways of making sense of mathematical problem posing. In B. Ubuz, (Ed.), *Proceedings of the 35th Conference of the International Group for the Psychology of Mathematics Education* (Vol. 2, pp 209-216). PME.

Chapman, O. (2013). Facilitating prospective secondary mathematics teachers' learning of problem solving. In A. M. Lindmeier & A. Heinze, (Eds.), *Proceedings of the 37th Conference of the International Group for the Psychology of Mathematics Education* (*Vol. 2*, pp. 153-160). Academic Press.

Cheng, D., Feldman, Z., & Chapin, S. (2012). Mathematical discussions in preservice elementary courses. In T. Tso, (Ed.), *Proceedings of the 36th Conference of the International Group for the Psychology of Mathematics Education* (Vol. 2, pp 139-146). PME.

Chissick, N. (2002). Factors affecting the implementation of reforms in school mathematics. In A. D. Cockburn, & E. Nardi, (Eds.), *Proceedings of the 26th Conference of the International Group for the Psychology of Mathematics Education* (Vol. 2, pp. 249-256). PME.

Christou, C., Philippou, G., Pitta-Pantazi, D., & Menn-Eliophotou, M. (2002). The effect of efficacy on teachers' concerns with regard to the implementation of a new mathematics curriculum. In A. D. Cockburn, & E. Nardi, (Eds.), *Proceedings of the 26th Conference of the International Group for the Psychology of Mathematics Education* (Vol. 2, pp. 257-264). PME.

Common Core State Standards Initiative. (2010). *Common core state standards for mathematics*. http://www.corestandards.org/assets/CCSSI_Math%20Standards.pdf

Dreher, A., Kuntze, S., & Lerman, S. (2012). Pre-service teachers' views on using multiple representations in mathematics classrooms - an inter-cultural study. In T. Tso, (Ed.), *Proceedings of the 36th conference of the international group for the psychology of mathematics education* (Vol. 2, pp 211-218). PME.

Elipane, L. E. (2011). Incorporating lesson study in pre-service mathematics teacher education. In B. Ubuz, (Ed.), *Proceedings of the 35th conference of the international group for the psychology of mathematics education* (Vol. 2, pp 305-312). PME.

Elipane, L. E. (2012). Infrastructures within the student teaching practicum that nurture elements of lesson study. In T. Tso, (Ed.), *Proceedings of the 36th conference of the international group for the psychology of mathematics education* (Vol. 2, pp 219-226). PME.

English, L. D., Ärlebäck, J. B., & Mousoulides, N. (2016). Reflections on progress in mathematical modelling research. In *The second handbook of research on the psychology of mathematics education* (pp. 383–413). Brill Sense. doi:10.1007/978-94-6300-561-6_11

Fernández, C., Llinares, S., & Valls, J. (2011). Development of prospective mathematics teachers' professional noticing in a specific domain: Proportional reasoning. In *Proceedings of the 35th Conference of the International Group for the Psychology of Mathematics Education* (Vol. 2, pp. 329-336). PME.

Franke, M. L., Kazemi, E., & Battey, D. (2007). Mathematics teaching and classroom practice. In F. K. Lester (Ed.), *Second handbook of research on mathematics teaching and learning* (pp. 225–256). NCTM.

Friesen, M., & Kuntze, S. (2016). Teacher students analyse texts, comics and video-based classroom vignettes regarding the use of representations—Does format matter. In C. Csikos, A. Rausch, & J. Szitanyi (Eds.), *Proceedings of the 40th Conference of the International Group for the Psychology of Mathematics Education* (Vol. 2, pp. 259-266). PME.

Gagatsis, A., & Monoyiou, A. (2011). Towards a comprehensive theoretical model of pre-service teachers' conceptual understanding of functions. In B. Ubuz (Ed.), *Proceedings of the 35th Conference of the International Group for the Psychology of Mathematics Education* (Vol. 2, pp. 369-376). PME.

Ho, S. Y., & Lai, M. Y. (2012). Pre-service teachers' specialized content knowledge on multiplication of fractions. In T. Tso (Ed.), *Proceedings of the 36th Conference of the International Group for the Psychology of Mathematics Education* (Vol. 2, pp. 291-298). PME.

Hoynes, S., Klemp, T., & Nilssen, V. (2018). Pre-service mathematics teachers' whole-class dialogs during field practice. In E. Bergqvist, M. Osterholm, C. Granberg, & L. Sumpter (Eds.), *Proceedings of the 42nd Conference of the International Group for the Psychology of Mathematics Education* (Vol. 3, pp. 123-130). PME.

Iskenderoglu, T. A. (2016). Multiplication and division problems posed by pre-service elementary mathematics teachers about fraction topic. In C. Csikos, A. Rausch, & J. Szitanyi (Eds.), *Proceedings of the 40th Conference of the International Group for the Psychology of Mathematics Education* (Vol. 2, pp. 35-42). PME.

Ivars, P., & Fernández, C. (2016). Narratives and the development of the skill of noticing. In C. Csikos, A. Rausch, & J. Szitanyi (Eds.), *Proceedings of the 40th Conference of the International Group for the Psychology of Mathematics Education* (Vol. 3, pp. 19-26). PME.

Ivars, P., Fernández, C., & Llinares, S. (2017). Pre-service teachers' uses of a learning trajectory to notice students' fractional reasoning. In *Proceedings of the 41st Conference of the International Group for the Psychology of Mathematics Education* (Vol. 3, pp. 25-32). PME.

Ivars, P., Fernández, C., & Llinares, S. (2019). Principles in the design of tasks to support preservice teachers' noticing enhancement. In M. Graven, H. Venkat, A. Essien, & P. Vale (Eds.), *Proceedings of the 43rd Conference of the International Group for the Psychology of Mathematics Education* (Vol. 2, pp. 408-415). PME.

Ivars, P., Fernández, C., Llinares, S., & Choy, B. H. (2018). Enhancing noticing: Using a hypothetical learning trajectory to improve pre-service primary teachers' professional discourse. *Eurasia Journal of Mathematics, Science and Technology Education, 14*(11), em1599. doi:10.29333/ejmste/93421

Jacobs, V., Lamb, L., & Philipp, R. (2010). Professional noticing of children's mathematical thinking. *Journal for Research in Mathematics Education, 41*(2), 169–202. doi:10.5951/jresematheduc.41.2.0169

Jeschke, C., Lindmeier, A., & Heinze, A. (2017). What do Mathematics pre-service teachers lack for mastering instructional demands? In *Proceedings of the 41st Conference of the International Group for the Psychology of Mathematics Education* (Vol. 3, pp. 33-40). PME.

Kar, T., Özdemir, E., Öçal, M. F., Güler, G., & Ipek, A. S. (2019). Indicators of prospective mathematics teachers' success in problem solving: The case of creativity in problem-posing. In M. Graven, H. Venkat, A. Eddien, & P. Vale (Eds.), *Proceedings of the 43rd Conference of the International Group for the Psychology of Mathematics Education* (Vol. 2, pp. 456-463). PME.

Karp, A. (2010). Analyzing and attempting to overcome prospective teachers' difficulties during problem-solving instruction. *Journal of Mathematics Teacher Education, 13*(2), 121–139. doi:10.100710857-009-9127-y

Kim, O., & Son, J. (2017). Preservice teachers' recognition of affordances and limitations of curriculum resources. In B. Kaur, W. K. Ho, T. L. Toh, & B. H. Choy (Eds.), *Proceedings of the 41st Conference of the International Group for the Psychology of Mathematics Education* (Vol. 3, pp. 57-64). PME.

Kuntze, S., Friesen, M., Krummenauer, J., Skilling, K., Fernandez, C., Ivars, P., Llinares, S., Samkova, L., & Healy, L. (2021). Multi-criterion noticing: Pre-service teachers' difficulties in analyzing classroom vignettes. In M. Inprasitha, N. Changsri, & N. Boonsena (Eds.), *Proceedings of the 44th Conference of the International Group for the Psychology of Mathematics Education* (Vol. 3, pp. 173-183). PME.

Lavy, I., & Shriki, A. (2012). Engaging prospective teachers in the assessment of geometrical proofs. In T. Tso (Ed.), *Proceedings of the 36th Conference of the International Group for the Psychology of Mathematics Education* (Vol. 3, pp. 35-42). PME.

Lim, W., Jiang, B., DeVillar, R. A., & Lee, J. (2017). Patterns of language use in two modes of writing mathematical solutions. In B. Kaur, W. K. Ho, T. L. Toh, & B. H. Choy (Eds.), *Proceedings of the 41st conference of the international group for the psychology of mathematics education* (Vol. 3, pp. 169-176). PME.

Lin, P. J., & Tsai, W. H. (2013). Enhancing pre-service teachers' knowledge of students' errors by using researched-based cases. In A. M. Lindmeier, & A. Heinze (Eds.), *Proceedings of the 37th Conference of the International Group for the Psychology of Mathematics Education* (Vol. 3, pp. 273-280). PME.

Livy, S., Downton, A., & Saito, E. (2017). Pre-service teachers' reflections of the summarise phase of a lesson study. In B. Kaur, W. K. Ho, T. L. Toh, & B. H. Choy (Eds.), *Proceedings of the 41st Conference of the International Group for the Psychology of Mathematics Education* (Vol. 3, pp. 185-192). PME.

Milinkovic, J. (2012). Pre-service teachers' representational preferences. In T. Tso (Ed.), *Proceedings of the 36th Conference of the International Group for the Psychology of Mathematics Education* (Vol. 3, pp 209-215). PME.

National Research Council. (2001). *Adding it up: Helping children learn mathematics.* National Academy Press.

NCTM. (2000). *Principles and standards for school mathematics.* Author.

NCTM. (2014). *Principles to action: Ensuring mathematical success for all.* Author.

Oh, Y. (2003). Applying theory of planned behavior model on studying teachers' change in mathematics instruction. In N. A. Payeman, B. J. Dougherty, & J. T. Zilliox (Eds.), *Proceedings of the 2003 Joint Meeting of PME and PMENA* (Vol. 4, pp. 405-412). PME.

Ozmantar, M. F., Akkoc, H., Bingolbali, E., & Demir, S. (2011). Development of preservice teachers in linking multiple representations via technology. In B. Ubuz (Ed.), *Proceedings of the 35th Conference of the International Group for the Psychology of Mathematics Education* (Vol. 3, pp 313-320). PME.

Pang, J. (2011). What do prospective teachers analyze when they watch a mathematics lesson? In B. Ubuz (Ed.), *Proceedings of the 35th Conference of the International Group for the Psychology of Mathematics Education* (Vol. 3, pp 329-336). PME.

Piñero, J. L., Castro-Rodriguez, E., & Castro, E. (2018). Prospective primary teachers' conceptual understanding of mathematical problems and problem solving. In E. Bergqvist, M. Osterholm, C. Granberg, & L. Sumpter (Eds.), *Proceedings of the 42nd Conference of the International Group for the Psychology of Mathematics Education* (Vol. 3, pp. 465-472). PME.

Popovic, G. (2011). *Informal settings: Implications for secondary mathematics teachers' ability to recognize and apply mathematics* (Doctoral dissertation). Illinois Institute of Technology.

Pyper, J. S., & MacGregor, S. (2018). Problem solving: How preservice teachers understand it during their preservice learning. In E. Bergqvist, M. Osterholm, C. Granberg, & L. Sumpter (Eds.), *Proceedings of the 42nd Conference of the International Group for the Psychology of Mathematics Education* (Vol. 4, pp. 19-26). PME.

Reinhold, S. (2015). Prospective elementary teachers' diagnostic proceeding in one-on-one diagnostic interviews: Facets of data collection and attention. In *Proceedings of the 39th Conference of the International Group for the Psychology on Mathematics Education* (Vol. 4, pp. 73-80). PME.

Ross, J. A., McDougall, D., & Hogaboam-Gray, A. (2002). Research on reform in mathematics education, 1993-2000. *The Alberta Journal of Educational Research, 48*(2), 122–138.

Royea, D., Nicol, C., & Osana, H. (2015). "Borrowing from the neighbour" – Preservice teacher' interpretations of student errors. In *Proceedings of the 39th Conference of the International Group for the Psychology on Mathematics Education* (*Vol. 4*, pp. 97-104). PME.

Sakonidis, H., Tzekaki, M., & Kaldrimidou, M. (2001). Mathematics teaching practices in transition: Some meaning construction issues. In M. van den Heuvel-Panhuizen (Ed.), *Proceedings of the 25th Conference of the International Group for the Psychology of Mathematics Education* (Vol. 4, RR248). PME.

Sánchez, V., & Garcia, M. (2011). Socio-mathematical and mathematical norms in pre-service primary teachers' discourse. In B. Ubuz (Ed.), *Proceedings of the 35th Conference of the International Group for the Psychology of Mathematics Education* (Vol. 4, pp 105-112). PME.

Sánchez-Matamoros, G., Fernández, C., Valls, J., & Callejo de la Vega, M. L. (2018). Noticing students' mathematical thinking: Characterization, development and contexts. In E. Bergqvist, M. Osterholm, C. Granberg, & L. Sumpter (Eds.), *Proceedings of the 42nd Conference of the International Group for the Psychology of Mathematics Education* (Vol. 2, pp. 91-98). PME.

Schoenfeld, A. H. (2006). Mathematics teaching and learning. In P. A. Alexander & P. H. Winne (Eds.), *Handbook of educational psychology* (pp. 479–510). Lawrence Erlbaum Associates Publishers.

Sevimli, E., & Delice, A. (2011). Is what you prefer what you do? Representations in definite integral. In B. Ubuz (Ed.), *Proceedings of the 35th Conference of the International Group for the Psychology of Mathematics Education* (Vol. 4, pp 153-160). PME.

Stein, M. K., Engle, R. A., Smith, M. S., & Hughes, E. K. (2008). Orchestrating productive mathematical discussions: Five practices for helping teachers move beyond show and tell. *Mathematical Thinking and Learning, 10*(4), 313–340. doi:10.1080/10986060802229675

Stockero, S. L. (2013). Student teacher noticing during mathematics instruction. In A. M. Lindmeier, & A. Heinze (Eds.), *Proceedings of the 37th Conference of the International Group for the Psychology of Mathematics Education* (Vol. 4, pp. 249-256). PME.

Vale, I., Pimentel, T., Cabrita, I., Barbosa, A., & Fonseca, L. (2012). Pattern problem solving tasks as a mean to foster creativity in mathematics. In T. Tso (Ed.), *Proceedings of the 36th Conference of the International Group for the Psychology of Mathematics Education* (Vol. 4, pp 171-178). PME.

Van den Heuvel-Panhuizen, M., & Fosnot, C. T. (2001). Assessment of mathematics achievements: Not only the answers count. In M. van den Heuvel-Panhuizen (Ed.), *Proceedings of the 25th Conference of the International Group for the Psychology of Mathematics Education* (Vol. 4, RR277). PME.

Vos, P., & Bos, K. (2001). Trends (1995-1999) in performances of Dutch grade 8 students in TIMSS against the background of the realistic mathematics curriculum. In M. van den Heuvel-Panhuizen (Ed.), *Proceedings of the 25th Conference of the International Group for the Psychology of Mathematics Education* (Vol. 4, RR045). PME.

Xenofontos, C., & Kyriakou, A. (2016). Prospective elementary teachers' talk during collaborative problem solving. In C. Csikos, A. Rausch, & J. Szitanyi (Eds.), *Proceedings of the 40th Conference of the International Group for the Psychology of Mathematics Education* (Vol. 4, pp. 395-402). PME.

Zapatera, A., & Callejo, M. L. (2013). Pre-service primary teachers' noticing of students' generalization process. In A. M. Lindmeier, & A. Heinze (Eds.), *Proceedings of the 37th Conference of the International Group for the Psychology of Mathematics Education* (Vol. 4, pp. 425-432). PME.

Chapter 2
Reform-Based Mathematics Teaching in the United States

Sumeyra Karatas
University of Texas Rio Grande Valley, USA

Reyhan Safak
The Ohio State University, USA

Gamze Bulut
Brooklyn College, USA

Veysel Karatas
University of Texas Rio Grande Valley, USA

ABSTRACT

This chapter examined the trends in reform-based mathematics teaching practices in the United States classrooms. The authors systematically analyzed the journal articles in the Mathematics Teacher: Learning and Teaching PK-12 (MTLT) in order to reveal the current practices that practitioners and experts in mathematics education deem significant and worthy. They found that the most trending reform practices were mathematical discourse, conceptual understanding, task selection, and real-life applications. They discussed each trending practice through sample strategies and provided examples from the reviewed articles. They also identified the least trending reform practices that need attention and discussed associated challenges.

INTRODUCTION

Past and current reform movements in mathematics education in the United States have been largely shaped by the National Council of Teachers of Mathematics (NCTM). State education standards were begun to be established in the early 1990s after the NCTM standard-based education movement in North America to support a systematic improvement in mathematics education (NCTM, 2014). Each state developed and adopted its own learning standards in the early 2000s (Common Core State Standards

DOI: 10.4018/978-1-7998-9422-3.ch002

Initiative [CCSSI], 2018). After about a decade, state leaders and commissioners of education emphasized the importance of having consistent goals all over the country through their membership in the National Governors Association Center for Best Practices (NGA Center) and the Council of Chief State School Officers (CCSSO). Then, the NGA and CCSSO (2010) released the Common Core State Standards (CCSS) for English Language Arts (CCSS-ELA) and Mathematics (CCSSM) in 2010 to ensure all students graduate from high school as college and career ready. Since then, 41 States, the District of Columbia, four territories, and the Department of Defense Education Activity (DoDEA) have adopted the CCSS (CCSSI, 2018).

Along with the CCSSM content standards, CCSSI (2010) developed Standards for Mathematical Practices (SMP) in the light of NCTM (2000) process standards (i.e., problem-solving, reasoning and proof, communication, representation, and connections) and based on the strands of mathematical proficiency specified in the National Research Council's (NRC, 2001) report (i.e., conceptual understanding, procedural fluency, strategic competence, adaptive reasoning, and productive disposition). The eight SMPs are listed as; (1) make sense of problems and persevere in solving them; (2) reason abstractly and quantitatively; (3) construct viable arguments and critique the reasoning of others; (4) model with mathematics; (5) use appropriate tools strategically; (6) attend to precision; (7) look for and make use of structure; and (8) look for and express regularity in repeated reasoning. The SMPs describe what students are expected to do as they learn mathematics. For example, mathematically proficient students can apply the mathematics they know to solve problems arising in everyday life, society, and the workplace; consider the available tools when solving a mathematical problem, and communicate precisely to others. In addition, each SMP supports the development of strands of mathematical proficiency. For example, conceptual understanding is developed when students are engaged in one or more SMPs (e.g., make sense of problems and persevere in solving them, construct viable arguments and critique the reasoning of others) (Stein et al., 2017).

BACKGROUND

With the adoption of CCSSM, the focus of mathematics standards moved from developing only skills efficiency to rigorous thinking, conceptual understanding, and reasoning (Stein et al., 2017). NCTM (2014) stated that the progress made, the challenges that have been left, and the necessary actions to ensure mathematics success for all students should be examined to realize the potential of these new standards. The percentage of fourth-graders and eighth-graders scoring "proficient" on the National Assessment of Educational Progress (NAEP) increased by 29 and 22 points, respectively, from 1990 to 2013. Yet, only 44% of United States high school graduates are college-ready according to Scholastic Aptitude Test (SAT) and American College Testing (ACT) scores (NCTM, 2014).

Although NCTM (2014) stated that the implementation of CCSSM is promising to address challenges and expand the progress, the CCSSM document does not indicate how to implement these standards in the classroom. Therefore, NCTM (2014) published the *Principles to Actions* (PtA) book to describe necessary conditions to implement CCSSM and ensure mathematical success for all students. The NCTM's PtA identified the eight research-informed high-leverage effective mathematics teaching practices to support the development of SMPs, based on NCTM's (2000) core principles. These eight effective mathematics teaching practices are: (1) establish mathematics goals to focus learning; (2) implement tasks that promote reasoning and problem solving; (3) use and connect mathematical representations; (4)

facilitate meaningful mathematical discourse; (5) pose purposeful questions; (6) build procedural fluency from conceptual understanding; (7) support productive struggle in learning mathematics; and (8) elicit and use evidence of student thinking. These effective mathematics teaching practices are necessary to promote deep mathematics learning and offer educators and school districts a standard lens to improve instructional practices and ensure mathematics success for all students (NCTM, 2014). They, hence, are one of the components that form reform-based teaching practices in the United States.

Reform-based mathematics teaching is broadly defined as integrating teaching practices that build on students` prior knowledge and involve interdisciplinary and real-life application activities (Joram & Gabriele, 2012). Mathematics teaching and curriculum reform emphasize mathematical thinking and reasoning rather than memorization and computation (Spangler & Wanko, 2017). In this respect, reform-based mathematics teaching practices focus on developing a conceptual understanding of mathematics that connects prior knowledge with new knowledge through problem-solving and inquiry-based learning that allows collaboration and hence student-centered instruction (Jong, 2016). Therefore, setting instructional goals to enhance students` engagement and conceptual and procedural understanding is considered essential for reform-based mathematics teaching (Hiebert et al., 2003). Accordingly, reform-based mathematics teaching emphasizes problem-solving, mathematical discourse, justification of the answers, and using multiple methods or representations to solve the problems (NCTM, 2014).

MAIN FOCUS OF THE CHAPTER

This chapter examines the trends of reform-based mathematics teaching practices that are focused on in the *Mathematics Teacher: Learning and Teaching PK-12* (MTLT) articles. The MTLT trends that experts in the field deem significant and worthy of writing and publishing would provide information about meaningful and effective reform-based mathematics teaching practices in the United States.

We reviewed the articles in the practitioner journal, MTLT, published by NCTM. We chose MTLT because NCTM is the world's most extensive mathematics education organization and a leading professional organization for mathematics teachers in the United States. MTLT's scope reflects the current practices of mathematics education and sustains an ongoing process of knowledge-based practice and policy regarding the future of the field. Therefore, MTLT articles reveal information about the most recent reform-based mathematics teaching practices emphasized by practitioners and experts in mathematics education. This chapter sought to answer the following question: What are the most trending reform-based mathematics teaching practices in the United States? In an attempt to describe the recent trends in the United States, we examined the themes and focus of the articles in MTLT.

NCTM combined its three practitioner journals—*Teaching Children Mathematics*, *Mathematics Teaching in the Middle School,* and *The Mathematics Teacher*—in a new journal that spans PK-12, MTLT, in January 2020. MTLT is a digital-first journal that allows authors to include screenshots, videos, or applets which allow the readers to understand the messages and implement the ideas in their classrooms (Barlow, 2020). The journal involves peer-reviewed and invited articles from preschool (PK) to 12th grade, which is the only practitioner journal that focuses on PK to 12 mathematics education published by NCTM since January 2020. Twelve issues were published in 2020 and nine issues in 2021 at the time of this writing. Each issue involves a front-and-center article that is designed to span the full PK–12 range. In addition, there is at least one feature and one focus article specified to grade bands.

The journal also involves tasks or problems specific to grade levels, asked and answered sessions, and other mathematics sections.

A systematic analysis of the trends in MTLT revealed how teacher educators and practitioners approach reform-based mathematics teaching. The trends analysis also revealed which reform-based practices received less attention and associated challenges and obstacles in implementing reform-based mathematics teaching practices. Thus, this chapter described the reform-oriented mathematics practices that have been emphasized in MTLT articles in recent years.

METHODOLOGY

This qualitative study relies on document analysis as a stand-alone methodology based on the research questions proposed. As being aware of the limitations of this methodology as "biased selectivity" (Bowen, 2009), we had a specific purpose of selecting NCTM articles since reforms in mathematics education for K-12 in the United States heavily focus on NCTM standards and publications.

A total of 21 issues of MTLT have been published at the time of this writing. The front-and-center article, feature articles, and focus articles were systematically reviewed to explore reform-based mathematics teaching trends by using open coding strategy (Corbin & Strauss, 2014). As we reviewed the front and center, feature, and focus articles of 21 issues of MTLT published between January 2020 and September 2021, we recognized that some of the reform-based teaching practices were emphasized more than others. Four authors independently created codes for 173 articles in 21 issues for the primary focus, grade level, and main practices. The authors coded several articles together in the beginning, resolved any inter-coder disagreement through discussion, and reached a consensus for coding. Two hundred ninety-five codes were created from the articles based on their focus and main ideas. Some of the articles were classified into more than one code. For instance, the article titled "Ready, Set, Launch! —The Engineering Cycle for Productive Struggle" from issue 14-feature article 3 was coded as both "productive struggle" and "problem-solving" practices.

The authors then merged codes into main themes and established a consensus on the coding using QSR International's NVivo qualitative data analysis software (Version 12) (2018). Relying on open coding methodology (Corbin & Strauss, 2014), 59 themes emerged from 173 articles. Then, these 59 themes were merged into bigger categories. For example, a cluster of themes including concept, sense-making, meaningful mathematics learning, making sense of mathematics, deeper level understanding was merged into "conceptual understanding." As displayed in Table 1, the most frequent themes were mathematical discourse, conceptual understanding, task selection, and real-life application, whereas equity, mathematical modeling, and using technology to solve problems were less frequent themes. Other themes that appeared less than or equal to 1% were anticipating student thinking, attending to precision, estimation, feedback, fluency, generalizing, graphing, lesson study, mathematical habits of mind, mathematical intuition, metacognition, objectives, professional noticing, professionalism, proficiency, project-based-instruction, proof teaching, quantitative reasoning, structural reasoning, student voice, and student-centered instruction. We discussed the four most frequent themes in the following sections.

Table 1. Frequency of common themes in NCTM P-12 articles

Themes	Frequency	Themes	Frequency
Mathematical Discourse	12%	Hands-On/Manipulatives	2%
Conceptual Understanding	10%	Culturally Relevant Pedagogy	2%
Task Selection	8%	Justification	2%
Real Life Applications	8%	Productive Struggle	2%
Technology	4%	Student Engagement	2%
Equity	4%	Tool Use	2%
Mathematical Modeling	4%	Problem Solving Strategies	2%
Interdisciplinary Teaching	3%	Productive Disposition	2%
Games	3%	Others	≤1%
Mathematical Language	3%	Total	100%

Mathematical Discourse

The results of the analysis showed that mathematical discourse was the most frequent focus across the MTLT articles. These articles often proposed various strategies that can foster mathematical discourse in the classroom. While some of the strategies were mainly about mathematical discourse, some of them discussed discourse as a secondary focus, with the main focus on another strategy. In our analysis, the mathematical discourse theme included the following codes: discourse action, meaningful discussion, facilitating discussion, productive discussion, purposeful questioning, assessment and advancement questioning, mathematics talk, representation talk, and chalk talk. The proposed strategies, in the reviewed articles, in support of mathematical discourse are discussed in the following sections.

Discourse Actions

One of the strategies proposed by the articles is applying discourse actions. Discourse actions are defined as the actions taken to promote mathematical discourse (Boston et al., 2019). The specific discourse actions which are highlighted in some articles are teacher's press (the press for reasoning), the student's providing, teacher's linking, student's linking. The articles identified the actions, addressed their benefits and challenges, and provided tools to improve the use of these actions.

Candela et al. (2020) identified the teacher's press as an act of following up on a student's contributions by asking students to say more about their strategy. Students' providing as an action of students to explain their thinking justifications and validate mathematical accuracy. The author described teacher's linking as an action in which a teacher asks students to restate, extend, compare and contrast or relate another student's contribution. Besides, student's linking was defined as an action in which students pick up and use ideas from their peers, such as explaining, relating to, or building on an idea offered by another student. Some tools were provided in the article to improve the implication of these actions, such as checklists and rubrics for measuring, analyzing, reflecting, and enhancing teachers' (and students') use of discourse actions in the classroom.

While the four actions are highlighted altogether by Candela et al. (2020), the press for reasoning is recognized explicitly by Anderson (2021). The author identified the press for reasoning as an instructional routine of responding to a student's explanation with additional questions to elicit student reasoning. One of the benefits of the strategy is to keep classroom talk focused on students' ideas instead of teacher-led explanations. One challenge of this strategy described in the article was asking questions that shape the students' ideas for them, canalize their thinking toward predetermined pathways, or reduce the cognitive demands of the task. To overcome these challenges and use the strategy for benefit, the author offered three routines: (1) anticipate student responses before the lesson begins, (2) target follow-up questions on specific parts of a more extensive explanation, and (3) use sequences of questions that unfurl the student's reasoning in its entirety.

Purposeful Questioning/Assessment and Advancement Questions

Another strategy that was pointed out by articles is purposeful questioning. This aligns with 'NCTM's (2014) vision that effective mathematics teaching relies on questions that encourage students to explain and reflect on their thinking as an essential component of meaningful mathematical discourse (p.35). Stohlmann (2020a) emphasized that assessment and advancement questions encouraged a rich mathematical discourse by helping students deepen their understanding and prepared teachers to provide informed answers to students' thinking. Besides, Hallman-Thrasher and Spangler (2020) developed strategies and planning tools to help teachers pose questions with specific purposes to facilitate mathematical discussion.

More specifically, Hallman-Thrasher and Spangler described how teachers could develop and pose purposeful questions to engage students in high-cognitive tasks that reveal their thinking, encourage students to communicate and justify their ideas, and maintain a high cognitive demand. They emphasized the use of cognitively demanding tasks, which provide more opportunities for higher-order questioning. In their article, some types of purposeful questions were illustrated by using examples of preservice teachers' work: questioning to elicit students' thinking, questioning to generate student ideas, questioning to clarify student explanation, and questioning to justify.

Stohlmann (2020a) defined the purposeful questions as assessment and advancement questions more specifically. He described how assessment and advancement questions could be planned and implemented based on anticipated solutions of students for a task that was focused on creating an equation to solve a real-world problem. He emphasized thinking about possible student strategies and the questions that could be asked to the students before implementing the task. Students watched a video about a world record for the longest distance traveled rolling across exercise balls and engaged in a real-world problem. He asked the questions that helped students assess their solutions and strategy and helped teachers figure out whether students understood the context. He anticipated possible students' strategies and created assessment and advancement questions for each possible strategy.

Facilitating Discussion

Teachers must determine how to build on and honor student thinking while ensuring that the mathematical ideas at the heart of the lesson remain prominent in class discussions (Engle & Conant, 2002, as cited in NCTM, 2014, p.29). NCTM (2014) emphasized that the teacher must decide how to facilitate the discussion by considering what approaches to share and their order. The teacher must also consider the questions that will help students connect the strategies and critical disciplinary ideas driving lessons to orchestrate a class discussion of student approaches to solving a task. We observed that some articles emphasized the ways of facilitating discussion to develop meaningful mathematical discourse by proposing some strategies. While Ahrendt et al. (2021) pointed out that productive discussion calls for rich tasks, Ghousseini et al. (2021) emphasized using classroom resources and representation to facilitate meaningful discussion. Bennett and Morgan (2020) defined the Chalk Talk as a nonverbal, student-centered, and interactive procedure of facilitating discussion. The article in which Brand et al. (2021) described a teacher's discussion strategies that promote mathematical discourse is another study that emphasized ways of facilitating discussion.

Ahrendt et al. (2021) pointed out that productive discussion calls for rich tasks with multiple solution pathways and gives students opportunities to share their reasoning in ways that inspire others by making them gain a deeper understanding. Ahrendt and colleagues presented a well-designed task for generating a meaningful discussion. They implemented Smith and Stein's (2011) five practices, viewed as crucial parts of orchestrating productive discussion, and a task that supported students to engage in meaningful mathematical discourse. The practices were: (0) choosing a mathematical goal and appropriate task, (1) anticipating student thinking, (2) monitoring student work, (3) selecting students to share work, and (4) sequencing students' work.

Ghousseini et al. (2021) suggested that children as early as kindergarten can reason about mathematics and justify their thinking using both oral language and gestures with the help of available representations in their classroom (e.g., number lines, pictures, and number charts). They highlighted teachers' crucial role in supporting students to engage in mathematical discourse by encouraging them to use classroom resources. Five lessons were observed and videotaped in a kindergarten classroom, and the teachers were interviewed after each lesson. By analyzing the videos and transcripts of class discussions and interviews, they shared three essential practices that yielded from the analysis: (1) establishing expectations that support mathematical talk, (2) eliciting student thinking, and (3) narrating student thinking.

Similarly, Bennett and Morgan (2020) defined the Chalk Talk as a nonverbal, student-centered, and interactive procedure to promote discussion by ensuring that all students have an equitable opportunity to share ideas and provide feedback (Roberts, 2013). They called the chalk talk a silent discussion. This discussion was composed of three steps: (1) student response for initial prompts, (2) student response for initial comment, and (3) whole-class discussion using student responses to highlight the intended outcomes and understandings. The goal was to share ideas and respond to each other. Students were required to express the reason for their agreement and disagreement with different responses. The teacher reminded and emphasized that disagreement is positive interaction that helps them understand others' mathematical ideas and perspectives better. During the first and second steps, teachers observed each station and tried to figure out common themes to lead a whole-class discussion effectively in step three. In other words, Chalk Talk was used as a way for students to engage in mathematical discourse in the form of writing communication.

Likewise, Brand et al. (2021) described a teacher's discussion strategies that promote mathematical discourse during a mathematical modeling activity. They named five notable strategies: (a) promoting common language, (b) encouraging students' perspectives and decision-making, (c) prompting a revision of students' thinking, (d) probing students' responses for assumptions, and (e) emphasizing students' reasoning for further discussion. They observed 6-8 grade students' mathematical modeling activities and categorized the outcome of students' responses associated with each strategy.

Purposefully Structured Lesson Techniques

Mathematical discourse includes the purposeful exchange of ideas through classroom discussion and other verbal, visual, and written communication forms (NCTM, 2014). In line with this, some articles presented instructional and assessment techniques which are purposefully structured to allow students to engage in meaningful mathematical discourse. For example, Thompson and Gay (2020) emphasized the impact of meaningful discourse with fact strategy; and Hicks and Bostic (2021) used a formative assessment technique named whole-class think aloud (WCTAs), which allows students to engage in meaningful mathematical discourse. Additionally, Gunter and Raymond (2020) presented a Spend a Buck strategy (Kagan, 2001) which fostered communication and metacognition through an effective mathematical discourse.

Thompson and Gay (2020) explored the impact of meaningful discourse with fact strategy by working with a group of third-grade teachers. They pointed out that using purposeful structure, instead of just showing the various techniques to students and expecting mastery, helped students engage with the reasoning and thinking behind the strategies. The authors argued that intentional use of discourse and fact strategy instruction could create the bridge students need to connect learning fact strategies to using fact strategies. The combination of using facts strategy and discourse will allow students to explore their thinking and to make sense of the reasoning of others for uncovering understanding and misunderstanding critical concepts.

Hicks and Bostic (2021) shared WCTAs, a formative assessment technique, for problem-solving instruction. They emphasized that high-quality formative assessments need to monitor students' progress toward learning targets and foster deep mathematical learning (NCTM, 2014). Students worked with their partners on one mathematical problem by writing their ideas and draft talk. While the students shared their rough draft thinking aloud, the teacher observed them and took notes about how they read and interpreted the problem, which strategies they applied to solve it, and how they reached a solution. In this technique, while teachers could attend to all students' thinking in real-time, they could monitor students' progress toward the learning target and foster their deep mathematical learning (NCTM, 2014). This technique allowed students to talk about the mathematics they learned and engage in meaningful mathematical discourse.

Similarly, Gunter and Raymond (2020) presented a technique that helped students better understand mathematics and developed their communicative and metacognitive skills. In the article, the Spend a Buck strategy (Kagan, 2001) was used to develop students' thinking strategies for solving equations. To apply this technique, the teacher asked students to "invest" a dollar in a set of strategies for solving equations by considering which strategy was most likely to be more effective. Students discussed their solutions strategy, shared their thinking aloud, and listened to the thoughts of their peers. Since they communicated their thinking during the activity, the Spend a Buck strategy also helped students engage in metacognitive behaviors, including *planning* solutions, *debugging* progress, and *evaluating* their results. As a result, all these activities fostered communication and metacognition through an effective mathematical discourse, which is crucial to help students build their ideas about mathematics.

To sum up, the review of the MTLT articles that focused on mathematical discourse showed that mathematical discourse is an essential element of any kind of mathematics class, regardless of the focus, content, and grade level. Some critical common strategies of facilitating mathematical discourse discussed in most articles were choosing high-cognitive tasks, assessing and advancing students' thinking by using purposeful questions, encouraging students to share their perspectives and respond to others' perspectives, and anticipating students' possible needs responses to lead a meaningful discussion.

Conceptual Understanding (Mathematical Thinking and Reasoning)

The analysis revealed that conceptual understanding was also one of the most common focuses across the MTLT articles. We grouped the following codes and developed the conceptual understanding category: leveraging students` thinking, meaningful mathematics learning, making sense of mathematics, strategic sense-making, eliciting, and using evidence of students` thinking. A variety of activities, strategies, tools, and teaching methods that promote conceptual understanding and leverage students` thinking was provided in the articles. Some of these tools and strategies that enable conceptual understanding could be listed as mathematical action technologies, real-life applications, exploration, scaffolding, gameplay, and choral counting.

Davis (2021) used TI-Nspire, and Harrow and Merchant (2020) used "Desmos" in order to explore how mathematical action technologies make the static more dynamic. Davis designed a lesson with the TI-Nspire calculators to teach the system of linear inequalities rather than using the test-point procedure that many textbooks have used. The graphing calculator let the students move the point in each region and even on the solid and dashed lines to see if the point was a solution for both inequalities. In this way, students could make conjectures and generalizations about when to use solid lines or dashed or what region was supposed to be shaded. Harrow and Merchant (2020) also found that the Desmos enhanced visualizations improved conceptual understanding across cartesian and polar coordinate systems.

On the other hand, Stephan et al. (2020) explored that real-life context plays a crucial role in learning mathematics conceptually by creating connections between abstract and concrete thinking and engaging students in conceptual mathematical discourse. Roy et al. (2020) also used a real-life application task in which students were asked how to order chicken nuggets from their favorite restaurant for the eighth-grade holiday party. Given that the restaurant sells the nuggets in packages of 6, 9, or 20, their goal was to determine any number of chicken nuggets from 1 to 100 that they could order. The teacher positioned the students as capable contributors and engaged students by using mathematical discourse during the instruction. By the end of the lesson, students were able to leverage their conceptual understanding of algebraic reasoning. Naanou and Rhodes (2020) also used real-life context, focusing on how culturally relevant experiences may promote cultural and conceptual understanding. They designed a lesson about folding (asafiri or crescent) qatayef (a special cultural dessert made on Saint Barbara's Day) to maximize the amount of filling in each qatayef.

Besides, Frank (2021) claimed that lessons that promote conceptual understanding let students learn mathematical structure, not just computations. In Frank's lesson implementation, exploring the quadratic formula allowed students to understand the structure of algebraic representations. Students derived the quadratic formula by quantifying the symmetry of a parabola. At the end of the activity, students were able to use the quadratic formula to solve quadratic equations and reason about the structure of the quadratic formula. Rakes et al. (2020) also showed that a procedural concept such as simplifying radical expressions could be taught in a way that helps students understand conceptually and create a stronger foundation for building on the knowledge rather than memorization. In their lesson, students explored radical simplification through geometry connections, in other words, through semi-concrete geometric representations. Besides, Lischka and Stephens (2020) showed how overarching concepts and models could build connections and sense-making. They provided examples of using the area model in different grade levels and enhancing students' conceptual understanding. They discussed that while the area model is used to multiply whole numbers in the third grade, it could also be used to multiply binomials or factors in Algebra. If students can use the area model across the grade levels and concepts, they could use it to make sense of mathematics (SMP1) and understand the structure of mathematical concepts (SMP7).

In addition to using technology, real-life applications, or geometric representations of procedures, scaffolding was also suggested as one of the strategies to promote conceptual understanding emphasized by Coleman (2020). Coleman (2020) designed a lesson for 6th-grade students who have significant gaps in the content knowledge and showed how scaffolding and building a bridge from a student's basic understanding of an intentionally selected grade-level topic to a strong foundation in the grade-level topic improve their conceptual understanding and procedural fluency.

Game-playing (MacDonald et al., 2020) and choral counting (McMillan & Sagun, 2020) were pointed out as other tools that promote conceptual understanding. MacDonald et al. (2020) showed how gameplays with dominoes could leverage students` quantitative and additive reasoning and how using hands-on tools and game-playing enhanced students` conceptual understanding of subitizing and unit coordination. McMillan and Sagun (2020) used choral counting to engage students in instruction. In their lesson implementation, students counted out loud; and the teacher charted each number in an organized way and asked students what they noticed about the numbers. Students recognized patterns and explored additive or multiplicative ideas using choral counting, which leveraged students' thinking and built conceptual understanding.

To sum up, the conceptual understanding was the second frequent theme emphasized in the MTLT articles over the last two years. Classroom stories presented various strategies and tools that can enhance conceptual understanding and leverage students` thinking. The main idea across the presented strategies and tools was helping students visualize mathematics by using technology or real-life or pictorial application; and assisting students in developing reasoning skills through games, interactions, and discussions.

Task Selection

Task selection has been highly emphasized in the reviewed MTLT articles. The authors presented diverse perspectives on task selection. Thus, four main codes emerged in the reviewed articles focusing on task selection: perseverance/curiosity/engagement/productive struggle; features of the quality tasks; and instructional approaches for implementing quality tasks. The following sections provide specific examples from some articles to illustrate how task selection was discussed in the reviewed articles.

Perseverance/Curiosity/Engagement/Productive Struggle

Task selection was often discussed and studied in relation to SMP1, making sense of problems and persevering in solving them. For example, Lischka et al. (2020) suggested implementing extended tasks to promote perseverance in students. These extended tasks must have multiple entry points and solution approaches within students' current understanding offering perspectives on future math topics. The swimmer task example was given as an extended task, and how it let two students persevere in solving the problem was reported in the article:

Two swimmers start at opposite ends of a 90 feet rectangular lap pool, each in their lane. One swims 30 feet per minute, and the other swim 20 feet per minute. If they swim for 30 minutes, how many times will they meet each other, and at what times will they meet? (p. 49).

This problem can be solved with both simple and complex strategies to promote perseverance in students. Similar to implementing extended tasks, Gilbertson (2021) presented "no-solution problems" to engage students in sense-making and perseverance in problem-solving. Gilbertson argued that "no-solution problems" provide students with multiple opportunities:

- Motivating meaningful discussion and positive engagement
- Establishing mathematical truth
- Sparking curiosity through open problems
- Motivating new mathematics
- No solution in the real world
- Engaging in mathematical practices (p.209-210)

Yee et al. (2020), Manousaridis (2021), and Taylor and Lee (2021) articulated the productive struggle, which is "a necessary facet of effective mathematics teaching" (p.124) via the example of STEM Rocket Launch Task. NCTM (2014) indicated that "Effective teaching of mathematics consistently provides students, individually and collectively, with opportunities and support to engage in productive struggles as they grapple with mathematical ideas and relationships" (p.10). STEM Rocket Launch Task required high school students to design a prototype rocket, fly it, and make revisions based on its flight. This task not only emphasized that "mistakes are valued" but also "persistence is valued" in classrooms engaging in productive struggle.

On the other hand, Manousaridis (2021) suggested using tasks promoting curiosity in students. The authors asked the following subtraction questions to the students:

54 – 45 = ?
43 – 43 = ?
98 – 89 = ?

Once students recognized the result of all problems equal to nine, they became curious about why it was always nine, which led to a great discussion in the classroom. Manousaridis (2021) confirmed 'Warshauer's (2015) claim that such tasks genuinely engage students in productive struggle.

Features of the Tasks

Features of the tasks that could lead to desired learnings and understandings in students were described in a variety of ways in MTLT articles. For example, Bostic et al. (2021) claimed that implementing mathematics tasks with high cognitive demand and multiple entry points (i.e., low-floor and high-ceiling tasks) is essential in fostering rich mathematics for each student, while O'Connor et al. (2021) described such tasks as simply high level. Similarly, Ahrendt et al. (2021) pointed to the significance of choosing high-level tasks that support students in meeting the goal of the lesson before implementing Smith and Stein's (2011) five practices for orchestrating productive discussions in the classroom.

Russo et al. (2020) provided a more comprehensive description of features of tasks: meaningful context, visualization, offering opportunities to make predictions, and comparison to other concepts. On the other hand, Wasserman et al. (2020) highlighted the desired task features as encouraging problem-solving, offering conceptual insights, and preparing students for future mathematics studies.

While Jung and Brand (2021) offered social-justice-oriented mathematical tasks to view mathematics as a way of understanding social justice issues, Barlow et al. (2020) relied on Substitution, Augmentation, Modification, and Redefinition. (SAMR) model (Puendetura, 2013) that intended to encourage educators to significantly enhance the quality of education provided via technology for selecting tasks from MTLT front-and-center articles to use in a virtual setting.

Instructional Approach

MTLT articles relating to task selection with desired features also proposed some instructional approaches for successful classroom implementation. For example, for word problems that are challenging to understand and solve, Melissa et al. (2021) offered reading comprehension strategies that could help students make sense of the context while solving: visualizing, retelling, making connections, and asking questions. By interviewing ten Algebra 2 teachers, Snider (2021) found that teachers employed a variety of considerations as general and specific examples, misconceptions, splitting a concept or procedure into cases or essential ideas, boundary cases, non-examples, using examples to cause cognitive dissonance, sequencing different types of examples, how students will engage with the examples, and coherence across lessons and units.

While Ivy et al. (2020) emphasized the importance of asking "why" questions while students are solving reversibility tasks which are planned to build flexibility in students' thinking, Lim (2020) offered Superficially Similar but Structurally Different (SS-SD) problems as a way of addressing hammer-and-nail phenomenon which is a tendency of students' using a proportion (hammer) to solve a missing-value problem even when it is non-proportional (mistaken as a nail) (p.488).

Brakoniecki et al. (2021) presented a common proportional reasoning problem, the orange juice task, which offers multiple solution strategies to solve the problem. They implemented the presentation of multiple strategies in response to the problem as an instructional strategy. The strategies were unitizing, partitioning, building up, norming, reasoning up and down, using multiplicative comparison, and percentage and decimal comparison.

Real-Life Applications

Real-life applications are also highly emphasized in the reviewed articles. The real-life application can be categorized as real-life contexts and multiple representations based on our coding criteria. The real-life context includes the following codes: Situation-oriented contexts, real-world contexts, real-world problems, real-life experiences, real-life applications, real-life problem solving, and real-life situations. The multiple representations include the following codes: Mathematical representations, visual representations, and using visual representations.

Real-Life Context

Effective mathematics teaching practices allow students to learn new mathematics by solving contextual problems (NCTM, 2014). Most mathematics teachers receive the question of "when/where will we use mathematics in real life?" from their students. Therefore, there is a need to implement activities that will allow students to make connections between school mathematics and real life. Cunningham (2021) claimed that mathematics needs to permeate the school day to assist students in making connections between mathematics at school and the mathematics they encounter in everyday life. Bartell (2020) stated that reimagining homework and assigning work that released information about students` backgrounds and lived experiences empowered the way of mathematical teaching and learning.

Stephan et al. (2020) used the work of several middle school teachers who used realistic contexts as both hooks and anchors to illuminate three features of real-world contexts: (1) support the transition from concrete to abstract reasoning; (2) engage students conceptually mathematical discourse; and (3) capitalize on real-world and experiential real-world situations. They concluded that real-life contexts increase students' interest and motivation and demonstrate how mathematics is used to make sense of their world. Also, Brown (2021) found that real-world tasks provide students with learning opportunities that allow mathematics to be used to understand the world, strengthen mathematical knowledge, and provide students with opportunities to use mathematics to analyze the world they live in.

Multiple Representations

Stohlmann (2020b) stated that teachers could support students' persistence in problem-solving by explaining how to be successful in an escape room, emphasizing and supporting productive teamwork, and combining multiple representations. The author chose the escape room challenges with these ideas in mind, so students solve the challenges by using multiple representations. The escape room challenge task had seven questions, and students were encouraged to use multiple representations to solve problems.

Students increase their problem-solving skills by choosing among many representations that they believe will most likely lead to the solution of a problem (NCTM, 2000). Monson et al. (2020) discussed that it is essential to represent increasingly complex ideas about fractional multiplication and division concepts, such as selecting problem-solving strategies and using mathematical representations to solve problems. Using multiple representations to teach key fraction ideas in learning mathematics expanded students' experience to develop a solid understanding of units, divisions, and fractions (Monson et al., 2020). The authors found the most effective teaching models for building fraction understanding, including fraction circles, paper-folding strips, and number lines.

CONCLUSION

The front and center, focus, and feature articles of 21 issues of MTLT were systematically reviewed by four researchers to examine the trends of reform-based instruction focused in MTLT articles. MTLT is the only practitioner journal that NCTM, one of the leading organizations in mathematics education, has been publishing since 2020. The analysis of the reviewed articles revealed information about the insight into what the authors of these articles and editors of MTLT journal feel is vital regarding reform-based instruction. The results showed that the most frequently published reform-based teaching practices were mathematical discourse, conceptual understanding, task selection, and real-life applications. Mathematical modeling, technology integration into teaching mathematics, questioning, gaming, interdisciplinary mathematics were the practices that were also paid attention to in the reviewed articles. Some of the infrequent practices were anticipating students` thinking, attending to precision, estimation, feedback, fluency, generalizing, and more.

MTLT articles provided many examples about reform-based mathematics teaching practices and how these practices promote deep learning of mathematics. The mathematical discourse was the most frequent practice across the reviewed articles over the last two years. Mathematical discourse is a set of tools and practices which give students opportunities to share their ideas and make their understandings clear, construct convincing arguments regarding why and how things work, develop a language for expressing mathematical ideas, and learn to see things from other perspectives by including purposeful exchange of ideas through classroom discussion and other forms of verbal, visual, and written communication (NCTM 1991, 2000, as cited in NCTM 2014 p.29). While some of the articles focused on discourse actions as *teacher's press student's providing*, *teacher's linking,* and *student's linking* (Anderson, 2021; Candela et al., 2020), some of them provided practices as choosing mathematical goals, anticipating students` thinking that will provide opportunities to have productive mathematical discussions (Ahrendt et al., 2021).

The second most frequent practice was conceptual understanding and leveraging students` thinking. The articles coded in this category provided strategies, tools, and practices that promote conceptual understanding. These strategies and tools could be listed as integrating mathematical action technologies, real-life application, activities that allow students to explore, scaffolding between grade levels, game-playing, and choral counting. Conceptual understanding is one of the five strands of mathematical proficiency and is defined as comprehension of mathematical concepts, relations, and operations (NRC, 2001). Conceptual understanding establishes the foundation needed for procedural fluency as it is emphasized in building procedural fluency from conceptual understanding effective mathematics teaching practice by NCTM (2014).

Task selection was the third most frequent focus in the reviewed articles. Roy et al. (2020) pointed out that one of the most critical features of effective mathematics teaching is task selection, allowing students to explore and discuss mathematics. While Gilbertson (2021) provided an example of "no solution problems" for sense-making and reasoning, Taylor and Lee (2021) presented the STEM Rocket Launch task to point out that mistakes and persistence are valued. Effective teaching of mathematics and reform-based mathematics teaching practices encourage teachers to engage students in solving and discussing tasks that promote mathematical thinking, reasoning, and problem-solving (NCTM, 2014). These tasks allow multiple-entry points, a variety of ways of representations, tools, and methods.

Last but not least, the fourth most frequent trend in the reviewed articles was real-life applications categorized as real-life context and multiple representations. Real-life applications allow students to connect school mathematics and everyday life (Cunningham, 2021). Mathematical concepts could be represented through multiple representations: symbolic, verbal, graphical, and numeric (NCTM, 2000). NCTM (2014) identified several key aspects of what students should do in the classroom: (a) using multiple representations to make sense and understand mathematics; (b) explaining and verifying their mathematical understanding and reasoning with drawings, diagrams, and other representations; and (c) contextualizing mathematical ideas by connecting them to real-world situations. By linking the mathematical content and impactful practices such as connecting mathematical representations and implementing tasks promote reasoning and problem-solving, powerful mathematics learning experiences, and meaningful classroom discourse (NCTM, 2014).

In summary, these four trending practices are central to reform-based mathematics teaching as mathematical discourse, conceptual understanding, mathematical thinking, integrating real-life application to mathematics teaching, using multiple representations are often pointed out as the prominent aspects of reform-based teaching by mathematics educators (e.g., Hiebert et al., 2003; Joram & Gabriele, 2012; Spangler & Wanko, 2017). On the other hand, other prominent aspects that are central to reform-based teaching practices, such as setting instructional goals (Hiebert et al., 2003) and problem-solving (NCTM, 2014), did not receive as much attention in the reviewed articles. It was surprising that only a few articles have focused on problem-solving per se in MTLT, although problem-solving is identified as one of the eight effective mathematics teaching practices; and making sense of problems and persevering in solving them is one of the eight SMPs in Common Core State Standards (2010). Although problem-solving was not often the main focus and goal of the articles, it was used as a tool for discourse and or conceptual understanding. We also noticed that each of the four most trendy practices is closely related to NCTM's (2014) eight effective mathematics teaching practices. For instance, while the mathematical discourse theme refers to facilitating meaningful mathematical discourse, the conceptual understanding theme directly links to building procedural fluency from conceptual understanding and has indirect links to elicit and use evidence of students` thinking. Besides, one can see explicit connections between the task selection theme and implementing tasks that promote problem-solving and reasoning, and between the real-life application theme and using and connecting mathematical representations. Other effective mathematics teaching practices, such as asking purposeful questions, were a stand-alone focus only in a few articles. Yet, some articles discussed it as a strategy supporting mathematical discourse. Anderson (2021) pointed to the challenges of implementing this strategy: reducing the task's cognitive demand and shaping students' reasoning. This is only one example of the challenges of implementing reform-based teaching practices mentioned in the reviewed articles.

Interestingly, only a few articles focused on productive struggle despite its popularity both in informal and formal education forums and presentations. This could be because of the challenge in observing productive struggle in students, and it's one of the most challenging practices among the eight effective mathematics teaching practices for teachers. When the United States is compared to other countries, students are rarely asked reasoning and thinking questions about mathematical ideas (Banilower et al., 2006; Heibert & Stigler, 2004). United States teachers want to rescue their students because they may perceive students' frustration as an immediate failure (NCTM, 2014). On the other hand, it is surprising that only a single article was found that focused on establishing focused mathematics goals. Although setting instructional goals is emphasized as an aspect of reform-based teaching, and also it is one of the NCTM's effective teaching practices, the reviewed articles paid no attention to this practice. Establishing clear goals helps teachers make decisions about the tasks and activities and provide opportunities for students to check their progress (NCTM, 2014). These eight effective mathematics teaching practices represent a core set of high-leverage practices that promote deep learning of mathematics. Therefore, each practice needs attention and should be observable in the MTLT articles. Problem-solving, supporting productive struggle, and establishing mathematical goals need more attention so that practitioners may better understand how to integrate these practices into teaching.

TAKE-AWAYS, FUTURE DIRECTIONS, AND LIMITATIONS

The trends that were focused on the most frequently by experts in mathematics education provide valuable information about the reform-based mathematics teaching practices and their impacts on students` performance. Classroom teachers may benefit from learning specific techniques in support of each practice that they can implement in their classrooms.

This chapter is limited to reviewing only 21 issues of MTLT. Other three journals of NCTM that are published until 2020 were not reviewed and included in this chapter. This is relatively small sample size when the number of journals is considered. Future researchers could review more articles from different journals and compare the most common trends in each journal. Besides, reviewing more articles may provide more information about experts` perspectives on reform-based mathematics teaching practices.

Future researchers may also group the articles according to the grade level as elementary, middle, high school and compare the most frequent reform-based mathematics teaching practices focused by experts.

REFERENCES

Ahrendt, S., Monson, D., & Cramer, K. (2021). Promoting discourse: Fractions on number lines. *Mathematics Teacher: Learning and Teaching PK-12, 114*(4), 284-289.

Anderson, N. (2021). Keep calm and press for reasoning. *Mathematics Teacher: Learning and Teaching PK-12, 114*(8), 591-597.

Banilower, E. R., Boyd, S. E., Pasley, J. D., & Weiss, I. R. (2006). *Lessons from a decade of mathematics and science reform: A capstone report for the local systemic change through teacher enhancement initiative.* Horizon Research, Inc.

Barlow, A. T. (2020). Well played. *Mathematics Teacher: Learning and Teaching PK-12, 113*(1), 2-3.

Barlow, A. T., Edwards, C. M., Robichaux-Davis, R., & Sears, R. (2020). Enhancing and transforming virtual instruction. *Mathematics Teacher: Learning and Teaching PK-12, 113*(12), 972-982.

Bartell, T. G. (2020). Reimagining homework to connect to home and community knowledge. *Mathematics Teacher: Learning and Teaching PK-12, 113*(12), 1023-1028.

Bennett, C. A., & Morgan, M. J. (2020). Developing norms with silent discussions. *Mathematics Teacher: Learning and Teaching PK-12, 113*(10), 841-844.

Bostic, J. D., Vostal, B., & Folger, T. (2021). Growing TTULPs through your lessons. *Mathematics Teacher: Learning and Teaching PK-12, 114*(7), 498-507.

Boston, M., Candela, A. G., & Dixon, J. K. (2019). *Making sense of mathematics for teaching to inform instructional quality*. Solution Tree Press.

Bowen, G. A. (2009). Document analysis as a qualitative research method. *Qualitative Research Journal, 9*(2), 27–40. doi:10.3316/QRJ0902027

Brakoniecki, A., Amador, J. M., & Glassmeyer, D. M. (2021). One task, multiple proportional reasoning strategies. *Mathematics Teacher: Learning and Teaching PK-12, 114*(1), 33-40.

Brand, S., Jung, H., Dorlack, A., & Gailliot, S. (2021). Launching a mathematical modeling lesson. *Mathematics Teacher: Learning and Teaching PK-12, 114*(4), 298-305.

Brown, J. P. (2021). Understanding the real world: Sweaty jo. *Mathematics Teacher: Learning and Teaching PK-12, 114*(7), 528-535.

Candela, A. G., Boston, M. D., & Dixon, J. K. (2020). Discourse actions to promote student access. *Mathematics Teacher: Learning and Teaching PK-12, 113*(4), 266-277.

Coleman, L. K. (2020). Build a bridge: Provide access to grade-level content for all students. *Mathematics Teacher: Learning and Teaching PK-12, 113*(7), 590-593.

Common Core State Standards Initiative. (2010). *National governors association center for best practices*. Council of Chief State School Officers.

Common Core State Standards Initiative. (2018). *National governors association center for best practices*. Council of Chief State School Officers.

Corbin, J., & Strauss, A. (2014). *Basics of qualitative research: Techniques and procedures for developing grounded theory* (4th ed.). Sage publications.

Cunningham, E. P. (2021). Infusing the day with mathematical moments. *Mathematics Teacher: Learning and Teaching PK-12, 114(*1), 70-73.

Davis, J. D. (2021). Constructing and adapting technology-infused lessons. *Mathematics Teacher: Learning and Teaching PK-12, 114*(4), 290-297.

Engle, R. A., & Conant, F. R. (2002). Guiding principles for fostering productive disciplinary engagement: Explaining an emergent argument in a community of learners classroom. *Cognition and Instruction, 20*(4), 399–483. doi:10.1207/S1532690XCI2004_1

Frank, K. (2021). The structure of the quadratic formula. *Mathematics Teacher: Learning and Teaching PK-12, 114*(5), 395-398.

Ghousseini, H., Lord, S., & Cardon, A. (2021). Supporting mathematics talk in kindergarten. *Mathematics Teacher: Learning and Teaching PK-12, 114*(5), 362-368.

Gilbertson, N. J. (2021). The opportunities of no-solution problems. *Mathematics Teacher: Learning and Teaching PK-12, 114*(3), 205-211.

Gunter, M. D., & Raymond, K. M. (2020). Communication and Metacognition in the Classroom. *Mathematics Teacher: Learning and Teaching PK-12, 113*(12), 1017-1022.

Hallman-Thrasher, A., & Spangler, D. A. (2020). Purposeful questioning with high cognitive–demand tasks. *Mathematics Teacher: Learning and Teaching PK-12, 113*(6), 446-459.

Harrow, C., & Merchant, N. (2020). Envelope curves unify sinusoidal graphing. *Mathematics Teacher: Learning and Teaching PK-12, 113*(4), 301-308.

Heibert, J., & Stigler, J. W. (2004). A world of difference: Classrooms abroad provide lessons in teaching math and science. *Journal of Staff Development, 24*(4), 10–15.

Hicks, T., & Bostic, J. D. (2021). Formative assessment through think alouds. *Mathematics Teacher: Learning and Teaching PK-12, 114*(8), 598-606.

Hiebert, J., Gallimore, R., Garnier, H., Givvin, K. B., Hollingsworth, H., Jacobs, J., Chui, A. M.-Y., Wearne, D., Smith, M., Kersting, N., Manaster, A., Tseng, E., Etterbeek, W., Manaster, C., Gonzales, P., & Stigler, J. (2003). *Teaching mathematics in seven countries: Results from the TIMSS 1999 Video Study*. PsycEXTRA Dataset.

Ivy, J. T., Bush, S. B., & Dougherty, B. J. (2020). Stacking the deck: Reversibility and reasoning. *Mathematics Teacher: Learning and Teaching PK-12, 113*(1), 65-68.

Jong, C. (2016). Linking reform-oriented experiences to teacher identity: The case of an elementary mathematics teacher. *The Journal of Educational Research, 109*(3), 296–310. doi:10.1080/00220671.2014.947398

Joram, E., & Gabriele, A. J. (2012). How mathematics teachers can help curb childhood obesity. Iowa Council of Teachers of Mathematics Journal, 39, 3.

Jung, H., & Brand, S. (2021). Student actions for social justice-oriented mathematical tasks. *Mathematics Teacher: Learning and Teaching PK-12, 114*(5), 378-385.

Kagan, S. (2001). *Kagan structures: Research and rationale*. Kagan Online Magazine.

Lim, K. H. (2020). Addressing the hammer-and-nail phenomenon. *Mathematics Teacher: Learning and Teaching PK-12, 113*(6), 487-495.

Lischka, A. E., Prince, K. M., & Reed, S. D. (2020). Implementing extended tasks: Developing perseverance. *Mathematics Teacher: Learning and Teaching PK-12, 113*(1), 47-52.

Lischka, A. E., & Stephens, D. C. (2020). The area model: Building mathematical connections. *Mathematics Teacher: Learning and Teaching PK-12, 113*(3), 186-195.

MacDonald, B. L., Moss, D. L., & Hunt, J. H. (2020). Dominoes: Promoting units construction and coordination. *Mathematics Teacher: Learning and Teaching PK-12, 113*(7), 551-557.

Manousaridis, T. (2021). Promoting inquisitiveness. *Mathematics Teacher: Learning and Teaching PK-12, 114*(5), 369-377.

McMillan, B. G., & Sagun, T. (2020). Extending choral counting. *Mathematics Teacher: Learning and Teaching PK-12, 113*(8), 618-627.

Monson, D., Cramer, K., & Ahrendt, S. (2020). Using models to build fraction understanding. *Mathematics Teacher: Learning and Teaching PK-12, 113*(2), 117-123.

Naanou, P., & Rhodes, S. (2020). Cultural connections: Exploring the mathematics of qatayef. *Mathematics Teacher: Learning and Teaching PK-12, 113*(12), 1034-1038.

National Council of Teachers of Mathematics. (1991). *Professional standards for teaching mathematics*. National Council of Teachers of Mathematics.

National Council of Teachers of Mathematics. (2000). *Principles and standards for school mathematics*. National Council of Teachers of Mathematics.

National Council of Teachers of Mathematics. (2014). *Principles to actions: Ensuring mathematical success for all*. National Council of Teachers of Mathematics.

National Governors Association Center for Best Practices (NGA Center) & Council of Chief State School Officers (CCSSO). (2010). Common core state standards for mathematics. Washington, DC: NGA Center and CCSSO.

National Research Council. (2001). *Adding it up: Helping children learn mathematics*. National Academies Press.

NVivo qualitative data analysis software (Version 12). (2018). *QSR International*. Retrieved on October 1, 2021, from https://www.qsrinternational.com/nvivo-qualitative-data-analysis-software/home

O'Connor, K., Dearborne, E., & Casa, T. M. (2021). Inquiry + math workshop model = success! *Mathematics Teacher: Learning and Teaching PK-12, 114*(3), 187-195.

Puendetura, R. (2013). *SAMR: Moving from enhancement to transformation*. Retrieved on October 1, 2021, from http://www.hippasus.com/rrpweblog/archives/2013/05/29/SAMREnhancementToTransformation.pdf

Rakes, C. R., Kirvan, R. J., & Witkowski, A. (2020). Discovering radical simplification through geometry connections. *Mathematics Teacher: Learning and Teaching PK-12, 113*(9), 737-744.

Roberts, S. L. (2013). The "Chalk Talk" 2.0: Using google docs to improve the silent discussion in social studies. *Social Studies, 104*(3), 130–136. doi:10.1080/00377996.2012.703972

Roy, G. J., Allen, J. S., & Thacker, K. (2020). Linking factors and multiples to algebraic reasoning. *Mathematics Teacher: Learning and Teaching PK-12, 113*(3), 216-222.

Russo, J., Russo, T., & Kalogeropoulos, P. (2020). Exploring exponential growth in elementary school. *Mathematics Teacher: Learning and Teaching PK-12, 113*(12), 989-994.

Smith, M. S., & Stein, M. K. (2011). *5 practices for orchestrating productive mathematics discussions*. National Council of Teachers of Mathematics.

Snider, R. B. (2021). Guiding questions for selecting mathematical examples. *Mathematics Teacher: Learning and Teaching PK-12, 114*(1), 63-68.

Spangler, D. A., & Wanko, J. J. (Eds.). (2017). Enhancing classroom practice with research behind principles to actions. NCTM, National Council of Teachers of Mathematics.

Stein, M. K., Correnti, R., Moore, D., Russell, J. L., & Kelly, K. (2017). Using theory and measurement to sharpen conceptualizations of mathematics teaching in the common core era. *AERA Open, 3*(1), 1–20. doi:10.1177/2332858416680566

Stephan, M. L., Reinke, L. T., & Cline, J. K. (2020). Beyond hooks: Real-world contexts as anchors for instruction. *Mathematics Teacher: Learning and Teaching PK-12, 113*(10), 821-827.

Stohlmann, M. S. (2020a). Planning questions to help surface understanding. *Mathematics Teacher: Learning and Teaching PK-12, 113*(4), 327-329.

Stohlmann, M. S. (2020b). Escape room math: Luna's lines. *Mathematics Teacher: Learning and Teaching PK-12, 113*(5), 383-389.

Taylor, C., & Lee, J. S. (2021). Ready, set, launch! -The engineering cycle for productive struggle. *Mathematics Teacher: Learning and Teaching PK-12, 114*(2), 117-124.

Thompson, D. M., & Gay, A. S. (2020). Fact strategy instruction + meaningful discourse = student understanding. *Mathematics Teacher: Learning and Teaching PK-12, 113*(7), 558-564.

Warshauer, H. K. (2015). Productive struggle in middle school mathematics classrooms. *Journal of Mathematics Teacher Education, 18*(4), 375–400. doi:10.100710857-014-9286-3

Wasserman, N. H., Weber, K., Fukawa-Connelly, T., & Mejía-Ramos, J. P. (2020). Area-preserving transformations: Cavalieri in 2D. *Mathematics Teacher: Learning and Teaching PK-12, 113*(1), 53-60.

Yee, S. P., Roy, G. J., & Graul, L. (2020). Conditional reasoning online with mastermind. *Mathematics Teacher: Learning and Teaching PK-12, 113*(5), 390-396.

Chapter 3
RETA Model for Teaching Mathematics:
From the United Kingdom to Turkey

Ipek Saralar-Aras
https://orcid.org/0000-0002-4942-4408
The University of Nottingham, UK

ABSTRACT

For a long time, research into students' geometry performance has been regarded as an essential research topic in mathematics. By outlining an innovative mathematics teaching model and offering sample lesson practices from geometry on how to implement the model in the Turkish setting, the purpose of this chapter is to give an international perspective on reform-based practice in mathematics teaching and learning. The chapter focuses on a model for teaching geometry classes that incorporates realistic, exploratory, technology-enhanced, and active (RETA) principles, as well as its implementations in Turkish middle schools. It presents the different approaches of teaching geometry, common geometry classroom practices in Turkey, the previous models leading to the RETA model, and finally, a review of the RETA model's principles together with their benefits and drawbacks followed by a discussion.

INTRODUCTION

This chapter provides an international perspective on a reform-based practice in mathematics teaching and learning by describing a novel mathematics teaching model, and presenting sample lesson practices on how to use the model in the Turkish context. Specifically, this chapter introduces a model to teach geometry lessons with realistic, exploratory, technology-enhanced and active (RETA) principles, and its practical applications in middle schools in Turkey. First, both the alternative models of teaching geometry and the typical geometry teaching approach in Turkey are discussed. Then, how these existing models led to the RETA model is described, followed by a discussion of the principles of the RETA model along with pros and cons. For the pros and cons of each principle of the model, practical applications of the principles are presented through sample lesson plans. These sample RETA-based lesson plans are on

DOI: 10.4018/978-1-7998-9422-3.ch003

orthogonal and isometric drawings of polycubes which were developed, implemented, and evaluated as a part of a design-based research project with four cycles. The lessons were intended to help middle school students learn about three-dimensional (3D) shapes and in particular orthogonal and isometric drawings of polycubes. Last, the chapter ends with a discussion of the solutions, recommendations, and future research directions.

The 3D geometry thinking is "the conception of thoughts and ideas about 3D geometry concepts by amalgamating various types of reasoning;" and reasoning in this concept refers to "a set of processes and abilities that act as a feasible tool in problem-solving and enable us to go beyond the information given" (Pittalis & Christou, 2010, p.192). Orthogonal and isometric drawings have a variety of names in the literature; for example, orthogonal drawings can be referred to as orthogonal projections (Jones et al., 2012), orthographic projections/drawings (Moyer-Packenham & Bolyard, 2002), plan/top view, and elevations/side views (Yeo et al., 2005), and isometric drawings are also known as isometric projections (Gambari et al., 2015) and perspective drawings (Oldknow & Tetlow, 2008), as well as sometimes very vaguely as a building or a representation/picture of a building (Ben-Haim et al., 1985). The differing titles for these 2D representations were deemed synonyms in this chapter, and it is opted to use orthogonal and isometric drawings, which are both among the earliest names for these sorts of representations in the literature (Cooper & Sweller, 1989). In a similar vein, polycubes are three-dimensional shapes that are constructed from unit cubes. Polycubes are also referred to as polycubical shapes/objects (Cooper & Sweller, 1989), (solid) cube constructions (Ben-Haim et al., 1985) and a solid or an object constructed from unit-sized cubes (Pittalis & Christou, 2010).

BACKGROUND

Models to Teach Geometry

Recently, there have been various attempts to develop frameworks to teach mathematics at the middle school level, as Usiskin (1987) did in the 1980s. Particularly, the focus of research on 2D representations of 3D shapes has been on developing frameworks that characterize and analyze 3D geometry thinking. Of this research, some helped to improve the RETA model which is introduced in this chapter.

For example, Yeh and Nason (2004) suggested and tested a framework for using technology to teach 3D geometry. They claimed that three separate components make up 3D geometry: communication, objects, and spatial thinking. In a technological context, communication refers to (a) spoken and written language to describe 3D geometric entities and (b) non-verbal 2D representation of objects. Taking these three components into account, they designed VRMath, a software program that displayed realistic representations of 3D geometry problems in a range of colours, along with a link to a discussion forum. The authors state that two primary school students in grades 6 and 7 found the tool effective in promoting the construction of knowledge about 3D geometry ideas and procedures.

Recently, Goodall and colleagues (2017) proposed a framework for teaching mathematics that many students in the United Kingdom can use at home or school (not specified in but including 3D geometry). In contrast to the tedious, isolated, rote, elitist, and depersonalized (TIRED) mathematics discovered by Nardi and Steward (2013), they believe mathematics teaching should be accessible, linked, inclusive, valued, and empowering (ALIVE). To comprehend ALIVE, we must first comprehend the TIRED framework. In a research study conducted by Nardi and Steward (2013) with 70 13- and 14-year-old pupils in the United Kingdom, they discovered that mathematics instruction can be experienced by students as TIRED:

- Tedious: Math was tedious for the majority of the pupils, who saw it as a dull and unimportant subject with no real-world applications. Furthermore, interviewed pupils stated that mastering mathematics provides little opportunities for practical life.
- Isolated: Students thought of mathematics as a solitary topic in which students had to work alone to solve a mathematics problem.
- Rote: Many children saw mathematics as a series of rules to follow, therefore there were indisputable and distinct answers to math problems for them.
- Elitist: Pupils saw mathematics as a difficult subject and came to believe that only very bright or gifted students could achieve in it.
- Depersonalized: The majority of the pupils in the research stated that their mathematics learning was not personalized, but that it could be.

Goodall and colleagues (2017) suggested ALIVE concepts to increase mathematics performance in opposition to this TIRED mathematics:

- Accessibility principle refers to the use of appropriate enactive tasks that allow pupils to develop their own knowledge. The concept is proposed in the premise that these activities provide pupils with practically no justification to be excluded from developing mathematical thinking.
- Linked principle entails referring to prior knowledge in order for pupils to connect new material to what they already know and understand.
- Inclusive principle implies that all students participate in the process of learning mathematics through numerous activities, rather than believing that only extraordinarily bright children can learn mathematics.
- Valued principle stresses the use of real-life examples in the classroom in order to comprehend the worth of mathematics. The researchers went on to say that the examples should be drawn from those that individuals see as valuable both individually and culturally.
- Empowering principle relates to the students' agency, which allows them to take charge of their own learning. The goal is to assist students to have a better grasp of lifelong learning while making as much progress as possible in school mathematics. This principle also provides ways for students to be more empowered in their learning through carefully planned mathematics exercises, while also acquiring 21st-century abilities like creativity and technological literacy.

When establishing the RETA principles for geometry education, this design-based research was inspired by these frameworks in order to meet the needs in the common mathematics teaching approach of geometry in Turkey.

Common Mathematics Teaching Approach of Geometry in Turkey

Exam-Focused Pedagogy

Many studies including Bayart et al. (2000), Duval (1998), and Fujita et al. (2017) discovered and documented difficulties in 2D representations of various 3D shapes. These studies revealed that students performed poorly on these representations, particularly when constructing orthogonal and isometric drawings. When the reasons for poor performance on such drawings were investigated, the reviewed literature showed teachers as one of the most influential factors. That is to say, the literature indicated that teachers' beliefs influence their actions in teaching. Particularly mathematics teachers in Turkey believed that students' high national exam scores indicate teacher quality; and hence, teachers "teach to the test," and avoid student-centred activities (Doruk, 2014; Saralar-Aras, 2022). There was a lack of teacher incentive to teach 3D shapes without exam-focused teaching. This exam-focused pedagogy might be a result of continuing teacher performance evaluations in Turkey, where teachers' performance was primarily assessed based on student achievement on national tests since 2016 (Konan & Yilmaz, 2017, 2018).

Limited Use of Technology

Furthermore, in various studies (e.g., Bayrakdar-Çiftçi et al., 2013; Ciftci & Tatar, 2015; Tekalmaz, 2019), Turkish mathematics teachers were observed using the available technology in restricted ways, mostly to present recorded educational videos and quizzes from the ministry's Moodle page while wrapping up their sessions. Despite the emphasis on integrating technology into the new mathematics curriculum (Ministry of National Education [MoNE], 2018), particularly the use of dynamic geometry software packages in mathematics lessons, none of the observed Turkish teachers in Saralar's (2020) study encouraged students to use such software to visualize 2D representations of polycubes in any of their lessons on 3D shapes. In this study, the observed teachers stated that they believed in the present mathematics program's efficacy, and support its reliance on technology. When it came to teaching, however, most teachers believed that memorizing their techniques was the only way to master 2D representations of 3D shapes, and they recommended practising and drilling with a larger number of questions for better and faster results, rather than learning with and through technology.

Teacher-Centered Lessons

Last but not least, lessons on teaching polycubes were found to be teacher-centred in many studies despite the emphasis on student-centred activities in primary and secondary education programmes in the Turkish education system (Birgin & Acar, 2020; MoNE, 2018). Studies on this issue discovered that mathematics teachers dominated the use of tools and manipulatives (Saralar, 2020). In these courses, children were given very limited chances to utilize manipulatives and express themselves. Earlier research (Christou et al., 2006; Widder & Gorsky, 2013) suggests that this pedagogy is connected to the requirement to perceive 3D shapes from their 2D representations since it creates hurdles for not only students' learning but also teachers' teaching. According to numerous researchers, teaching 3D shapes in middle schools is regarded to be difficult among instructors, and as a result, teachers either do not teach it or utilize direct instruction rather than student-centred activities (e.g., Bakó, 2003).

Based on these findings and wider research, a model was created that proposes mathematics education, particularly teaching of geometry, may be realistic, exploratory, technology-enhanced, and active, hence the RETA principles. These principles led to the development of four lessons on orthogonal and isometric polycube drawings, one of which is available in Appendix 1.

METHODS

How the Work on this Approach Led to the RETA Model

The initial goal in developing the RETA model was to examine how national exam performance may be enhanced without relying on direct instruction or the repetition of national exam questions. Based on the wider literature and the findings of Saralar's (2020) study the RETA model was developed as an alternative to exam-focused pedagogy, limited use of technology and teacher-centred lessons. The following section briefly summarizes the RETA model, whilst explaining how the RETA model differs from more common approaches to teaching in Turkey.

First of all, rather than emphasizing the importance of the topic with the national exam questions, the RETA-based lessons were designed using real-life examples (R) and contexts. Secondly, the courses were supplemented with worked examples that were quite similar to national exam questions, but instead of drill and practice, they were meant to include mistakes for students to detect and remediate, which highlights the exploratory (E) principle. Thirdly, given that the Turkish government provided tablets to all students in the sample and interactive whiteboards to all classrooms, it was decided to strategically integrate dynamic mathematics tools such as GeoGebra and Cabri in the study of polycubes to provide multiple visual and dynamic representations using the technology-enhanced (T) principle. Finally, in response to teacher-centred teaching and teachers' dominance in manipulative use, the RETA's active (A) principle emphasized learning environments in which students have autonomy over manipulative use rather than passively following teacher constructions and replicating drawings on the board; hence, active refers to students' involvement as active participants.

RETA MODEL

Realistic (R), exploratory (E), technology-enhanced (T), and active (A) are the four principles of the RETA model. Even though these terms are polysemic, each principle is described in the context of the model, and the meanings connected with these principles are explained in the following sections. This section describes these principles more explicitly for their use in the context of the RETA model. For each principle, positive and negative aspects, as well as sample lesson applications, are discussed.

Realistic

Being the first principle of the RETA model, *realistic* lessons refer to the inclusion of real-life examples and situations in the teaching. Real-life examples demonstrate how the knowledge and skills gained in the mathematics course may be used in the real world (Gravemeijer, 1994; Van den Heuvel-Panhuizen, 2003). This is meant to raise students' awareness of the topic's relevance in their everyday lives, allow them to draw inferences about mathematical ideas' real-world connections, increase motivation for studying mathematics (three-dimensional shapes in this specific case), and assist students in solving real-world issues in the future. The immediate goal of this research was to improve on the Turkish government's national geometry exam, however, it is hoped that there is more to it than just passing the Turkish national exam. The following examines the spatial and mathematical aspects (importance) of realistic lessons before explaining how they are applied in the sample lessons prepared.

When we think of how three-dimensional shapes are related to real life, spatial thinking comes into conversation as the first aspect. Spatial thinking is employed in a variety of ways in everyday life. To illustrate, we utilize spatial thinking automatically while preparing luggage and packing it into a car trunk (Liben, 2007). It is also required for education programs, ranging from the use of molecular models in chemistry (Barke & Engida, 2001; Pribyl & Bodner, 2006) to the comprehension of mountain strata in geography (Lee & Bednarz, 2009; Robertson et al., 2009). Spatial thinking is also seen as a vital skill in many disciplines including science, engineering, and mathematics, particularly in geometry.

Many occupations, from radiologists to product designers, require the ability to think in two and three dimensions. Doctors, for example, study two-dimensional scans of the body generated by radiologists to identify the patient's problems. Architects who sketch their construction designs on paper or on a computer screen to depict them in two dimensions are another and a closer approximation to geometrical drawings. They create architectural plans and elevations to reduce three-dimensional forms into a sequence of two-dimensional components for a variety of purposes, including solar energy conservation, and heating and electrical structures (Matusiak, 2017). Elevations are crucial to understand key dimensions such as wall lengths and heights, as well as to show openings such as doors and windows; and, plans are handy depictions to understand proximity and spatial relations between the rooms of a building, thus, they are both useful in real-life contexts. In middle school geometry curricula, plans are mostly called top views, while elevations are termed views from the front, rear/back, left, and right. The author designed lessons on teaching elevations. Incorporating such real-life examples into lesson plans might pique students' attention and motivate them to study more about the subject (Fredricks et al., 2017).

The second (mathematical) aspect of realistic education of the RETA model considers English realistic mathematics education as an example. That is to say, the author took the studies that are conducted in the United Kingdom into consideration with the knowledge that it is one of the first countries to embrace and implement realistic mathematics education (De Lange, 1996) – which started early in the 1970s to enhance the quality of mathematics instruction in Dutch schools (Freudenthal, 1971, 1973). According to Freudenthal (1987), there is a global need to apply mathematics, and realistic mathematics education is one approach to do so. According to him, by starting and staying in reality while teaching mathematics, realistic mathematics education creates mathematics of human worth. It relates mathematics to reality in order to keep the children's attention and encourage them to study mathematics (van den Heuvel-Panhuizen, 2003).

It is much easier for children to make this connection (subject and its real-life relation) when it is more noticeable as in the study of population in geography or the study of motion of objects in physics. However, many children and even adults struggle to make this connection when it comes to mathematical topics (Cornell, 1999; Larkin & Jorgensen, 2016). They either see mathematics as an abstract discipline or only link specific aspects of mathematics to real life, but they fail to explain the broader application of these elements in actual life (e.g., Mulero et al., 2013; Reid et al., 2003). For instance, Mulero and friends (2013) showed that more than half of 94 university students did not associate architecture with mathematics and did not even know any architect who is pioneering a mathematical contribution to architecture. While this is the situation, it is critical to give a realistic mathematics education that allows children to recognize the connection between mathematics and real life as early as elementary school.

Children learn mathematics based on tasks they may face in their daily lives in *realistic mathematics education*, and they have the chance to create their own knowledge via group work, debate, and reflection (van den Heuvel-Panhuizen & Drijvers, 2014). These criteria are consistent with constructivist theories (Cobb, 1994; Cobb & Yackel, 1995; Gravemeijer, 1994; Simon, 1995) and correspond to the RETA model's other principles, such as exploratory and active, as detailed in the following sections.

An excellent example of an effective intervention that complements the author's approach to teaching mathematics is English realistic mathematics education, especially geometry (Cooper & Harries, 2009; Dickinson & Eade, 2005; Dickinson & Hough, 2012). For example, in 2003, a local Manchester school trialled realistic mathematics education in English classes with Year 7 children, and the response to the created realistic education materials was extremely favourable (Dickinson & Eade, 2005; Eade & Dickinson, 2006). This study took place in over 20 schools, and data revealed that children's comprehension of mathematics and approach to problem-solving improved. Furthermore, not only did students who actively engaged with the material improve but so did lower attaining ones. Over the course of three years, there have been consistent findings concerning increases in mathematics proficiency in different areas, including geometry in mathematics (Blum et al., 2019; Dickinson et al., 2010; Hough et al., 2017; van den Heuvel-Panhuizen, 2019). Since then, this realistic approach to mathematics instruction has been implemented into middle school mathematics classes in the United Kingdom (Dickinson & Hough, 2012), and it has served as the foundation for the realistic principle of the RETA Model.

Together with all the positive responses, there are also several criticisms of realistic mathematics education. To begin with, there is a criticism that it has been exaggerated, and that abstract mathematical concepts are far superior to realism (Keune, 1998, 1999). Regarding this, in realistic mathematics education, "there is a need for giving more attention to abstraction and logical reasoning to better make use of the Dutch mathematical talent that would be lost because of the realistic approach," said Keune (1999, p. 365). However, it is essential to highlight that Keune's (1999) speech, which outlines his perspective on realistic mathematics education, focused on what realistic mathematics education has become in the Netherlands rather than what it is meant to be. As a response to this criticism, the author agrees with Gemert (2015) that just because something is practical does not mean it cannot be abstracted or taught through examples. Considering this, in addition to including realistic mathematics, the RETA model included worked examples for practice, explained in the Exploratory principle.

Secondly, realistic mathematics education has been criticized as being reductionist since it gives realistic but not actual settings (Verstappen, 1994). According to Verstappen (1994), issues in realistic mathematics education give simplified actual situations that may subsequently pose difficulty in managing real problems formally in mathematics and later in life. Gravemeijer (2001) reacted to this by highlighting that the issues in realistic mathematics education can, but do not have to, deal with actual everyday life situations that are more difficult for children and involve many variables than what realistic mathematics education recommends. According to him, providing a familiar setting in which students can act intelligently in order to fully understand the mathematics in it is critical in realistic mathematics.

Finally, some studies demonstrate that just presenting things in a realistic setting does not always make things simpler to understand for children. For instance, Chandler and Kamii's (2009) experiment with 98 children revealed that using coins as a real-life example to teach place value makes it more difficult for children to grasp the topic. They discovered that while it is easy for teachers to conceive of one dime as ten pennies as adults, it is difficult for children to do so, and it is considerably more difficult when teachers incorporate monetary examples in the teaching of ones and tens. This is just one example showing that real world applications may make teaching certain concepts more difficult, thus real-life examples should be carefully chosen to meet the needs of the children. Considering both advantages and disadvantages, the next paragraph discusses how to include realistic mathematics into the sample lessons.

In the designed lessons, videos and photos are utilized to bring the concepts of orthogonal and isometric drawing of polycubes to life (see Appendix 1 for a sample lesson). Many studies recommend real-life videos and explanations showing mathematical material in action. However, some critics have claimed that videos are perceived by children as entertainment rather than information, resulting in children not benefitting from the videos as much as they could (Salomon & Perkins, 2005). As a result, in the sample RETA-based lessons, the design decision was made to not only present important material in the videos but also to create a student-centred atmosphere for discussion. Peer discussions, as well as a subsequent whole-class discussion, were meant to assist students to establish links between classroom knowledge and skills and their real-life applications. Furthermore, nearly every unit cube construction made by children is a depiction of a realistic image of a building (e.g., a castle and a school).

Exploratory

The term *exploratory* describes the second principle, which relates to the use of worked examples in classes to assist students in investigating the topic. Worked examples are pedagogical devices that give students someone else's answer to study (Evans & Swan, 2014). The exploratory principle is aimed at providing students with worked examples of the topic with some of these examples designed to include mistakes to explore and remediate. These mistakes could be chosen from the most common mistakes of the students, reported by available research. The aim for including these worked examples is, in a way, taking advantage of problem-solving opportunities, as many argue that problem-solving affords exploration (e.g., Carreira & Jacinto, 2019; Schoenfeld, 1985, 2013), and students who study geometry through exploration with problem-solving achieve higher scores than those who study with traditional methods in middle school geometry (see Klančar et al., 2021).

Geometry is mostly taught via examples, which have a place in many teaching and learning theories (Bruner, 2017; Marton et al., 2013; Marton et al., 2004; Skemp, 2012; Watson et al., 2006; Wilson, 1986). Students are frequently asked to work on examples and solve problems in geometry classes. Solving problems, nonetheless, may not be particularly useful when children are just beginning to study and have just basic knowledge of the subject (Renkl, 2011; Salden et al., 2009). When compared to solving problems at the beginning of the lessons, worked examples were shown to be more successful for initial skill acquisition (Kalyuga et al., 2001; Renkl, 2014, 2017). Hence, in the design of lessons with the exploratory principle of the RETA, it may be suggested to include worked examples when introducing the topic.

Moreover, learning geometry through criticising, comparing, and debating various solutions has several advantages, ranging from increasing student engagement with examples to supporting students in successfully integrating previous knowledge into present learning processes (Pierce et al.,, 2011; Silver et al., 2014). "Mathematical discourse has long been shown influential in supporting students' learning of mathematics" (Bennett, 2010, p.79). Many mathematics education researchers support the idea that dialogic teaching is beneficial (Bakker et al., 2015; Hofmann & Mercer, 2016; Kazak et al., 2015; Mercer & Sams, 2006; Ruthven et al., 2017; Warwick et al., 2016). Some also argue that dialogic processes aid conceptual development in mathematics (e.g., Kazak et al., 2015). Yet, dialogue during work with examples is not without its criticisms. Some researchers have argued that implementing dialogic mathematics teaching might be difficult, not just for newly qualified teachers (Bennett, 2010) but also for more experienced ones (Wegerif & Scrimshaw, 1997). More recent research also found evidence indicating that not only teachers but also students, frequently oppose discourse in a mathematics lesson. Regarding this, Bennett (2010) asserted that "it's hard getting kids to talk about math" (p.79). Particularly, in the country of Turkey, except for occasional experiments for research purposes (e.g., Gürbüz & Agsu, 2017), dialogic mathematics education is not a prevalent teaching approach. Teachers of mathematics are still lecturing in front of the class, and there is little room for student discussion and interaction in mathematics classes in Turkey; therefore, there may be resistance to dialogic teaching.

In RETA-based lesson plans, similar to Durkin and Rittle-Johnson (2012) and Evans and Swan (2014), some of the planned worked examples have intentionally designed mistakes for students to diagnose, remediate, and debate probable causes for them. This is trialled previously and found to enhance students' understanding of geometry topics. For example, Evans and Swan (2014) worked with eight secondary school mathematics teachers in the United Kingdom and over twenty in the United States to integrate worked examples with intentional errors into their lessons. The teachers prepared lessons using the worked examples provided by the researchers. These examples are available at http://map.mathshell.org/ on the Mathematics Assessment Project's official website. When the children failed to solve a geometry problem, these teachers offered them to work with worked examples. The teachers highlighted as a constraint that some children were more concerned with correcting mistakes than making holistic comparisons. Nonetheless, all students managed to improve their performance on the mathematics test. This could be related to the productive failure of Kapur (2014) and Sinha and Kapur (2021) who argue that failure in problem-solving, particularly when followed by discussion, is productive and results in learning (Hedge's $g = 0.87$).

To help students develop their conceptual understanding of holistic concerns related to the topic (2D representations of polycubes), the designed lessons featured peer discussions followed by a whole-class discussion. The author was aware that conversations were a) more demanding for both teachers and students due to the abilities required to evaluate and analyse the thinking behind solutions, and b) significantly different from how teaching takes place in the Turkish setting. To conclude, the model's second principle attempts to provide an exploratory mathematical education in which students engage with worked examples, questions, and mathematics dialogue.

Technology-Enhanced

The third principle advocates for *technology-enhanced* learning, which refers to the strategic use of dynamic tools in teaching mathematics. There may be various dynamic geometry tools to integrate into mathematics lessons including Cabri, GeoGebra and SketchPad. The aim of integrating these tools into the mathematics lessons through the technology-enhanced principle is to present various visual and dynamic representations of mathematics topics. The following examines two elements of technology-enhanced education: spatial and mathematical aspects, as well as their implementations in the designed sample lessons.

People many find it challenging to reason about three-dimensional shapes when working with two-dimensional representations, thus making this a more complex task (Reisberg & Heuer, 2005). In order to impart meaning to three-dimensional shapes, people frequently create two-dimensional representations. Two-dimensional representations can be incorporated into the human mind in a variety of ways, such as using landmarks based on the properties of a shape or by reference to other shapes (Tversky, 2005). Reisberg and Heuer (2005) point out that "mental images seem to be represented from a determinate viewing angle and distance … since visibility from a perspective and occlusion seem to play a role in those data" (p.39). This statement is supported by their depiction study, which shows that when people are asked to describe a cat in a picture, their responses are faster than when they are asked to describe specific parts of a cat, and that they are even faster to describe large and visible parts like the cat's head in the picture than small and hidden parts like whiskers and claws. Cubes that lie behind other cubes in geometry courses on polycubical shapes may be partially visible or invisible, making it difficult to express them in two-dimensional forms, such as orthogonally or isometrically.

Although numerous studies have shown that dynamic geometry tools can assist students in expressing three-dimensional structures in two dimensions (e.g., Oldknow & Tetlow, 2008; Simpson et al., 2006), the introduction of technology to geometry instruction has historically been met with opposition (Bolt, 1991). The Royal Society's working group on teaching and learning geometry (2001) advised that students pay more attention to studying three-dimensional shapes and that they learn them more effectively through the use of digital tools (such as virtual manipulatives and dynamic geometry tools) in the classroom. Taking this into account, Oldknow and Tetlow (2008) tested the efficiency of a 3D geometry program in small-scale pilot schools before expanding their research to a larger project in a set of Hampshire schools in the United Kingdom. Their research found that working with such software — which allows students to build two-dimensional representations of three-dimensional shapes — gives students a deeper understanding of three-dimensional shapes as well as a wealth of opportunities for active participation, collaboration, and confidence-building. According to Widder and Gorsky (2013), who investigated students' utilization of mathematics software in three-dimensional shape lessons, students utilized these tools according to their needs. That is to say, students with different pre-test scores used the software for different purposes: those with low spatial skills used it primarily for measuring their constructed representations, while those with high spatial abilities used it primarily for self-examination (e.g., 50 per cent more constructions and operations than their peers with lower spatial skills) and for speeding up mental processes like (re)constructing and rotating.

Despite the fact that all representations have benefits and drawbacks (Friedlander & Tabach, 2001), many researchers agree that learning from an appropriate combination of representations with the help of technology is more beneficial than learning from a single representation (e.g., single-use of verbal, numeric, symbolic, or graphical representations), and geometry software packages provide an environment where this can happen (Ainsworth, 2006; Hoyles & Noss, 2003; Kaput, 1992; Pape & Tchoshanov, 2001; Pierce et al., 2011). However, not all of the results/applications of incorporating dynamic technologies are beneficial. Lessons with these technologies are more difficult to plan (Grandgenett, 2007), especially since they demand the use of more student-centred approaches, and they are more challenging for teachers to manage in general (Bates, 2005). Furthermore, a lack of teacher knowledge in teaching geometry with technology, as well as inconsistent beliefs and goals regarding the use of technology, make it much more difficult to develop technology-enhanced lessons. Teachers' knowledge, beliefs, and goals all influence how they teach with technology (Ball et al., 2008; Mishra & Koehler, 2006; Niess, 2008, Shulman, 1986). While some teachers regard dynamic geometry tools as distractions rather than learning aids and find them time-consuming, others see them as a highly useful facilitator and successful teaching method (Saralar & Ainsworth, 2017; Saralar-Aras, 2022).

In the study of sample lessons, teaching was strategically supported with GeoGebra, free dynamic software for manipulating 3D shapes that may be used individually on tablets or collectively on interactive boards. This decision was partially pragmatic, as 1.5 million tablets were handed to children and all middle school classes in Turkey were equipped with interactive whiteboards from 2011. To conclude, by using dynamic geometry software, the third principle of this approach intends to provide students with *technology-enhanced* experiences of an acceptable combination of 2D representations of 3D shapes.

Active

The fourth principle relates to *active* learning settings in which students manage the use of concrete manipulatives rather than watching teacher constructs and reproducing drawings on the board, as seen in various studies (e.g., Saralar, 2020). Active lessons in the RETA model are more than merely a contrast to passive learning of Chi (2009) and Schank (1994). Although the activities involved could be described as constructive (e.g., students building cube constructions for themselves) and sometimes interactive (e.g., looking at students' cube-constructions and solutions and receiving feedback on them) in the literature (see Chi, 2009), it was chosen to use the term active because this principle refers to the involvement of learners as active participants. With the active principle, students are intended to engage in the learning process through the control of manipulatives. This is intended to increase students' motivation as well as their academic performance in the intended topics. The next paragraphs detail how concrete manipulatives were used to make students active, review assertions about their use and advantages in geometry, and examine how they were used in the sample lessons.

To deliver middle school mathematics in the past, teachers depended on workbooks and memory. Nevertheless, educators have argued for over two decades that these techniques are "ineffective and outdated" (Cain-Caston, 1996, p.271). Many researchers who compared standard teaching techniques to alternative methods, such as concrete manipulatives, discovered that children perform better with alternative methods (An & Tillman, 2015; Driscoll, 1983; Kong & Mohd-Matore, 2022; Suydam, 1984). Concrete manipulatives are "objects that can be handled by an individual in a sensory manner during which conscious and unconscious mathematical thinking will be fostered" (Swan & Marshall, 2010, p.14). Concrete manipulatives are also previously defined as "models that incorporate mathematical concepts, appeal to several senses and can be touched and moved around by students", which incorporates the author's notion of making students active (Hynes, 1986, p.11). Unit cubes, such as multilink cubes or linking cubes, and unifix cubes are referred to as concrete manipulatives in the sample lesson plans.

In the literature, there is no unanimity on the use of concrete manipulatives. While some researchers argue against manipulatives (e.g., Ross, 1989; Uttal et al., 1997), others argue in favour of them (e.g., Moch, 2001; Van de Walle et al., 2010).

The author agrees with proponents who believe based on their empirical study that manipulatives work if specific circumstances are met. Plastics cannot teach mathematics, as Ball (1992) emphasizes with the following words: "Understanding does not travel through the fingertips and the arm… Mathematical ideas really do not reside in cardboard and plastic materials." (p.47). Using manipulatives does not help students learn new mathematics quickly. Many academics have identified specific circumstances that make manipulatives helpful, such as effective lesson plans and students' desire to learn meaningfully (Carbonneau et al., 2013; Furner & Worrell, 2017; Moch, 2001). Instruction, for example, is stressed as a critical element. The way teachers teach with manipulatives determines how effective manipulatives are. Carbonneau and colleagues (2013) conducted a meta-analysis using 7237 students from 55 studies to determine the usefulness of teaching mathematics with concrete manipulatives. Their results showed that manipulatives offer greater learning than standard approaches that simply supply abstract mathematical symbols. Instruction was shown to regulate the connection between students' learning and their use of concrete manipulatives. It also depends on students' understanding of what manipulatives represent, and on the teacher's directions and assistance throughout the classroom exercise (Uttal et al., 1997). According to theories in the literature, concrete manipulatives aid learning when they promote any of the following instructional characteristics:

- Providing chances for student-centered research on the subject (Kirschner et al., 2006; Mayer, 2004),
- Assisting students in their abstract reasoning (Bruner, 1964; Montessori, 1964; Piaget, 1962),
- Providing physical enactment (Biazak et al., 2010; Engelkamp et al., 1994; Kormi-Nouri et al., 1994),
- Encouraging students to apply what they've learned in class to real-life situations (Baranes et al., 1989; Rittle-Johnson & Koedinger, 2005; Tindall-Ford & Sweller, 2006).

Environments in which manipulatives are utilized for meaningful learning by building on prior knowledge and challenging students to reflect and think are another example of these circumstances (Baroody, 1989; Furner & Worrell, 2017). Both Baroody (1989) and Furner and Worrell (2017) stress that manipulatives aid students when they use inquiry to connect their prior knowledge to the desired learning goals. When students have prior experience with the subject (Sowell, 1989) and use it regularly over time (Marzolf & DeLoache, 1994), the use of manipulatives in the classroom leads to even higher learning results.

Teachers frequently use unit cubes to teach three-dimensional shapes in geometry classes. According to Swan and Marshall's (2010) study of 249 teachers from New South Wales, cubes are the most commonly used manipulative in introducing three-dimensional shapes and the third most commonly used manipulative in mathematics classes (after blocks and polyhedrons) followed by unifix cubes and multilink cubes. Although there are pros and cons to utilizing these manipulatives, many researchers agree that they are helpful for teaching and learning mathematics and that they improve students' understanding of mathematics by enabling them to explore the topic actively (e.g., Canny, 1984; Clements & Battista, 1990; Skemp, 1987; Suydam, 1984). Children's use of concrete manipulatives can help them visualize shapes better and so improve their mathematics understanding. Regarding this, "The relevant application of manipulatives to … classroom situations helps students visualize and develop problem-solving strategies", says Moch (2001, p.83). Furthermore, concrete manipulatives, particularly cubes, can help children acquire more meaningful mathematical thinking and reasoning by allowing them to build and compare amounts, and allowing them to create interlinked understandings of mathematical ideas (Stein & Bovalino, 2001). Stated differently, via their experience with concrete manipulatives, children may integrate and link a range of concepts and develop a profound knowledge of them. Concrete manipulatives may also give students concrete and exploratory experiences for two-dimensional representations, allowing them to embody the problem scenario by touching, manipulating, and determining the proper structure (e.g., see Carroll & Porter, 1997).

However, if teachers monopolize the use of these manipulatives and students are not given opportunities to actively engage with them by touching and moving them around, teaching polycubes with them may not be particularly useful. The way teachers use concrete manipulatives in their lessons on polycubes, as well as how they teach other mathematics concepts, is crucial to students' success (Alfieri et al., 2011; Wearne & Hiebert, 1988). Saralar (2020) observed that teachers dominated the use of manipulatives in all observed Turkish classes, and as a result, children appeared to be disengaged with the instructional material. This might be a major factor in children's poor performance on three-dimensional shape comprehension tests. Three-dimensional shapes education must become more student-centred, with classrooms where students actively engage in rich mathematical activities during which they have control over manipulatives, to boost students' engagement and improve their knowledge of shapes.

Furthermore, the disparity in the usage of manipulatives in the literature indicated a very particular rationale for why they might be useful in the situations investigated in this study in Turkey. Most people who argue against manipulatives utilize blocks to represent abstract ideas, such as place value and percentages, which are represented by addition and subtraction (e.g., Bartolini-Bussi & Mariotti, 2002; Chandler & Kamii, 2009; Fuson & Briars, 1990). The active principle of the RETA model, on the other hand, is quite similar to how tangible manipulatives are employed in chemistry. Students in chemistry create a model containing atoms and structures so that they do not have to continually envision 3D structures; instead, they externalize these into a 3D model (see Hegarty et al., 2013). In a similar vein, in the sample RETA-based lessons, students construct polycubes from concrete unit cubes to externalize them so that they don't have to envision them while drawing orthogonal and isometric views. Constructions will not reflect anything else in mathematics (for example, abstract symbols), but will simply be externalized to allow students to draw them themselves. As a result, the researcher's goal in utilizing manipulatives is to teach symbolic relations rather than traditional mathematical facts. To summarize, active lessons strive to create learner-centered environments in which students interact with concrete manipulatives. The designed lessons (one of which is in Appendix 1) are meant to allow students to explore polycubical forms not just via student-centred activities with unit cubes but also through chances for reflection through teacher-led conversations.

SOLUTIONS AND RECOMMENDATIONS

The RETA model's principles may be applied to a wide range of topics in mathematics. The work on this has already started, and Turkish mathematics teachers started to prepare RETA-based lesson plans for various topics (other than 3D shapes' orthogonal and isometric drawings) in mathematics from polygons to types of triangles (e.g., Esen & Saralar-Aras, 2021). Other teachers and/or teacher-researchers than the author could continue designing RETA-based lessons and evaluate the effectiveness of these lessons by measuring students' academic performance through various types of assessments.

The RETA principles are not claimed to be sufficient; additional principles can be created or implemented, but the presented principles are required for teaching three-dimensional shapes in the studied context in Turkey. Furthermore, lesson plans do not have to employ all of the model's principles in a single lesson plan; rather, they may use them in their overall teaching of the objectives from mathematics.

Moreover, such research in real classrooms can provide valuable insights for teachers, educators, programme developers and policymakers. Teachers may gain self-awareness about their practices while applying RETA principles, whilst educators can use and trial the model in other contexts to see whether it meets the needs of their specific contexts. Finally, policymakers and programme developers can benefit from the findings. In particular, the Turkish Ministry of National Education and programme developers working for the ministry might be interested in the study as the sample lesson plans were developed specifically for the Turkish context.

FUTURE RESEARCH DIRECTIONS

Each classroom has a different student profile, particular traditions and contextual needs with its own atmosphere. While there are various ways of teaching mathematics, there can be particular ways that work for specific contexts. The RETA model was developed for the Turkish context, and sample lesson plans were developed and found to be effective in teaching orthogonal and isometric drawings of polycubes. In a quasi-experimental study with more than 200 students, scores of the RETA intervention classes improved almost up to 100% in orthogonal drawings and almost 90% in isometric drawings that are equivalent to national exam, whilst the final performance was about 60% in orthogonal drawings and 45% in isometric drawings in the traditional classes (Saralar, 2020; published elsewhere). Although it was developed as a response to the need in the Turkish context, clearly, there are various ways that the need could be met. This model is only one of many possibilities that might have been developed.

The chapter outlined the need for student-centred environments, the need to motivate teachers to teach 3D shapes, the need to integrate realistic scenarios to increase student engagement, and the need to provide technologies that can help both students' learning and teachers' teaching. Therefore, the RETA was created to address the specific needs of middle school children in the Turkish context. Overall, the results show that RETA-based lessons helped students score better on the national exam. However, this does not imply that the RETA model and RETA-based teachings are flawless, prompting the issue of how the RETA model and RETA-based lessons may be improved. For example, for future studies, it is possible to supplement other principles to the RETA principles according to the needs in other contexts. For example, in retrospect, the author wonders why, as a researcher who appreciates conversation in her classes, she didn't explicitly list dialogue as a principle. It is suspected that one explanation is as a Turkish mathematics teacher herself, she resisted having it as a principle while knowing it would be challenging in the Turkish context.

Last but not least, there is considerable scope for this model to be tested and further developed in a range of contexts. The RETA model is trialled as a quasi-experiment only for one topic, orthogonal and isometric drawings of polycubes, which is from a middle school programme in Turkey. It could be utilized for other topics, as well as other levels of education. There is a much greater geometric world that the RETA model could be instantiated, in other years in middle schools, and other towns and countries. Moreover, in addition to the RETA model itself, the RETA-based lesson plans can be tested and improved. Having a greater sample of teachers that have met specific criteria such as years of experience and background of teaching would have been ideal when testing RETA-based lessons.

CONCLUSION

To conclude, study on students' geometry proficiency has long been considered an important research issue in mathematics (Clements, 2003; Clements & Battista, 1992). This chapter introduced a reform-based mathematics teaching model, the RETA model, as well as design decisions made during its application in sample lessons to teach geometry objectives. It is anticipated that this model, as well as the lesson plans that were developed utilizing the RETA principles, would be useful in the future for teachers teaching in Turkey, and maybe beyond.

To note, the chapter was derived from the author's PhD thesis titled "Designing Lessons to Help Middle School Students Learn about Orthogonal and Isometric Drawings of Three-dimensional Shapes".

ACKNOWLEDGMENT

This research was supported by the Ministry of National Education in Turkey [YLSY-1416- 2014/11]; and the University of Nottingham [NOT-24/03/2017].

REFERENCES

Ainsworth, S. (2006). DeFT: A conceptual framework for considering learning with multiple representations. *Learning and Instruction*, *16*(3), 183–198. doi:10.1016/j.learninstruc.2006.03.001

Alfieri, L., Brooks, P. J., Aldrich, N. J., & Tenenbaum, H. R. (2011). Does discovery-based instruction enhance learning? *Journal of Educational Psychology*, *103*(1), 1–18. doi:10.1037/a0021017

An, S. A., & Tillman, D. A. (2015). Music activities as a meaningful context for teaching elementary students mathematics: A quasi-experiment time-series design with a randomly assigned control group. *European Journal of Science and Mathematics Education*, *3*(1), 45–60. doi:10.30935cimath/9420

Bakker, A., Smit, J., & Wegerif, R. (2015). Scaffolding and dialogic teaching in mathematics education: Introduction and review. *ZDM*, *47*(7), 1047–1065. doi:10.100711858-015-0738-8

Bakó, M. (2003). Differing projecting methods in teaching spatial geometry. *Proceedings of the Third European Research in Mathematics Education*, 1–9. http://www.mathematik.tu-dortmund.de/~erme/CERME3/Groups/TG7/TG7_Bako_cerme3.pdf

Ball, D. L. (1992). Magical hopes: Manipulatives and the reform of math education. American Educator. *The Professional Journal of the American Federation of Teachers*, *16*(2), 14–18.

Ball, D. L., Thames, M. H., & Phelps, G. (2008). Content knowledge for teaching: What makes it special? *Journal of Teacher Education*, *59*(5), 389–407. doi:10.1177/0022487108324554

Baranes, R., Perry, M., & Stigler, J. W. (1989). Activation of real-world knowledge in the solution of word problems. *Cognition and Instruction*, *6*(4), 287–318. doi:10.12071532690xci0604_1

Barke, H.-D., & Engida, T. (2001). Structural chemistry and spatial ability in different cultures. *Chemistry Education Research and Practice*, 2(3), 227–239. doi:10.1039/B1RP90025K

Baroody, A. J. (1989). Manipulatives don't come with guarantees. *The Arithmetic Teacher*, *37*(2), 4–5. https://eric.ed.gov/?id=EJ405968. doi:10.5951/AT.37.2.0004

Bartolini-Bussi, M. G., & Mariotti, M. A. (2002). Semiotic mediation in the mathematics classroom: Artifacts and signs after a Vygotskian perspective. In L. English, M. G. Bartolini-Bussi, R. Jones, & D. Tirosh (Eds.), *Handbook of international research in mathematics education* (2nd ed., pp. 746–783). Lawrence Erlbaum. http://math.unipa.it/~grim/YESS-5/Semiotic%20mediation%20Ch%2028%20Mariotti.pdf

Bayart, C., Gos, C., Hindelang, C., & Keyling, M. A. (2000). Voir et raisoner: À laconquête de l'espace au collège [See and reason: The survey of space at the college]. *Repères IREM*, *33*(1), 51–73. https://publimath.univ-irem.fr/numerisation/PX/IGR14026/IGR14026.pdf

Bayrakdar-Çiftçi, Z., Akgün, L., & Deniz, D. (2013). Teachers' opinions and solution suggestions regarding encountered issues on the ninth-grade mathematics curriculum. *Anadolu University Educational Sciences Institute Journal, 3*(1), 2–21.

Ben-Haim, D., Lappan, G., & Houang, R. T. (1985). Visualizing rectangular solids made of small cubes: Analyzing and affecting student performance. *Educational Sciences in Mathematics, 16*(4), 389–409. doi:10.1007/BF00417194

Bennett, C. A. (2010). It's hard getting kids to talk about math: Helping new teachers improve mathematical discourse. *Action in Teacher Education, 32*(3), 79–89. doi:10.1080/01626620.2010.10463561

Biazak, J. E., Marley, S. C., & Levin, J. R. (2010). Does an activity-based learning strategy improve preschool children's memory for narrative passages? *Early Childhood Research Quarterly, 25*(4), 515–526. doi:10.1016/j.ecresq.2010.03.006

Birgin, O., & Acar, H. (2020). The effect of computer-supported collaborative learning using GeoGebra software on 11th-grade students' mathematics achievement in exponential and logarithmic functions. *International Journal of Mathematical Education in Science and Technology*, 1–18. doi:10.1080/0020739X.2020.1788186

Blum, W., Artigue, M., Sträßer, R., & Van den Heuvel-Panhuizen, M. (2019). European didactic traditions in mathematics: Introduction and overview. In W. Blum, M. Artigue, M. A. Mariotti, R. Sträßer, & M. Van den Heuvel-Panhuizen (Eds.), *European Traditions in Didactics of Mathematics* (pp. 1–11). Springer. doi:10.1007/978-3-030-05514-1_1

Bolt, B. (1991). *Mathematics meets technology*. Cambridge University Press.

Bruner, J. S. (1964). The course of cognitive growth. *The American Psychologist, 19*(1), 1–15. doi:10.1037/h0044160

Bruner, J. S. (2017). *A study of thinking*. Routledge., doi:10.4324/9781315083223

Cain-Caston, M. (1996). Manipulative queen. *Journal of Instructional Psychology, 23*(4), 270.

Canny, M. E. (1984). The relationship of manipulative materials to achievement in three areas of fourth-grade mathematics: Computation, concept development and problem-solving. *Dissertation Abstracts International, 45*, 775–776. http://hdl.handle.net/10919/106313

Carbonneau, K. J., Marley, S. C., & Selig, J. P. (2013). A meta-analysis of the efficacy of teaching mathematics with concrete manipulatives. *Journal of Educational Psychology, 105*(2), 380–400. doi:10.1037/a0031084

Carreira, S., & Jacinto, H. (2019). A model of mathematical problem solving with technology: The case of Marco solving-and-expressing two geometry problems. In *Mathematical problem solving* (pp. 41–62). Springer. doi:10.1007/978-3-030-10472-6_3

Carroll, W. M., & Porter, D. (1997). Invented strategies can develop meaningful mathematical procedures. *Teaching Children Mathematics, 3*(7), 370–374. doi:10.5951/TCM.3.7.0370

Chandler, C. C., & Kamii, C. (2009). Giving change when payment is made with a dime: The difficulty of tens and ones. *Journal for Research in Mathematics Education, 40*(2), 97–118. doi:10.2307/40539328

Chi, M. T. H. (2009). Active-constructive-interactive: A conceptual framework for differentiating learning activities. *Topics in Cognitive Science, 1*(1), 73–105. doi:10.1111/j.1756-8765.2008.01005.x PMID:25164801

Christou, C., Jones, K., Mousoulides, N., & Pittalis, M. (2006). Developing the 3DMath dynamic geometry software: Theoretical perspectives on design. *The International Journal for Technology in Mathematics Education, 13*, 168–174. https://www.learntechlib.org/p/107486/

Ciftci, O., & Tatar, E. (2015). Teacher opinions about the updated secondary mathematics curriculum. *Turkish Journal of Computer and Mathematics Education, 6*(2), 285–298. doi:10.16949/turcomat.15375

Clements, D. H. (2003). Teaching and learning geometry. In J. Kilpatrick, W. G. Martin, & D. Schifter (Eds.), *A research companion to principles and standards for school mathematics* (pp. 151–178). National Council of Teachers of Mathematics.

Clements, D. H., & Battista, M. T. (1990). Constructive learning and teaching. *The Arithmetic Teacher, 38*(1), 34–35. doi:10.5951/AT.38.1.0034

Clements, D. H., & Battista, M. T. (1992). Geometry and spatial reasoning. In D. A. Grouws (Ed.), *Handbook of research on mathematics teaching and learning* (1st ed., pp. 420–464). Macmillan.

Cobb, P. (1994). An exchange: Constructivism in mathematics and science education. *Educational Researcher, 23*(7), 4. doi:10.2307/1176932

Cobb, P., & Yackel, E. (1995). Constructivist, emergent, and socio-cultural perspective in the context of development research. In *Proceedings of the Annual Conference of North American Chapter of the International Group for Psychology of Mathematics Education.* Annual Conference of North American Chapter of the International Group for Psychology of Mathematics Education. https://eric.ed.gov/?id=ED389535

Cooper, B., & Harries, T. (2009). Realistic contexts, mathematics assessment, and social class: Lessons for assessment policy from an English research programme. In Words and worlds (pp. 93–110). doi:10.1163/9789087909383_007

Cooper, M., & Sweller, J. (1989). School students' representations of solids. *Journal for Research in Mathematics Education, 20*(2), 202–212. doi:10.2307/749283

Cornell, C. (1999). I hate math! I couldn't learn it, and I can't teach it! *Childhood Education, 75*(4), 225–230. doi:10.1080/00094056.1999.10522022

De Lange, J. (1996). Using and applying mathematics in education. In A. J. Bishop, K. Clements, C. Keitel, J. Kilpatrick, & C. Laborde (Eds.), *International handbook of mathematics education: Part 1* (pp. 49–97). Springer Netherlands. doi:10.1007/978-94-009-1465-0_4

Dickinson, P., & Eade, F. (2005). Trialling realistic mathematics education (RME) in English secondary schools. *Proceedings of the British Society for Research into Learning Mathematics, 25*(3), 1–14. http://www.bsrlm.org.uk/wp-content/uploads/2016/02/BSRLM-IP-25-3-1.pdf

Dickinson, P., Eade, F., Gough, S., & Hough, S. (2010). Using realistic mathematics education with low and high attaining pupils in secondary schools. In M. Joubert & P. Andrews (Eds.), *Proceedings of the British Congress for Mathematics Education* (pp. 73–80). https://www.researchgate.net/publication/266487229_Using_Realistic_Mathematics_Education_with_low_to_middle_attaining_pupils_in_secondary_schools

Dickinson, P., & Hough, S. (2012). Using realistic mathematics education in UK classrooms. In P. Nickolson (Ed.), *Realistic mathematics education impact booklet.* https://www.hoddereducation.co.uk/media/Documents/Maths/Using-Realistic-Maths-Education-in-UK-classrooms.pdf

Doruk, B. K. (2014). The educational approaches of Turkish pre-service elementary mathematics teachers in their first teaching practices: Traditional or constructivist? *The Australian Journal of Teacher Education, 39*(10), 113–134. doi:10.14221/ajte.2014v39n10.8

Driscoll, M. J. (1983). *Research within reach: Secondary school mathematics.* National Council of Teachers of Mathematics & CEMREL, Inc. https://eric.ed.gov/?id=ED225842

Durkin, K., & Rittle-Johnson, B. (2012). The effectiveness of using incorrect examples to support learning about decimal magnitude. *Learning and Instruction, 22*(1), 206–214. doi:10.1016/j.learninstruc.2011.11.001

Duval, R. (1998). Geometry from a cognitive point of view. In *Reasoning in geometry* (pp. 37–52). https://ci.nii.ac.jp/naid/10020806198/

Eade, F., & Dickinson, P. (2006). Exploring realistic mathematics education in English schools. In J. Novotná, H. Moraová, M. Krátká, & N. Stehlíková (Eds.), *Proceedings of the 30th conference of the international group for the psychology of mathematics education* (*Vol. 3*, pp. 1–8). Academic Press.

Engelkamp, J., Zimmer, H. D., Mohr, G., & Sellen, O. (1994). Memory of self-performed tasks: Self-performing during recognition. *Memory & Cognition, 22*(1), 34–39. doi:10.3758/BF03202759 PMID:8035683

Esen, B., & Saralar-Aras, I. (2021). Designing lesson plans for sustainable development focused education in the context of teaching polygons to middle school students. *Proceedings of the Second International Academician Studies Congress*, 1–8.

Evans, S., & Swan, M. (2014). Developing students' strategies for problem-solving in mathematics: The role of pre-designed "sample student work". *Educational Designer, 2*(7), 1–31. https://www.educationaldesigner.org/ed/volume2/issue7/article25/pdf/ed_2_7_evans_swan_14.pdf

Fredricks, J. A., Hofkens, T., Wang, M., Mortenson, E., & Scott, P. (2017). Supporting girls' and boys' engagement in maths and science learning: A mixed-methods study. *Journal of Research in Science Teaching, 55*(2), 271–298. doi:10.1002/tea.21419

Freudenthal, H. (1971). Geometry between the devil and the deep sea. In H.-G. Steiner (Ed.), *The teaching of geometry at the pre-college level: Proceedings of the second CSMP international conference co-sponsored by Southern Illinois University and Central Midwestern Regional Educational Laboratory* (pp. 137–159). Springer Netherlands. 10.1007/978-94-017-5896-3_10

Freudenthal, H. (1973). *Mathematics as an educational task.* Reidel. doi:10.1007/BF00121085

Freudenthal, H. (1987). Mathematics starting and staying in reality. In I. Wirszup, & R. Street (Eds.), *Proceedings of the USCMP conference on mathematics education on development in school mathematics around the world.* NCTM.

Friedlander, A., & Tabach, M. (2001). Promoting multiple representations in algebra. In A. Cuoco (Ed.), *The roles of representation in school mathematics* (pp. 173–184). NCTM.

Fujita, T., Kondo, Y., Kumakura, H., & Kunimune, S. (2017). Students' geometric thinking with cube representations: Assessment framework and empirical evidence. *The Journal of Mathematical Behavior, 46*(1), 96–111. doi:10.1016/j.jmathb.2017.03.003

Furner, J., & Worrell, N. (2017). The importance of using manipulatives in teaching math today. *Transformations, 3*(1), 1–25. https://nsuworks.nova.edu/transformations/vol3/iss1/2

Fuson, K. C., & Briars, D. J. (1990). Using a base-ten blocks learning/teaching approach for first- and second-grade place-value and multidigit addition and subtraction. *Journal for Research in Mathematics Education, 21*(3), 180–206. doi:10.2307/749373

Gambari, I. A., Balogun, S. A., & Alfa, A. S. (2014). Efficacy of interactive whiteboard on psychomotor skills achievement of students in isometric and orthographic projection. *Contemporary Educational Technology, 5*(4), 316–330. doi:10.30935/cedtech/6133

Gemert, S. B. (2015). *All positive action starts with criticism: Hans Freudenthal and the didactics of mathematics.* Springer. doi:10.1007/978-94-017-9334-6

Goodall, J., Johnston-Wilder, S., & Russell, R. (2017). The emotions experienced while learning mathematics at home. In U. X. Eligio (Ed.), *Understanding 284 emotions in mathematical thinking and learning* (pp. 295–313). Academic Press. doi:10.1016/B978-0-12-802218-4.00011-X

Gravemeijer, K. P. E. (1994). *Developing realistic mathematics education.* CD-ß Press/Freudenthal Institute.

Gravemeijer, K. P. E. (2001). Developmental research: Fostering a dialectic relation between theory and practice. In J. Anghileri (Ed.), *Principles and practice in arithmetic teaching* (pp. 147–161). Open University Press.

Gürbüz, M. Ç., & Agsu, M. (2017). Dialogic teaching model for ninth class students to conceptualize inequalities. *Journal of Education and Practice, 8*(1), 171–187. https://files.eric.ed.gov/fulltext/ED582086.pdf

Hegarty, M., Stieff, M., & Dixon, B. L. (2013). Cognitive change in mental models with experience in the domain of organic chemistry. *Journal of Cognitive Psychology, 25*(2), 220–228. doi:10.1080/20445911.2012.725044

Hofmann, R., & Mercer, N. (2016). Teacher interventions in small group work in secondary mathematics and science lessons. *Language and Education, 30*(5), 400–416. doi:10.1080/09500782.2015.1125363

Hough, S., Solomon, Y., Dickinson, P., & Gough, S. (2017). Investigating the impact of a realistic mathematics education approach on achievement and attitudes in post-16 GCSE resit classes. Manchester Metropolitan University/The Nuffield Foundation.

Hoyles, C., & Noss, R. (2003). What can digital technologies take from and bring to research in mathematics education? In A. J. Bishop, M. A. Clements, C. Keitel, J. Kilpatrick, & F. K. S. Leung (Eds.), *Second international handbook of mathematics education* (pp. 323–349). Springer Netherlands. doi:10.1007/978-94-010-0273-8_11

Hynes, M. C. (1986). Selection Criteria. *The Arithmetic Teacher, 33*(6), 11–13. doi:10.5951/AT.33.6.0011

Jones, K., Fujita, T., & Kunimune, S. (2012). Representations and reasoning in 3-D geometry in lower secondary school. In T. Y. Tso (Ed.), *Proceedings of the 36th conference of the international group for the psychology of mathematics education* (Vol. 2, pp. 229–346). The Psychology of Mathematics Education. https://eprints.soton.ac.uk/445627/1/Jones_etc_represent_reason3D_PME36_2012.pdf

Kalyuga, S., Chandler, P., Tuovinen, J., & Sweller, J. (2001). When problem-solving is superior to studying worked examples. *Journal of Educational Psychology, 93*(3), 579–588. doi:10.1037/0022-0663.93.3.579

Kapur, M. (2014). Productive failure in learning math. *Cognitive Science, 38*(5), 1008–1022. doi:10.1111/cogs.12107 PMID:24628487

Kaput, J. J. (1992). Technology and mathematics education. In *Handbook of research on mathematics teaching and learning* (pp. 515–556). https://www.learntechlib.org/primary/p/5474/

Kazak, S., Wegerif, R., & Fujita, T. (2015). The importance of dialogic processes to conceptual development in mathematics. *Educational Studies in Mathematics, 90*(2), 105–120. doi:10.100710649-015-9618-y

Keune, F. J. (1998). Naar de knoppen [Inaugural lecture]. In *NRC/Handelsblad.* Catholic University. https://lib.ugent.be/en/catalog/rug01:000433729

Kirschner, P. A., Sweller, J., & Clark, R. E. (2006). Why minimal guidance during instruction does not work: An analysis of the failure of constructivist, discovery, problem-based, experiential, and inquiry-based teaching. *Educational Psychologist, 41*(2), 75–86. doi:10.120715326985ep4102_1

Klančar, A., Starčič, A. I., Cotič, M., & Žakelj, A. (2021). Problem-based geometry in seventh grade: Examining the effect of path-based vs. conventional instruction on learning outcomes. *International Journal of Emerging Technologies in Learning, 16*(12), 16–35. doi:10.3991/ijet.v16i12.21349

Konan, N., & Yilmaz, S. (2017). Teachers' views on teacher performance assessment: A mixed-method research. *Proceedings of the 26th International Educational Sciences Conference,* 1–3. https://dergipark.org.tr/download/journal-file/11372

Konan, N., & Yilmaz, S. (2018). Ogretmen iliskileri performans degerlendirmeye iliskin ogretmen gorusleri: Bir karma yontem arastirmasi. *Milli Egitim, 219*(1), 137–160. https://dergipark.org.tr/tr/pub/milliegitim/issue/39923/473428

Kong, S. F., & Mohd-Matore, M. E. E. (2022). Can a Science, Technology, Engineering, and Mathematics (STEM) approach enhance students' mathematics performance? *Sustainability, 14*(1), 379. doi:10.3390u14010379

Kormi-Nouri, R., Nyberg, L., & Nilsson, L. G. (1994). The effect of retrieval enactment on recall of subject-performed tasks and verbal tasks. *Memory & Cognition*, *22*(6), 723–728. doi:10.3758/BF03209257 PMID:7808281

Larkin, K., & Jorgensen, R. (2016). 'I hate maths: Why do we need to do maths?' Using iPad video diaries to investigate attitudes and emotions towards mathematics in year 3 and year 6 students. *International Journal of Science and Mathematics Education*, *14*(5), 925–944. doi:10.100710763-015-9621-x

Lee, J., & Bednarz, R. (2009). Effect of GIS learning on spatial thinking. *Journal of Geography in Higher Education*, *33*(2), 183–198. doi:10.1080/03098260802276714

Liben, L. S. (2007). Education for spatial thinking. In Handbook of child psychology (Vol. 4). Wiley. doi:10.1002/9780470147658.chpsy0406

Marton, F., Booth, S., & Booth, S. (2013). *Learning and awareness*. Routledge., doi:10.4324/9780203053690

Marton, F., Tsui, A. B. M., Chik, P. P. M., Ko, P. Y., & Lo, M. L. (2004). *Classroom discourse and the space of learning*. Routledge., doi:10.4324/9781410609762

Marzolf, D. P., & DeLoache, J. S. (1994). Transfer in young children's understanding of spatial representations. *Child Development*, *65*(1), 1–15. doi:10.2307/1131361 PMID:8131641

Matusiak, B. S. (2017). Daylighting is more than an energy-saving issue. In Energy efficient buildings. doi:10.5772/65866

Mayer, R. E. (2004). Should there be a three-strikes rule against pure discovery learning? *The American Psychologist*, *59*(1), 14–19. doi:10.1037/0003-066X.59.1.14 PMID:14736316

Mercer, N., & Sams, C. (2006). Teaching children how to use language to solve maths problems. *Language and Education*, *20*(6), 507–528. doi:10.2167/le678.0

Mishra, P., & Koehler, M. J. (2006). Technological pedagogical content knowledge: A framework for teacher knowledge. *Teachers College Record*, *108*(6), 1017–1054. doi:10.1177/016146810610800610

Moch, P. L. (2001). Manipulatives work! *The Educational Forum*, *66*(1), 81–87. doi:10.1080/00131720108984802

MoNE. (2018). *Middle school mathematics program (Grades 5-8)*. MoNE. https://mufredat.meb.gov.tr/Dosyalar/2018813171044420-1-2018-80%20Matematik%20Uygulamalar%C4%B1%20Dersi%20(5-8.%20S%C4%B1n%C4%B1flar).pdf

Montessori, M. (1964). *The Montessori method* (A. E. George, Trans.). Robert Bentley.

Moyer-Packenham, P., & Bolyard, J. (2002). Exploring representation in the middle grades: Investigations in geometry with virtual manipulatives. *Australian Mathematics Teacher*, *58*(1), 19–25. https://digitalcommons.usu.edu/teal_facpub/52/

Mulero, J., Segura, L., & Sepulcre, J. M. (2013). Is maths everywhere? Our students respond. *International Association of Technology, Education and Development (IATED)*, 1–10. http://rua.ua.es/dspace/handle/10045/27248

Nardi, E., & Steward, S. (2013). Is mathematics T.I.R.E.D? A profile of quiet disaffection in the secondary mathematics classroom. *British Educational Research Journal*, *29*(3), 345–366. doi:10.1080/01411920301852

Niess, M. (2008). Mathematics teachers developing technology, pedagogy and content knowledge (TPACK). *Society for Information Technology and Teacher Education Proceedings*, 5297–5304. https://www.learntechlib.org/primary/p/28121

Oldknow, A., & Tetlow, L. (2008). Using dynamic geometry software to encourage 3D visualisation and modelling. *The Electronic Journal of Mathematics & Technology*, *2*(1), 1–8. https://link.gale.com/apps/doc/A178451802/AONE?u=anon~501ead4c&sid=googleScholar&xid=695b9a4a

Pape, S. J., & Tchoshanov, M. A. (2001). The role of representation(s) in developing mathematical understanding. *Theory into Practice*, *40*(2), 118–127. doi:10.120715430421tip4002_6

Piaget, J. (1962). *Play, dreams, and imitation in childhood* (C. Gattegno & F. M. Hodgson, Trans.). Norton.

Pierce, R., Stacey, K., Wander, R., & Ball, L. (2011). The design of lessons using mathematics analysis software to support multiple representations in secondary school mathematics. *Technology, Pedagogy and Education*, *20*(1), 95–112. doi:10.1080/1475939X.2010.534869

Pittalis, M., & Christou, C. (2010). Types of reasoning in 3D geometry thinking and their relation with spatial ability. *Educational Studies in Mathematics*, *75*(2), 191–212. doi:10.100710649-010-9251-8

Pribyl, J. R., & Bodner, G. M. (2006). Spatial ability and its role in organic chemistry: A study of four organic courses. *Journal of Research in Science Teaching*, *24*(3), 229–240. doi:10.1002/tea.3660240304

Reid, A., Petocz, P., Smith, G., Wood, L., & Dortins, E. (2003). Mathematics students' conceptions of mathematics. *New Zealand Journal of Mathematics*, *32*(Supplementary Issue), 163–172. http://www.sun.ac.za/english/learning-teaching/ctl/Documents/Reid%20et%20al%202003.pdf

Reisberg, D., & Heuer, F. (2005). Visuospatial images. In A. Miyake & P. Shah (Eds.), *The Cambridge handbook of visuospatial thinking* (pp. 35–80). Cambridge University Press. doi:10.1017/CBO9780511610448.003

Renkl, A. (2011). Instruction based on examples. In R. E. Mayer & P. A. Alexander (Eds.), *Handbook of research on learning and instruction* (pp. 272–295). Routledge. https://www.routledgehandbooks.com/doi/10.4324/9780203839089.ch3

Renkl, A. (2014). Toward an instructionally oriented theory of example-based learning. *Cognitive Science*, *38*(1), 1–37. doi:10.1111/cogs.12086 PMID:24070563

Renkl, A. (2017). Learning from worked-examples in mathematics: Students relate procedures to principles. *ZDM*, *49*(4), 571–584. doi:10.100711858-017-0859-3

Rittle-Johnson, B., & Koedinger, K. R. (2005). Designing knowledge scaffolds to support mathematical problem-solving. *Cognition and Instruction*, *23*(3), 313–349. doi:10.12071532690xci2303_1

Robertson, C., Nelson, T. A., Jelinski, D. E., Wulder, M. A., & Boots, B. (2009). Spatial-temporal analysis of species range expansion: The case of the mountain pine beetle, dendroctonus ponderosae. *Journal of Biogeography*, *36*(8), 1446–1458. doi:10.1111/j.1365-2699.2009.02100.x

Ross, S. H. (1989). Parts, wholes, and place value: A developmental view. *The Arithmetic Teacher*, *36*(6), 47–51. doi:10.5951/AT.36.6.0047

Ruthven, K., Mercer, N., Taber, K. S., Guardia, P., Hofmann, R., Ilie, S., Luthman, S., & Riga, F. (2017). A research-informed dialogic-teaching approach to early secondary school mathematics and science: The pedagogical design and field trial of the epiSTEMe intervention. *Research Papers in Education*, *32*(1), 18–40. doi:10.1080/02671522.2015.1129642

Salden, R. J. C. M., Aleven, V. A. W. M. M., Renkl, A., & Schwonke, R. (2009). Worked examples and tutored problem-solving. *Topics in Cognitive Science*, *1*(1), 203–213. doi:10.1111/j.1756-8765.2008.01011.x PMID:25164806

Salomon, G., & Perkins, D. (2005). Do technologies make us smarter? Intellectual amplification with, of, and through technology. In R. J. Sternberg & D. D. Preiss (Eds.), *Intelligence and technology: The impact of tools on the nature and development of human abilities* (pp. 71–86). Erlbaum. doi:10.4324/9780203824252-11

Saralar, İ. (2020). *Designing lessons to help middle school students learn about orthogonal and isometric drawings of three-dimensional shapes* [Doctoral Thesis, The University of Nottingham]. Nottingham eThesis.

Saralar, İ., & Ainsworth, S. (2017). *An exploration of middle school mathematics teachers' beliefs and goals regarding GeoGebra: Four cases from the Turkish Republic* [Paper Presentation]. European Conference on Educational Research (ECER) 2017, Copenhagen, Denmark. https://eera-ecer.de/ecer-programmes/conference/22/contribution/39865/

Saralar-Aras, İ. (2022). An exploration of middle school mathematics teachers' beliefs and goals regarding a dynamic mathematics tool: GeoGebra. *Journal of Research in Science Mathematics and Technology Education, 5*(SI), 41–63. doi:10.31756/jrsmte.113SI

Schank, R. C. (1994). Active learning through multimedia. *IEEE MultiMedia*, *1*(1), 69–78. doi:10.1109/93.295270

Schoenfeld, A. H. (1985). *Mathematical problem-solving*. Academic Press.

Schoenfeld, A. H. (2013). Reflections on problem-solving theory and practice. *The Mathematics Enthusiast*, *10*(1), 9–34. doi:10.54870/1551-3440.1258

Shulman, L. (1986). Those who understand: Knowledge growth in teaching. *Educational Researcher*, *15*(2), 4–14. doi:10.3102/0013189X015002004

Simon, M. A. (1995). Reconstructing mathematics pedagogy from a constructivist perspective. *Journal for Research in Mathematics Education*, *26*(2), 114–145. doi:10.2307/749205

Simpson, G., Hoyles, C., & Noss, R. (2006). Exploring the mathematics of motion through construction and collaboration. *Journal of Computer Assisted Learning*, *22*(2), 114–136. doi:10.1111/j.1365-2729.2006.00164.x

Sinha, T., & Kapur, M. (2021). When problem-solving followed by instruction works: Evidence for productive failure. *Review of Educational Research*, *91*(5), 761–798. doi:10.3102/00346543211019105

Skemp, R. R. (1987). *The psychology of teaching mathematics* (Revised American Edition). Lawrence Erlbaum. doi:10.4324/9781315621128

Skemp, R. R. (2012). The psychology of learning mathematics (Expanded American Edition). Routledge. doi:10.4324/9780203396391

Sowell, E. J. (1989). Effects of manipulative materials in mathematics instruction. *Journal for Research in Mathematics Education*, *20*(1), 498–505. doi:10.2307/749423

Stein, M. K., & Bovalino, J. W. (2001). Manipulatives: One piece of the puzzle. *Mathematics Teaching in the Middle School*, *6*(6), 356–359. doi:10.5951/MTMS.6.6.0356

Suydam, M. N. (1984). Research report: Manipulative materials. *The Arithmetic Teacher*, *31*(5), 27. doi:10.5951/AT.31.5.0027

Swan, P., & Marshall, L. (2010). Revisiting mathematics manipulative materials. *Australian Primary Mathematics Classroom*, *15*(2), 13–19. https://files.eric.ed.gov/fulltext/EJ891801.pdf

Tekalmaz, G. (2019). Teacher reviews about reformed high school mathematics curriculum. *Kocaeli University Education Journal*, *2*(1), 35–47. doi:10.33400/kuje.548562

The Royal Society. (2001). *Teaching and learning geometry 11-19: Report of a Royal Society / Joint Mathematical Council working group*. The Royal Society. https://royalsociety.org/-/media/Royal_Society_Content/policy/publications/2001/9992.pdf

Tindall-Ford, S., & Sweller, J. (2006). Altering the modality of instructions to facilitate imagination: Interactions between the modality and imagination effects. *Instructional Science*, *34*(4), 343–365. doi:10.100711251-005-6075-5

Tversky, B. (2005). Functional significance of visuospatial representations. In P. Shah & A. Miyake (Eds.), *The Cambridge handbook of visuospatial thinking* (pp. 1–34). Cambridge University Press. doi:10.1017/CBO9780511610448.002

Usiskin, Z. (1987). Resolving the continuing dilemmas in school geometry. In M. M. Lindquist & A. P. Shulte (Eds.), *Learning and Teaching Geometry, K12: 1987 Yearbook* (pp. 17–31). National Council of Teachers of Mathematics.

Uttal, D. H., DeLoache, J. S., & Scudder, K. V. (1997). Manipulatives as symbols: A new perspective on the use of concrete objects to teach mathematics. *Journal of Applied Developmental Psychology*, *18*(1), 37–54. doi:10.1016/S0193-3973(97)90013-7 van de Walle, J. A., Karp, K. S., & Bay-Williams, J. M. (2010). Geometric thinking and geometric concepts. In *Elementary and middle school mathematics teaching developmentally* (7th ed., pp. 402–433). Pearson Education. http://xn--webducation-dbb.com/wp-content/uploads/2019/09/John-A.-Van-de-Walle-Karen-S.-Karp-Jennifer-M.-Bay-Williams-Elementary-and-Middle-School-Mathematics_-Teaching-Developmentally-Pearson-2012.pdf

van den Heuvel-Panhuizen, M. (2003). The didactical use of models in realistic mathematics education: An example from a longitudinal trajectory on percentage. *Educational Studies in Mathematics, 54*(1), 9–35. doi:10.1023/B:EDUC.0000005212.03219.dc

van den Heuvel-Panhuizen, M. (2019). Didactics of mathematics in the Netherlands. In W. Blum, M. Artigue, M. A. Mariotti, R. Sträßer, & M. van den HeuvelPanhuizen (Eds.), European traditions in didactics of mathematics (pp. 57–94). Springer. doi:10.1007/978-3-030-05514-1_3

van den Heuvel-Panhuizen, M., & Drijvers, P. (2014). Realistic Mathematics Education. In S. Lerman (Ed.), *Encyclopedia of mathematics education* (pp. 521–525). Springer Netherlands. doi:10.1007/978-94-007-4978-8_170

Verstappen, P. (1994). Het dogma: Van de leefwereld naar de wiskundewereld [The dogma: From the daily life world to the world of mathematics]. *Tijdschrift Voor Didactiek der Bètawetenschappen, 12*(2), 104–129. http://www.fi.uu.nl/publicaties/literatuur/199402-verstappen.pdf

Warwick, P., Vrikki, M., Vermunt, J. D., Mercer, N., & van Halem, N. (2016). Connecting observations of student and teacher learning: An examination of dialogic processes in Lesson Study discussions in mathematics. *ZDM, 48*(4), 555–569. doi:10.100711858-015-0750-z

Watson, A., Mason, J., & Mason, J. (2006). *Mathematics as a constructive activity: Learners generating examples.* Routledge. doi:10.4324/9781410613714

Wearne, D., & Hiebert, J. (1988). A cognitive approach to meaningful mathematics instruction: Testing a local theory using decimal numbers. *Journal for Research in Mathematics Education, 19*(5), 371–384. doi:10.2307/749172

Wegerif, R., & Scrimshaw, P. (Eds.). (1997). *Computers and talk in the primary classroom.*

Widder, M., & Gorsky, P. (2013). How students solve problems in spatial geometry while using a software application for visualizing 3D geometric objects. *Journal of Computers in Mathematics and Science Teaching, 32*(1), 89–120. https://eric.ed.gov/?q=solve+AND+in&pg=11&id=EJ1006172

Wilson, P. S. (1986). Feature frequency and the use of negative instances in a geometric task. *Journal for Research in Mathematics Education, 17*(2), 130–139. doi:10.2307/749258

Yeh, A., & Nason, R. A. (2004). Toward a semiotic framework for using technology in mathematics education: The case of learning 3d geometry. *ICCE2004 Proceedings,* 1–10.

Yeo, K. K. J., Goh, S., & Koh, E. Z. F. (2005). Exploring plan and elevation geometry with prodesktop. *Journal of Science and Mathematics Education, 28*(1), 125–137. https://repository.nie.edu.sg/bitstream/10497/15471/1/JSMESEA-28-1-125.pdf

ADDITIONAL READING

Ben-Chaim, D., Lappan, G., & Houang, R. T. (1989). Adolescents' ability to communicate spatial information: Analyzing and effecting students' performance. *Educational Studies in Mathematics*, *20*(2), 121–146. doi:10.1007/BF00579459

Fennema, E. H. (1973). Manipulative in the classroom. *The Arithmetic Teacher*, *20*(1), 350–352. doi:10.5951/AT.20.5.0350

Furner, J., & Worrell, N. (2017). The importance of using manipulatives in teaching math today. *Transformations, 3*(1), 1–25. https://nsuworks.nova.edu/transformations/vol3/iss1/2

Saralar, İ., Ainsworth, S., & Wake, G. (2018). BSRLM Report: Helping students learn two-dimensional representations of polycubical shapes. *Research in Mathematics Education Journal*, *21*(1), 107–111. doi:10.1080/14794802.2019.1587668

Saralar, İ., Ainsworth, S., & Wake, G. (2019). *Working with a mathematics teacher to teach with technology* [Paper Presentation]. The European Association for Research in Learning and Instruction Conference (EARLI) 2019, RWTH Aachen University, Aachen, Germany. https://bit.ly/3KZdgRj

Saralar, İ., Işıksal, M., & Akyuz, D. (2018). The evaluation of a pre-service mathematics teacher's TPACK: Case of 3D objects. *The International Journal for Technology in Mathematics Education*, *25*(2), 3–22. doi:10.1564/tme_v25.2.01

Saralar-Aras, İ., & Ainsworth, S. (2020). *A categorisation of middle school students' errors in representing three-dimensional shapes* [Paper Presentation]. The European Association for Research in Learning and Instruction (EARLI) JURE 2020 Conference. https://bit.ly/3ulAH13

Thompson, A. G. (1984). The relationship of teachers' conceptions of mathematics and mathematics teaching to instructional practice. *Educational Studies in Mathematics*, *15*(2), 105–127. doi:10.1007/BF00305892

Zilkova, K., & Partova, E. (2019). Virtual manipulatives with cubes for supporting the learning process. *International Symposium of Elementary Mathematics Teaching Proceedings: Opportunities in Learning and Teaching Elementary Mathematics*, 427–438.

KEY TERMS AND DEFINITIONS

Isometric Drawing: A two-dimensional representation of a 3D object with three primary lines which are angled away from the observer in the same way with vertical lines staying vertical whereas horizontal lines at 30-degree angle to the horizontal plane. Isometric projections; perspective drawings.

Orthogonal Drawing: A two-dimensional representation of a 3D object from the top, bottom, front, back, left and right. Orthogonal projections; orthographic projections/drawings; plan/top view, and elevations /side views.

Polycube: A three-dimensional shape which is constructed from unit cubes.

RETA Model: A model for teaching mathematics that supports realistic, exploratory, technology-enhanced, and active learning.

APPENDIX

The process of designing the following lesson plan was based on the RETA model and the reviewed literature. Turkish mathematics teachers had a very tight curriculum to cover with a fixed time for teaching the topic. It should also be noted that the lesson was designed to meet the needs of the children in the sample and does not claim to be the ideal option for different educational settings. The researcher developed the design decisions based on the RETA model and her ability to construct lesson plans as a mathematics teacher and a researcher. As a result, other researchers would make various decisions to meet the needs of the students in their classrooms; yet, the decisions were able to meet the needs of the current teaching in Turkey.

Each lesson plan comprised a lesson abstract, a lesson structure that specified the amount of time allotted for each activity, and activity descriptions in the form of a teacher's guide. It is worth noting that different classrooms had varied dynamics and student profiles, making it difficult to forecast how the lesson would proceed in different classes, such as how students would respond to a specific activity. As a result, the mathematics teachers who used these plans did not carefully adhere to the time allotted for each task. The first of the four developed lesson plans to teach 2D representations of 3D shapes follows.

Sample Lesson Plan

Lesson abstract: Students focus on the issue of why we need two-dimensional drawings (orthogonal and isometric drawings) of three-dimensional objects. They develop an awareness of how drawing views from the top, front and sides and isometric drawing are related to real-life practices. They engage with several real-life examples and consider how these may be represented mathematically (realistic). After engaging with the real-life examples, they construct polycubical shapes corresponding to pictures of 3D objects (realistic) with linking cubes (active). They create their concrete polycubical shapes in the authoring tool, created through GeoGebra. They explore the view from the front by manipulating their representation in GeoGebra (technology-enhanced). They develop an awareness of how the front views change when they manipulate the shape. They diagnose and remediate and discuss possible reasons for worked examples with designed mistakes (exploratory). Lesson 1 focused particularly on a realistic principle of the RETA model where students articulate real-world experiences through videos and photographs on engineering, architecture and a drawing tool AutoCAD.

Lesson Structure

- Agenda and aims (5 minutes)
- Real-life examples: Engineering, Architecture (10 minutes)
- Tools for drawing: AutoCAD, GeoGebra (15 minutes)
- Activity 1: Constructing buildings and drawing their front views (15 minutes)
- Activity 2: Finding the mistakes (5 minutes)
- Conclusion and feedback (5 minutes)

Agenda and aims: Introduce yourself, if necessary. Explain to students that during the next four lessons we are going to discover different types of drawings and drawing tools, and relate our learning with their use in real-life situations. Ask whether students have any question before starting to the first lesson, and answer their questions if any. Explain to students that in this lesson we are going to consider some real-life examples to understand why people need two-dimensional representations of three-dimensional objects.

Real-life example 1 - Engineering: Show a two-minute part of a technical drawing video. Explain why we might need to learn such drawings and how these can be related to mathematics. Explain that this is how engineers start to draw multi-views of the shapes. They use different views to design different parts of the machines. All of the machines we use in daily life, such as computers, mobile phones, hairdryers and fruit squeezers, are composed of small parts, which were designed and drawn by engineers. Ask whether anyone's mother or father is an engineer or whether anyone has seen such a drawing before. Invite students to share their ideas as well.

Real-life example 2 - Architecture: Show the whole class an architecture photo and repeat the same procedure in Engineering activity. Discuss why architects need to learn these drawings and how these can be related to mathematics. Invite students to share their ideas as well. Summarize the discussion with a few sentences.

Explain that the drawing in the video is the first step to draw plans of the houses, or draw new interior designs of the houses. All architects similar to engineers learn how to draw multi-views and prepare projects based on this knowledge. Such drawings are natural parts of their jobs. Architects also try to draw these shapes more clearly as their drawing should be easy to understand by the people who ask for their help or their customers.

Tools for drawing 1 - AutoCAD: Show whole class a two-minute part of an AutoCAD drawing video. Explain that this is how engineers use a tool to construct their shapes.

Discuss how such tools might help them draw three-dimensional polycubical shapes in two-dimension. Invite students to share their ideas as well.

Possible Prompt Questions

- What are the potential advantages and disadvantages of using a tool like this?
- What about accuracy?
- What about getting the detail right?
- What about time spent on construction?
- Why do you think we both need tools and pen and pencil drawings? – A possible answer: Sometimes it is easier to sketch, sometimes tools help us to visualise the shape so that we can draw with pen and pencil.
- Do you think using a similar tool in this class help us visualise different views of a shape? How? – An expected answer: with the help of the manipulations the tool allows

Tools for drawing 2 - GeoGebra: Explain to students that there are some tools which help us in drawing shapes similar to engineers. Introduce the authoring tool. Ask students to turn on their tablets and run the authoring tool created through GeoGebra. Give them some time to explore how it works. Move to construction examples after the discovery of GeoGebra.

Open the slide – Exploring the Authoring Tool in Figure 1. Give an example (use a simple construction similar to the one on the slide) to show how buttons work and how we can manipulate the shape constructed on the authoring tool. Ask students to construct the same shape on their tools. Continue asking questions to stimulate their exploration of the tool.

Possible Guiding Questions

- 2 x 2 x 1 (Length x Breadth/Width x Height)
- 3 x 3 x 3

Then, ask them to remove cubes so that they have 3D shapes which have the following dimensions:

- 3 x 2 x 2
- 2 x 2 x 2
- 2 x 1 x 2

Move to construction examples after the discovery of GeoGebra.

Figure 1. Exploring the authoring tool

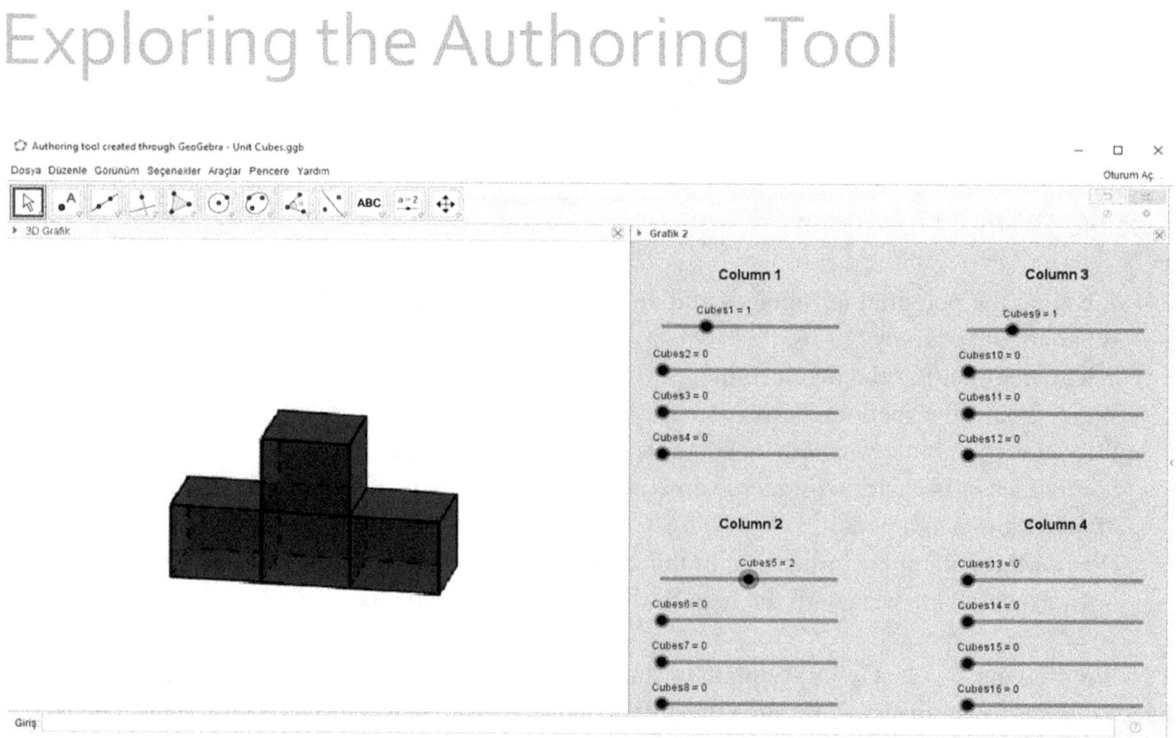

Activity 1

Organize students into groups of 2-3. Give a box of linking cubes to each group. Show the school picture and ask them to construct the school using linking cubes. Check their constructions and discuss their answers. Repeat the same procedure for the castle. Possible correct answers are included in Figure 2.

Figure 2. School and castle pictures – I

Students may construct the castle totally different than each other as only one view (front view) of the castle is seen from the pictorial representation. Here, explain that we need more than one view to construct the exact shape. Then, ask at least how many views we need and why?

Figure 3. Possible correct constructions from linking cubes – I

Point to the castle in Figure 2. Ask them to construct the castle from linking cubes. Following that ask them to construct the same castle on the tool (by now they are familiar with both the polycubical shape and its pictorial representation).

Note: If there are students who constructed with a different depth, remind the point discussed before: Having only one view of a shape is not enough to decide all of its dimensions, therefore we may not construct the exact shape only having its one view. Ask them to remind you at least how many views we need to construct the exact shape and why?

After they all construct the shapes on the authoring tool, give a copy of the Activity Sheet Castles and focus on the front view of the first representation. Invite the group of students to manipulate their GeoGebra constructions to decide how to represent the front view. Ask all students to draw the front view on the dotted paper individually.

Do not forget to ask students to save their files before moving to the next question on the sheet. Please note that it might be useful or easier for students to see the depth if we use the angles of the isometric paper on the tool (30°-60°), so before giving the authoring tool to students set the angle accordingly. After they manipulate the shape they may decide which angle they would like to use. Figure 4 shows a possible correct GeoGebra construction of the castle and its front view.

Figure 4. A possible correct GeoGebra construction and its front view – I

Show the next slide having pictures of the buildings which require relatively more complex constructions in Figure 5. Point to the school picture and ask them to construct the school using linking cubes or/and GeoGebra. Check their constructions and discuss their answers. Repeat the same procedure for the castle. Collect students' constructions to give them back in the next lesson. Some of the possible correct answers are presented in Figure 6.

Figure 5. School and castle pictures – II

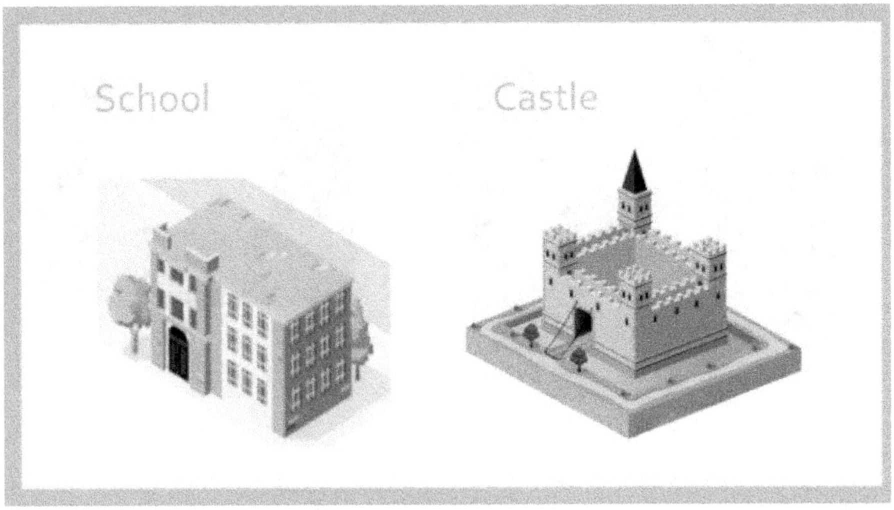

Figure 6. Possible correct constructions from linking cubes – II

Point to the Castle II in Figure 5. Ask students to construct the castle, which they constructed from the linking cubes, on the tool. After they all construct the shapes on the authoring tool, invite them to focus on the front view of their Castle II representation on the tool. Invite the groups of students to manipulate their GeoGebra constructions to decide how to represent the front view. Ask all students to draw the front view on the dotted paper individually. Figure 7 presents a possible correct construction of the castle and its front view.

Figure 7. A possible correct GeoGebra construction and its front view – II

Show the slide – Question – III in Figure 8. Follow the same procedure for the third question of the worksheet as well. This time, do not ask students to construct the shape with linking cubes. However, students who need concrete construction may continue constructing with them. Do not forget to ask students to save their files in GeoGebra before moving to the next question on the sheet as they will use them during the next lessons while exploring the views from top and sides. Figure 9 presents a possible correct construction of the third representation and its front view.

Figure 8. Question – III

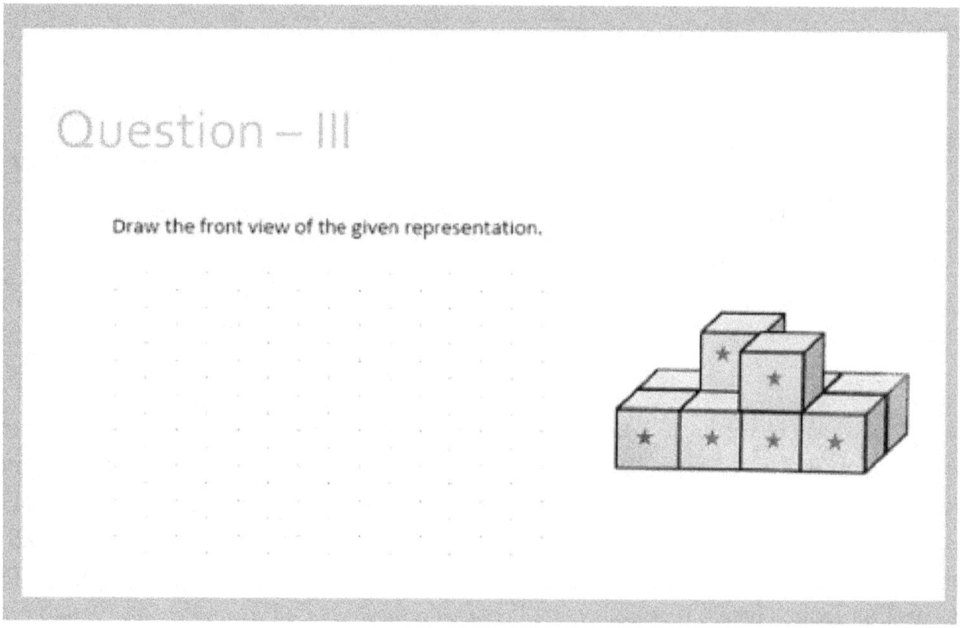

Figure 9. A possible correct GeoGebra construction and its front view – III

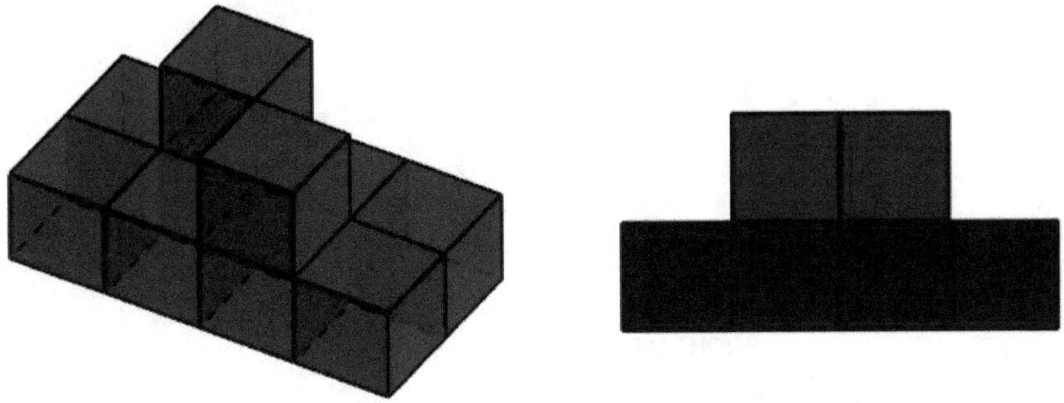

Ask students to construct other shapes having the same front view. Say that students can choose to use linking cubes or GeoGebra to construct the shape. Figure 10 presents possible answers.

Figure 10. Possible correct GeoGebra constructions

Activity 2

Provide one activity sheet to each student. Activity sheets could have five worked examples, specifically designed with mistakes. Explain to students that now we will explore common student errors in orthogonal and isometric drawings.

Show the slide – Find the mistake and discuss why – I in Figure 11. Say that here is Deniz's work. Ask them the following question. What is the mistake in the drawing and why is that?

Note that it is better to use "what was he thinking" question instead of asking "what do you think he thought" question to push students to step into the other person's shoes a bit more.

Figure 11. Worked example

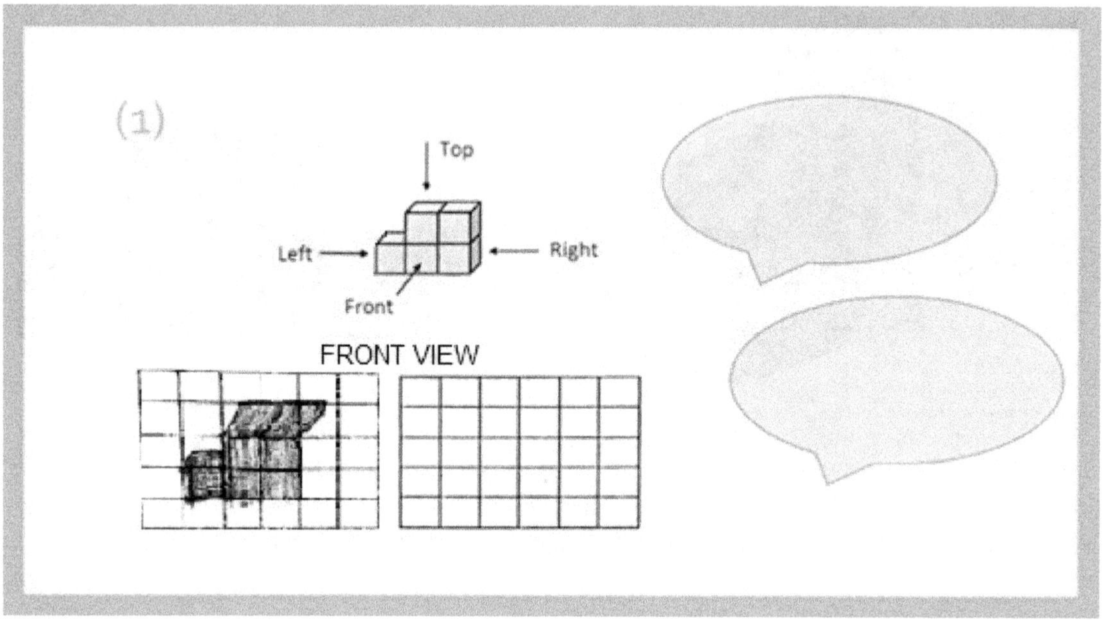

Ask students to discuss why it's wrong in their groups and to note the reason for it to their worksheets. Invite students to share their ideas with the whole class. Ask them to draw the correct representation individually. The correct drawing for the first question is represented in Figure 12. Repeat the procedure for the other four questions.

Figure 12. A possible correct drawing for the worked example - I

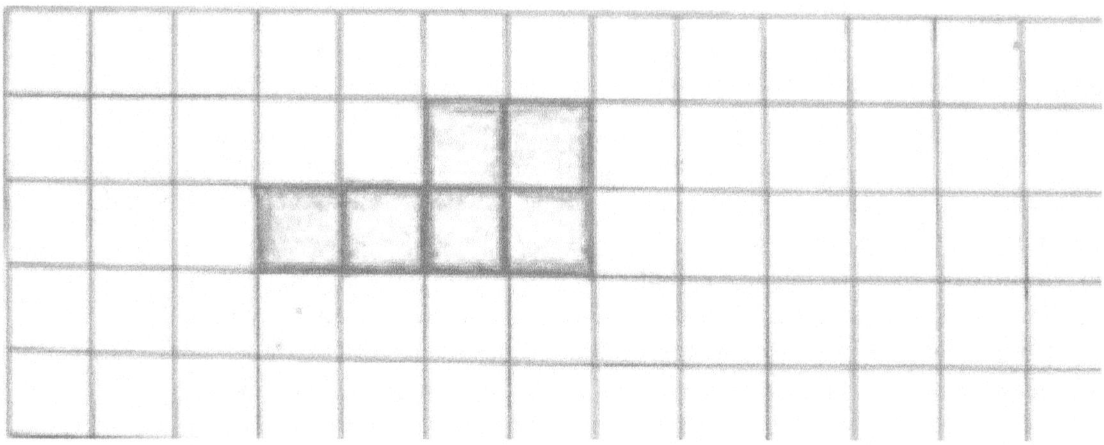

Conclusion and feedback: Conclude the lesson with a question. Show the slide What if in Figure 13, and distribute the sheets of paper to students and ask them to draw the front view of the given shape on the dotted paper.

Figure 13. What if question

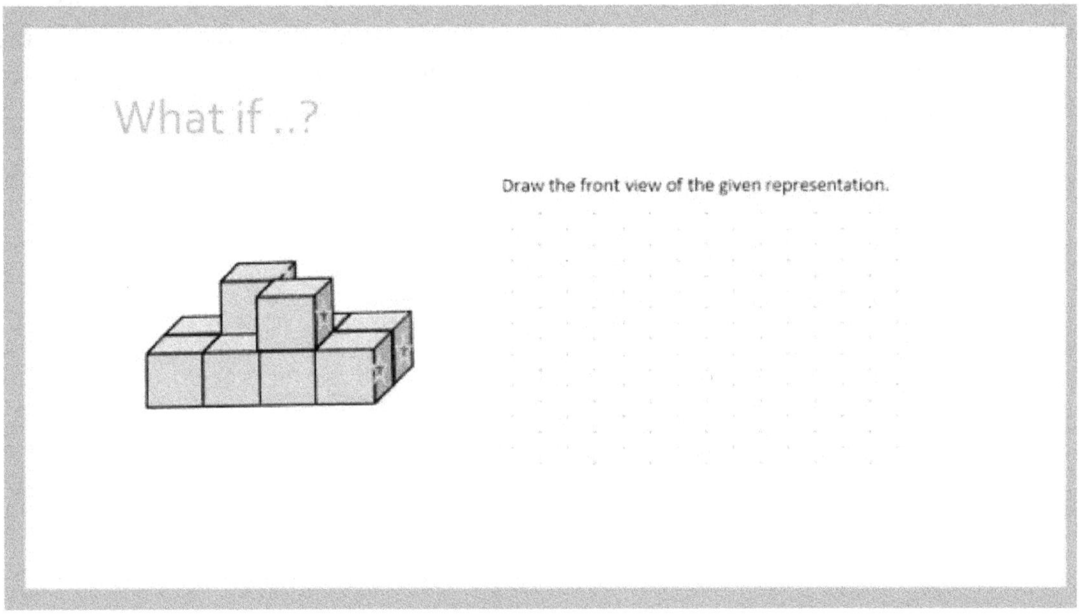

This question is different than the other questions and the stars indicating the front view are not in the front perspective. The aim is to raise awareness that front views of polycubical shapes can change according to the perspective we look at. Some students might tend to draw the front view similar to the front view of the third question since the shape actually is the same and the only change is that the stars indicating the front view. Ask them to go back to their GeoGebra constructions and manipulate their constructed shapes to indicate the front view. A possible correct construction of the given representation and its front view are represented in Figure 14.

Figure 14. A possible correct GeoGebra construction and its front view – IV

Explain to students that in this lesson we looked at real-life examples where orthogonal drawings are used, and a real-life example where a dynamic tool was used for drawing. Then, we constructed 3D shapes from linking cubes to represent some buildings in the given pictures mathematically. We explored an authoring tool and represented polycubical shapes on it, and used the tool to draw the view from the front on the dotted paper. The next lesson will be about the views from the top and sides.

Chapter 4
A Holistic Approach to Teaching the Process of Problem Solving: A Curriculum of Nonroutine Problems

Robert (Bob) H. London
California State University, San Bernardino, USA

ABSTRACT

This chapter concerns an approach to teaching mathematics that facilitates the student seeing the connection between the process of solving nonroutine mathematics problems and solving significant problems in the student's life. In this chapter, the author will define a nonroutine problem; discuss the concept of a curriculum of nonroutine problems; discuss a few examples of nonroutine problems, including some that facilitate the student seeing the connection between the process of solving nonroutine mathematics problems and solving significant problems in the student's life; and suggest guidelines for implementing a curriculum of nonroutine problems consistent with improving the Common Core Mathematical Practices and the concept of reform-based mathematics education.

INTRODUCTION

Any reform-based mathematics curriculum needs to address the basic question, "What is essential in mathematics education?". Certainly, such a mathematics curriculum must provide the average student with the tools to effectively address significant problems in that student's life. This chapter concerns an approach to teaching mathematics appropriate for secondary and post-secondary students that facilitates the student seeing the connection between the process of solving nonroutine mathematics problems and solving significant problems in the student's life. The approach is embedded in a curriculum of nonroutine problems field-tested from 1984 to 1995 in a secondary school setting, and later applied in an undergraduate mathematics course in problem-solving. The development and field-testing of the curriculum was based on a methodology for addressing significant problems in one's context that require a transformation of understanding (London, 2016) — Perhaps an appropriate label for this methodology would be spiritual pragmatism, "A practical way of solving problems" (DeVinne, 1985, p. 973) without

DOI: 10.4018/978-1-7998-9422-3.ch004

a fixation on views and theories, and with an understanding of the dynamics of the relationship between our inner work and our functional activities in the process of transformation. For this curriculum the development process consisted primarily of these steps: (1) The initial development of a curriculum appropriate for a high school Calculus course consisting of eight nonroutine problems, each requiring about two weeks work, primarily outside of class. The curriculum was viewed initially by the author as enrichment for the students with three of the problems connected to the Calculus curriculum (see London, 1989). The unexpected significance of the curriculum was described thusly,

The significance of the curriculum can probably be best described by reporting the observed effect on the mathematical maturity of the students. It is as if the student has been transformed mathematically! Instead of acting in all the ways that we normally attribute to most high school students, the student acts similarly to a 'mathematically mature' person. For example, instead of stopping when an obstacle is encountered, the student will persist. Instead of ignoring obvious contradictions or inaccuracies, the student will actively examine them. Instead of being intimidated by ambiguity, the student will tolerate the ambiguity. Instead of being satisfied with the first solution to a problem, the student will work on a problem until a more satisfactory solution is reached. Instead of staring at a problem that seems unsolvable or confusing, the student will try something until the problem naturally becomes clearer. (London, 1989, p. 1)

(2) The development of a four-year curriculum by the author, consisting of 60 nonroutine problems in seven strands, fifteen per year, suitable to be integrated into a four-year secondary academic mathematics curriculum. Development of the curriculum was primarily completed during a ten-week Alden B. Dow summer residential creativity fellowship. (3) The field-testing and revision of the curriculum in all the mathematics courses at Old Saybrook High School in Old Saybrook, Connecticut (an upper middle-class shoreline community) for which the author was the mathematics department chair. This field-testing included the author's one-year elective course Problem Solving, which allowed for more in-depth field-testing. (4) The further development and revision of the curriculum as a professor of education at California State University, San Bernardino, including the teaching of an undergraduate liberal studies one-quarter mathematics course in Problem Solving and field-testing of the curriculum by four graduate students (e.g., DeLeon, 2003). It should be clear that the methodology is primarily classroom-based research versus an empirical research study. Specifically, the methodology was appropriate for the development of a curriculum effective for the author's professional context, but not necessarily generalizable to other classrooms. In fact, it is the author's opinion that the curriculum requires an educator that is able to adjust the curriculum to be appropriate for the educator's context, including being sensitive to unique nonroutine problems meaningful for the educator's students.

In the rest of this chapter, the author will define a nonroutine problem; discuss the concept of a curriculum of nonroutine problems; discuss a few examples of nonroutine problems, including some that facilitate the student seeing the connection between the process of solving nonroutine mathematics problems and solving significant problems in the student's life; and suggest implications for implementing a curriculum of nonroutine problems consistent with the concept of reform-based mathematics education.

DEFINITION OF A NONROUTINE PROBLEM

Recent years have seen the improvement of problem-solving skills and the Common Core Standards of Mathematical Practice (SMPs) identified as an important focus of the National Association of Teachers of Mathematics (NCTM) and the Mathematical Association of America (MAA). This chapter is concerned with a type of problem, labeled as a nonroutine problem, not typically included in even the improved materials based on this focus. For instructional purposes, a nonroutine problem at an appropriate level of difficulty is defined to have the following characteristics (London, 2004, 2007, & 2021): (a) The problem requires three steps to complete: problem recognition and orientation, trying something, and persistence; (b) the problem is open-ended; that is, the problem allows for various solutions; (c) the problem requires students to evaluate a variety of potential strategies and approaches to the problem such as finding a pattern and generalizing, generating and organizing data, and reducing a problem to an easier equivalent problem; and (d) every student is able to "solve the problem;" that is, each student will be able to confront the problem and generate a solution consistent with their ability and efforts. Typically, each problem requires at least a few hours work over at least a week's time and written justification and description of the solution. Given this specific definition, in this chapter, the author will use the spelling "nonroutine" versus the dictionary spelling, "non-routine." Other works (London 1976, 1989, 1993, 2004, 2021) have discussed the theoretical and pedagogical basis for this definition. It has been argued that the definition of the three steps of problem-solving is consistent with other definitions of the steps of mathematical problem-solving (e.g., Polya, 1962). In addition, a properly structured curriculum of nonroutine problems is consistent with holistic education and constructivist theory (e.g., Brooks & Brooks, 1993; Hickey et al., 2001; Ward, 2001), as well as being consistent with the guidance from professional organizations such as the Association of Supervisors and Curriculum Developers (e.g., ASCD, Schmoker, 2018) and the NCTM (e.g., NCTM, 2014, 2018). Previous research (London, 1989, 1993, 2021) supports the hypothesis that: (a) The ability to solve nonroutine problems is an essential skill in mathematics education in the sense that these skills can be applied by students to solve significant problems in their life. It is argued that a curriculum which focuses primarily on the process of problem solving involved in solving a nonroutine problem, versus the content of mathematics (e.g., adding fractions, solving a first-degree equation) or strategies of problem solving (e.g., guess and check, problem reduction), focuses on what is an essential skill in mathematics education. (b) Even most above average ability students in mathematics need significant instruction in this objective (London, 1993, 2021). (c) Most students need an integrative approach emphasizing problems from a variety of fields, including problems meaningful to them, to insure transfer to significant problems in their personal life. (d) Successful instruction is feasible within a curriculum consistent with the content requirements of traditional content-focused courses. In addition, a curriculum of nonroutine problems can be easily integrated into other contemporary approaches to curriculum such as project-based learning (Curtis, 2002), place-based learning (Anderson & Gurnee, 2016), creating communities of mathematical inquiry (Peressini & Knuth, 2000), or rich tasks (Moulds, 2004). Finally, according to Tekkumru-Kisa et al.'s (2020) classification of levels of demand of tasks, a nonroutine problem is at the highest of four levels, "higher-level demands (doing mathematics)" (p. 610).

Before proceeding it is necessary to clarify the fourth characteristic of a nonroutine problem: Every student can "solve the problem." Given the realities of the level of understanding of mathematical content of the typical average mathematics student, it is inappropriate to initially include nonroutine problems that require significant mathematics content prerequisite skills, <u>if your objective is to improve the student's</u>

ability to solve nonroutine problems. For example, the following two problems from Schoenfeld (1985, pp, 15-16) are examples of problems requiring the three steps of nonroutine problems, but not satisfying the criteria that every student is able to "solve the problem": (1) Given two intersecting lines and a point P marked on one of them, show how to construct, using straightedge and compass, a circle that is tangent to both lines and that has the point P as its point of tangency to one of the lines; and (2) Given triangle ABC, show that it is always possible to construct, with straightedge and compass, a straight line that is parallel to line segment AB and that divides triangle ABC into two parts of equal area. It is not implied that these two problems are inappropriate for students, but that, because of the content prerequisite skills assumed, these problems would not provide a good context for focusing on the skills involved in solving nonroutine problems. In contrast, the curriculum of nonroutine problems was constructed so as to minimize the mathematical content prerequisite skills required of the student. Therefore, using Schoenfeld's terminology, it should be clear that this curriculum does not directly attempt to improve the category he labels resources ("the facts, procedures, and skills — in short, the mathematical knowledge — that the individual is capable of bringing to bear on a particular problem" (p.17)). It needs to be noted that typically in a content-oriented mathematics course the curriculum of nonroutine problems only accounts for approximately 20% of the curriculum. In addition, field-testing has supported the hypothesis that the curriculum of nonroutine problems can be effectively integrated into a typical content-oriented mathematics curriculum (London, 2021). This characteristic of a curriculum of nonroutine problems especially addresses the issue of accessibility, specifically, by definition and in practice each problem needs to be appropriately challenging, meaning that there needs to be an obstacle to the solution, yet not an overly frustrating obstacle, such as mathematics content not yet mastered. In fact, the pedagogy for the curriculum provides seven levels of structure (e.g., scaffolding; London, 2012, 1993) that allow the teacher to adjust the difficulty level of each problem so that it is appropriately challenging for most students, with additional guidance how to adjust the problem for individual students or groups that need additional scaffolding or need additional challenge.

A CURRICULUM OF NONROUTINE PROBLEMS

This section will discuss what is meant by a curriculum of nonroutine problems. Before giving examples of a sequence of nonroutine problems, some of the general characteristics of a curriculum of nonroutine problems will be briefly described, as well as some guidelines in implementing such a curriculum — topics more thoroughly discussed in previous works (London, 1993, 2021). For a traditional four-year academic mathematics curriculum for grades 9 through 12, it is recommended that the curriculum consist of eight to fifteen nonroutine problems for each academic course. Each problem typically requires at least one week to complete, involving one to three hours of actual class time with additional time outside of class. It is recommended that approximately two-thirds of the problems are solved in cooperative groups. Students would be required to, orally and/or in writing, document the process for solving each problem, including describing all methods that were considered. Approximately forty percent of the problems would involve content not typically considered mathematical, including problems that students individually define and solve. The curriculum should include an introductory unit that describes the three steps of a nonroutine problem, gives several examples of the steps applied in a variety of fields, and has the students research additional examples. The curriculum is primarily designed as one component of a grades 9-12 mathematics curriculum (approximately twenty percent of the commonly allotted time for mathematics instruction) and is constructed to easily integrate with a curriculum consistent with the NCTM standards. In addition, the curriculum can be implemented as an elective secondary mathematics course in problem solving, an undergraduate liberal studies mathematics course, or as a significant component of an alternative secondary program (London, 2021).

To clarify: A brief example of a sequence of nonroutine problems in teaching an undergraduate mathematics problem solving course included: (1) solving one of three practical problems: planning a trip to a nearby resort, buying a car, or selecting a restaurant for a special occasion, (2) calculating the area of an irregular shape on graph paper, (3) determining the distribution of four different color marbles based on random samples, (4) determining grades based on raw test scores and some minimal information, (5) identifying significant improvements in one's home consistent with an ecological perspective, (6) constructing three solid figures without a "lid" that maximize volume, using one piece of construction paper for each solid, (7) arrange for a speaker to increase the class' appreciation of diversity, (8) constructing the most pleasing (aesthetically) rectangle, with one dimension 6", from a piece of construction paper – the rectangle judged most pleasing usually has a ratio of sides close to the Golden Mean, (9) graphing nonlinear equations without a graphing calculator (e.g., plotting points and looking for patterns), and (10) two individual problems that are meaningful in the student's life.

To clarify the individual problems, a few examples that undergraduate students selected include planning a vacation (e.g., planning a cruise for less than $1000); selecting the best engagement ring; finding a new job; relieving my frequent headaches; losing weight effectively; and developing a profitable tutoring service. Perhaps the most interesting problem was to find a humane way to reduce the number of squirrels on the student's property – the student discovered a somewhat complicated, but effective solution.

SAMPLE NONROUTINE PROBLEMS

In this section, a few sample nonroutine problems will be discussed, some that deal directly with mathematics content, and others that concern other fields. The primary purpose of this section is to give the reader a better sense of the characteristics of a curriculum of nonroutine problems, inductively through concrete examples. However, first, in this section the author will describe two nonroutine problems used effectively with both beginning average secondary students and more experienced calculus students, both problems appropriately engaging for most students, requiring only minor prerequisite skills. The two examples will be followed by additional examples both in the content areas of mathematics, as well as integrative examples in other fields. Finally, examples addressing the SMPs will be discussed, focusing on the pedagogical approach embedded in a curriculum of nonroutine problems that addresses the SMPs, as well as the connection to reform-based mathematics education.

The first problem is "Five calculations." For this problem, each group of three students prepares for the following task: for each of 20 numbers, between 100 and 900, provided by the teacher, reduce each number to 0 with five or fewer calculations involving addition, subtraction, multiplication, and/or division by the numbers 1 to 9. Each group is given two calculators and twenty minutes to complete the task. Each group may bring one 3 by 5 card with writing on one side. Groups are given one point if they reduce a given number to 0 in five calculations; two points if reduced in four calculations; and three points if reduced in three calculations. The groups are given some time over two weeks to prepare for the task.

A discussion of one solution will clarify the problem. There is a straightforward method to generate a solution in five or fewer steps for the numbers 100 - 819: (a) if not divisible by 9, subtract a number less than 9, resulting in a multiple of 9, (b) divide by 9, resulting in a number less than 91, (c) if not divisible by 9, subtract a number less than 9, resulting in a multiple of 9, (d) divide by 9, resulting in a number less than 10, and (e) subtract to equal 0. The numbers from 819 - 900 require you to find numbers in the 819 - 900 range that permit reduction in four steps. Then you can generate an interval of solutions in five steps by first adding or subtracting to get that number. For example, the numbers from 855 to 873 can be reduced to 0 by first getting to 864 by adding or subtracting, then 864/9 = 96, 96/8 = 12, 12/6 = 2, and 2 - 2 = 0. This process "allows" all numbers from 811 to 900 except 850 to 854 to be reduced in five steps. 850, 852, and 854 can be reduced by individual solutions. There is no solution in five steps for 851 and 853. Investigations of situations that can be solved in less than five steps and consideration of how to work effectively as a group will further improve a team's score. If a group discovers the solution in five steps, there is plenty of room on the 3 by 5 card to write the solutions for 811 – 900, as well as information that facilitates solutions in three or four steps. Students are provided the rules for divisibility, and it is indicated that there is no more "mathematics content" that will be useful – that is, they need to develop their own methods. By including a field-test of their solutions, the problem lends itself naturally to a transition from trying something to persisting. For example, most groups believe they have developed a good strategy before the first field-test, wondering why a field-test is even necessary, and are surprised that their strategies are not nearly as effective as expected. The structure of requiring a field-test helps the students focus on the need for persistence. Also, students are not informed that there are two numbers that cannot be reduced in five steps, and the two numbers are not included on the field-test, but one is included on the final test, providing an interesting twist to the problem, and a good basis for discussing persistence during processing.

The second problem is "Maximizing volume." The students are informed: "Your group is a research team for a company. The company sends out many containers and the company wants to save money by fitting more material in each container. The containers are made from material the size of a piece of construction paper, not having a lid. You can design any shape as long as the top open section is flat. The company wants your three best designs for a container; each design should be a different shape." The students are provided with material to test their volumes, such as a bucket of sand and a measuring cup.

This problem provides an opportunity for students to see directly what it means to persist through additional exploration after their initial solutions. For example, after the processing of their initial work, you can give them a concrete example of persisting by asking how we can be confident that we have picked the rectangular solid (or cone, or cylinder) that holds the most volume. You can adjust the instructions depending on your assessment of their understanding. For example, as a class you can systematically try rectangular solids with square bases with sides of 1", 2", ..., 8" by having each group construct one, then look for a pattern and persist.

Mathematics Problems

Initially, most students need a careful introduction to the three steps. Therefore, initial problems need to concretely give most students a successful experience with the three steps. For example, the problem "Expensive Tape" is a nonroutine problem that can be used early in the curriculum to introduce the students to the process of solving a nonroutine problem. Students complete the following task: determine three boxes that hold the most volume, given that the box is taped with exactly 30 inches of expensive tape, covering one length and one girth (two widths and heights) of the box, as illustrated in Figure 1.

Figure 1. "Expensive Tape" problem illustrated

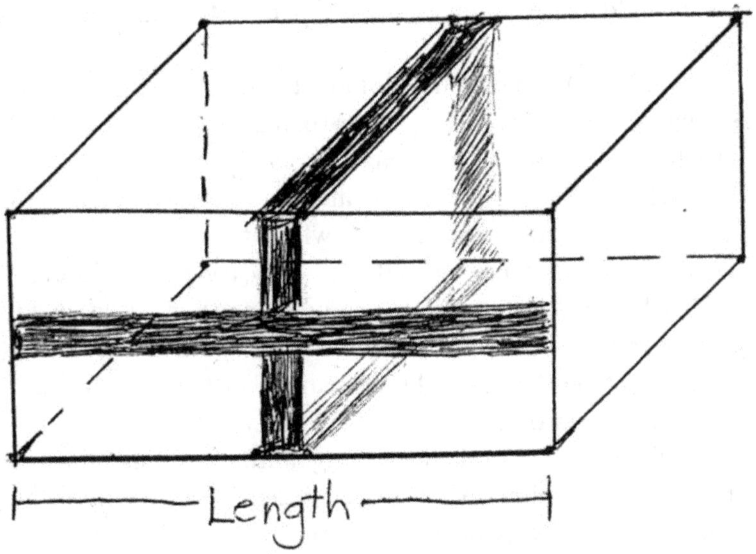

Two of the three dimensions (length, width, and height) must be whole numbers — the reason for this restriction is clarified in the discussion of the solution. For students with little experience solving nonroutine problems, a procedure to generate data is demonstrated and the students are encouraged to generate a good amount of data and to look for two patterns that will help them solve the problem. In addition, student's calculations are checked for accuracy only — this prevents students from generating invalid data such as more than 30" of tape, or incorrect multiplication. It is not indicated whether their volumes are good or not. It should be noted that if this problem was given to more experienced students, the teacher would not correct their data since discovering the errors would be an important part of a good solution for those students. When students generate data, they typically notice two patterns in the data. Before reading on, see if you can spot the two patterns in the data shown in Table 1.

Table 1. Table of data for the "Expensive Tape" problem

Length	Width	Height	Volume
10	5	5	250
10	6	4	240
10	7	3	210
10	8	2	160
10	9	1	90
16	4	3	192
14	4	4	224
12	5	4	240
10	5	5	250
8	6	5	240
6	6	6	216
4	7	6	168

The two patterns illustrated above are: Girths closest to squares yield higher volumes, and lengths closer to ten inches tend to yield higher volumes. Most groups notice these two patterns, thereby finding good solutions to the problem, such as: 10" by 5" by 5" (250 cubic inches); 8" by 6" by 5" (240 cubic inches); and 12" by 5" by 4" (240 cubic inches). Students that persist realize that 9" and 11" lengths should provide better solutions than 8" and 12" lengths (being closer to 10"), and discover the following two solutions, each of which includes one fraction: 9" by 5" by 5 1/2" and 11" by 5" by 4 1/2" (both with a volume of 247.5 cubic inches). Many students initially assume that lengths of 9" and 11" will not work because the sum of the width and height is not a whole number.

The processing usually helps students not only see how trying something results in uncovering patterns that make the problem easier, but also notice how persistence can lead to the best three solutions. Given the help described, most groups of students are successful with this problem, meaning not necessarily that all the groups find the best solution, but rather that most of the students are engaged in working on a problem at the appropriate level of difficulty and participate in the processing of the problem — a good indication that their problem-solving abilities will naturally improve (e.g., Caine et al., 1994). An important part of the directions is that they, as a group, decide which three solutions are the best, and write a short paragraph why they are convinced they found the best solutions. The teacher does not indicate to them whether their solutions are the best or not until the processing. This provides an excellent focus in the processing on the step of persistence.

"$Y=e^x-\sin x$" is an example of a nonroutine problem that is appropriate for more advanced students with experience solving nonroutine problems. The task is to find the fiftieth root less than 0 for the equation $y=e^x-\sin x$. You cannot solve the equation for x through normal algebraic manipulation, rather the student needs to generate data using a graphing calculator and look for patterns. In working with calculus students with experience solving nonroutine problems, a typical student solution is as follows: (a) problem recognition: the student realizes the problem is not solvable by algebraic manipulation, sees the need to generate data, and has some confidence that trying something and persisting will lead to a solution; (b) trying something: the student uses a graphing calculator to generate the first few roots and looks for patterns, noticing a fairly linear decrease in the value of the roots; and (c) persistence: in trying to make sense of the pattern, the student realizes that the roots change by a value approaching pi (perhaps believing the value approaches 3 first; the first three roots are approximately -3.1830, -6.2813, and -9.4249) and realizes that e^x contributes close to zero to the value $e^x-\sin x$ as x decreases in value; therefore, the appropriate root of $y= -\sin x$ (i.e., $x= -50pi$) is an excellent approximation of the actual root. In fact, further persistence may convince the student that a better approximation for the root is not possible with a graphing calculator. When tested with Calculus students with experience solving nonroutine problems, approximately 75% of the students found an excellent solution.

Integrative Problems

A major objective of the curriculum consistent with a holistic approach is that students will apply the three steps to solve meaningful problems that might not appear to be mathematical in nature but require the three steps for a good solution. A representative problem of this type is "Planning a Trip," which requires groups to plan an actual class trip that (a) will increase the class' appreciation of diversity, (b) is interesting and enjoyable from the students' point of view, and (c) is inexpensive. In the field-testing, one high school class planned a trip to Boston. Their research uncovered many good ideas including: (a) visiting a variety of relevant museums such as the Museum of the National Center of Afro-American Artists, Cambridge Multicultural Arts Center, and the African Meeting House; (b) seeing relevant sights in culturally rich sections of Boston; and (c) interacting with Cultural Survival, an organization that focuses on helping indigenous cultures survive. Perhaps the most interesting idea was to plan an afternoon with a group of inner-city students trained to be peer leaders for an after-school program. The idea was to share skills — we would lead a problem-solving session and they would lead a community building activity.

How is this problem connected with the mathematics curriculum? First, the students see how the three steps of solving nonroutine problems are useful in solving this practical problem. Students might demonstrate the following type of evidence of applying the three steps: (a) Problem recognition and orientation: Students might realize that some of the criteria for evaluating their project are contradictory, and part of a good solution will be finding possibilities that satisfy all three criteria. In addition, they might realize the need to gather data from classmates rather than relying solely on their own ideas. (b) Trying something: Students might brainstorm possible sources for ideas and decide which resources to actually investigate. They might also decide to develop a questionnaire to gather data. (c) Persistence: Students might realize that there is a need for additional data to finish the problem well. For example, most questionnaires give some clear data, but also generate some data that suggests the need to ask follow-up questions.

After students have demonstrated a reasonable understanding of how to solve nonroutine problems, students are required to formulate and solve meaningful nonroutine problems in their own life. It should be noted that even most talented mathematics students require a curriculum of at least eight to ten nonroutine problems to be adequately prepared for this portion of the curriculum. Typically, students select problems involving a major purchase (e.g., used automobile, stereo system, or bicycle), obtaining employment (e.g., a well-paying, interesting summer job), budgeting (e.g., budgeting to be able to save money to buy an automobile), or selecting an appropriate postsecondary institution. Two examples indicating the potential of this type of curriculum are taken from two vocational students of poor mathematics ability who took the author's one year Problem Solving course to fulfill their mathematics requirement for high school graduation. The first student had a lawn care business with three clients with large lawns. He was having much difficulty organizing the business and was ready to give it up. He defined as his problem to effectively organize the business. To gather information, he interviewed his clients and three other lawn care professionals (including two teachers at the high school). His initial conclusions included: (1) plan a schedule that allows you to complete the lawns in four days — this allows for bad weather and other complications, and (2) schedule your largest lawn early in the week. Based on these conclusions and the other data, he devised a tentative schedule, which he checked with his clients and a fourth lawn care professional. The plan was acceptable. He successfully implemented the plan and was very satisfied with the results.

The second student picked a very personal problem — he was concerned about his inability to relate well with other students, particularly his inability to make close friends. He investigated several options and decided to see a therapist working with the school. The next fall the therapist indicated that he continued seeing her through the summer and had made significant progress.

These two examples indicate the power of a curriculum of nonroutine problems. These two students were able to define a significant problem in their life and apply the steps a mathematician uses to solve a difficult problem to effectively address their problem. It is claimed that mastery of this type of problem solving is what is essential in mathematics education and most useful to the great majority of students.

Additional general problems in this category include: (1) Improving the Class. Students generate a list of ways to improve the class based on collected data. (2) Having Fun in Your Community. Students determine ways to have fun in their local area. (3) Home and the Environment. Students determine how to make their homes more ecologically sound. (4) Persistence. Using questionnaires, students investigate topics (e.g., How do students define a loving relationship?) that require persistence to research satisfactorily. (5) Interesting Rituals and Customs. Students find at least five interesting rituals or customs practiced in their community. (6) Comparing Products. Students compare the effect on the environment of certain products (e.g., cloth diapers versus disposable diapers). (7) Improving the School. Students prepare a list of five to ten ways to improve the school. They develop questionnaires to help them with the project.

Examples: Common Core Standards of Mathematical Practice

What is the connection between the SMPs and a curriculum of nonroutine problems? The purpose of this section is not only to outline how the approach to instruction for a curriculum of nonroutine problems promotes the SMPs, but also to describe how this instructional approach represents a reform-based, holistic approach to improving the SMPs. This will be discussed in two steps, firstly, by describing a few specific examples of the connection using specific nonroutine problems that well illustrate the connection, and, secondly, describing the difference in implications of the approach to the SMPs in a curriculum of nonroutine problems and the approach implied by much of the current material concerning implementing curriculum that has the purpose of improving the SMPs of students. First, it is appropriate to make a few introductory remarks: The pedagogy described for a curriculum of nonroutine problems was developed before the SMPs were adopted and issued; that is, the pedagogy described was not developed in response to the adoption of the SMPs. Basically, a curriculum of nonroutine problems is one of a variety of approaches to mathematics education prior to the formal definition of the SMPs that facilitates the improvement of students' ability to do mathematics, including their ability to integrate the SMPs into their work in mathematics classes, as well as in their daily life. Of course, there will also be new approaches in years to come in response to the formal emphasis on the SMPs.

In this section, a few individual nonroutine problems that illustrate the connection will be discussed. More than the other SMPs, common practice 1, "Make sense of problems and persevere in solving them," is most consistent with the definition of a nonroutine problem and is therefore a common practice addressed in every nonroutine problem. Specifically, "Make sense of problems" corresponds well with the first step of solving a nonroutine problem, "problem recognition", and "persevere in solving them" corresponds well with the third step of solving a nonroutine problem, "persistence." The second step of solving a nonroutine problem, "trying something," is certainly assumed or implied in the wording of the first common practice.

The primary example for this practice not only clarifies the step of persisting or persevering in solving a problem, but also illustrates that even the best secondary mathematics students have difficulty (without instruction) with this step. The step of persistence implies having a connection with what a satisfactory solution looks like. For example, persistence implies that the problem solver realizes that a particular solution is not satisfactory. A clear example of the lack of this awareness is the typical response of beginning AP Calculus students (generally the best mathematics students) to the nonroutine problem of calculating the area under the curve $y=x^2+2$ between $y=0$, $x=0$, and $x=3$ as illustrated in Figure 2 (the problem is given before the concept of integration and is the first nonroutine problem they attempt to solve). Despite directions to determine the area as best they can, the great majority of the students calculate the area as 15 1/2 square units (actual area is 15 square units) by "replacing" the curve with the straight lines connecting (0,2) and (1,3), (1,3) and (2,6), and (2,6) and (3,11) and then dividing the area into triangles and rectangles.

Figure 2. Graph of $y=x^2+2$ between $y=0$, $x=0$, and $x=3$

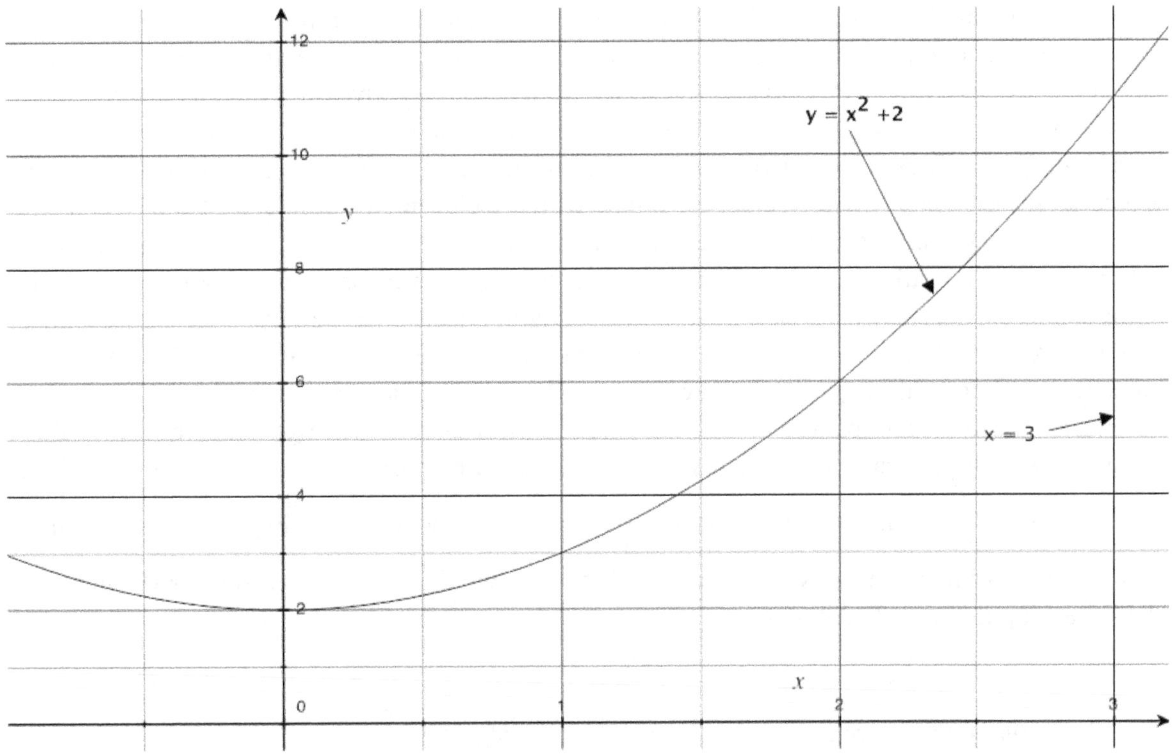

These students fail to recognize that their calculations could be more accurate by simply increasing the number of divisions (e.g., nine divisions yield a solution of 15.125 square units). Notice their solution is inadequate even though the strategy they picked is excellent and, in fact, is the basis of the method of integration in Calculus that gives an exact answer to the problem. From having worked with these students, most of the students, typically high achievers, would have been willing to spend the time necessary to persist but simply did not have experience solving problems which require them to recognize that a problem does not always have a straightforward solution (or is insolvable by them). Certainly, typical textbook problems do not require this skill. In practice, typically approximately 75% of Calculus students determine the best approximation of the area to be 15.5 square units. Of the remaining students a few even determine the area to be 15 square units (actual area).

In terms of the first step, problem recognition, the students fail to make an initial connection with the solution, that is, they have no clear sense at the beginning of the problem what an adequate solution (the best approximation of the area) looks like. Notice their solution is inadequate even though the strategy they picked (basically the trapezoid method) in step two is excellent. In fact, it could be argued that the inadequacy of their solution is connected to the first step (e.g., a lack of connection with step three, i.e., a sense of what a good solution might look like and some sense of the obstacles to a solution) rather than the third step. Typically, a person cannot persist in the sense of step three if there is not an adequate sense of what a good solution might look like in step 1.

This example provides a good opportunity to discuss one approach to improving the students' ability to effectively practice the SMPs. Specifically, after a curriculum of eight nonroutine problems over three quarters of a school year, integrated into a Calculus course, the students demonstrated a good mastery of this common practice (London, 1989). It is hypothesized that if given this area problem to Calculus students after they completed the curriculum, at least 75% would determine an answer better than 15.5 square units (versus 25%) — that hypothesis has not been tested directly with Calculus students since by that point in the school year they have learned how to determine the exact area quickly by integration. However, other data supports that hypothesis: (1) from similar problems requiring the three steps given to Calculus students later in the year, and (2) data from an eleven-week undergraduate liberal studies problem-solving course (for non-mathematics majors) involving a curriculum of nonroutine problems in which students who did not take a Calculus course were given this problem as part of a posttest.

The eighth SMP is, "Look for and express regularity in repeated reasoning." This practice includes the following skills (1) notice regularities in data and use those regularities to help solve problems; (2) draw conclusions concerning patterns or regularities based on adequate data and examples of the pattern; (3) generate data to look for patterns or regularities; and (4) being careful in generalizing and evaluating the implications of patterns and regularities. For example, one more advanced problem, "Making an area of 1," explores the area bounded by $y = \frac{1}{x}$, y=0, x=1, and x=b (see Figure 3) and requires the student to answer three questions — what values of b will yield an area of 1, 2 and 1000 (answer: e, e^2 and e^{1000}; e is approximately 2.718)?

Figure 3. Graph of area bounded by $y = \frac{1}{x}$, y=0, x=1, and x=b

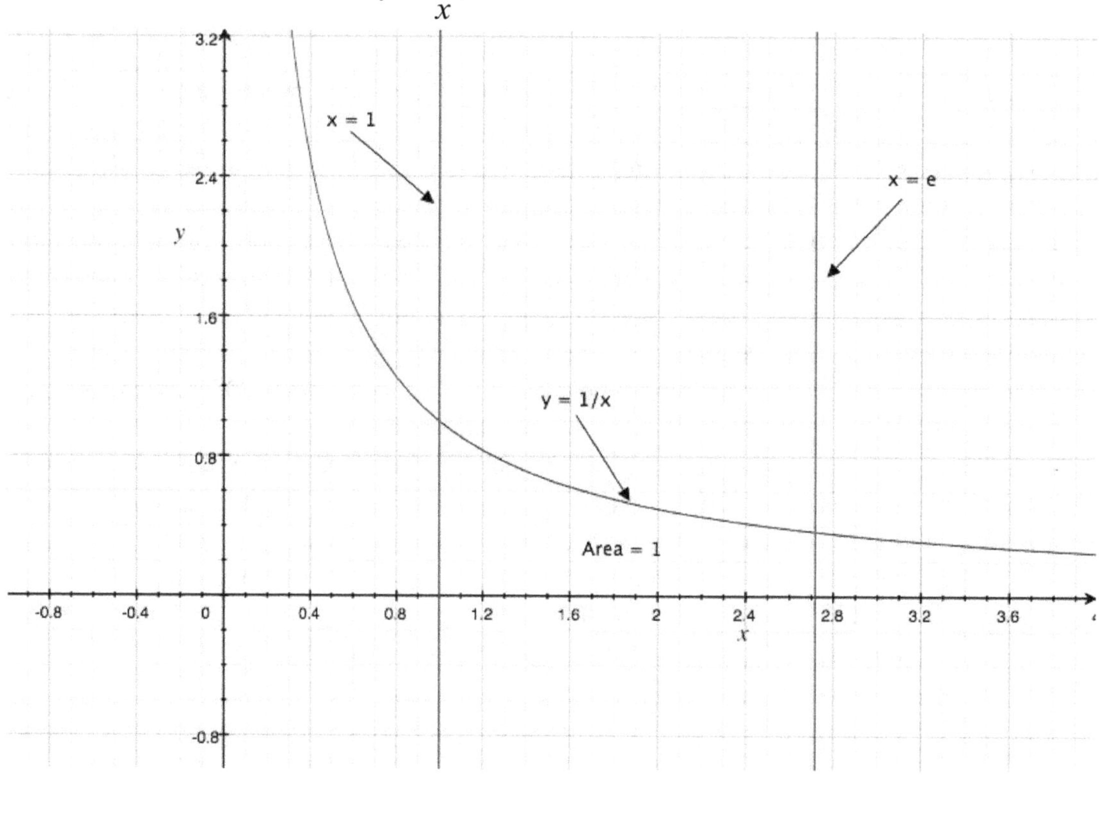

The directions for this problem make it clear that for all practical purposes it would be impossible to directly determine the value of b that would give an area of 1000; it is necessary to discover a pattern. There are two aspects of this problem that make it particularly valuable as an inductive problem. First, the student is required to generate the data, and this is certainly not an easy task — the student is not familiar with the number e and calculating the areas well is a long process. Second, in processing the problem it is emphasized that full credit is given for an answer of approximately e^{1000} (e.g., 2.7^{1000}) only if the student includes direct calculations (or educated estimates) for at least three values of b (intentionally, this specific requirement is not specified in the rubric for the problem). For example, if only two values of b (1 and 2) are calculated, a student who incorrectly guesses that the pattern is $e, e^2, e^4, e^8, \ldots$ is given the same credit as one who correctly guesses the pattern because both students made the same mistake from a problem-solving point of view, generalizing with too little data.

Students who generalize too quickly in this problem learn from their experience and the processing, and generally do not make that type of error on subsequent nonroutine problems. This fact indicates something about the value of nonroutine problems, and how they can provide a good context for exploring the SMPs. When the author emphasized the need for enough examples of a pattern to generalize in the context of a lesson in the regular curriculum, most students continue to make that error in lessons in subsequent months. What helps the transfer from the nonroutine problems is the amount of time and effort invested in the solution — after spending two weeks on a problem the student is more interested in the processing and more likely to retain the substance of the processing.

"Making an area equal to one" is a nonroutine problem recommended for more advanced students after a good amount of experience with nonroutine problems. In contrast, a second example, "The most pleasing rectangle," is one of the ten introductory nonroutine problems involving mathematics content suggested in implementing a curriculum of nonroutine problems (London, 2021), and well illustrates the connection between instruction in this common practice (as well as the other common practices), "look for and express regularity in repeated reasoning," and the three steps of solving a nonroutine problem. "The most pleasing rectangle" requires students to construct the rectangle that will be judged most pleasing by other people. To insure a reasonable basis for comparing rectangles, there is a restriction that the rectangle must be made from one piece of black (or any consistent color) construction paper with no designs on the paper, and have one side equal to six inches. These restrictions help ensure that factors other than the shape of the rectangle do not effect whether the rectangle is judged pleasing or not. The restriction of one side being six inches helps keep the number of possibilities manageable. Initially it seems to most students not to have a solution and to have nothing to do with mathematical problem solving. However, after the problem is completed and processed it becomes clear that the three steps of solving nonroutine problems result in fairly consistent good solutions to this problem. Basically, the solutions judged to be most pleasing tend to have a ratio of adjacent sides close to the Golden Mean. What is needed for the typical student is scaffolding to make this a problem at the appropriate level of difficulty to allow for the student to be actively engaged and likely to improve in the appropriate SMPs. A typical student given a problem requiring these skills without scaffolding would probably either be inappropriately frustrated from an instructional viewpoint and/or quickly generate a poor solution. The scaffolding can include: (1) The suggestion that they start the problem by remembering the second step of solving a nonroutine problem — trying something. Specifically, it can be suggested that they cut out a variety of rectangles consistent with the restrictions, show them to a variety of people, and see if any patterns emerge concerning which are judged pleasing. (2) The students are reminded of the third step of persistence, noting that the first set of rectangles may suggest some patterns but that an excellent solution will require the student to refine the experiment until the results are consistent and convincing.

The specifics of the directions need to be adjusted to be consistent with the primary purpose of this introductory problem: To introduce the student to the SMPs and the three steps of a nonroutine problem in a way that facilitates them being successful in the future with nonroutine problems with less scaffolding. Remember, for an introductory problem most students are likely to need help seeing what is appropriate for the steps of trying something and persisting. For example, in addition to the scaffolding suggested above, you might give them specific directions such as: for the first step, create at least five different rectangles, with a difference of at least one inch for the second dimension of the rectangle, and gather data from at least 10 people. You might indicate what type of data to gather (for example, two most pleasing rectangles; rank order the rectangles; etc.). You might process this first round of data as a class by having each student identify their most pleasing rectangle (or two if a close vote) and looking for patterns. For example, rectangles with a second dimension between 9 and 10 should be judged most pleasing more than other dimensions (people generally prefer the shorter dimension being 6"). This type of outcome provides a good basis to discuss the importance of persistence. Specifically, you might indicate that the results imply that the best answer is between 9 and 10, and ask how can you persist so that you feel more confident and precise in your results? Depending on your assessment of your class, you might either (1) come to a class decision, for example, compare additional rectangles with given suggestions or directions such as compare using 1/8" intervals (9 1/8; 9 2/8; 9 3/8; 9 4/8; etc.) or (2) let them decide how to persist, perhaps allowing them to work on this step in small groups.

If this is a problem given later in the curriculum, much less, if any, scaffolding would be appropriate. This general principle of gradually reducing the scaffolding for the problems is not only essential in a curriculum of nonroutine problems, but also an important principle in facilitating students' progress in developing the eight common core mathematical practices. Specifically, a variety of teachers and curricula that do give students problems involving the common practices and provide scaffolding for students, tend not to move the students in the direction of solving similar problems with less scaffolding. For example, we can compare three different levels of scaffolding for this problem, (1) the directions suggested above, (2) only the initial scaffolding of requiring them to gather data from at least 10 people before determining their final conclusion, and (3) give them only the statement of the problem with the restriction that the rectangle must be made from a piece of black construction paper and have one side equal to six inches. Generally, the first set of directions would be appropriate for students with little or no experience solving nonroutine problems; the second set for students with some experience; and the third set for students later in the year already with experience solving nonroutine problems with the level of scaffolding suggested in the first two sets of scaffolding. Typically, the students that persist in their experiments are more likely to construct a rectangle with proportions close to the Golden Mean (e.g., 6" and approximately 9.7"). Therefore, in the processing of this task, it is emphasized how the steps of trying something and persisting allow the problem solver to find a good solution to what seems like an impossible problem to most students.

These three examples provide a good illustration of the connection between a curriculum of nonroutine problems and instruction in the SMPs. In closing this section, two nonroutine problems already discussed in this chapter that address the fifth common practice, "Use appropriate tools strategically," will be briefly related to the SMPs. First, the problem, "$y=\sin x-e^x$" is appropriate for more advanced students with experience solving nonroutine problems and asks students to estimate as accurately as possible the fiftieth root of $y=\sin x-e^x$ (e » 2.714 and is an important number in calculus) to the left of x = 0 (y-axis). Second, "Five Calculations" is a nonroutine problem that is particularly appropriate as an introductory problem with appropriate scaffolding. Both problems require the use of calculators and the ability to make sound decisions concerning how and when the tool might be appropriate, as well as recognizing the limitations of the tool. For both problems the scaffolding needs to move the problem into the appropriate level of difficulty versus making the problem straightforward. Notice that this approach differs from most approaches to teaching content objectives for which you generally want to eliminate student mistakes and dissonance.

In summary, this type of approach to improving the SMPs helps facilitate the type of coherent focus that is needed to significantly improve students' ability to integrate the SMPs into their problem-solving process both in mathematics education and in solving significant real-life problems.

IMPLICATIONS AND CONCLUSION

In summary, field-testing (e.g., London, 1993, 2021) indicates that the following guidelines are important when implementing a curriculum of nonroutine problems:

1. The most effective way to improve students' ability to solve nonroutine problems is to repeatedly put them in a situation in which they work on a nonroutine problem at an appropriate level of difficulty, and then discuss or process the problem as a class.
2. In an effective sequence it is normal that each problem is difficult. It is normal that the student would not be clear about how to solve the problem initially; that at times it would feel as if the problem is not solvable; that the first attempts at solution would seemingly lead nowhere; and that one would need to persist until the problem eventually became clear.
3. The teacher needs to provide a structure to allow students to work on problems at an appropriate level of difficulty. Perhaps the most significant factor affecting the level of difficulty of the problem that we can manipulate is the level of support and assistance we provide the student during the process of solving the problem. In previous work (London, 1993), seven levels of support and assistance are defined that provide an instructional sequence to help students move effectively to the objective of being able to solve nonroutine problems without scaffolding. For example, at the fourth level, the students are given a nonroutine problem to solve (either in groups or individually), and each group or individual brainstorms ideas for each of the three steps for teacher review before attempting each step. At the seventh level, the students generate and define nonroutine problems important to them individually (or as a group) and solve each problem. Within each level, the difficulty of a problem increases as you move from a format involving whole class discussion, to group work, to individual work. An implication of this guideline is that each student can be provided with a problem at an appropriate level of difficulty to maximize learning of the process of problem solving.
4. A cooperative group model is consistent with the purposes of this curriculum. For example, students benefit from being exposed to the strategies and thinking of other students, and cooperative group work provides a supportive atmosphere for dealing with the natural difficulty of the nonroutine problems (e.g., Leikin & Zaslavsky, 1997; Johnson & Johnson, 2016). In addition, the proper use of cooperative groups has been found to be an effective method in classes with students of diverse backgrounds.

5. The curriculum needs to emphasize problems from a variety of fields and problems that directly affect or are meaningful to the student's life to facilitate transference of the skills to other fields and daily life. Field-testing indicates that without variety and relevance most students are unlikely to transfer the problem-solving skills to their day-to-day life. The need to integrate a variety of problems from fields not generally considered mathematical only became apparent after the first few years of field-testing. Specifically, after a few years of field-testing with the problem solving class, a few strands were integrated that emphasized "nonmathematical" problems such as problems which attempt to increase the student's appreciation of diversity, especially diversity of ethnic groups and cultures; ecology problems with an emphasis on problems directly affecting their lives; and practical applications affecting their lives, including a sequence of four problems (starting in the third year) affecting their lives that each student defines and solves. These strands account for approximately 40% of the problems in the suggested sample four-year secondary curriculum. Posttest data supports the hypothesis that the students were able to apply the problem-solving skills to meaningful "nonmathematical" problems (London, 2021). It should be noted that many mathematics educators have emphasized the importance of integrating instruction in the SMPs with instruction in the content standards. This point is not inconsistent with the intent of that emphasis. It is argued that this approach increases the ability of the mathematics educator to effectively integrate the understanding of the content with an emphasis on the SMPs. In addition, a focus only on "mathematics content" reduces the likelihood of transfer to other types of content.
6. Nonroutine problems that do not require significant content prerequisite skills are more effective initial vehicles to teach the process of problem solving than nonroutine problems that require significant content prerequisite skills. For example, two major obstacles to solving a nonroutine problem that requires significant content prerequisite skills are: (a) students tend to focus on the content aspects of the problem, lessening the focus on the process of problem solving, and (b) students that do not have the content prerequisite skills necessary to understand the content involved in the problem are usually not able to be successful in solving the problem. A nonroutine problem can provide an excellent introduction to significant content objectives, and, after students have experience solving nonroutine problems, nonroutine problems with significant content objectives can become an integral part of the curriculum. The point is that appropriate instruction in the process of problem solving does not have to be delayed because of lack of mastery of content (e.g., Bottge, 2001). For example, in the field-testing process, the author was able to successfully instruct a class of ninth grade students, considered of poor mathematics ability by previous coursework and/or testing, in the process of problem solving by using nonroutine problems that did not require significant content prerequisite skills. In the author's experience integrating a sequence of nonroutine problems (e.g., 8 to 12 problems) into traditional academic mathematics courses, most of the nonroutine problems did not directly relate to content objectives of the course; however, the academic achievement of students was consistently as much as, if not more, than other comparable sections (London, 2021). This may seem counterintuitive; however, in the author's experience effective instruction focusing on the process of problem solving naturally improves the student's ability to learn and to do mathematics. In addition, such instruction can be integrated in a way that minimizes disruption of classroom instructional time devoted to content objectives (e.g., London, 2021). The "Most pleasing rectangle" problem discussed earlier is a good example of the type of problem that allows improvement in the ability to solve nonroutine problems but does not create unnecessary obstacles connected to a focus on content objectives.

7. A curriculum of nonroutine problems is holistic in the sense that, by definition, each nonroutine problem requires the same three steps to solve, and processing is organized around the whole picture of the three steps and how the steps were implemented in the process of solving each problem. Similarly, it is suggested that activities that are meant to improve the SMPs need to be primarily organized around the whole picture of the direction of the SMPs versus an emphasis on discrete questions for each of the eight practices.
8. The focus of the processing for each problem should naturally emerge from the students' solutions and the processing of their work; that is, while there may be some general questions that are suggested for processing most problems, the most significant questions or focuses naturally emerge from the students' solutions and the processing of their work.
9. The process of improving students' ability in the SMPs and in solving nonroutine problems should be framed as an ongoing conversation with the students. That conversation can include addressing the fact that each problem is difficult and helping students become comfortable with feelings generated by that difficulty, helping the students notice improvements in their abilities and discussing what contributed to the improvement, noticing patterns of difficulty and discussing how to address the difficulties, and helping them realize that improvement will gradually occur over the course of the year.

What characteristics of this approach warrant the label of reform-based mathematics education? First and foremost, the focus of the curriculum is on what the author believes is essential for mathematics education, providing the average student with the tools to effectively address significant and meaningful problems in that student's life. For the author, this implied a curriculum that is focused on the process of problem solving (i.e., the three steps to solve a nonroutine problem), versus a primary focus on the content/concepts of the traditional mathematics curriculum. Note that it has been argued that such an approach can be effectively integrated into a traditional content-oriented mathematics course or curriculum. Second, the approach is holistic in the sense as mentioned above that, "by definition, each nonroutine problem requires the same three steps to solve, and processing is organized around the whole picture of the three steps and how the steps were implemented in the process of solving each problem." In addition, the approach is holistic with a multidisciplinary emphasis on meaningful problems, including problems identified as meaningful by the students. Third, the approach requires dissonance in the sense of obstacles to solutions that are an integral characteristic of the curriculum. In fact, the curriculum includes guidance how to adjust the problems so that most, if not all, students experience obstacles at the appropriate level of difficulty to be challenging, but not overly frustrating – considered essential in facilitating students' understanding of the process of problem solving.

Finally, we can connect these problems to the theme of this book, reform-based mathematics education, by noting the connections to the call for the book. Specifically, the approach addresses "using instructional practices that build on students' informal knowledge of mathematics"; "provid[ing] opportunities for students to contribute and invent their own methods"; "expect[ing] students to provide explanations for their answers" through oral presentation, written documentation and/or participation in the processing of the problem; "emphasiz[ing] classroom discourse and analysis of multiple methods for solving a given problem;" and "emphasiz[ing] instructional goals related to student engagement in and deep understanding of mathematical reasoning, concepts, and procedures."

Hopefully, the author has conveyed much enthusiasm for this type of approach to teaching the process of problem-solving. One last suggestion in implementing such a curriculum — allow yourself to be open to the richness of the problems, to the variety of solutions, and to the opportunities for growth that the student efforts will create. If you have any questions or comments, please contact me at rlondon@csusb.edu.

REFERENCES

Anderson, S., & Gurnee, A. (2016, March). Home-grown citizens. *Educational Leadership, 73*(6), 72–75.

Bottge, B. (2001). Using intriguing problems to improve math skills. *Educational Leadership, 58*(6), 68–72.

Brooks, J., & Brooks, M. (1993). *In search of understanding: The case for constructivist classrooms*. Association for Supervision and Curriculum Development.

Caine, G., Caine, R., & Crowell, S. (1994). *Mindshifts: A brain-based process for restructuring schools and renewing education*. Zephyr Press.

Curtis, D. (2002). The power of projects. *Educational Leadership, 60*(1), 50–53.

DeLeon, A. (2003). *A curriculum of nonroutine problems in the middle school* [Unpublished masters project]. California State University.

DeVinne, P. B. (Ed.). (1985). The American heritage dictionary, Second college edition. Houghton Mifflin Co.

Hickey, D., Moore, A., & Pellegrino, J. (2001). The motivational and academic consequences of elementary environments: Do constructivist innovations and reforms make a difference? *American Educational Research Journal, 38*(3), 611–652. doi:10.3102/00028312038003611

Johnson, D., & Johnson, F. (2016). *Joining together: Group theory and group skills*. Pearson.

Leikin, R., & Zaslavsky, C. (1997). Facilitating student interactions in mathematics in a cooperative learning situation. *Journal for Research in Mathematics Education, 28*(3), 331–354. doi:10.2307/749784

London, R. (1976, April). *Process-oriented curriculum* [Paper presentation]. *American Educational Research Association Annual Meeting*, San Francisco, CA.

London, R. (1989). *Nonroutine problems: Doing mathematics*. Janson Publications.

London, R. (1993, April). *A curriculum of open-ended nonroutine problems* [Paper presentation]. *American Educational Research Association Annual Meeting*, New York, NY.

London, R. (2004). What is essential in mathematics education. *Encounter: Education for Meaning and Social Justice, 17*(3), 30–36.

London, R. (2007). What is essential in mathematics: A holistic view. *MSOR Connections, 7*(1), 30–34. doi:10.11120/msor.2007.07010030

London, R. (2016). The enneagram: A spiritual perspective for addressing significant problems through research. In J. Lin, T. Culham, & R. Oxford (Eds.), Toward a spiritual research paradigm: Exploring new ways of knowing, researching, and being (pp. 97 – 126). Information Age Publishing.

London, R. (2021). *A curriculum of nonroutine problems: A holistic approach to process-oriented mathematics education* [manuscript in preparation]. College of education, California State University, San Bernardino, CA.

Moulds, P. (2004). Rich Tasks. *Educational Leadership, 61*(4), 75–78.

National Council of Teachers of Mathematics. (2014). *Principles to actions: Ensuring mathematical success for all.* National Council of Teachers of Mathematics.

National Council of Teachers of Mathematics. (2018). *Catalyzing change in high school mathematics: Initiating critical conversations.* National Council of Teachers of Mathematics.

Peressini, D., & Knuth, E. (2000). The role of tasks in developing communities of mathematical inquiry. *Teaching Children Mathematics, 6*(6), 391–397. doi:10.5951/TCM.6.6.0391

Polya, G. (1962). *Mathematical discovery* (Vol. 1). John Wiley & Sons.

Schmoker, M. (2018). *Focus: Elevating the essentials to radically improve student learning.* Association of Supervisors and Curriculum Developers.

Schoenfeld, A. H. (1985). *Mathematical problem solving.* Academic Press.

Tekkumru-Kisa, M., Stein, M. K., & Doyle, W. (2020). Theory and research on tasks revisited. *Educational Researcher, 49*(8), 606–617. doi:10.3102/0013189X20932480

Ward, C. (2001). Under construction: On becoming a constructivist in view of the Standards. *Mathematics Teacher, 94*(2), 94–96. doi:10.5951/MT.94.2.0094

ADDITIONAL READING

Caine, G., Caine, R., & Crowell, S. (1994). *Mindshifts: A brain-based process for restructuring schools and renewing education.* Zephyr Press.

Johnson, D., & Johnson, F. (2016). *Joining together: Group theory and group skills.* Pearson.

London, R. (2007). What is essential in mathematics: A holistic view. *MSOR Connections, 7*(1), 30–34. doi:10.11120/msor.2007.07010030

National Council of Teachers of Mathematics. (2014). *Principles to actions: Ensuring mathematical success for all.* National Council of Teachers of Mathematics.

National Council of Teachers of Mathematics. (2018). *Catalyzing change in high school mathematics: Initiating critical conversations.* National Council of Teachers of Mathematics.

Polya, G. (1962). *Mathematical discovery* (Vol. 1). John Wiley & Sons.

Schmoker, M. (2018). *Focus: Elevating the essentials to radically improve student learning*. Association of Supervisors and Curriculum Developers.

Schoenfeld, A. H. (1985). *Mathematical problem solving*. Academic Press.

Tekkumru-Kisa, M., Stein, M. K., & Doyle, W. (2020). Theory and research on tasks revisited. *Educational Researcher*, *49*(8), 606–617. doi:10.3102/0013189X20932480

KEY TERMS AND DEFINITIONS.

Cooperative Groups: A cooperative group has the following characteristics: (a) provides for individual accountability; (b) an emphasis on group skills; (c) fosters interdependence; and (d) includes a processing component. Note: These are research-based characteristics found to facilitate effective group work.

Curriculum of Nonroutine Problem: A curriculum of at least 8 nonroutine problems with the goal of improving the student's ability to solve nonroutine problems. The curriculum may or may not be integrated into a content-oriented curriculum (e.g., Geometry).

Holistic: An approach to mathematics education that implies a focus on the whole child, body, mind, emotions, and spirit. This type of focus implies a focus on transferring the process of solving nonroutine problems to fields other than pure mathematics that are meaningful to the student.

Nonroutine Problem: A nonroutine problem is defined to have the following characteristics: (a) The problem requires three steps to complete: problem recognition and orientation, trying something, and persistence; (b) the problem is open-ended; (c) the problem requires students to evaluate a variety of potential strategies and approaches to the problem; and (d) every student is able to "solve the problem."

Optimal Disparity: The level of difficulty of a nonroutine problem given to students to solve needs to be appropriately challenging, meaning that the problem cannot be solved straightforwardly, nor is the problem unnecessarily frustrating. Theoretically, this is considered the type of instructional problem (with appropriate processing) likely to improve the student's ability to solve nonroutine problems.

Process-Oriented: An approach to teaching problem solving implies an emphasis on the process of problem solving versus either a focus on mathematics content (e.g., adding fractions) or individual problem-solving strategies (e.g., guess and check).

Processing: After students complete their work on solving a nonroutine problem, processing means that you take instructional time to assist the students in making sense of the implications of their work on the problem.

Chapter 5
When Multidisciplinarity Meets Mathematics Teaching

Tamar Apel Campo
CimeH Institute, Israel

ABSTRACT

The great expectations put in reform-based mathematics education to solve the existing conflict between the teaching constructum and the learning agents empower "some kind" of different thinking, not vertical, not lateral, but multidimensional. This situation needs to be explored using appropriate tools that enable a clear vision of the complex educational environment where it takes place. The chapter presents in a systematic sequence the various dimensions identified through the use of complex system analysis and the fuzzy set approach. Examples of implementation of the multidisciplinary approach support the theoretical aspects of the phenomena studied and reinforce a didactic change in the normative teaching model. This change complies to the use of differentiation skills in the planning stage enabling a wider perspective of the participants in the teaching situation: mathematics, teacher, and students.

INTRODUCTION

We could start this chapter with a description of classrooms where mathematics is being taught. The diversity of the descriptions would be the same as the descriptions of what we encounter at restaurants in the area at lunch time. Opening one door and another will enable us to see different settings, different people, and different dishes. The important part of the visit is probably the impression you get of the satisfaction of those who are participating in the event, eating, cooking, or serving. Yet, this impression is not enough for a restaurant critic, who incognito will sit at a table and ask for those dishes that can give some clue about the quality of the place. These clues include the look of the dish, the taste, the relation between the food and the serving size, the atmosphere of the restaurant, the price, and probably the service. The results, summarized in an evaluative report, can shorten or not the distance to the Michelin star.

The many dimensions included in the restaurant evaluation are an example of those that, similarly, are present in the educational scenario. The mathematics that is served in school needs to be presented in the proper way to the students, taking in account the changes that actual knowledge management

DOI: 10.4018/978-1-7998-9422-3.ch005

contribute with. One of them is the multidisciplinary approach for teaching planning which results in an enriched view of the learning scenario. The mathematics classroom is hereby depicted from different disciplinary perspectives which include complex system analysis, fuzzy set theory, Hologramic sensitivity, organizational theory, social sciences and other. They will be presented in the following sections in the specific context of education.

The tangent points between the teaching of mathematics in a formal setup and the contributions of the various disciplines mentioned, produce a multidimensional presentation of the situation. We will move from the classic concept of "learning to teach" to a wider conception of the role of the teacher, as part of a learning organization emphasizing respectful and effective actions and the creation of an "ideal-type" (Easterby-Smith, 1997) of educational organization. This standing point meets the recommendations that: "due to the diversity of the purpose and perspective, it is suggested that is better to consider organizational learning as a multidisciplinary field containing complementary contributions and research agenda" (Easterby-Smith, 1997). The teacher role is from now on viewed as the expression of a developmental process in the organization called school, college, or higher academic institutes. The product of this development is a better learning scenario: secure, interesting, enjoyable, and effective. This takes us to reform-based programs.

The great expectations put in reform-based mathematics education to solve the existing conflict between the teaching constructum and the learning agents, empower "some kind" of different thinking, not vertical, not lateral, but multidimensional (Morin, 1999). Multidimensional mental movement, involving more than three dimensions is beyond the human capacity of concrete representation but still accessible enough to maintain the connections between different planes of knowledge. In these planes, the relation knowledge/thought converges towards option/action, pendulating between the complex and the simple. This natural oscillation is embedded in the educational environment we are interested in developing. An environment where "transformative learning communities could emerge, grow, and sustain themselves as a result of the interactions of constituents within boundary conditions, or parameters, which enable those interactions to develop and flourish" (Mason, 2008). This description of the learning environment is not only a must, but also a possibility, if and only if, critical factors are investigated through the prism that differentiates between the dimensions involved.

Throughout history, in more than one occasion, people trying to solve simple problems have made situations more complicated than they really are, mainly for one reason: the difficulty to see the whole picture. The mathematics' classroom falls in this category. Teaching is simple, but only if it doesn't castrate the act, assuming that all learn in the same way, are attracted positively to the same areas of knowledge, and have the same mental ability to deal with mathematical questions. Mathematics teaching offers the possibility to connect with, investigate, express, and formulate ideas, topics, and emotions. All these lead us to understand that education is suitable for a redefinition inside the margins of knowledge management including the consideration that effective people behaviour, required to achieve success, implies that the person must be considered. We must integrate cognition, motivation, personal satisfaction, feeling of security, and many other factors. This leads naturally to multidisciplinarity in the teaching plans.

To build effective dynamic models of educational institutions, we will need to know not just what people do, but why they do it, how they might imagine things being different, and what they would really want to do (Lemke & Sabelli, 2008). The classic teaching strategy based on cognitive activation strategies, teacher-directed instruction, student orientation and use of formative assessment must be updated and replaced by actions in affinity to social and research development. For example: 1-Differential planning; 2-Inclusion of interrelated topics; 3-Establishment of rutinary team communication frameworks.

Figure 1. Dimensions involved in Educational Activities

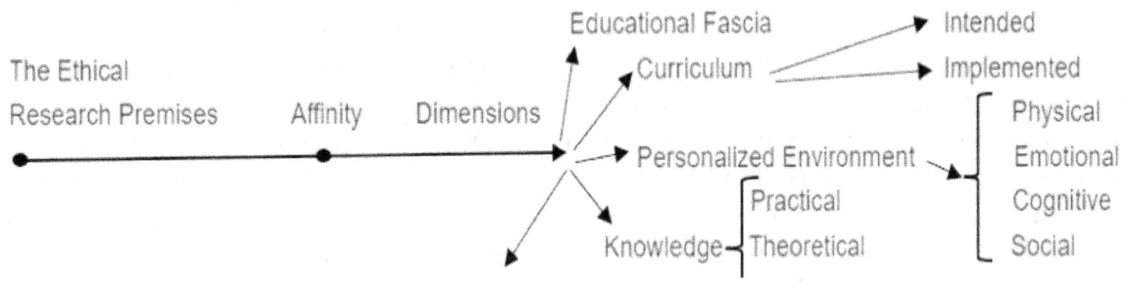

The differential planification model (DPM), used in the Teaching Mathematics with a Humanistic Approach program (Apel, 1996), gives the teacher a step-by-step planning protocol embedded in a cyclic structure where three areas of teaching interlap: theory, personalized didactics, and administrative issues, including evaluation and assessment. This planning tool positions the teacher in a new point of view, with a realistic vision of the complex situation and enables the design of units of the curriculum in ways that provide the conditions for interesting, enjoyable, and respectful learning of mathematics.

Interrelated topics emerge from flexible boundaries between disciplines and the introduction of phrasing that automatically connects among them. When Maria, a teacher in Argentina was worried that her students, who came to school after working hours, weren't paying attention to her statistics lessons, and requested advice, the suggestion given to her solved the situation. "Ask your students to make a list of topics related to their route from work to school that they would like to find information about. Use them as raw material to develop concepts and skills related to statistical analysis." From there on, the situation changed. Students started to participate in the class, contributing with questions and issues that interested them without putting curriculum boundaries. The gathering of data to answer the questions and the analysis came immediately as a proper way of looking into real world in an organized way. The artificial walls built in education when related to the mathematics spectrum which is present in every bit of real life, create an invisible distance between the contents and the participants, teachers, and students. On this basis, multidisciplinarity serves a reconciliation purpose and provides the opportunity to shorten the gap between theory and practical implementation at the learning levels. More examples are presented in the next section.

Enabling team communication is another way of linking social disciplines. Teamwork provides opportunities to choose, get involved, participate and be active during the learning process. All these actions of the teacher, open the involvement gates for the students and provide the space for formalization of ideas, concepts and tools required in the mathematics' classroom. We are going to proceed presenting the different dimensions involved in the teaching scenario and their interactional influences during the educational activities. The following graph (Figure 1) presents a proposal of the dimensions involved.

Components are presented not only with the usual academic terminology which deals with learning, concepts, didactics, curriculum, and research features, but from the topological aspect that reflects educational reality. "Education itself is 'coming to display a proliferation of surfaces' that are becoming-topological. Multitudes of multiplicitous topologies interact, overlap, and interpret each other in education" (Thompson & Cook, 2014).

These surfaces appear to the expert eye through the micro-telescope of the assemblage as presented by Deleuze and Guattari (2006), in their "rhizome" metaphor: "An assemblage is precisely this increase in the dimensions of a multiplicity that necessarily changes in nature as it expands its connections" and yet, it is not enough for the understanding of the teaching-learning situation, which requires adding to it the dynamic and constant interaction of all the components involved.

This multidimensional point of view leads to a perception of the phenomena from a perspective that enables seeing the dimensions of the teaching-learning situation, one by one. To get acquainted with these dimensions, we will define and state the feature of our intrusion before taking any steps in that direction. From there on, we will move in a sequential modality to present examples of multidisciplinary situations involved in the mathematics' teaching-learning scenario.

To have success in our enterprise some premises need to be made. The first of them is that the phenomena we want to study is grounded in the mental component of the mind. The second is that we have the proper tools to observe it. The third premise is that our observation will not interfere with the phenomena itself. Based on these premises, we proceed to analyze the dimensions of the mathematics teaching-learning situation. Altering any of the above will lead us to a different axiomatic constructum, contradicting the affinity among its statements. As much as this affinity seems negligible, it turns out to be one of the mostly influential and vulnerable factors present in the didactic spectrum, requiring our attention and definition.

The affinity element that guides us through our proposal will be considered in a similar context as in James Paul Gee's (2017) description of affinity spaces: "Affinity spaces are primarily defined by an affinity for solving certain sorts of problems." This definition comes as a payoff from the involvement of students in videogames, which offer a perfect setting for those willing to participate in an adventure attracted by a simple enjoyable teaser, the unexpected in the known scenario. "A video game is just a set of well-designed problems to solve, where the very design of the game teaches and mentors players to solve the problems, using good principles of teaching and learning" (Gee, 2017).

Video games are not the only place where we encounter the affinity idea. In qualitative research, interview techniques demand the creation of a relationship that must be nurtured and sustained, through proper rapport. This rapport implies "getting along with each other, a harmony with, a conformity to, an affinity for one another" (Seidman, 2005, p. 96). Whether we refer to a couple of participants in an interview or a group in the video game, or in a math's class, the main element present is the motivation of all to succeed in the joint enterprise. There is affinity in the intentions.

LEARNING WITHOUT BOUNDARIES

The following examples, taken from onsite situations, introduce possible approaches for the teaching of mathematics based on a *complex-fuzzy-multidisciplinary-humanistic* model. The examples are presented in a sequenced and concise modality, from low to high levels of education.

The 1st example comes from a preschool, in Netanya, during the project "Mathematics with a Humanistic approach for Preschool." In this project, four to five-year-old children in a kindergarten school, with cultural diversity, in the city of Netanya, Israel, were invited to choose from eight different mathematical activities offered in the preschool site. One of them was the Flower Shop. The layout included pots and artificial flowers of four different colours. The teacher told the group of children: "You are the owners of a flower shop. There are flowers of different colours. If you want, you can prepare bouquets to sell to your clients." The first part of the activity provided the space needed to get acquainted with the materials giving the children the opportunity to play with the pots and flowers and making bouquets. After this period, the first math question was posed: How many different bouquets could you make if in every bouquet you put four flowers of two different colours? The "making of bouquets" game, through trial and error, continued, leading to the production of bouquets. The quantity of flowers in each bouquet limited the solutions which were recorded in the form of drawings that changed afterward to traces of coloured lines representing the flowers, and then to coloured numbers symbolizing quantities and colour of flowers.

In the next phase the second math question was posed: "If you have three colours of flowers, how many bouquets can you prepare?" Or, "if you are asked to prepare bouquets with five flowers with only two colours, how many could you prepare?" The simulation game went on, the same day and the next one, with more level of depth. Intuitive counting, comparison of results, documentation (which leads to symbols written under each bouquet composition: 2+2, 3+1, 1+3, etc.). (Apel, 2001)

In this activity mathematics emerges from a need to understand and resolve a situation. The transmergent[1] didactic approach used by the instructional designers provided an environment in which students created new knowledge by blending principles of experiences with properties of complexity theory (Honebein, 2009). In this preschool, algebraic thinking was introduced through sophisticated games, in a learning environment that values the actions of the students who are involved in inquiry procedures, exploring, arguing, and testing ideas to solve problems (Blanton & Kaput, 2005).

The evaluation of the implementation of the "Mathematics with a Humanistic Approach" program, using the multidisciplinary approach in preschools in Netanya showed "statistically significant reduction in the percent of poor students in mathematical thinking: 8 percentage points: from 22% who perform poorly to 14%" (Lifshitz, Dolev et al., 2004, p. 8).

The second example relates to the implementation of the project: Knowing my School, in a group of elementary schools in the coastal area of Israel, This project included all the students from grade 1 to six for a period of two weeks at the beginning of the school year. I would like to remind the reader that the beginning of the school year is usually used by mathematics' teachers to "refresh" knowledge from the past year, since during the vacation period students "forgot" what they had learned. For the students, this has negative effects. Boredom is there for those that know the topics, frustration for those that learned and forgot, a general deception from all, expecting to get in contact with new and interesting topics. This project changed the point of view, assuming that students' knowledge is *sleeping* and would only need a good trigger to *wake up*.

The mathematics´ teachers opened the year in all the classes with the same question: *Do we really know our school?* (Apel, 1993). Teachers moderated the dialogue that rose in the class, giving an opportunity to find out what the question means and raise new questions about the school. The clarification dealt with what knowledge is and how can we achieve it, how information helps to learn more about any topic. In the following step, students were invited to ask questions that would help them to know their school *well*. This part enabled students to ask questions on topics that interested them, ignoring boundaries of disciplines. Every question was written on the board. Mathematic questions were highlighted; students were invited to write their names near the question they wanted to find an answer to. Thus, teams of researchers were formed. Each team prepared a plan to obtain the information required to answer their question. Questions with no link to mathematics were addressed in separate opportunities.

The students' research plans were submitted to the teacher for comments and approval and returned for implementation. In cases where there was a need to widen a topic, the required content was taught in the mathematics class. The question *Do school bags in our class weigh the same every day of the week?* required knowledge of weight measurement. The topic was subsequently taught to the whole class, linking with the curriculum of the grade. Data of family members in the school, number of boys and girls, arrival hours, ice cream preferences of students, salary of teachers, distance to toilets from each classroom, number of trees in the backyard, size of the rooms and buildings, number of windows and doors, designs of floors, etc. were collected by the students during the following days. Special hours were allocated for going around and entering classes to ask questions, measure, and count. Guidance on how to record observation, formulate questionnaires and represent data through diagrams was part of the teaching plan. The teams organized the information, wrote conclusions, and prepared a presentation including the question, their method of research and their answers. A summarizing session in each class gave the opportunity for each team to present their work and answer questions that came up. Some groups used posters, other a musical presentation or even a short theatrical scene. After the final session, with their *colleagues,* each team summarized the work in an A3 size paper that was posted in the corridors for three days, as a newspaper wall item. The whole school could enjoy the results of their collaborative work.

Two weeks had ended. The students refreshed their mathematics knowledge applying it to a topic that interested them, actively. The teachers made sure that there would be a correlation between the questions and topics of the curriculum by making enrichening suggestions. In grade five, the question *How many hours do we study each subject during the week?* got an addendum of, *what part of the whole week schedule do we study each subject?* coordinating with the topic of fractions in the curriculum. The same approach was implemented for the whole school. The ambiguous question posted at the beginning led to a network of information coming from varios areas of knowledge. Students, teachers, and parents were surprised with the results. Statistics was the connecting tissue in this enterprise.

The third example takes us to grade 10 in the Environmental School at Sde Boker Academy in south Israel. The curriculum of the Environmental School in the Sde Boker Academy included in every grade one-week outdoor studies, three times a year. The outdoor studies, called "Field Workshops", were planned by a team of teachers focusing on environmental issues. During the trip, theory was connected to practice through a precise elaborated syllabus. The connection with the official curriculum was created through the formal channels in each subject. The *Field Workshops* enabled the students to take a glance at complex phenomena not encountered in classroom.

Two teachers made a change in the rules of the game: Instead of having teachers plan the field workshop, the students were given the assignment to plan it. They organized themselves around common interests or social interaction without the intervention of the conducting teachers of the project: the English and mathematics teachers. Each team submitted a proposal for revision and approval. Logistics for the three days on site research program were included. After approval, the requirements of knowledge for the topic of research were submitted to the teachers of the class. Preparatory classes from different subject matters were given by the teachers of each subject and incorporated by the groups in their project portfolio. After three days on site study, each team organized the information collected, explanations, and new questions emerging from the study. The "paper" of their study was submitted to the teachers for revision.

The topics chosen by the students, although correlative with the curriculum frame, did not appear always in the same sequence as the official curriculum. Instead, generic titles emerged: Pesticides, Yarkon River pollution, Eilat sewage problem, Recycling: Haifa and Hedera Area, etc. Statistics, finances, measurement, and geometry got together with chemistry, biology, history, geography, engineering, and ecology in this adventure. (Apel & Rave, 1979). Students manifested great involvement, responsibility, and motivation during the whole project.

The fourth and last example comes from a higher level of education, the Ben Gurion University in Beer-Sheva Israel. The Teacher Training Department of the Faculty of Humanities in the Ben Gurion University in the years 1979-1984 offered courses designed in a non-conventional format. The courses were bi-disciplinary: geography and mathematics, biology and literature, history, and English, etc. After two months new alliances were formed. In one of the periods, the teachers of the mathematics and geography group posed a research question as the starting point: *Why were pyramids built in Mexico and Egypt?* The takeoff took place with enthusiasm. The question was scientifically relevant and created immediate engagement in the research. Groups, including students from both disciplines, designed research plans and collected data from geography, astronomy, sociology, mathematics, history, and geology. Digging in the topic revealed new knowledge. At the end of the project, the answers from the groups were presented for discussion, questioning and evaluation. The outcome of the course was an interesting document where all the disciplines merged. The manipulation of information collected from different disciplines in each group was possible due to the weak framing instaured, following Basil Bernstein's statement: 'frame refers to the degree of control teacher and pupil possess over the selection, organization, pacing and timing of the knowledge transmitted and received in the pedagogical relationship' (Sadovnik, 2001). Weak framing implies more freedom to move in the topic. The curricular change in the classification of knowledge permeated in the pedagogical arena due to the integrated code implemented. The mechanical listing of subjects moved towards and harmonic organization of knowledge. The teachers served as facilitators in an effective and natural way (instead of the classic "guiding and questioning" discourse) leading to a "questioning and finding" alumni discourse involving mathematical issues.

The above examples present a multidisciplinary perspective of the teaching-learning situation based on two principles: 1- Inclusion of new dimensions present in the situation 2- Adding points of view. With these two in mind, the perspective of the teaching role changed. In the following sections we will analyze them and their contributions to reform-based education.

FROM LISTS TO CHAOS

Let's dive into the curriculum dimension. The mathematics curriculum can be considered as the "plan for the experiences that learners will encounter, as well as the actual experiences they do encounter, that are designed to help them reach specified mathematics objectives" (Remillard & Heck, 2014). Educational institutions use this definition as a means to focus on what will be offered to mathematics' students. The offer is then translated by departments of education into a more specific and detailed list of topics to be expounded upon by mathematics educators. In a document from the Department of Education of England, dated January 2021, we find that "Mathematics is an interconnected subject in which pupils need to be able to move fluently between representations of mathematical ideas. The programs of study are, by necessity, organized into apparently distinct domains, but pupils should make rich connections across mathematical ideas to develop fluency, mathematical reasoning, and competence in solving increasingly sophisticated problems. They should also apply their mathematical knowledge to science and other subjects" (Department for Education, Gov.uk, 2013). The precise and elaborated document expands the theoretical vision towards the practical topics to be taught, providing clear guidance for teachers. This important contribution of the official government of England is not a unique piece of art. It is an example of how responsible education authorities around the world present the "what to teach" content in a step-by-step manner, making public their expectations from the teaching staffs in coherency with the policy maker approach of the *intended* curriculum (Hume, 2010).

The intended curriculum and the implemented curriculum have been referred to along the years as components of the theory-practice enterprise linking policymakers with teachers. While there is agreement on the need of a curriculum as a guideline, the influence of its design must be considered if we want to move to a reform-based mathematics' teaching. The curriculum is presented as a list of objectives, contents, and expectations of the teaching in a specific discipline, suffering from the illness of artificiality, since reality is castrated in favour of a comfortable formulation of guidelines for the teaching, classifying the contents according to areas of knowledge. Bernstein defines this classification as "the degree of boundary maintenance between contents and is concerned with the insulation or boundaries between curricular categories" (Berenstein cited in Moore, 2001). Within this framework, strong classification refers to a curriculum that is highly differentiated and separated into traditional subjects; weak classification refers to a curriculum that is integrated and in which the boundaries between subjects are fragile. (Bernstein, 2003, 2005)

The consequences of this classification are seen in the teaching praxis. The classic 'one discipline' feature, which is the chosen alternative among most curriculum planners, seems to have a contradicting impact in nowadays global educational approach, interfering with the outdoor dynamic environment. This needs to be taken in consideration in any reform-based education, since the starting point in the route influences teaching mathematics. The chaotic feature of knowledge spreading in the surroundings through electronic media maintains a stable dynamic and effective pattern of behaviour which makes it accessible and comfortable to users. The curriculum can be presented not as a list of concepts and skills to be taught, but as an organized system of relations between questions to be dealt, ideas to be shared and historical references that can highlight mathematical knowledge in an evolving feature, giving space for information on social influences and communication elements. The presence of "outsiders" in the mathematics' curriculum enriches the contents with supporting contributions from real life facts that are understandable by all, facilitating a connection channel with the abstract ideas involved. A curriculum of this type will link topics from different disciplines in conjunction without cutting aspects of the reality involved.

As stated before, the list of topics in the curriculum is in contradiction with the dynamic order existing in the chaotic featured curriculum. These two versions of the curriculum, based on a binary codification, do not contribute to a proper allocation of the reform-based approach curriculum which would be positioned more effectively using the "fuzzy sets" approach. In the "fuzzy sets" approach, continuity would be implied in the range, enabling transition from the "list" curriculum to the "chaotic" curriculum in a gradual manner. The varied classes of curricula appear when we introduce concepts from various disciplines and merge them through diversified routes of accessibility; this can be represented with the use of helicoidal concept maps, in which different aspects of mathematical concepts—their meaning and use—are positioned at different levels. This can provide a document where routes of relations can be chosen with freedom and responsibility. The movement in the helicoidal map is regulated by problem solving requirements and personal preferences which filter the steps logically in the way to the solution. The step-by-step sequence is replaced by a 'choose between alternatives' technique aimed at optimization of mental investment. This approach supports the complex system view, maintaining the affinity principle stated before, but is not enough for its implementation in the classroom. The helicoidal model of connection of knowledge presented in the Let's Track Methodology uses complex system approach for research in mathematics learning based on ethical postulates. (Apel, 2022).

POINTS OF VIEW

Looking at the math class from multiple angles, as shown in Figure 2, illuminates the agents and components in the occurring dimensions: structural organization (including rules and physical setup), pedagogic atmosphere, social aquarium, individual personality, collective objectives, and political postulates.

Figure 2. Multiple Points of View

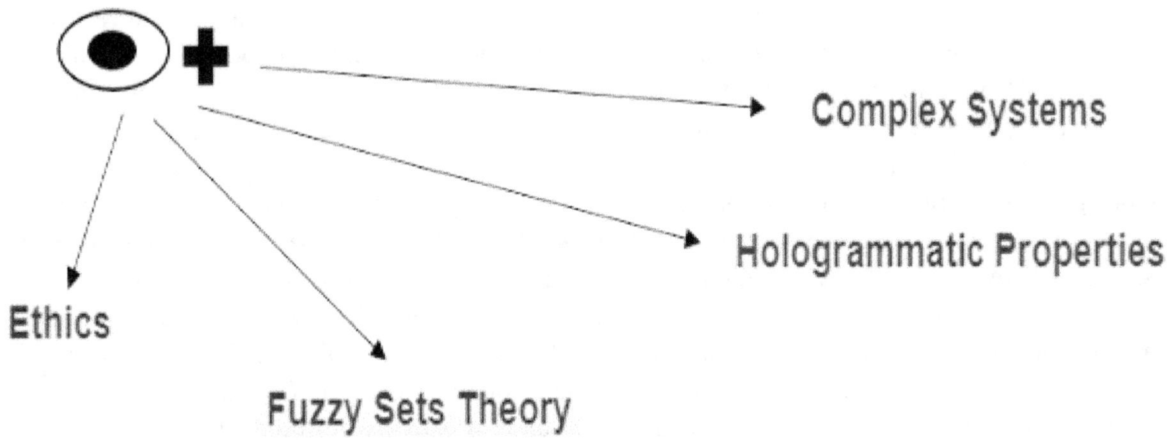

This view, springing from complex system analysis adds information on the occurrence in the pedagogical milieu "such as self-organization, feedback loops, equilibrium, nonlinearity, and emergence" (Jacobson et al., 2019). The emergence component is considered by many scientists as the most important complexity construct (Moore, 2001). This component is of great importance in education, introducing us to two different aspects of the emergence manifestation: local and global. The local emergence component takes us to a place where we can identify collective behaviour in a small part of the system. Thus, a child sitting on the floor in a preschool class in the city of Netanya in Israel, arranging flowers on a pot in ways that meet a special request presented by the teacher exemplifies the pedagogic atmosphere, interrelations, and objectives in the preschool class. The global emergence component magnifies the behaviour of the system as a whole. For example, when students at a school in Jerusalem systematically did not leave their mathematics class when recess bell rang. They manifested involvement and interest in the activities offered in the class. The global component emerged in the system, spontaneously.

There is also a somewhat counterintuitive aspect of emergence, which Jacobson et al. (2019) describe as the "formation of collective properties at a macroscopic level of a system from simple behaviours of the parts, with properties which frequently are not found in the parts." This last commentary is of major importance in the teaching scenario where didactic guidance, aimed to foster movement in one direction, can produce a "knowledge traffic jam" that directs the participants towards side routes in their efforts to arrive to the goal. This "out of the route" movement gives the opportunity to use creativity to find alternative knowledge independently. "In complex systems every causal chain is mediated, and many chains branch and loop back on themselves in complicated webs of mutual interdependent, self-regulation, and amplification of effects. This conceptualization is consistent with Fullan's (2009) 'systems at the edge of chaos' view of education" (Lemke & Sabelli, 2008).

The reductionist approach to the teaching mathematics phenomenon encountered in normative classrooms manifests itself through the linear approach of the classic "industrial" education. The triangular teacher-content-student model, which has not been demolished yet by technology or intelligent insight of educators, is still taught widely in teacher training institutes. An important contribution for its replacement comes from the work of Ball (2017), who presents a dynamic model[3] of the teaching-learning situation adding components that have been ignored until now. The appearance of many triangles in Ball's model is a proof of the complex situation we are involved in when teaching mathematics, leading us to an hologrammatic view of the teaching situation.

The new understanding of the teaching-learning scene scatters light in a range of directions, allowing us to view the scene from different angles, enrichening the original understanding with added information. "This has fundamental implications for the operation of educational systems, given that holographically and semiotic fractality education entails the recursive presence of the sociocultural fabric that it generates and to which it belongs" (Lavanderos Gallardo & Oliva Figueroa, 2012). This declaration brings us to inquire about the Hologramic characteristics present in the educational site. Since any point of a hologram contains the information of the whole of which it is part, we can extrapolate to the classroom where each participant contains hologrammatically the whole of which he is part and this is at the same time part of him. (Morin, 1999). The hologram of the teaching learning situation gives place to the phenomena letting us know that the educational society with its culture is inside the spirit of the individual member of it (Parra Sabaj, 2005). The student copying a problem from the board is a member of a learning community doing abstract reasoning which is already part of himself.

This hologrammatic property introduces the relationship and co-production of pieces of information joining the local and the global, the rustic and the formal, the isolated and the merged. The figure and the background are separated and connected again in a dynamic network. Teaching connects between theoretical sources and individuals through the basic principles of the hologram essence embedded in the classroom situation.

Mathematics is blessed with this hologram-like characteristic, too. The symbolism used in mathematics subsumes in a single idea the specific and the general, the concept and the process. Symbols are used throughout all the levels with a dual meaning. "The function notation $f(x)= x^2A= \pi r2^{-}3$ simultaneously tells both: how to calculate the value of the function for a particular value of x and encapsulates the complete concept of the function for a general value of x; The symbol -7 stands for both the process of "subtract seven" and shift seven units in the opposite direction along the number line, and the concept of the negative number -7" (Gray & Tall, 1994). This ambiguity is nothing more than freedom to move around in the mental arena, leading to choices emerging from personal investment focused in understanding the situation. If the learning environment permits flexibility it will lead to a supported involvement of the participants, enhancing motivation to deal with mathematic issues in a simple and yet formal academic feature.

The motivational aspect is not trivial, especially academic intrinsic motivation, taking us to the personalized dimension of the phenomena we are looking into. This drive or desire of the student to engage in learning 'for its own sake' results from the connection between the learner and the setting in the classroom. "Students who are intrinsically motivated engage in academic tasks because they enjoy them. They feel that learning is important with respect to their self-images, and they seek out learning activities for the sheer joy of learning" (Middleton & Spanias, 1999). This factor can't be ignored in teaching plans if we want to witness personal serious involvement in the topics presented. "When individuals engage in tasks in which they are motivated intrinsically, they tend to exhibit a number of pedagogically desirable behaviours including increased time on task, persistence in the face of failure, more elaborative processing and monitoring of comprehension, selection of more difficult tasks, greater creativity and risk taking, selection of deeper and more efficient performance and learning strategies, and choice of an activity in the absence of an extrinsic reward" (Lepper, 1988, as cited in Middleton & Spanias, 1999).

The motivational connection between mathematics and the student takes place through different features: characteristic of contents, skills involved, modality of presentation of topic, social preferences, tone of phrasing and possible rewards of success. An optimization of the learning conditions would enhance intrinsic motivation. The provision of a secure environment that enables comfortable involvement depends on the ethical platform which will support the avoidance of imposed, compulsory, instructions, which ignore personal interests and modality of learning. Instead, gentle, and friendly invitations to take part in "something" interesting, pleasant, and significant, can enhance motivating learning energies. This takes us to the final dimension in the teaching-learning situation.

Multidisciplinarity is a natural fact. Separation of knowledge is not. The isolation of a topic requires skills based on analysis of reality and a strong conviction that there is a firm justification for the limits created. The fact that epistemology is a separated domain in philosophy indicates that it is not easy to deal with knowledge. In classrooms, and not only there, "we discover a great deal by seeking explanations and positing one or another hypothesis to explain the puzzling data. What we discover in this way commonly constitutes knowledge" (Audi, 2011). Knowledge is also acquired by investigating at different levels and from different points of view, given its multidimensional nature (Mazzocchi, 2019). In this dichotomy, mathematics is part of both. The preparation of a meeting between the student, the teacher and mathematics in the multidimensional environment can rely on the dimensions and points of view presented in a multidisciplinary feature.

MULTIDISCIPLINARY TEACHING UNITS - A PREAMBLE

Disciplinary versus multidisciplinary approaches to teaching commonly stand on a binary platform, the 0 and 1 signals. In a disciplinary framework, we share a specific vocabulary along with a set of basic means for the acquisition of knowledge (Mazzocchi, 2019). This organized discrimination of information leads to the creation of sets of knowledge and the creation of barriers in research, teaching, social interaction, and creativity. In theory, it may happen. In practice it doesn't exist. Any event in the human physical or mental dimension is a result of non-discrimination of information. The *Person* teaching or learning is a multidimensional entity with easy access to knowledge, regardless of the label glued to it (Cropley, 2015). The more affinity there is between the grade of multidisciplinarity of the topic and the person, the easier it will be to get in contact with the knowledge.

The use of a fuzzy set of tools to position grades of multidisciplinarity enables us to get closer to the situation in the mathematic class but does not shorten the distance to our understanding. The continuum *one to many* makes our task almost impossible. On which variables will we focus to decide that we are viewing a disciplinary, bi-disciplinary, tri-disciplinary, transdisciplinary, multidisciplinary phenomena? What makes the change from one value to the other? Is it the language? The content? The structure? What is the flexibility of the fence in the separation of knowledge? Is there a possibility to find a *zero* value in coherence with disciplinarity?

Any attempt to simplify the complexity of the teaching-learning situation will imply a redefinition of the boundaries between disciplines and its consequences in research and implementation in mathematics teaching. This action has been taken in consideration, in a different context, yet meaningful, separating mathematics education into four ingredients for research: The students, the teachers, the curriculum and the pedagogical models. The conclusion from this definition was not optimistic: "Our position is that educational research cannot consider all of these variables. The result we must live with is acceptance that educational research cannot answer definitively all the questions we might ask about mathematics education" (Dean, 2019). This brings us to some examples that can raise more questions, without forgetting that the priority of any educational enterprise is to support a happy learning environment.

Moving from the analytical scientific media into the classroom gives us the opportunity to remind ourselves that we are dealing with "people". Persons that feel, not only think. The emotional and physical components of the person are inseparable from the human constitution. Mathematics learning is done by persons who have mental, physical, and emotional components, present every moment of their life, integrated in a woven net with one objective: happiness. Happiness and joy are linked everywhere, especially in education. One way of promoting happiness in the classroom is through the achievement of goals. This tricky truth can cause great damage in mathematics class if it is nurtured by pressure and competition, segregating the winners from the rest, the best from the others, the strong from the weak. According to Mirvis, "Challenges of competition were found to be stimulating and enjoyable. But when beating the opponent takes precedence in the mind over performing as well as possible, enjoyment tends to disappear. Competition is enjoyable only when it is a means to perfect one's skills; when it becomes an end, it ceases to be fun" (Mirvis, 1991).

The use of the competition factor is thus an unbearable component for the students, in any teaching approach, unless the design of learning situations is compatible with the improvement of the pupil's skills without relation to another pupil. Competing with oneself would be the proper terminology adhered to a situation of the kind. Another way of achieving the goals would be by group cooperation or by designing "sequences of activities that are goal-directed and bounded by rules –activities that require the investment of psychic energy (attention) and that could not be done without skills" (Mirvis, 1991). Taking this in account, we will join the perspective given by Csikszentmihalyi (2009), connecting teaching with happiness in the "flow" theory and repairing the situation where children "have to sit in school, doing abstract tasks they don't care about, without moving their bodies –which were selected by evolution for movement and action" (p. 209). The power to create "beautiful settings" that enable looking for new connections among ideas is in the hands of the teacher who needs to be also creative. As Csikszentmihalyi (2014) says, "it is essential to have perspectives on issues we are dealing with, i.e., to have a "prepared mind" (p. 9). The multidisciplinary teaching we are looking into connects mathematics with other fields of knowledge taking in account the mental, physical and emotional dimensions of the participants involved, as well as those of the educational environment. This is the real essence of multidisciplinary teaching, making meaningful connections between the disciplines in the learning-teaching realm: mathematics, sciences, humanities, communication, and technology.

CONCLUSION

Knowledge boundaries in curriculum create a culture of learning adapted to the limits defined for each dimension. "All normative curricula risk becoming sites of irresolvable ideological clashes over which or whose values should be advanced, and in which oppressive and exclusionary power relations flourish. However, complexity offers ways to think about educational inputs and outcomes that do not assume that causal relationships between them are, or should only be, instrumental" (Gough, 2013). The use of complex system analysis in education allows the creation of cultures of learning, furnishing "raw" materials for proper elaboration of topics by all, questioning and answering (Sträßer, 2017). Formal expository mathematics will not be thrown to the garbage can, instead it will enrichen knowledge through a supporting angle. History, geography, engineering, philosophy, and ethics are involved in the mathematics we teach.

The Relational Theory of Mathematics Teaching (Apel, 2022), using complex system analysis, fuzzy sets understanding and ethical values, presents a multidimensional model of the didactic actions involved before, during, and after the lesson starts, aiming to create a learning environment where all participate in harmonious frequency. The multidisciplinarity-complex attitude of the designers takes in account knowledge, participants, and conditions for learning in the n-dimensions present, in a conscious modality where the alumni discourse emerges naturally. The ethic element opens the door to a planning modality where the conditions for learning are taken in account using the Differential Planification model included in the theory, merging theoretical analysis with the Human Dynamics (Seagal, 1997) typology. Based on the ethic postulates of Levinas, (Joldersma, 2002) the teacher-student relationship moves towards respect to the "other," an exteriority existing between both. Learning thus takes place coming from the posture of welcoming something genuinely other, new, not in the possession of the student or the teacher leading to new thematization and promoting participation and collaboration. Rights, respect, and responsibility programs have proved to increase participation in classroom and enhanced academic engagement (Covell & Howe, 2001). The inclusion of the ethics component in the teaching schema adds the confidence and involvement required in mathematics learning as a concomitant to rights and responsibility to deal with knowledge issues.

Lowering boundaries between disciplines will enable the flow of knowledge in the educational institutes in affinity with the human characteristics of the persons participating in the enlarging mathematics knowledge adventure. The challenge for education and research is to find the proper procedure for the implementation in wide scales of the process. This remains to be investigated.

REFERENCES

Apel, T. (1993). *Opening an account, Tools for Arithmetic teaching using an integrative approach.* Reches Publishers.

Apel, T. (2001). *Mathematics in Preschool.* CimeH Institute, Biniamina.

Apel, T. (2012). *Differential Planning.* CimeH Institute. https://members.tripod.com/math_human/id1.html

Apel, T. (2022). *The Relational Theory of Mathematics Teaching.* Las Huellas Matemáticas del Éxito, Cimeh Institute. https://www.amazon.com/LAS-HUELLAS-MATEM%C3%81TICAS-DEL-%C3%89XITO-ebook/dp/B09RSFWXM1

Apel, T., & Rave, A. (1979). *Pupil's Workshop: A learning approach to an interdisciplinary unit in environmental education.* National Congress of Education.

Audi, R. (2011). *Epistemology: A Contemporary Introduction to the Theory of Knowledge. Taylor & Francis Group.* Routledge. doi:10.4324/9780203846469

Ball, D. L. (2017). Uncovering the Special Mathematical Work of Teaching. In *Proceedings of the 13th International Congress on Mathematical Education* (pp. 11-34). Springer Open. 10.1007/978-3-319-62597-3_2

Bernstein, B. (2003). *Class, Codes and Control.* Routledge.

Bernstein, B. (2011). Elaborated and Restricted Codes: Their Social Origins and Some Consequences. *The Ethnography of Communication, 66*(6), 55–69. https://www.jstor.org/stable/668161

Blanton, M. L., & Kaput, J. J. (2005). Characterizing a classroom practice that promotes algebraic reasoning. *Journal for Research in Mathematics Education, 36*(5), 412–446. https://mathed.byu.edu/kleatham/Classes/Fall2010/MthEd590Library.enlp/MthEd590Library.Data/PDF/BlantonKaput2005CharacterizingAClassroomPracticeThatPromotesAlgebraicReasoning-1974150144/BlantonKaput2005CharacterizingAClassroomPracticeThatPromotesAlgebraicReasoning.pdf

Covell, K., & Howe, R. B. (2001). Moral education through the 3 Rs: Rights, respect and responsibility. *Journal of Moral Education, 30*(1), 29–41. doi:10.1080/03057240120033794

Cropley, D. H. (2015). *Teaching engineers to think creatively: Barriers and challenges in STEM disciplines. In Handbook of Research on Teaching Thinking. Routledge.*

Csikszentmihalyi, M. (2009). The promise of positive psychology. *Psihologijske Teme, 18*(2), 203–211. https://www.researchgate.net/publication/43170663_The_Promise_of_Positive_Psychology

Csikszentmihalyi, M. (2014). *CREATIVITY Flow and the psychology of discovery and invention.* Harper Perennial.

Dean, P. G. (2019). *Teaching and Learning Mathematics.* Routledge. doi:10.4324/9781315788753

Deleuze, G., & Guattari, F. (2006). *A Thousand Plateaus: Capitalism and Schizophrenia.* University of Minnesota Press. https://libcom.org/files/A%20Thousand%20Plateaus.pdf

Department for Education. (2013). *National curriculum in England. Mathematics programmes of study* (Publication No. DFE-00180-2013, Stages 1 and 2). https://www.gov.uk/government/publications/national-curriculum-in-england-mathematics-programmes-of-study/national-curriculum-in-england-mathematics-programmes-of-study

Easterby-Smith, M. (1997). Disciplines of Organizational Learning: Contributions and Critiques. *Human Relations, 50*(9), 1085–1113. https://www.researchgate.net/publication/241002680_Disciplines_of_Organizational_Learning_Contributions_and_Critiques/link/56a7f9f908ae0fd8b3fe44a0/download

Fullan, M. (2009). Large-scale reform comes of age. *Journal of Educational Change, 10*(2-3), 101–113. doi:10.100710833-009-9108-z

Gee, J. P. (2017). Affinity Spaces and 21st Century Learning. *Educational Technology, 57*(2), 27–31. https://www.jstor.org/stable/44430520

Gough, N. (2013). Towards deconstructive nonalignment: A complexivist view of curriculum, teaching and learning. *South African Journal of Higher Education, 27*(5), 1213–1233. https://www.researchgate.net/publication/274375251_Towards_deconstructive_nonalignment_a_complexivist_view_of_curriculum_teaching_and_learning

Gray, E. M., & Tall, D. O. (1994). Duality, Ambiguity, and Flexibility: A "Proceptual" View of Simple Arithmetic. *Journal for Research in Mathematics Education, 25*(2), 116–140. doi:10.2307/749505

Hincapie, S., & Hincapie, D. (2013). *La Fascia: Sistema de Unificación Estructural y Funcional del Cuerpo*. https://repository.ces.edu.co/bitstream/handle/10946/2630/Fascias.pdf?sequence=1&isAllowed=y

Honebein, P. C. (2009). Transmergent Learning and the Creation of Extraordinary Educational Experiences. *Educational Technology, 49*(1), 27–34. https://eric.ed.gov/?id=EJ829871

Hume, A., & Coll, R. (2010). Authentic student inquiry: The mismatch between the intended curriculum and the student-experienced curriculum. *Research in Science & Technological Education, 28*(1), 43–62. doi:10.1080/02635140903513565

Jacobson, M. J., Levin, J. A., & Kapur, M. (2019). Education as a Complex System: Conceptual and Methodological Implications. *Educational Researcher, 48*(2), 112–119. doi:10.3102/0013189X19826958

Joldersma, C. W. (2002). Pedagogy of the Other: A Levinasian Approach to the Teacher-Student Relationship. *Philosophy of Education*, 181-188. https://www.researchgate.net/publication/314152336_Pedagogy_of_the_Other_A_Levinasian_Approach_to_the_Teacher-Student_Relationship

Lavanderos Gallardo, L., & Oliva Figueroa, I. (2012). Desde la distinción a la configuración: complejidad, evaluación y aprendizaje desde la visión relacional. *Estudios Pedagógicos (Valdivia), 38*(1), 273–283. .273 doi:10.4067/s0718-07052012000100016

Lemke, J. L., & Sabelli, N. H. (2008). Complex systems and educational change: Towards a new research agenda. *Educational Philosophy and Theory, 40*(1), 118-129. https://www.tandfonline.com/doi/abs/10.1111/j.1469-5812.2007.00401.x

Lepper, M. R. (1988). Motivational Considerations in the Study of Instruction. *Cognition and Instruction, 5*(4), 289–309. doi:10.12071532690xci0504_3

Lifshitz, C., Dolev, T., Margolin, T., & Agur, M. (2004). *The PACT Project for Ethiopian Israeli Immigrant Pre-schoolers and their Families Selected Findings from the Evaluation of the Second Year of Implementation in Netanya*. Academic Press.

Mason, M. (Ed.). (2008). Complexity Theory and the Philosophy of Education. Wiley-Blackwell. .124 doi:10.1002/9781444307351

Mazzocchi, F. (2019). Scientific research across and beyond disciplines. *EMBO Reports, 20*(6), 1–6. doi:10.15252/embr.201947682 PMID:31040110

Middleton, J. A., & Spanias, P. (1999). Motivation for Achievement in Mathematics: Findings, Generalizations, and Criticisms of the Research. *Journal for Research in Mathematics Education, 30*(1), 65–88. doi:10.2307/749630

Mirvis, P. H., (1991). Flow: The Psychology of Optimal Experience. *Academy of Management Review, 16*(3), 636–640. .3 doi:10.5465/amr.1991.4279513

Moore, R. (2001). Basil Bernstein 1924-2000. *British Educational Research Journal, 27*(3), 367–370. doi:10.1080/01411920120048359

Morin, E. (1999). *Seven complex lessons in education for the future*. UNESCO. https://unesdoc.unesco.org/ark:/48223/pf0000117740

Remillard, J. T., & Heck, D. J. (2014). Conceptualizing the curriculum enactment process in mathematics education. *ZDM - Mathematics Education, 46*(5), 705–718. .707 doi:10.1007/s11858-014-0600-4

Seagal, S., & Horne, D. (1997). *Human Dynamics. A New Framework for Understanding People and Realizing the Potential in Our Organizations*. Pegasus Communications.

Seidman, I. (2005). Interviewing as Qualitative Research: A Guide for Researchers in Education and the Social Sciences. Teachers College Press.

Sträßer, R. (2017). Cultural Framing of Teaching and Learning Mathematics. *Didactics of Mathematics as a Scientific Discipline*, 399-455. https://www.researchgate.net/publication/227113904_Cultural_Framing_of_Teaching_and_Learning_Mathematics

Thompson, G., & Cook, I. (2014). Becoming-topologies of education: Deformations, networks and the database effect. In *Discourse: Studies in the Cultural Politics of Education*. Taylor & Francis.

ENDNOTES

1 Transmergent learning is a macro instructional strategy that increases the likelihood of exceptional educational experiences, where creativity and innovation reign. By blending the principles of transformational experiences with evolutionary and emergent properties of complexity theory, instructional designers are able to craft an educational experience weaving ten instructional tactics that together simulate a neural network. The result is an environment in which students can create new knowledge and tangible results that are more than the sum of the parts, that is, extraordinary learning (Honebein, 2009).

2 The algebraic reasoning embedded in finding, describing, justifying, and symbolizing mathematical relationships between quantities that vary is crucial to elementary school mathematics because it creates conceptual underpinnings for the more formalized functional thinking that occurs in later grades (Blanton & Kaput, 2005, p. 426).

3 Model Figure can be seen in Ball (2017) p. 16. Uncovering the Special Mathematical Work of Teaching. 11–34. https://doi.org/10.1007/978-3-319-62597-3_2

Section 2

Reform-Based Teaching Practices in Teacher Education Programs

The chapters in this section present studies that are implemented in mathematics methods or content courses in teacher education programs in support of preservice teachers' development of reform-based teaching skills. There are implications and recommendations for teacher educators.

Chapter 6
Shifting Preservice Teachers' Sources of Mathematics Teaching Efficacy Through Scaffolded Reflection:
Fostering Commitment to Reform-Based Mathematics

Brooke Krejci
Winona State University, USA

Elana Joram
University of Northern Iowa, USA

Anthony J. Gabriele
University of Northern Iowa, USA

ABSTRACT

Efforts to promote reform-based mathematics instruction in schools are pervasive, yet many teachers continue to use more traditional, procedural methods. In this chapter, the authors suggest that to achieve sustained adoption of reform-based instruction in mathematics, it is essential for professional development programs to draw teachers' attention to sources of mathematics teaching efficacy afforded by reform-based teaching, such as enhanced student thinking. Educator preparation programs are ideal settings in which to begin this process, and the authors describe how scaffolded reflection activities within a mathematics methods course were used to focus preservice teachers' attention on sources of mathematics teaching efficacy that are congruent with reform-based mathematics instruction. The authors suggest that engendering such a shift in the sources for mathematics teaching efficacy may lead to more widespread and sustained use of reform-based teaching methods by preservice teachers in their future practice.

DOI: 10.4018/978-1-7998-9422-3.ch006

INTRODUCTION

Despite widespread efforts by teacher educators and those providing professional development to help teachers integrate reform-based mathematics approaches into their everyday instruction (McGee, Wang & Polly, 2013; Paolucci, 2015), traditional methods of teaching mathematics are still common (Hill et al. 2018). Teachers may focus classroom instruction on procedural computations without making deeper connections, and emphasize fact and skill recall instead of reasoning and problem solving even after professional development in reform-based mathematics instruction (NCTM, 2014; Paolucci, 2015). Indeed, research reveals that sometimes even teachers who appear to embrace reform-based mathematics principles may utilize more traditional approaches of instruction in their practice (e.g., Lane & Ríordáin, 2020).

Although many factors—including institutional, socio-economic, and cultural—may contribute to this puzzling pattern, the authors suggest that one important element largely missing from the discussion is "mathematics teaching efficacy," defined as teachers' or preservice teachers' beliefs that they can effectively teach mathematics (Briley, 2012). This chapter explores mathematics teaching efficacy, explains why it is central to supporting and sustaining reform-based mathematics teaching, and describes some of the special challenges faced by mathematics educators who attempt to support its development. The authors argue that an opportune time to support and facilitate the development of mathematics teaching efficacy is during teacher preparation (Giles et al., 2016), when teacher beliefs may be more malleable than after years of classroom practice. This chapter describes the unique opportunities and challenges of encouraging the development of mathematics teaching efficacy beliefs for teaching reform-based mathematics with preservice elementary teachers. Examples of one of the author's initial efforts to do so in an undergraduate mathematics methods course are included. The purpose of this description and accompanying examples is two-fold: it is intended to help focus the attention of both educational researchers and teacher educators on a potential roadblock to the adoption and sustained use of reform-based instructional methods in mathematics, and also to provide an illustration of how mathematics teaching efficacy can be facilitated within teacher education programs.

Background

Since the late 1980s, marked by the introduction of the National Council of Teachers of Mathematics (NCTM) *Curriculum and Evaluation Standards for Mathematics* (Romberg, 1989), mathematics education in the United States changed from emphasizing teacher-centered instruction to promoting more student-centered approaches (Hill, 2021; NCTM, 2014). More recently, the NCTM updated their recommendations of student-centered and inquiry-based teaching practices ("reform-based") in *Principles and Standards for School Mathematics* (NCTM, 2000). Further, in the *Common Core State Standards Initiative* (CCSSI), eight standards for mathematical practice are provided that describe the ways in which students should engage with mathematical content standards (CCSSI, 2010). As noted in the description of each mathematical practice standard, students play an active role in learning content rather than replicating the action of the teacher (CCSSI, 2010). In line with these student-centered practice standards and specifically targeting effective mathematics teaching practices for student-centered mathematics, NCTM (2014) released *Principles to Actions: Ensuring Mathematical Success for All.* A common thread in these practices is for teachers to incorporate student thinking and allow it to guide their instructional moves, with the goal of developing deep conceptual understanding in students (NCTM, 2014).

The acknowledged benefits and deepened understanding that students gain from reform-based mathematics instruction raises the question of why these standards are not currently implemented everywhere. This is the question addressed in this chapter, and the authors suggest that one factor that may play a role in the implementation and sustained practice of reform-based mathematics instruction is "mathematics teaching efficacy." More generally, "teaching efficacy" refers to a teacher's belief in their ability to effectively teach and positively influence their students' learning (Tschannen-Moran & Hoy, 2001), which has been credited with increased teacher effectiveness (von Suchodoletz et al., 2018). In a general sense, higher efficacy is linked to increasing likelihood of teachers' willingness to try new instructional strategies, according to a meta-analyses over 40 years of literature (Zee & Koomen, 2016); for this reason, "teacher efficacy" is relevant to those hoping to promote reform-based mathematics instruction.

Beliefs about one's ability to teach can be further specified according to content domain (Tschannen-Moran & Hoy, 2001); thus "mathematics teaching efficacy" is based on the beliefs a teacher holds about their ability to effectively teach mathematics (Briley, 2012). With respect to mathematics teaching efficacy, along with others, the authors argue that shifting to teaching reform-based mathematics requires significant changes to mathematics teaching efficacy, and that this shift is necessary for the sustained adoption of reform-based mathematics instruction (Gabriele & Joram, 2007; Smith III, 1996). Below, specific aspects of these changes to mathematics teaching efficacy are described.

Sources of Mathematics Teaching Efficacy and Its Role in the Shift to Reform-Based Teaching

In the mathematics education literature, mathematics teaching efficacy has most often been discussed with respect to enhancing teachers' content and pedagogical content knowledge, with the underlying assumption that mathematics teaching efficacy naturally grows as teachers gain confidence in their own mastery of mathematical content and knowledge of teaching strategies (e.g., Kleickmann et al., 2017; Thomson et al., 2017). Although this assumption intuitively makes sense, and may in fact be true with some mathematical teaching practices linked to traditional mathematics instruction, prior research suggests that this is likely not the case with reform-based mathematics instruction. Instead, the authors have previously argued that active, intentional support is required for its development (Gabriele & Joram, 2007). In part, this may be because the sources of mathematics teaching efficacy related to traditional and reform-based mathematics instruction are different and require distinct kinds of support (Althauser, 2018; Gabriele & Joram, 2007; Smith III, 1996).

In a study comparing teachers with extensive experience using reform-based mathematics instruction (*veteran*) to those with less experience (*novice*), teachers who had used reform-based methods for five or more years reported pleasure and enjoyment when they observed a student showing partial or deep understanding of a problem or concept (Gabriele & Joram, 2007). In contrast, those teachers who had used reform-based methods for fewer than five years tended to evaluate their own success as a mathematics teacher with respect to their lesson coverage and students getting correct answers to problems, suggesting they were still transitioning to fully embracing reform-based methods. In other words, with respect to reform-based mathematics instruction, the "veteran" and "novice" teachers in this study drew on different criteria to judge the success of their teaching practices even though both were implementing mathematical tasks considered to be reform-based in nature. This study suggests, as illustrated in Figure 1, that important distinctions need to be made with respect to how mathematics teaching efficacy develops and is sustained in reform-based and more traditional mathematics instructional contexts (Gabriele

& Joram, 2007; Smith III, 1996). This, in turn, points to the critical role that sources of mathematics teaching efficacy may play in the shift to reform-based mathematics instruction.

Thus, this chapter focuses on the relatively unexplored dimension of teachers shifting to a greater focus on student thinking as evidence for their mathematics teaching efficacy and valuing the observed changes in student thinking that align with reform-based teaching. In this sense, the authors suggest that the very concept of "mathematics teaching efficacy" needs to be understood within the context of reform-based mathematics instruction as continuously cultivated through specific observations of student thinking that serve as evidence for teachers that their teaching methods result in deeper understanding of mathematics by their students. This evidence can then serve as an important source of their mathematics teaching efficacy for teachers using reform-based methods (see Figure 1).

Figure 1. Congruence within different mathematics teaching approaches and success criteria drawn on as sources of mathematics teaching efficacy

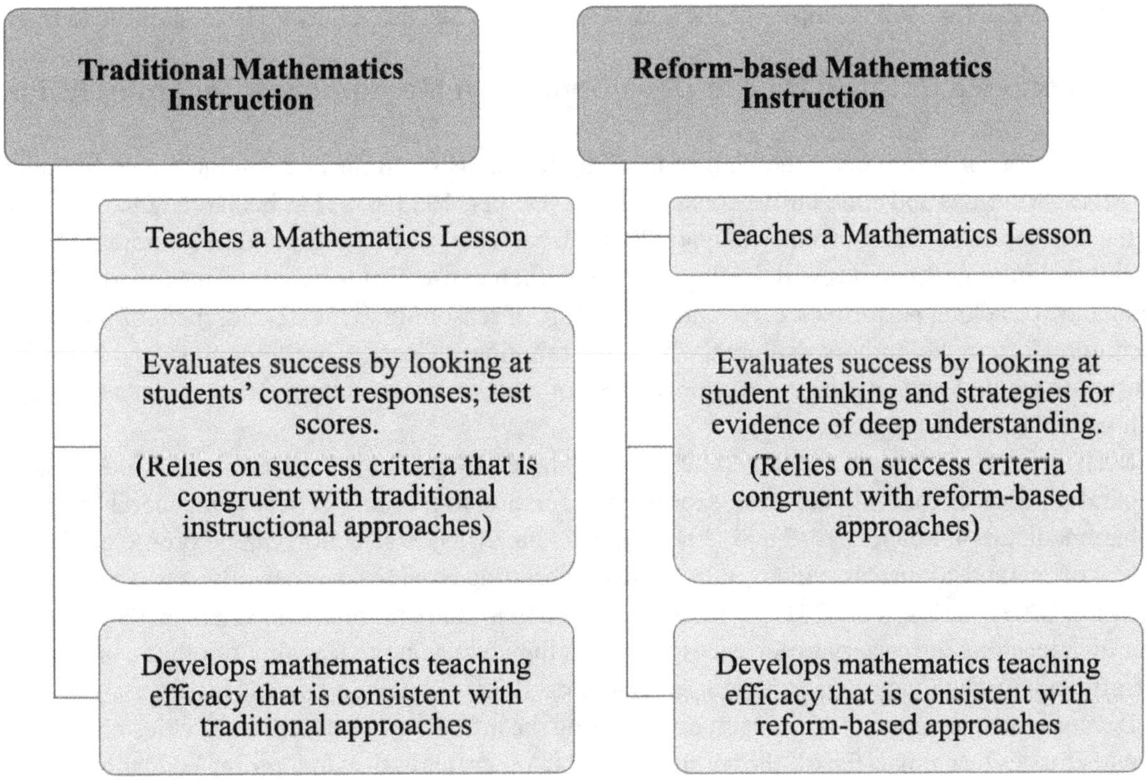

Preservice Teacher Education as an Ideal Context to Foster Mathematics Teaching Efficacy

Teacher beliefs about effective instruction are often developed through their own past experiences (Althauser, 2018; Fives et al., 2015). Many teachers' past experiences as students of mathematics likely involved primarily traditional and/or procedural methods (e.g., Jao, 2017; Paolucci, 2015), and the authors suggest teacher education programs may present an opportune time in which to target beliefs about the nature of effective mathematics instruction (Giles et al., 2016). Wyatt (2016) notes that teacher self-efficacy beliefs may be easily impacted during teacher education, and recommends that teacher educators target the growth of teacher self-efficacy in relation to specific tasks rather than more generally. Thus, methods courses within teacher education programs seem well suited to doing so.

In U.S. educator preparation programs, mathematics methods courses are typically a primary context in which preservice teachers (referred to hereafter as "*PSTs*") encounter reform-based approaches to mathematics instruction. Exposing PSTs to reform-based mathematics instruction in these methods courses and encouraging reflection on these instructional methods gives them an opportunity to embrace these approaches and shift their beliefs about effective instruction. As suggested above, it may be that in the absence of efforts by instructors to intentionally scaffold PSTs' observations and reflections, mere exposure to reform-based teaching methods is not sufficient to shift PSTs' beliefs about effective instruction in mathematics. Therefore, to draw attention to salient aspects of reform-based methods, the authors propose that teacher educators make intentional use of scaffolds when working with PSTs in mathematics methods courses to guide their observation and reflection. Later in this chapter the authors provide a description and examples of the use of such scaffolds, as well as PSTs' responses to them.

Reflection as a Catalyst for the Development of Mathematics Teaching Efficacy

The process of reflection or "...the kind of thinking that consists in turning a subject over in the mind and giving it serious and consecutive consideration" (Dewey, 1933, p. 3) has been recognized as critical for the development of self-efficacy (Wyatt, 2016). Uzun et al. (2013) suggest that an increase in PSTs' reflective thinking tendencies will result in growth in their self-efficacy beliefs, which in turn may lead to enhanced teaching performance. Extending this line of reasoning to the development of mathematics teaching efficacy, the authors explored whether promoting reflection in the context of reform-based instruction would lead to a greater focus on sources of mathematics teaching efficacy that are congruent with reform-based mathematics.

Reflective thought that is facilitated by professional development for teachers and through educator preparation courses for PSTs has been examined in relation to teaching efficacy in general (e.g., Noormohammadi, 2014; Phan, 2014; Tavil, 2014; Uzun et al., 2013). Reflection can be promoted through a variety of methods: journals, videos, talk-alouds, debriefing sessions, frameworks, etc. (e.g., Gabriele & Joram, 2007; van Es et al., 2017; Wilkerson et al., 2018). In some cases, teachers or PSTs reflect on their own teaching or mastery experiences; in others they focus on the teaching of others, thus building self-efficacy through vicarious experiences (Bandura, 1977).

Because there is a paucity of research investigating the influence of reflective activities on the growth of mathematics teaching efficacy, the authors looked to other fields also undergoing educational reforms in which researchers have examined the relationship between reflective activities and growth in teaching efficacy. For example, in science education, the action-reflection cycle was found to be productive for promoting PSTs' science teaching efficacy across instructional activities (Naidoo & Naidoo, 2021). Specifically, Naidoo and Naidoo (2021) recommend the use of successful teaching episodes to develop PSTs' perceived abilities, reflection in a collaborative environment, and recognition that challenges are part of the development of PSTs' own teaching competence. These studies suggest that similar intentional reflective opportunities following instructional tasks for PSTs in the context of reform-based mathematics teaching could be beneficial for increasing their mathematics teaching efficacy.

Using Reflective Prompts to Shift PSTs' Attention to Student Thinking and Understanding

Although researchers have focused on professional development for practicing teachers to encourage a greater focus on student thinking, less attention has been directed to such efforts for PSTs (Coffey & Edwards, 2015). There is an indication in the literature, however, that prompting and focusing questions in reflection activities can guide PSTs toward focusing on specific aspects of instruction such as student learning and understanding (e.g., Santagata & Yeh, 2014; van Es et al., 2017). This could be particularly beneficial for PSTs because they tend to focus on teaching rather than on learning (e.g., Chamoso et al., 2012; Gelfuso & Dennis, 2014; Seung et al., 2014); in other words, on a teacher's actions rather than students' thinking and strategies. This focus conflicts with the goals and aims of reform-based mathematics instruction which places greater emphasis on students' learning and understanding. Reflective prompts can guide PSTs away from an exclusive focus on the teacher, and more towards attending to student learning and understanding (Santagata & Yeh, 2014; van Es et al., 2017). Through reflective activities, it may be possible to draw PSTs' attention to student thinking and understanding as a way of gauging a teacher's (and eventually their own) success in teaching mathematics. This would also shift them away from looking to summative test scores and number of correct responses by students as the most important (or sole) indicators of a teacher's impact.

To summarize, the authors suggest that teacher education provides an ideal context in which to scaffold the development of mathematics teaching efficacy through the provision of opportunities to reflect on reform-based instructional tasks that PSTs either observe or carry out. More specifically, the authors suggest that it will be critical during teacher education courses to provide activities that direct PSTs to shift to relying on sources of mathematics teacher efficacy that align with reform-based mathematics instruction (see Figure 2). This contrasts with those sources that have typically been relied on to build mathematics teaching efficacy in traditional contexts. The authors argue that this shift to drawing on sources that are consistent with the culture of reform-based mathematics instruction is key to building mathematics teaching efficacy in a reform-based context. In turn, this should, lead to sustained and consistent use of reform-based mathematics instruction.

Figure 2. Instructional activities in a mathematics methods course designed to facilitate mathematics teaching efficacy in the context of reform-based instruction

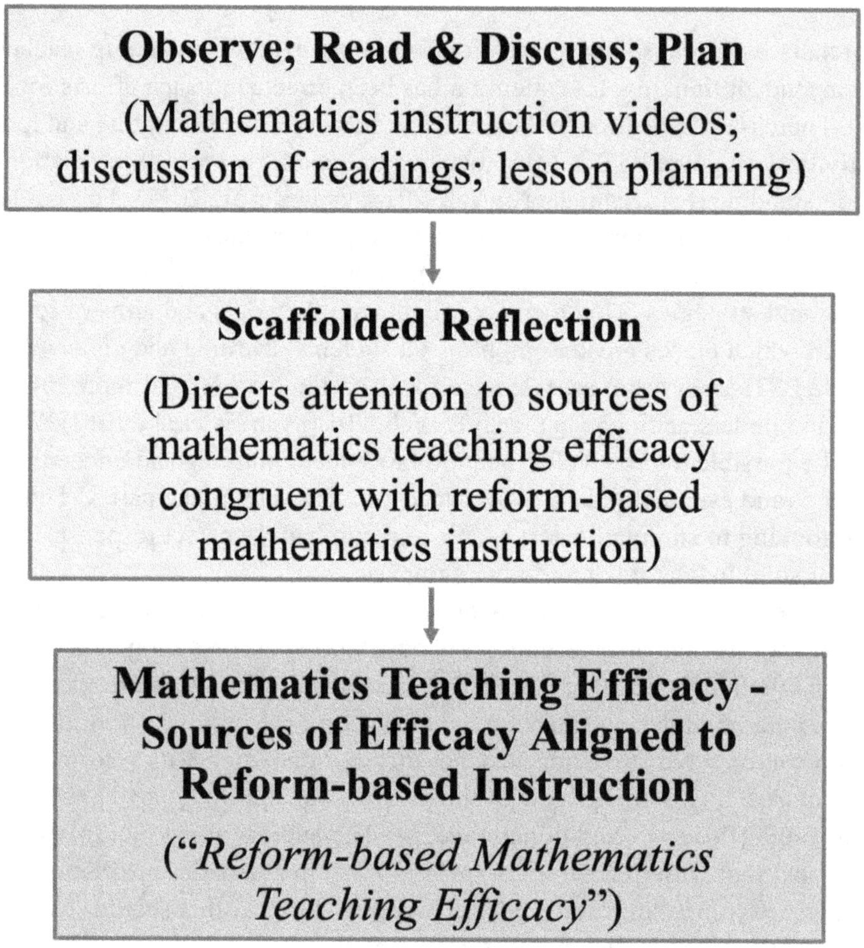

THEORETICALLY MOTIVATED ACTIVITIES DESIGNED TO PROMOTE REFORM-BASED MATHEMATICS INSTRUCTION

As suggested above, in a reform-based mathematics culture, it is critical for teachers to seek out windows into their students' thinking about mathematical concepts and problems, value the insights their students may have, and take satisfaction in these. These sources provide evidence for the teacher's own positive impact on student mathematical thinking and understanding; in other words, their mathematics teaching efficacy. To achieve these goals, one of the authors designed a series of instructional tasks that were introduced in an elementary mathematics methods course, described below.

Prior to engaging in the instructional activities outlined in this methods course, PSTs had completed a mathematical content course for elementary teachers, typical of the sequence of coursework in many teacher preparation programs in the United States. PSTs were enrolled in a four-year undergraduate teacher education program in a medium-size university in the midwestern United States. The sequence of courses in this program consisted of general education courses (e.g., English composition, elective humanities), followed by coursework for developing content knowledge and pedagogical practices across main subject areas and core professional knowledge courses such as assessment, diverse learners, etc. In addition, PSTs engaged in field experiences throughout the program, culminating with student teaching.

The design of the mathematics methods course described here was based on the assumption that the intentional facilitation of mathematics teaching efficacy in PSTs should be integrated into opportunities to further develop their content and pedagogical knowledge in mathematics. This contrasts with the assumption described above that it will emerge naturally on its own after PSTs acquire content and pedagogical content knowledge. Another rationale for targeting PSTs in a methods course is that they can be given tasks in which they comment on someone else's teaching, thus freeing up their attention to focus on the task at hand. If they were delivering a lesson in a field experience or student teaching, PSTs' attention would instead likely be devoted to the logistics of delivery, making it difficult to introduce another focus of attention.

Throughout the eight-week long mathematics methods course, PSTs' mathematics teaching efficacy was intentionally facilitated through activities including watching instructional videos, planning lessons, and discussing course readings. The lesson planning activity was structured around *Thinking Through a Lesson: Successfully Implementing High-Level Tasks* (Smith et al., 2008), which encourages the use of student-centered instruction. All tasks included were immediately followed by reflective prompts (see Table 1 for examples).

Table 1. Instructional tasks and examples of reflection prompts

Scaffolded Reflection on Reform-Based Teaching (videos)	Weeks 1 and 8 (General reflection prompts): 1. "Please reflect on the lesson you just observed." 2. "How do you think the lesson went today? What stood out for you?" Weeks 2-7 (Prompts targeting aspects of reform-based instruction): 1. "Specific to the mathematical concept of ________, how did the student(s) in the video demonstrate their understanding? Provide specific examples." 2. "What did the teacher say or do to develop student understanding of ______? Provide specific examples." 3. "What could the teacher do, additionally, to increase student understanding of ______?" 4. "How does reflecting on this video inform your own practice? What will you take away from this video, or what connections can you make to your own teaching or future teaching?"
Scaffolded Reflection on Reform-Based Lesson Planning	1. "What are the mathematical concepts you are teaching in this lesson?" 2. "How confident are you that this lesson would be effective in helping students understand the mathematical concepts? Please explain your reasoning for your response." 3. "How will you know that students understand the mathematical concept targeted in your lesson?" 4. "How confident are you that you could successfully teach this lesson? Please explain your reasoning for your response."
Scaffolded Reflection on Reform-Based Readings (Example)	After reading "Building a Discourse Community: Initial Practices" by Hodge and Walther (2017): 1. "As you think about your own classroom, how will you incorporate the '4 Initial Practices to Build a Foundation of Productive Discourse' in your teaching?"

Task 1: Scaffolded Reflection of Reform-Based Teaching

In the first instructional task, PSTs were asked to observe videos showing a teacher demonstrating reform-based mathematics instruction. They were then prompted with reflection questions focused on student thinking and understanding as evident in the video, with the intent of drawing the PSTs' attention to student thinking. Researchers have conducted studies centered around the use of videos for reflection, demonstrating the potential for PSTs to shift the focus of their reflection in this context (e.g., Chamoso et al., 2012; van Es et al., 2017; Wilkerson et al., 2018). A common theme in these studies is the use of intentional prompts focused on student learning and thinking, and this was incorporated into the design of the activities in the present course. As suggested by van Es and colleagues (2017), videos should include cognitively demanding tasks that lend themselves to reflective prompts focused on student understanding, consistent with reform-based instructional practices. Thus, videos were selected for the methods course that showed teachers using reform-based instructional methods that involved tasks with a scope that allowed for reflective prompts about student understanding.

Video reflection prompts from weeks 1 and 8 were general in nature to allow any shifts in PSTs' focus that may have occurred throughout the course to be observed. In contrast, prompts used in weeks 2 through 7 directed PSTs' attention to specific aspects of instruction. These scaffolded reflection prompts were adapted from two different sources. The first three scaffolded prompts (see Table 1) were drawn from a pilot study conducted by one of the authors of a small number of PSTs in a mathematics methods course, and focused PSTs' attention on student understanding and the teachers' role in this development during the intervention. As recommended by Lee and Ertmer (2006), prompts did not narrowly focus PSTs' attention (which could potentially limit the scope of their reflections). Instead, PSTs were invited to process the information and select examples they felt demonstrated student understanding. The fourth prompt, which encouraged PSTs to focus on making connections to their own practice, was adapted from a framework used by Wilkerson et al. (2018) in their study of teachers reflecting on written vignettes.

Reflections were prompted by questions tailored to each mathematical concept exhibited in the video (see Table 1, weeks 2-7, prompts 1 & 2). For example, one prompt was: "Specific to the mathematical concept of fraction representation of equal parts, how did the student(s) in the video demonstrate their understanding? Provide specific examples." In addition, to direct PSTs to attend to student understanding as a source of self-efficacy, another prompt focused on the observed teacher's actions as related to aspects of reform-based mathematics, "What did the teacher say or do to develop student understanding of equivalent fraction representation? Provide specific examples."

The prompts provided an opportunity to evaluate an important component of reform-based mathematics instruction: developing student understanding and recognizing the teacher's actions that promoted this understanding. Scaffolded reflection following observation provides an opportunity for PSTs to engage in observing an instructor model reform-based mathematics. It also invites PSTs to take the time to reflect on important characteristics of reform-based mathematics instruction to inform their future teaching (e.g., Wilkerson et al., 2018), which may then facilitate the development of their beliefs in their ability to teach mathematics in a student-centered way.

Task 2: Scaffolded Reflection of Reform-Based Lesson Planning

Lesson planning activities directed PSTs' attention to important components of reform-based mathematics and also provided insights into what PSTs intend to enact in their instruction. Similar to recommendations by van Es and colleagues (2017) regarding video selection, lesson plan templates were designed to be consistent with the aims of reform-based mathematics, specifically based on components of student-centered instruction outlined in *Thinking Through a Lesson: Successfully Implementing High-Level Tasks* (Smith et al., 2008). In contrast to more traditional methods of mathematics instruction in which the teacher models procedures for mathematical computations, Smith et al.'s (2008) protocol encourages PSTs to support student understanding by implementing tasks with multiple entry or access points, asking questions to develop and advance student thinking, and facilitating a classroom discussion by sequencing predicted student strategies. "Multiple entry/access points" is a term commonly used in reform-based mathematics instruction, and means that students can begin or access a problem in many different ways, regardless of different levels of prior knowledge.

After PSTs completed their written lesson plans both at the beginning and end of the course, they engaged in reflection using prompts designed to focus their attention on specific aspects of reform-based teaching such as developing and assessing student understanding. Specifically, the third lesson plan reflection prompt asked PSTs to describe student understanding based on their anticipated instruction, allowing PSTs to identify their own examples based on their interpretation of student understanding (see Table 1). The focus on student understanding can be observed in several of the prompts used in the lesson planning task, for example, "How confident are you that this lesson would be effective in helping students understand the mathematical concepts? Please explain your reasoning for your response." By directing PSTs' attention to key themes and aspects of reform-based instruction such as the development student understanding as a process, these prompts encouraged PSTs to reflect on their anticipated instruction and its alignment with reform-based instruction.

Task 3: Scaffolded Reflection of Reform-Based Readings

The final instructional task designed to shift PSTs' sources of efficacy was scaffolded reflection following course readings that embody reform-based mathematics teaching principles and give examples of reform-based practices. Similar to video selections, readings that are aligned with reform-based mathematics instruction offer PSTs an opportunity to learn about key aspects of reform-based instruction that may be unfamiliar to them (van Es et al., 2017). Readings were selected from the NCTM's practitioner journals and reflected the aims of the eight *Mathematics Teaching Practices* outlined in NCTM's *Principles to Actions: Ensuring Mathematical Success for All* (2014). They included questioning strategies teachers could use to move their students' thinking forward, developing productive classroom discourse, sequencing and connecting strategies to develop student understanding, etc. Readings were selected with the intent to both inform PSTs about important components and examples of reform-based instructional practices and to provide PSTs with opportunities to reflect on how they might integrate components of reform-based instruction into their future classrooms.

As noted above, one important aspect of reform-based instruction is using tasks with multiple entry points to allow all students access and to engage them in the development of their own thinking and understanding. To focus on this aspect, the following was used as a course reading: *Building A Discourse Community: Initial Practices* (Hodge & Walther, 2017) which discusses how to encourage student participation through discourse in the mathematics classroom. Consistent with the other instructional activities, PSTs' reflection on course readings was prompted, this time in an online forum. Prompts were also tailored to the selected readings to direct PSTs' attention to specific instructional components that embody reform-based mathematics instruction (van Es et al., 2017) and their implementation in their future classroom (Wilkerson et al., 2018) (see Table 1).

OBSERVED SHIFTS ACROSS INSTRUCTIONAL TASKS: CONSIDERATIONS FOR TEACHER EDUCATORS

In this section, the authors explicitly link the theoretical model of mathematics teaching efficacy outlined above with a description of PSTs' responses to the instructional activities in the mathematics methods course. The discussion of these responses is organized according to three shifts observed across PSTs' observations and reflections that the authors suggest may constitute new sources of mathematics teaching efficacy in the context of reform-based mathematics instruction.

Increased Focus on Teacher Actions to Facilitate Student Thinking

At the beginning of the course, PSTs tended to focus on traditional aspects of instruction, that is, teacher modeling and student performance of procedural steps. Following the course and instructional activities, in contrast, PSTs' reflections suggest that they came to view the teacher as more of a facilitator of the classroom and developer of student understanding. Although PSTs still focused on the teacher near the end of the course, the role they saw teachers playing in mathematics instruction better reflects the goals of reform-based instruction. This change in the perceived role of the teacher was observed across all instructional tasks.

For example, prior to the use of scaffolded video prompts, when asked to reflect on a video-taped lesson, PSTs focused on general student engagement and on the teacher's overall management of the classroom, as characteristic of traditional mathematics instruction. In contrast to reflections at beginning of the course, by the end of the methods course, PSTs focused more often on specific examples of student understanding and what the teacher did in the video to address or further student understanding. This shift in focus when watching video-taped lessons was noted in all PSTs who were enrolled in the course, and is illustrated below by one PST when prompted generally about the video-taped lesson they observed at the end of the course:

As the lesson progressed, the students were modeling their thinking through different strategies. The teacher highlighted many different opinions throughout the classroom and those students either discussed or lead their thinking at the front of the classroom. The teacher never discussed the idea of the commutative property but the students highlighted the idea of the concept through their ideas.

Similarly, a change in the perceived role of the teacher in mathematics instruction was also observed in PSTs' lesson plans and reflections on course readings. In lesson plans constructed prior to the intervention, PSTs described the teacher's role as central; specifically, as a modeler of the mathematics students were intended to replicate. For example, one PST launched a task, assuming students knew multiplication, and used the exploratory phase as a way for the teacher to model a single strategy and have students practice that one strategy, stating, "The strategy for solving multiplication by 8 is the double, double, double strategy. The teacher will use 6 x 8 for example. We take the factor (that is not 8) and we are going to double it." The lesson continued by asking students to replicate this specific strategy. Despite the nature of the lesson planning template, PSTs' initial lesson plans appeared to be designed according to traditional approaches to mathematics instruction.

At the end of the course when PSTs wrote their final lesson plans, the role of the teacher depicted in these plans seemed to shift from that of a central lecturer to more of a facilitator of instructional conversations. In these final lesson plans, teachers' actions, as reflected in the lesson plans, appeared to be designed to support the development of student thinking and understanding, thus more clearly embodying the goals of reform-based mathematics instruction. As evident in the following example, the actions of the teacher in PSTs' lesson plans shifted from a modeler of mathematics to a facilitator of student thinking, as would be expected in the actions of a teacher in a reform-based mathematics instructional setting. The PST below did not funnel students to a singular method of solving the task, but instead incorporated students' methods to orchestrate a whole class discussion:

The teacher will make notes on each student/group and the strategy they are using. This information will help rank the strategies used, clear up common misconceptions, and will help when students are asked to share their strategies under the document camera to the class and participate in whole group discussion.

After outlining the use of student strategies to guide the classroom discussion, the PST included the order in which they would present anticipated students' responses and strategies, building in complexity:

I would first share the misconception, only if others are also making similar mistake. I would then have students who solved the problem by drawing a picture, array, or table share first. All students could connect to this thinking as we have worked with arrays in the past. All students could clearly understand the problem and clear up misconceptions if seeing it displayed as a visual representation.

This continued with three more strategies, leading to the desired strategy of using an "unknown" factor:

Last, I would have the students who knew that 12 x 8 =96 go last because they set up their division problem by using the unknown factor in a multiplication problem which is what we want to focus on. I would then go back to strategy 3 and show how division and multiplication can both help solve the same problem

As demonstrated by this progression, the PST became the facilitator of student strategies, connecting them in a way that helped advance student thinking from their current level of understanding, and orchestrating students' ideas in a way that was mathematically productive.

Following the course reading, *Building A Discourse Community: Initial Practices* (Hodge & Walther, 2017), students were prompted to reflect, "As you think about your own classroom, how will you incorporate the '4 Initial Practices to Build a Foundation of Productive Discourse' in your teaching?" In response to this prompt, one PST focused on facilitating a classroom discussion, specifically centered around students' strategies and understanding:

Building this discourse in the classroom begins by providing the students with more opportunities to develop knowledge. In my classroom, engaging students in a more broad task seems beneficial... As the discussion builds, the students can discuss the strategies and techniques they used to solve their problems and learn from other students.

Here, the PST is clearly focusing on student thinking that is facilitated by a class discussion and further developed by engaging with other students.

Similarly, when reflecting on another course reading, *Differentiating Instruction in Mathematics for the English Language Learner* (Murrey, 2008), several PSTs mentioned the role of the teacher as the facilitator of the classroom and in moving all students' understanding forward. Specifically, one PST mentioned: "When solving a math task the teacher should encourage students to further their understanding by listening, reading, writing and discussing. It may be helpful for students to be in groups to really discuss the math with peers…" Again, the PST's focus here is on the teacher providing opportunities for students to develop understanding of concepts rather than on their ability to lead students through step-by-step procedures. These examples of PSTs' reflections suggest that the reflective activities in the methods course provided an opportunity for PSTs to consider some of the key elements of reform-based mathematics instruction to inform their future teaching.

Evidence of Student Thinking and Understanding

Another change noted in PSTs' reflections, lesson plan designs, and comments about readings was that they placed a greater emphasis on student thinking and understanding following the use of these instructional tasks than at the beginning of the course. Prior to the intervention, PSTs tended to focus on students' ability to replicate procedures or their general engagement. However, following the implementation of the instructional tasks in the course, PSTs focused more on student thinking as illustrated by their written comments, use of multiple strategies, and perseverance when solving tasks.

Evidence of student thinking was lacking in PSTs' initial observations and reflections across the three instructional tasks. However, consistent with the aims of reform-based instruction, PSTs came to emphasize the idea of embracing productive struggle and connecting to students' current level of understanding throughout the instructional tasks in the course. At times, PSTs' comments seemed to suggest a growing sense confidence in their ability to successfully deliver the lesson described in their plan in relation to their ability to successfully develop student understanding. For example, one PST stated: "I feel very confident if I were to teach this lesson. I feel the task is cognitively demanding but as we have learned--we want students to struggle... I believe I set up assessing and advancing questions to benefit students' understanding." Another PST focused on the use of multiple strategies and class discussion to model student thinking when prompted about a lesson: "As the lesson progressed, the students were modeling their thinking through different strategies. The teacher highlighted many different opinions throughout the classroom and those students either discussed or lead their thinking at the front of the classroom."

This shift is also illustrated by the following PST when reflecting on a course reading, *Strategies to Support the Productive Struggle* (Warshauer, 2015). PSTs were prompted to think about how one of the four strategies mentioned in the article would look in their future classroom. In response, one PST focused on questioning and student understanding:

When you question students you are encouraging them to use their voice and explain... When you question students you [are] getting them to think about the why and how. When students start thinking about why and how they can ask questions and have the potential of full understanding…

Understanding, according to this PST's reflection, includes the ability to explain, not just replicate procedures, as recommended by reform-based mathematics approaches. As illustrated in both examples above, following their experiences with the instructional tasks, PSTs tended to focus more on student thinking and understanding, specifically on the ways in which students can demonstrate their understanding. The changes observed across PSTs' reflections and observations suggest that these activities helped focus their attention on student thinking about mathematical content.

Evaluating Student Thinking and Understanding

What is considered to be evidence of student understanding varies with the instructional styles used. In traditional mathematics instruction, student understanding is often considered to be evident in students' ability to replicate procedures or compute correct answers. However, in the context of reform-based mathematics instruction, evidence for student understanding requires more than a demonstration of a correct procedure; instead, teachers must look for evidence of deeper, conceptual understanding such as explanation, application, or interpretation. To develop conceptual understanding, teachers need to devote attention to students' current level of understanding, and allow them to make meaningful connections. Across all tasks, PSTs demonstrated a shift in their ability to gauge student understanding through a variety of instructional actions in contrast to their initial reflections.

Prior to engaging in the instructional activities, the PSTs designed their first lesson plan, and were then asked "How will you know that students understand the mathematical concept targeted in your lesson?". One PST responded: "the students are able to compute the answer and show their work by using the new strategy [as demonstrated by the teacher]." This PST's comment suggests they initially believed that student understanding is demonstrated through computations or written work, consistent with the intentions of traditional mathematics teaching which emphasizes the accuracy of a solution only. Similarly, another PST stated that: "The 'exit ticket' assessment piece of this lesson will show me if students are successfully understanding this concept," indicating that the exit ticket would serve as a summative assessment. At the end of the course, however, one PST also mentioned the exit ticket but also included a description of how they would check for understanding throughout the lesson in order to prepare them for the class discussion:

The exit ticket is the grand summation of the lesson, but it's at the end. I would walk around during the explore phase to see which of my students was A) understanding the information and B) using the different strategies. The basic to more complex strategies offers some insight into student understanding and helps me know where I need to funnel my emphasis for later.

The PST in the quotation directly above voices their sense of the importance of evaluating the level of student understanding to develop their thinking throughout the lesson versus assessing only the students' ability to compute the correct answer. This contrasts with the earlier the example in which the PST uses the exit ticket only as a summative assessment, Another PST also referred to evaluating students' level of understanding throughout the lesson: "The variety of different assessments I add to the lesson will show if they are understanding the concept. Discussion and explanation of their strategies is where I will learn the most about their understanding during the lesson…" Again, this PST emphasizes evaluating student understanding through the use of multiple methods throughout the lesson, not only by means of a summative assessment at the end of the lesson that focuses on the attainment of correct answers.

Similarly, at the end of the course, when asked "How do you think this lesson went?" following a video observation, one PST focused on the development of student understanding throughout the observed lesson. This PST evaluated students' level of understanding through their verbal responses and ability to explain their thinking: "The students were verbal with their understanding of the concept and explained their thinking when asked too. The teacher highlighted several strategies that the students used and the class acknowledged their understanding of those strategies."

In placing a greater emphasis on the use of multiple sources to evaluate student understanding throughout a lesson, PSTs' responses better reflect the goals and aims of reform-based instruction, which center on student thinking and understanding. In the next section, the authors discuss how shifting PSTs to an increased focus on student understanding may promote the development of their mathematical teaching efficacy.

IMPLICATIONS FOR TEACHER EDUCATORS

The three themes across instructional tasks demonstrate PSTs' increased ability to focus on key components of reform-based instruction, serving as potential sources to draw on for their mathematics teaching efficacy. Instead of gauging their ability to teach mathematics effectively by means of examining students' performance on standardized assessments and their ability to replicate procedures for correct answers, PSTs can develop beliefs about their own mathematics instruction in light of how it supports their students' growing understanding. In addition, the instructional tasks help focus PSTs' attention on their own role in facilitating this understanding. As observed across the instructional activities described above, shifting PSTs' focus to student understanding offers a potential new source to draw on for developing their mathematics teaching efficacy that fits with the culture of reform-based mathematics instruction.

In making this transition, confidence in PSTs' own mathematics teaching becomes redefined. For example, prior to the mathematics methods course, one PST referenced a single strategy they would use in teaching their lesson plan as the sole basis for their confidence in teaching it effectively, stating: "I am fairly confident because I have taught this lesson while subbing 4th grade before. This helps students with figuring out their x8 math facts." This confidence was based on their prior experience (substitute teaching) with this specific content, using a single strategy (double-double-double). This limited prior experience seems to fit with traditional approaches of mathematics instruction, placing greater emphasis on procedural computations without facilitating students making deeper connections to the concept, similar to findings of another study with a larger group of PSTs (Althauser, 2018).

In contrast, following the course, PSTs expressed confidence in their ability to teach mathematics that better reflects the goals of reform-based instruction. This entailed gauging their own ability to effectively facilitate a reform-based mathematics classroom: predicting student strategies, developing questions for students in a way that attended to their level of thinking and understanding, and using anticipated student strategies to orchestrate discussions.

CONCLUSION

In summary, throughout this chapter the authors argue that the lack of uptake and consistent, sustained use of reform-based methods in mathematics (Hill, 2021) may stem to some extent from the mismatch between use of reform-based methods and the criteria teachers seek out to evaluate their own teaching success. As depicted in Figure 3, when a teacher adopts reform-based teaching methods yet continues to evaluate their success by drawing on sources of mathematics teaching efficacy that are more suited to traditional teaching methods, this may lead them to feel a sense of failure and revert back to traditional methods of teaching mathematics. The presence of this kind of mismatch reflects gaps in professional development programs aimed at reform-based mathematics instruction, as well as in mathematics methods courses in teacher education programs.

Figure 3. Congruent and incongruent sources of evidence for lesson success with subsequent impact on development of mathematics teaching efficacy in the context of reform-based mathematics

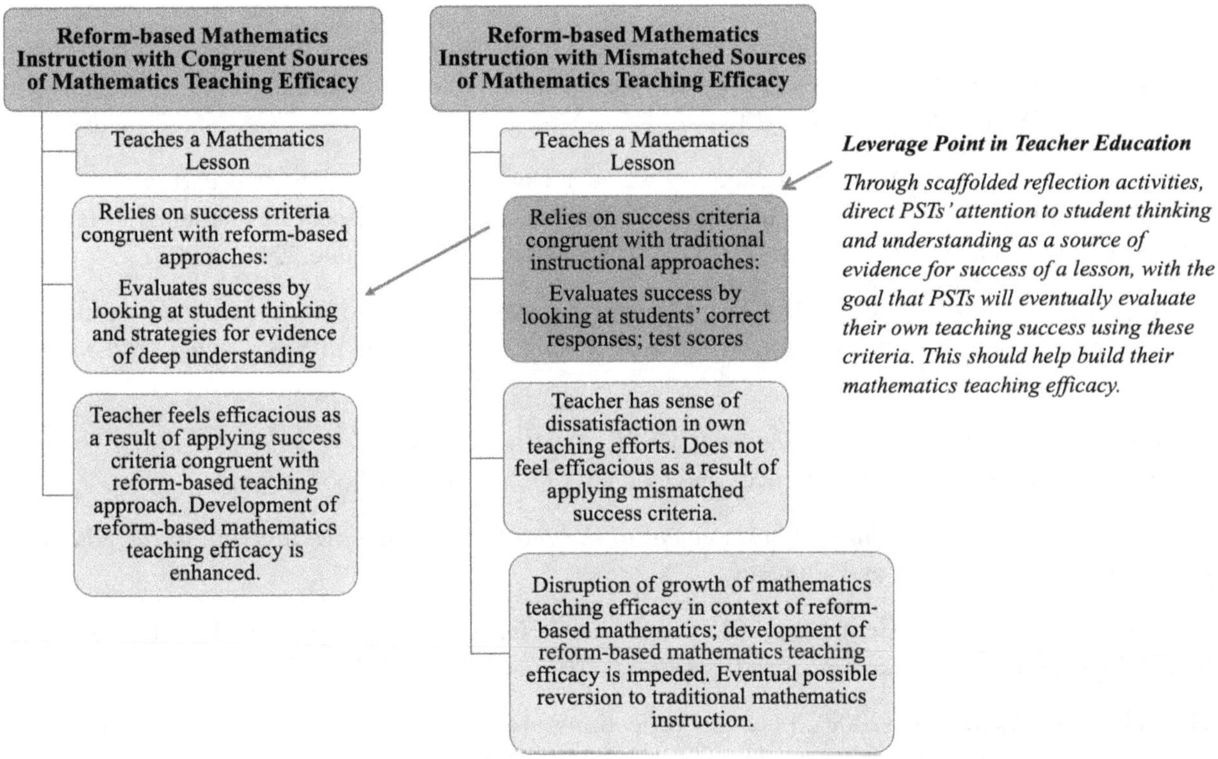

As shown in Figure 3, a key leverage point for change may lie in shifting PSTs to draw on those sources of mathematics teaching efficacy that match the goals of reform-based mathematics. The authors have suggested that teacher education is an opportune time to target the development of PSTs' mathematics teaching efficacy, and the examples in this chapter of activities in a mathematics methods course illustrate one way that teacher educators can facilitate a fuller adoption of reform-based methods. Each of the three reflective components of this mathematics methods course were designed to facilitate PSTs' developing ability to attend to student thinking and understanding and use it to evaluate the effectiveness of their mathematics teaching. This may ultimately lead to increased mathematics teaching efficacy and continued, lasting use of reform-based mathematics instruction. The authors have argued here that it is only when teachers look to these sources to evaluate their own success in teaching, will they fully embrace reform-based teaching by building the kind of mathematics teaching efficacy that sustains the use of reform-based mathematics teaching.

REFERENCES

Althauser, K. L. (2018). The emphasis of inquiry instructional strategies: Impact on preservice teachers' mathematics efficacy. *Journal of Education and Learning, 7*(1), 53-70.

Bandura, A. (1977). Self-efficacy: Toward a unifying theory of behavioral change. *Psychological Review, 84*(2), 191–215. doi:10.1037/0033-295X.84.2.191 PMID:847061

Briley, J. S. (2012). The relationships among mathematics teaching efficacy, mathematics self-efficacy, and mathematical beliefs for elementary pre-service teachers. *Issues in the Undergraduate Mathematics Preparation of School Teachers, 5*, 1–13.

Chamoso, J. M., Cáceres, M. J., & Azcárate, P. (2012). Reflection on the teaching-learning process in the initial training of teachers. Characterization of the issues on which pre-service mathematics teachers reflect. *Teaching and Teacher Education, 28*(2), 154–164. doi:10.1016/j.tate.2011.08.003

Coffey, J. E., & Edwards, A. R. (2015). The role subject matter plays in prospective teachers' responsive teaching practices in elementary math and science. *Responsive Teaching in Science and Mathematics*, 163-179.

Common Core State Standards Initiative (CCSSI). (2010). *Common core state standards*. National Governors Association for Best Practices and Council of Chief State School Officers.

Dewey, J. (1933). *How we think: A restatement of the relation of reflective thinking to the educative process* (Vol. 8). DC Heath and Company.

Fives, H., Lacatena, N., & Gerard, L. (2015). Teachers' beliefs about teaching (And learning). In International Handbook of Research on Teachers' Beliefs (pp. 249-265). Taylor and Francis.

Gabriele, A. J., & Joram, E. (2007). Teachers' reflections on their reform-based teaching in mathematics: Implications for the development of teacher self-efficacy. *Action in Teacher Education, 29*(3), 60–74. doi:10.1080/01626620.2007.10463461

Gelfuso, A., & Dennis, D. V. (2014). Getting reflection off the page: The challenges of developing support structures for pre-service teacher reflection. *Teaching and Teacher Education, 38*, 1–11. doi:10.1016/j.tate.2013.10.012

Giles, R. M., Byrd, K. O., & Bendolph, A. (2016). An investigation of elementary preservice teachers' self-efficacy for teaching mathematics. *Cogent Education, 3*(1), 1160523.

Hill, H., Litke, E., & Lynch, K. (2018). Learning lessons from instruction: Descriptive results from an observational study of urban elementary classrooms. *Teachers College Record, 120*(12), 120304. doi:10.1177/016146811812001207

Hill, H. C. (2021, February 4). *After 30 years of reforms to improve math instruction, reasons for hope and dismay*. Brookings Institute. https://www.brookings.edu/blog/brown-center-chalkboard/2021/02/04/after-30-years-of-reforms-to-improve-math-instruction-reasons-for-hope-and-dismay/

Hodge, L. L., & Walther, A. (2017). Building a discourse community: Initial practices. *Mathematics Teaching in the Middle School, 22*(7), 430–437. doi:10.5951/mathteacmiddscho.22.7.0430

Jao, L. (2017). Shifting pre-service teachers' beliefs about mathematics teaching: The contextual situation of a mathematics methods course. *International Journal of Science and Mathematics Education, 15*(5), 895–914. doi:10.100710763-016-9719-9

Kleickmann, T., Tröbst, S., Heinze, A., Bernholt, A., Rink, R., & Kunter, M. (2017). *Teacher knowledge experiment: Conditions of the development of pedagogical content knowledge. Competence Assessment in Education*. Springer.

Lane, C., & Ríordáin, M. N. (2020). Out-of-field mathematics teachers' beliefs and practices: An examination of change and tensions using Zone Theory. *International Journal of Science and Mathematics Education, 18*(2), 337–355. doi:10.100710763-019-09964-5

Lee, Y., & Ertmer, P. A. (2006). Examining the effect of small group discussions and question prompts on vicarious learning outcomes. *Journal of Research on Technology in Education, 39*(1), 66–80. doi:10.1080/15391523.2006.10782473

McGee, J. R., Wang, C., & Polly, D. (2013). Guiding teachers in the use of a standards-based mathematics curriculum: Teacher perceptions and subsequent instructional practices after an intensive professional development program. *School Science and Mathematics, 113*(1), 16–28. doi:10.1111/j.1949-8594.2012.00172.x

Murrey, D. L. (2008). Differentiating instruction in mathematics for the English language learner. *Mathematics Teaching in the Middle School, 14*(3), 146–153. doi:10.5951/MTMS.14.3.0146

Naidoo, K., & Naidoo, L. J. (2021). Designing teaching and reflection experiences to develop candidates' science teaching self-efficacy. *Research in Science & Technological Education*, 1–21. doi:10.1080/02635143.2021.1895098

National Council of Teachers of Mathematics (NCTM). (2000). *Principles and standards for school mathematics*. National Council of Teachers of Mathematics.

National Council of Teachers of Mathematics (NCTM). (2014). *Principles to actions: Ensuring mathematical success for all*. NCTM.

Noormohammadi, S. (2014). Teacher reflection and its relation to teacher efficacy and autonomy. *Procedia: Social and Behavioral Sciences, 98*, 1380–1389. doi:10.1016/j.sbspro.2014.03.556

Paolucci, C. (2015). Changing perspectives: Examining the potential for advanced mathematical studies to influence pre-service teachers' beliefs about mathematics. *Teaching and Teacher Education, 49*, 97–107. doi:10.1016/j.tate.2015.03.002

Phan, H. P. (2014). Self-efficacy, reflection, and achievement: A short-term longitudinal examination. *The Journal of Educational Research, 107*(2), 90–102. doi:10.1080/00220671.2012.753860

Romberg, T. A. (1989). *Curriculum and Evaluation Standards for School Mathematics*. The National Council of Teachers of Mathematics.

Santagata, R., & Yeh, C. (2014). Learning to teach mathematics and to analyze teaching effectiveness: Evidence from a video-and practice-based approach. *Journal of Mathematics Teacher Education, 17*(6), 491–514. doi:10.100710857-013-9263-2

Seung, E., Park, S., & Jung, J. (2014). Exploring preservice elementary teachers' understanding of the essential features of inquiry-based science teaching using evidence-based reflection. *Research in Science Education, 44*(4), 507–529. doi:10.100711165-013-9390-x

Smith, J. P. III. (1996). Efficacy and teaching mathematics by telling: A challenge for reform. *Journal for Research in Mathematics Education, 27*(4), 387–402. doi:10.5951/jresematheduc.27.4.0387

Smith, M. S., Bill, V., & Hughes, E. K. (2008). Thinking through a lesson: Successfully implementing high-level tasks. *Mathematics Teaching in the Middle School, 14*(3), 132–138. doi:10.5951/MTMS.14.3.0132

Tavil, Z. M. (2014). The effect of self reflections through electronic journals (e-journals) on the self efficacy of pre-service teachers. *South African Journal of Education, 34*(1), 1–20. doi:10.15700/201412120931

Thomson, M. M., DiFrancesca, D., Carrier, S., & Lee, C. (2017). Teaching efficacy: Exploring relationships between mathematics and science self-efficacy beliefs, PCK and domain knowledge among preservice teachers from the United States. *Teacher Development, 21*(1), 1–20. doi:10.1080/13664530.2016.1204355

Tschannen-Moran, M., & Hoy, A. W. (2001). Teacher efficacy: Capturing an elusive construct. *Teaching and Teacher Education, 17*(7), 783–805. doi:10.1016/S0742-051X(01)00036-1

Uzun, M. S., Yüksel, N. S., & Dost, Ş. (2013). Investigation of pre-service teachers' mathematics teaching efficacy beliefs in terms of their reflective thinking tendencies. *Procedia: Social and Behavioral Sciences, 106*, 1595–1602. doi:10.1016/j.sbspro.2013.12.180

van Es, E. A., Cashen, M., Barnhart, T., & Auger, A. (2017). Learning to notice mathematics instruction: Using video to develop preservice teachers' vision of ambitious pedagogy. *Cognition and Instruction, 35*(3), 165–187. doi:10.1080/07370008.2017.1317125

von Suchodoletz, A., Jamil, F. M., Larsen, R. A., & Hamre, B. K. (2018). Personal and contextual factors associated with growth in preschool teachers' self-efficacy beliefs during a longitudinal professional development study. *Teaching and Teacher Education, 75*, 278–289. doi:10.1016/j.tate.2018.07.009

Warshauer, H. K. (2015). Strategies to support productive struggle. *Mathematics Teaching in the Middle School, 20*(7), 390–393. doi:10.5951/mathteacmiddscho.20.7.0390

Wilkerson, T., Kerschen, K., & Shelton, R. (2018). PreService teachers' critical connections to effective mathematical teaching practices: An instructional approach using vignettes. *Action in Teacher Education, 40*(4), 358–373. doi:10.1080/01626620.2018.1512430

Wyatt, M. (2016). "Are they becoming more reflective and/or efficacious?" A conceptual model mapping how teachers' self-efficacy beliefs might grow. *Educational Review, 68*(1), 114–137. doi:10.1080/00131911.2015.1058754

Zee, M., & Koomen, H. M. Y. (2016). Teacher Self-Efficacy and Its Effects on Classroom Processes, Student Academic Adjustment, and Teacher Well-Being: A Synthesis of 40 Years of Research. *Review of Educational Research, 86*(4), 981–1015. doi:10.3102/0034654315626801

ADDITIONAL READING

Gelfuso, A., & Dennis, D. V. (2014). Getting reflection off the page: The challenges of developing support structures for pre-service teacher reflection. *Teaching and Teacher Education*, *38*, 1–11. doi:10.1016/j.tate.2013.10.012

Naidoo, K., & Naidoo, L. J. (2021). Designing teaching and reflection experiences to develop candidates' science teaching self-efficacy. *Research in Science & Technological Education*, 1–21. doi:10.1080/02635143.2021.1895098

Smith, M. S., Bill, V., & Hughes, E. K. (2008). Thinking through a lesson: Successfully implementing high-level tasks. *Mathematics Teaching in the Middle School*, *14*(3), 132–138. doi:10.5951/MTMS.14.3.0132

Uzun, M. S., Yüksel, N. S., & Dost, Ş. (2013). Investigation of pre-service teachers' mathematics teaching efficacy beliefs in terms of their reflective thinking tendencies. *Procedia: Social and Behavioral Sciences*, *106*, 1595–1602. doi:10.1016/j.sbspro.2013.12.180

Wyatt, M. (2016). "Are they becoming more reflective and/or efficacious?" A conceptual model mapping how teachers' self-efficacy beliefs might grow. *Educational Review*, *68*(1), 114–137. doi:10.1080/00131911.2015.1058754

KEY TERMS AND DEFINITIONS

Mathematics Teaching Efficacy: One's belief in their ability to successfully teach mathematics and influence their students' learning of mathematics.

Pedagogical Content Knowledge: Specialized knowledge around the how to instruct in a particular subject area.

Reflection: Conscientious thought about actions or ideas.

Scaffolded Reflection: Directing attention to specific considerations for reflection.

Self-Efficacy: One's belief in their ability to successfully perform a task.

Student-Centered Mathematics: Mathematics instruction centered around students' thinking and current level of understanding, often explorative in nature.

Teaching Efficacy: One's belief in their ability to successfully teach and influence their students' learning/performance.

Traditional Mathematics Instruction: Mathematics instruction centered around direct and often, procedural instruction from the teacher who disseminates this knowledge to students.

Vicarious Experiences: Experience through observation of others.

Chapter 7
Supporting Preservice Elementary Teachers in Planning for Mathematical Discussions

Kathryn Mary Rupe
https://orcid.org/0000-0002-9467-182X
Western Washington University, USA

Rebecca S. Borowski
Western Washington University, USA

ABSTRACT

One goal for teacher preparation programs is to develop preservice teachers' ability to plan student-centered lessons that include meaningful mathematical discourse. The five practices for orchestrating productive mathematical discussions provides one framework for planning this critical component of mathematics instruction. This chapter discusses several strategies the authors implemented in undergraduate mathematics education courses. Mathematics teacher educators can use these strategies to support preservice teachers as they develop their understanding of mathematical discussions and their mathematical knowledge for teaching. The culminating activity, a three-part lesson planning sequence, prompts preservice teachers to apply knowledge of the five practices to plan responsive lessons in which student thinking is centered.

INTRODUCTION

Mathematics teacher educators are tasked with supporting preservice teachers (PSTs) to develop the knowledge and skills necessary to be effective mathematics teachers. For mathematics teacher educators (MTEs) working with elementary PSTs, this can be especially challenging for many reasons, including the limited number of mathematics content and methods courses students take, the common hesitance

DOI: 10.4018/978-1-7998-9422-3.ch007

with math content, and the emphasis on student-centered learning that conflicts with the way many PSTs were taught (Banilower et al., 2013). Engaging in meaningful mathematical discourse around rich problem-solving tasks in elementary content or methods courses can provide important experiences, but PSTs also need opportunities to practice planning for and facilitating these types of conversations. In this chapter, the authors provide a series of activities based on the *Five Practices for Orchestrating Productive Mathematical Discussions* (Stein et al., 2008) that they implemented with undergraduate elementary and special education PSTs during the second of three blended mathematics content and pedagogy courses at a mid-sized public university in the Pacific Northwest. The activities—which were designed to operationalize the learning of the different areas of Mathematical Knowledge for Teaching, or MKT, (Ball et al., 2008) by building a bridge to the Five Practices—have undergone several iterations, with changes made to address challenges and build on the successes of each implementation. Authors' experiences with the activities are shared not to engage in data analysis for the sake of research, but rather to reflect upon implementation of the activities. We encourage other mathematics educators to adopt the approach as they support PSTs in developing their knowledge and skills to plan for and facilitate mathematical discussions with their own students.

BACKGROUND

Five Practices for Orchestrating Productive Mathematical Discussions (Stein et al., 2008) offers a popular framework that includes actions that occur during planning as well as during classroom implementation of a lesson. Stein et al. suggest that meaningful mathematical discussions based on students' ideas can be attained through this process of five steps, each step depending on the previous one. Teachers must anticipate students' responses to a task that will be taught, monitor students' thinking as they are engaged in the task, purposefully select students to present based on their choice of representations, sequence those representations in a purposeful way, and then make connections among the representations so that students are able to understand key mathematical concepts. To facilitate a student-centered discussion, teachers predict how students may engage in the task, plan key questions, consider scaffolds students may need, and think about how big ideas will be summarized at the end of the lesson. Without careful planning, teachers who use rich tasks can feel out of control of the learning, reducing teacher efficacy (Schoenfeld, 1998; Smith, 1996). Stein et al. (2008) noted that while some expert teachers may be able to successfully improvise in the classroom, most cannot. As such, making lesson planning a public and reflective practice between MTEs and PSTs allows instructors to reinforce the idea that intentional planning versus improvisation is the norm for well-designed lessons that center students' thinking. The Five Practices provide a process for PSTs to use when planning to implement rich mathematical tasks that value students' ideas and strategies.

Preparing PSTs to plan lessons in which they will facilitate meaningful mathematical discussions requires MTEs to carefully consider the knowledge and skills required to do so. Ball et al. (2008) provided a framework that distinguishes among several types of Mathematical Knowledge for Teaching. In this framework, both mathematical content knowledge and pedagogical knowledge include three subsets of math teacher knowledge. In addition to *common content knowledge*, teachers use other types of knowledge including *specialized content knowledge, knowledge of content and students,* and *knowledge of content and teaching*. These distinctions help conceptualize *what* PSTs need to know and be able to do, while also supporting math teacher educators in considering *how* to develop these areas of knowledge.

One area of content knowledge developed in teacher preparation programs is *specialized content knowledge*, which includes content knowledge that is not "common" mathematical knowledge but does not directly require knowledge of students. For example, being able to use mathematical content knowledge to evaluate if a non-traditional algorithm "works" is important for assessing student responses and requires application of mathematical understanding in ways distinct to teaching. Also included in specialized content knowledge is the ability to respond to "why?" questions and select particular representations for specific purposes. *Knowledge of content and students* includes knowledge of likely student strategies and ideas students might find challenging. Knowledge of content and students can impact teacher decisions about task choice as well as their interpretation of students' written and verbal responses that include emerging mathematical ideas. A third area of MKT, *knowledge of content and teaching,* requires teachers to make intentional choices about the sequencing of mathematical ideas in a lesson, including the advantages and disadvantages of various tools, strategies, and procedures. These three areas of MKT speak directly to the knowledge and skills that teachers draw upon to facilitate productive mathematical discussions that focus on their students' strategies, early conceptions, and connections to important mathematical ideas. The lesson planning process shared in this chapter considers these areas of MKT both as distinct areas of knowledge that need attention, and as interconnected ways of thinking about teaching mathematics. Participating in learning experiences that attend to these areas of MKT gives PSTs opportunities to develop their knowledge in these areas and supports the culminating lesson planning activity discussed in this chapter.

Drawing upon and developing preservice teachers' MKT can be done in many ways. This chapter provides examples of several learning activities and practices that led to the formal lesson planning of a mathematical discussion around a rich task in a blended mathematics content and pedagogy course for future K-8 teachers. These learning activities were designed using a practice-based approach that included modeling, rehearsal, and providing feedback to PSTs (Shaughnessy et al., 2016). This allowed PSTs to develop their understanding of student-centered mathematics learning through experience. Courtney et al. (2015) found that many PSTs see lesson planning as a guide or an accountability tool, while many practicing teachers view lesson planning as a reflective tool. They suggested that specifically emphasizing student thinking, versus "covering" content, can positively impact PSTs' thoughts about lesson planning (Courtney et al., 2015). It is possible for PSTs to create mathematics lesson plans that follow a student-centered *launch, explore, summarize* format, attend to important mathematical ideas, and explicitly make connections among different student representations (Lee, 2019). To make lesson planning meaningful and responsive to student thinking, the authors suggest that PSTs need time to practice and develop their understanding of different representations and tools, to experience mathematical discussions facilitated by the instructor as a learner, and to explicitly debrief lessons with a focus on teacher moves and insight into the instructor's thinking about planning and facilitation of the discussion.

MAIN FOCUS OF THE CHAPTER

This chapter describes a series of learning experiences coupled with a formal lesson planning assignment that has been implemented in university classrooms to engage PSTs in learning to plan with consideration of the Five Practices. This lesson planning assignment was meaningful for PSTs because several foundational learning experiences built into the course gave attention and time to developing separate skills (e.g. knowledge of different representations and tools for solving problems) followed by opportunities for application of those skills when formally planning a lesson. The activities described in this chapter were integrated throughout a blended mathematics content and pedagogy course designed for undergraduate preservice elementary and special education teachers. At the university, PSTs take three quarters of blended content and pedagogy courses in their teacher preparation program. The courses are taught by former elementary and middle school teachers, with personal connections and expertise incorporated into the courses. The second course, during which the activities were implemented, enrolls approximately 50 students (across two sections) each term. The PSTs have been formally accepted into the education program and are working toward certificates in elementary education, special education, or both, but have not yet begun student teaching. One of the content areas of focus in the course is fractions, including exploration of all four operations with fractions, an area that provides a particular challenge for many PSTs (Son & Lee, 2016; Thurtell et al., 2019).

The activities provided in this chapter model a practice-based approach to reimagine lesson planning and prepare PSTs to facilitate mathematical discussions that are responsive to students' thinking. The authors present four practices mathematics teacher educators can use to support PSTs' lesson planning:

- Use rich tasks and ask PSTs to solve them in multiple ways.
- Develop collective knowledge about the Five Practices.
- Explicitly facilitate discussions that show a variety of strategies
- Develop a lesson plan using the Five Practices.

Several teaching practices, such as providing specific and timely feedback, taking an asset-based approach to looking at student work, and reflecting on the reasoning of others, are included throughout the model. The sequence of activities described in this chapter can be taken up by mathematics teacher educators in a variety of settings. A description of each activity includes examples of artifacts from PSTs such as discussion highlights, solution strategies, and reflective writing. Preservice teachers can experience, learn about, and plan for facilitating mathematical discussions in meaningful and important ways when given opportunities to practice and reflect upon the process.

Use Rich Tasks and Ask PSTs to Solve Them in Multiple Ways

This section illustrates how allowing students to solve rich tasks in ways that are meaningful to them is essential to facilitating meaningful mathematical discussions. Specifically, it explores how PSTs can incorporate rich tasks in lesson plans for mathematical discussions that center students' thinking.

Content and methods courses that regularly incorporate rich tasks with multiple possible solution strategies (and, potentially, a variety of "correct" solutions) can help build PSTs' capacity for anticipating student answers. The implementation of such tasks in the authors' courses supports PSTs in building their *specialized content knowledge*, sharing their solutions with their classmates, justifying their solutions, reasoning about the solutions of their peers, and finding connections among various solutions.

Before PSTs can begin to anticipate and respond to students' thinking, it is important for them to practice generating a variety of strategies and solutions using their own reasoning. Engaging in regular number talks is one foundational practice for building preservice teachers' quantitative reasoning and flexibility of solution strategies. During a number talk, students are asked to mentally solve problems in a variety of ways, share their thinking orally, and reason about the strategies of their classmates (Parrish, 2010). Number talks are a regular practice in the authors' classes and reinforce the philosophy that mathematical reasoning is valued over "correctness." PSTs are expected to use a variety of strategies flexibly and evaluate others' strategies. These experiences, along with reflective discussions after the number talks, support PSTs as they develop their own content knowledge and reasoning about various solution strategies.

For example, consider the task *Half of a Recipe* (Illustrative Mathematics, 2016).

Caitlin is making ½ of a recipe. The full recipe calls for 3 ¼ cups of flour. How many cups of flour should Cailtin use? Explain your thinking.

PSTs solved this problem in a number of ways, some of which are illustrated in Figure 3. In one common strategy, PSTs drew a picture showing three wholes and one fourth, eliminated half of each, then added the remaining parts. In another strategy, PSTs represented 3 ¼ on a number line, cut the length in half, then used the number line to find the value of that half. A third strategy was based on an oft-used algorithm; PSTs renamed 3 ¼ as 13/4, then multiplied it by ½ and simplified. Some PSTs got stuck and were unable to find a solution, and some PSTs had incorrect solutions. Giving PSTs opportunities to hear each other's solutions to a task such as this one and to reason about each other's work helped them develop an understanding of the wide range of potential solutions to the task. Further, it exposed them to solution strategies that had not occurred to them as well as student conceptions that might lead to incorrect answers. PSTs developed their *common content knowledge* as they became familiar with a range of fraction representations (number lines, pictures of circles, bar models, etc.). They also developed *specialized content knowledge* as they explored concepts of iterating, equipartitioning, and the concepts underling algorithms for fraction computation.

Figure 1. Student solution using partitioning each section in half
If we split each part of 1+1+1+¼ into half, then we get ½+½+½+⅛ which is 1 ⅝.

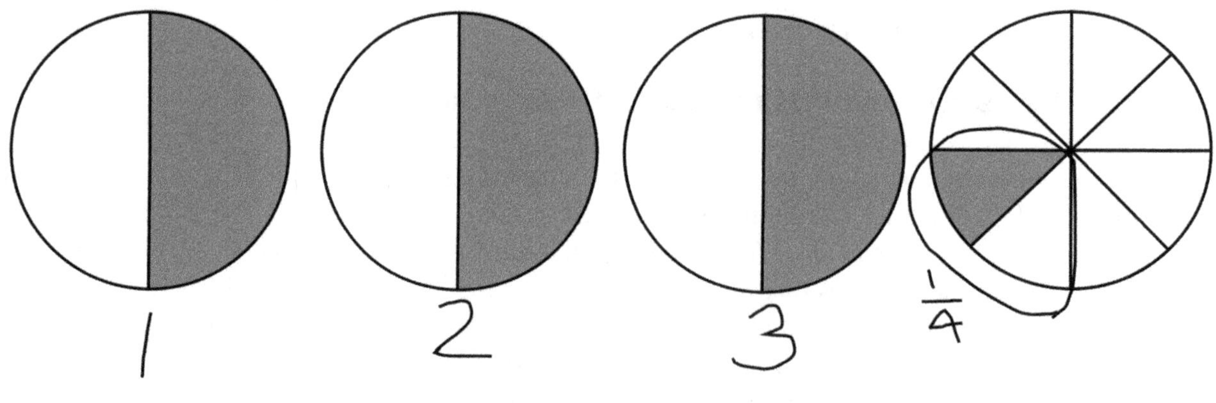

Figure 2. Student solution using a number line

I started making equal sized jumps from 0 and 3 ¼ until I found the halfway point. It's between 1 ½ and 1 ¾, which is 1 ⅝.

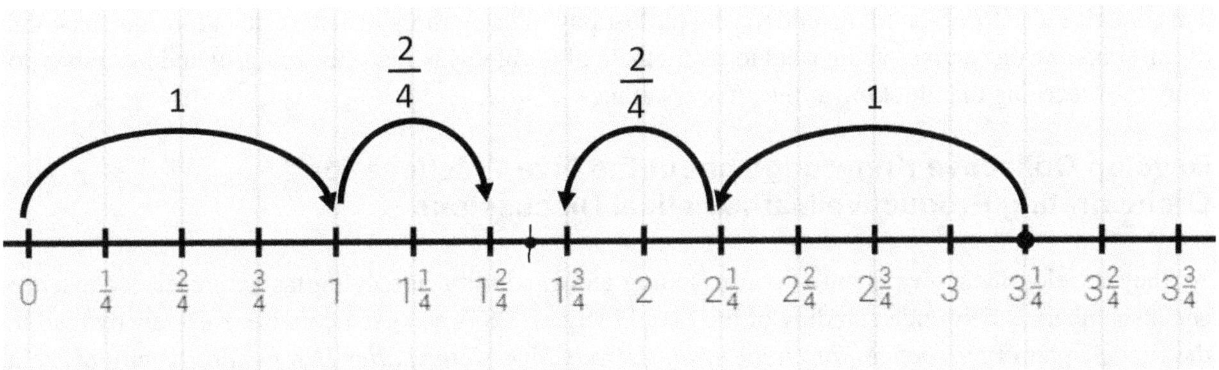

Figure 3. Student solution using a multiplication algorithm

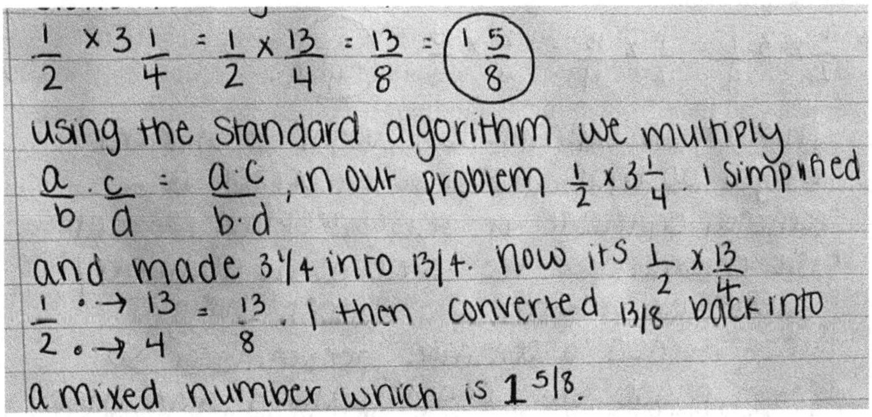

Once PSTs had a chance to solve the task and see multiple solution strategies for such tasks, they were expected to solve subsequent tasks in *multiple* ways. Rather than solving the task to get an answer, they were encouraged to find the answer using a variety of strategies. In this way, PSTs developed their ability to use different strategies and become more comfortable with strategies that were previously unfamiliar to them. Thus, the purpose of the task shifted from finding and justifying an answer to developing PSTs' ability to use multiple representations and solution strategies.

After PSTs had engaged in solving a task in multiple ways and discussed it with their classmates, they were given opportunities to analyze samples of student (children's) work on the same or similar tasks. In some cases, PSTs watched videos of students solving similar tasks. This experience prepared them to engage with and analyze work done by children, which contributed to their *knowledge of content and students*.

Building classroom experiences around rich mathematical tasks encourages PSTs to develop their MKT. They build their *common content knowledge* by solving the tasks themselves. They deepen their *specialized content knowledge* by analyzing their own solutions and the work of their peers. Finally, they build their *knowledge of content and students* by analyzing children's solutions to similar tasks. This practice also foregrounds an underlying philosophy in the authors' teacher education courses—that discussing *how* one arrives at answers to mathematical problems is more meaningful and has more potential for learning than looking at the answers alone.

Develop Collective Knowledge about the Five Practices for Orchestrating Productive Mathematical Discussions

As they develop their own quantitative reasoning and flexibility in solving tasks, preservice teachers can also enhance their understanding of the Five Practices. PSTs can get a concise overview by reading the first chapter of *5 Practices for Orchestrating Productive Mathematics Discussions* (Smith & Stein, 2018). NCTM's practitioner journals also provide numerous, recent descriptions of the Five Practices. Reading practice-based ideas written by teachers can help connect PSTs to a professional community and envision their own future classrooms. Articles that emphasize particular practices can be introduced at appropriate times to support course content such as selecting or sequencing. To help teacher educators select readings, Table 1 in the Appendix summarizes writings on the Five Practices and notes which practices are addressed in each.

The authors engaged PSTs with readings on the Five Practices in a variety of ways over various iterations. Providing PSTs with a variety of resources and allowing them to choose among them led to increased engagement and more robust understanding. It is important for PSTs to see teachers across grade levels and in different instructional contexts using an intentional and replicable process to be responsive to students' thinking during mathematical discussions. Engaging with multiple sources, PSTs could see the various contexts in which these practices are implemented and the different perspectives teachers take when they use them.

At other times, the authors used a jigsaw approach. Groups of PSTs were assigned to read different articles about the Five Practices and share their new learning with the others. They connected and synthesized ideas presented by their peers, and, as a group or class, developed a collective understanding of the practices and important discourse strategies. One benefit of the jigsaw approach is to expose PSTs to a larger body of work than can be assigned to individual students. A second is that sharing knowledge with fellow PSTs builds expertise. In some implementations, PSTs were asked to create drawings, tables, or concept maps of a common reading to compare and reflect upon their own conceptualization of the Five Practices.

PSTs' written reflections showed insight into the meaning they gained from reading various articles. One PST made explicit connections among the Five Practices and other experiences in the class, stating, "When I think about this article, I think a lot about number talks, because both are remarkably similar in their approach to talking about problems." Another student reflected on the importance of framing the questions asked of students, "because wording could be crucial here." They also shared that "thinking about students' responses could even help the teacher think about the problem in an entirely different way," showing that they understood how important it is to understand and center student thinking. Another PST reflected that they valued discussions about all types of answers, including incorrect ones, because "wrong answers are the catalyst for a meaningful discussion." Finally, one PST reflected, "Meaningful classroom discussions stem from teachers' mindsets and planning… When groups present ideas, you as the teacher can put them in a specific order to direct the conversation." This final comment demonstrates an understanding that good discussions don't happen as a result of luck, but, rather, are the result of careful planning and close attention to students' thinking.

Explicitly Facilitate Discussions that Show a Variety of Strategies

To support preservice teachers' ability to plan for and facilitate meaningful mathematical discussions with their future students, instructors should model discussions that are responsive to student thinking in their courses. PSTs need multiple opportunities to engage in lessons and discussions as learners, and to explicitly debrief and reflect on those experiences, as a foundation for their own planning. Mathematics teacher educators must, therefore, be intentional in their task choices and draw upon several areas of MKT. Choosing tasks that allow for multiple entry points and possible solution strategies can expose PSTs to several new ways to solve a problem and allow them to apply strategies or tools that have been introduced (e.g., using a number line or fraction circles to model fraction addition). Solving problems in a variety of ways and critiquing others' strategies makes use of and develops PSTs' *knowledge of content and teaching* (Ball et al., 2008).

Whole class discussions that include intentionally selected and sequenced student solutions that employ different strategies allow PSTs to broaden their understanding of the possibilities of mathematical discussions. Discussing the Five Practices in an undergraduate blended content and pedagogy math education course, PSTs recognized the importance of selecting and sequencing student responses versus calling on volunteers. Reasons for selecting and sequencing provided by the PSTs included hearing from students that do not share often, avoiding repeated strategies, and planning ahead for questions. A selection of PSTs' strategies and discussion artifacts follows, demonstrating how a common fraction task and selected student solution methods led to a discussion that featured several important understandings for PSTs.

You are making lemonade. You have 2 ½ cups of sugar at home. If a batch of lemonade needs ¾ cup of sugar, how many batches of lemonade can you make? Explain your thinking

This task led to three common solutions—three batches of lemonade with ¼ cup of sugar left over, 3 ¼ batches of lemonade, and 3 ⅓ batches of lemonade—in addition to some less common incorrect solutions. Discussion of the question *What is the correct answer to this problem?* allowed PSTs to consider the idea of *correctness* as well as contextualize the solutions. *Is it possible to make partial batches? Could both three batches and 3 ⅓ batches be correct solutions?* One PST commented, "The answer that the student gives helps me think about what questions to ask them." Making sense of solutions to division problems, including whether fractional answers are appropriate and how to interpret remainders, is a common challenge for students (Silver et al., 1993). Tasks like these highlight the idea that teachers should attend to students' thinking and strategies rather than focusing on "correctness" and following procedures. If one of the purposes of mathematical discussions using the Five Practices is to make connections among representations as well as important mathematical ideas, uncovering a range of student thinking and solutions is a positive outcome.

Preservice teachers' solutions, discussion questions, and selected comments from a class discussion provide insight into the implementation of the task, important understandings from the discussion, and key takeaways of the PSTs. In this example, four students' solutions were shared during the whole class discussion. Maria's solution used fraction circles and provided a division number sentence. Alex's and Jaime's solutions both included the extra ¼ cup of sugar in their description of the number of batches, although in different ways. Last, Erica created a division number sentence and used a standard algorithm to find the solution of 3 ⅓ batches of lemonade.

When facilitating mathematical discussions that are responsive to students' thinking, teachers should use an asset-based approach focused on students' strengths, informal knowledge, culture and experiences (Gutstein et al., 1997). The asset-based approach counters deficit narratives that focus on what students, families, schools, and communities are lacking. Deficit thinking is pervasive and can show up in classrooms in ways including a focus on correctness, applying rules without conceptual understanding, avoiding discussions of incorrect solutions, and providing few opportunities to engage in problem solving with excuses like "students lack basic skills" (Adiredja, 2019). When implemented, PSTs began to confront these ideas and develop an asset-based approach to looking at student work during classroom discussion experiences.

A Notice and Wonder routine framed the discussion of student work. This routine can be used to engage learners in a problem situation before a formal question is posed, which provides an accessible way for all learners to engage with the mathematics (Rumack & Huinker, 2019). The strategy can also be used with teachers in a variety of ways to support professional conversations around students' written responses (Dobie & Anderson, 2021). *What do you notice about this person's work? What do you wonder about this person's thinking or solution method?* When given time to think individually and then discuss with their small group, PSTs had time to make meaning of each other's thinking and practice the skill of analyzing student work. It was an intentional choice to give ownership of the ideas to the PSTs and allow them to reflect on the variety of thinking that was developed. When looking at Maria's work, PSTs shared:

PST 1: I see that they are using fraction circles and were thinking of 2 ½ as two and two-fourths.
PST 2: I notice that they use the red pen to circle three sections in their visual model.
PST 3: The student wrote the number sentence two and one-half divided by three-fourths equals question mark, so they know it's division.
PST 4: We wondered if they knew what to do with the extra part that is left over.
PST 5: I wonder why they wrote a division number sentence but then didn't get 3 ⅓ as their answer.

Figure 4. Maria's solution to the lemonade task

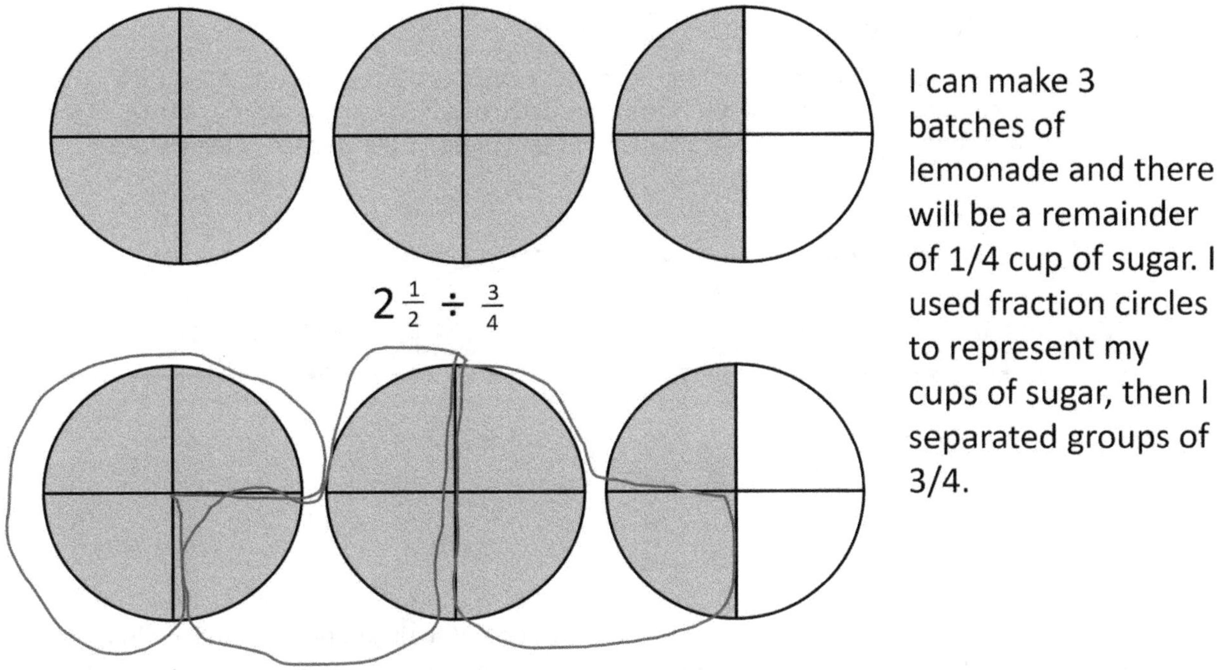

When looking closely at Alex and Jaime's solutions, the PSTs talked about whether it was possible for the problem to be modeled with addition or subtraction. Additionally, PSTs focused on similarities and differences between the two models and what might be concluded about student thinking. PSTs decided it would be best to ask questions of the students because "it's sometimes hard to write what you are really thinking." Using a Notice and Wonder routine for looking at student work, PSTs developed their agency as teachers of math.

Figure 5. Alex's solution to the lemonade task

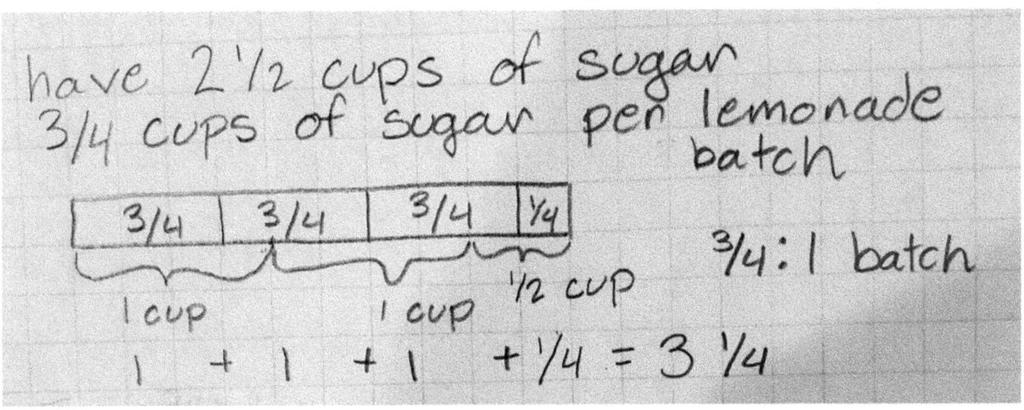

Figure 6. Jaime's solution to the lemonade task

Number Sentence: $2\frac{1}{2} \div \frac{3}{4}$

I used a number line to solve the problem. I started at 2 2/4 since 2/4 is equivalent to 1/2. I then jumped back 3/4 three times until I only had 1/4 left. Since once batch needs 3/4 cup of sugar, 1/4 is 1/3 of the amount needed for one batch. Therefore, 2 1/2 cups of sugar makes 3 1/3 batched of lemonade.

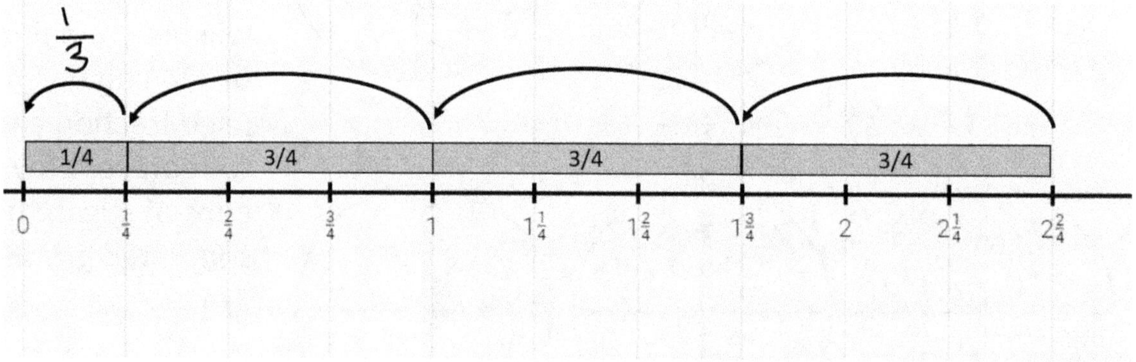

When discussing Erica's work, PSTs commented on aspects of their solution method, including "they know it's division," "they use the rule," and "they have a correct label, so they probably understand what the answer means." The PSTs wondered if the student only knew the algorithmic strategy or find the answer another way. In addition to their responses to the Notice and Wonder prompts, PSTs talked about contextualizing and decontextualizing the numbers in the solutions, iterating fractions, connecting repeated addition and subtraction number sentences, and whether multiplication would be appropriate to use to solve the problem. PSTs also raised the issue "Does using an algorithm show conceptual understanding?" Several PSTs mentioned questions they would want to ask the students such as, in reference to Maria's work, "I would ask them what the fourth represented, or, if they were to include the fourth in the answer, how that would look."

Figure 7. Erica's solution to the lemonade task

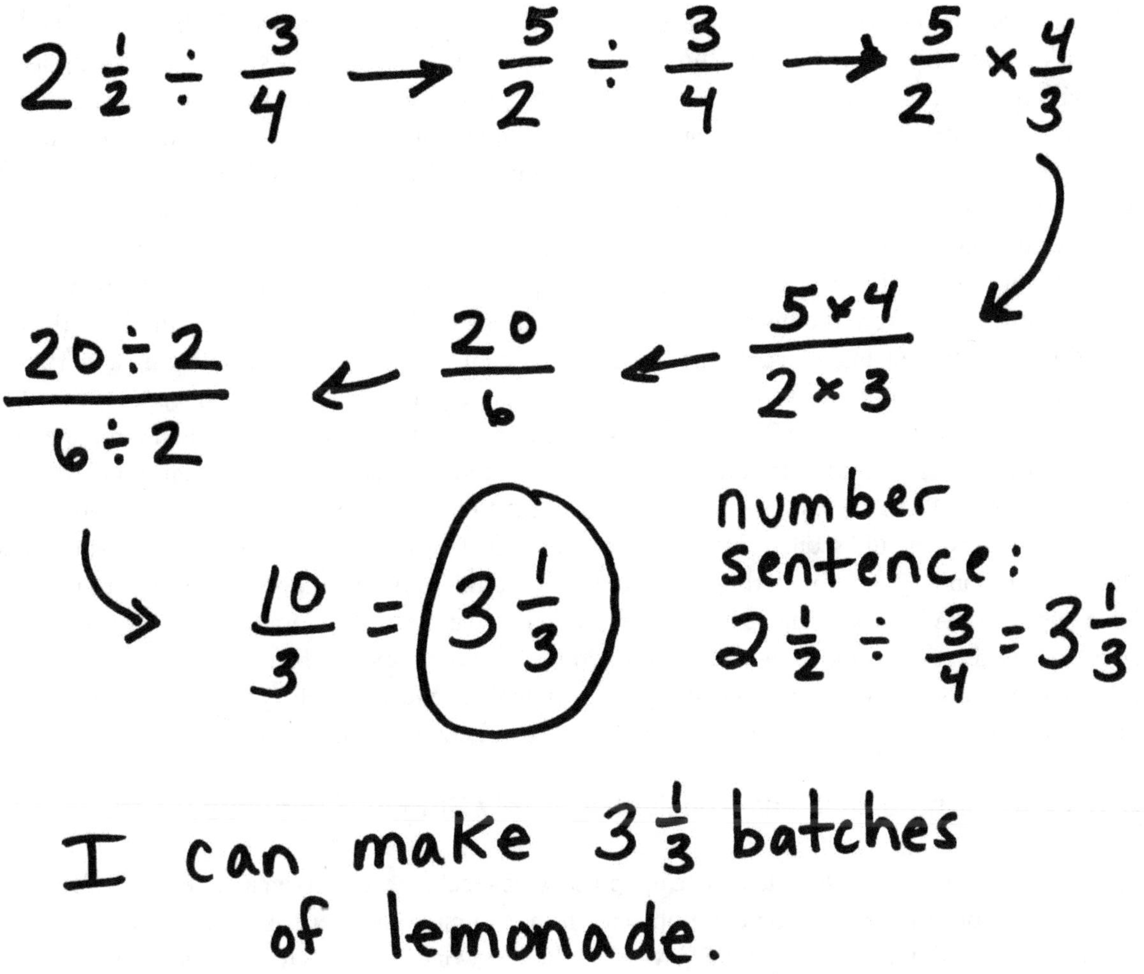

After PSTs engaged in class discussions such as these, they explicitly debriefed about the teacher moves and viewed artifacts related to planning the lesson. This provided insight into the instructor's planning process and thinking. Questions posed for debriefing the lesson included:

Why do you think I chose to include these students' work in the discussion?
What important mathematical ideas do you think came up in this lesson and discussion?
How would you describe the instructors' role and the students' role during the discussion?

The instructor shared their lesson plan, list of anticipated student responses, and the monitoring sheet (Figure 8 in the Appendix) used to keep track of student ideas while they worked on the task. Reflecting upon and connecting planning practices and classroom actions to the ideas presented in the articles about the Five Practices helped PSTs recognize that these practices ensure student thinking is at the center of the discussion.

Develop a Lesson Plan Using the Five Practices

Five Practices for Orchestrating Productive Mathematical Discussions (Stein et al., 2008) calls for teachers to select a worthwhile task, anticipate student responses to the task, monitor students' thinking as they engage with the task, select and sequence students' strategies to highlight during a class discussion that will illuminate important mathematical concepts, and use questioning to help students make connections among representations and to important mathematical concepts. Lessons that follow this structure require thoughtful choices in both the planning and classroom instruction phases of lesson implementation.

Preservice teachers who are not able to implement lessons with students in field experiences can still engage meaningfully with the *Five Practices*. A three-phase lesson planning assignment, characterized by consistent feedback from the instructor and opportunities for PSTs to revise their thinking throughout the process, provided an opportunity for such engagement. Phase 1 involved selecting a worthwhile task, considering the important mathematical ideas, and anticipating ways in which students may solve or engage with the task. In Phase 2, students began to complete the formal lesson plan template. They considered how the task would be introduced, and planned student questions for each anticipated solution method. In Phase 3, PSTs selected from their anticipated student strategies and provided a rationale for the way in which they would sequence the student ideas in a classroom discussion. PSTs envisioned how the class discussion might go, including questions that would be posed and possible student responses to those questions. After each phase, PSTs received extensive feedback from the instructor which was taken into consideration as they made revisions and built on the previous phase.

Next, more detailed descriptions of each phase of planning, with examples from implementation, are provided to illustrate the depth of thinking that PSTs exhibited during this process.

Phase 1: Choosing a Task and Anticipating Likely Student Responses

Before beginning a formal lesson plan write up, preservice teachers were given a large bank of rich tasks focused on fractions (an area of content emphasis in the course), and were asked to choose a task that they would like to think deeply about. PSTs were invited to work in groups to discuss the mathematical task and lesson plan, while also expected to submit their own unique writing.

After selecting a task, PSTs considered the important mathematical ideas addressed and anticipated at least three correct and two partially correct or incorrect ways of solving the task. This practice of anticipating ways students may solve a problem was woven throughout the course and practiced often. PSTs also submitted two to five important mathematical ideas addressed by the task.

In this beginning phase of the lesson planning process, PSTs were asked to reflect upon questions such as: *What strategies do you expect to see? What do you expect to be the most common approach?* Many provided thoughtful details and connected to important representations and mathematical ideas. For example, one PST anticipated:

Strategies that I expect to see students use include number lines, fraction circles, and the idea of repeated addition. I would expect that students would approach this task by breaking down visual models to represent each fruit on its own, (i.e., a number line for apples, another for lemons, and a third for peaches). Once students solve for the amount of each individual fruit needed, I would expect that they would then relate each fractional amount back to the context of the problem. Whether they do this through the use of number lines, fraction circles, grouping, repeated addition, or another strategy is up to them. With fraction circles students would be able to work deeply with the concepts behind the multiplication of fractions, and procedurally what it looks like.

PSTs' writings about such ideas gave the authors considerable insight into their thinking.

Instructor Feedback on Phase 1

Feedback was provided early in the planning process to allow the instructor to support PSTs in thinking deeply about the task before moving on to formally writing in a lesson planning template. Much of the feedback from the instructor was related specifically to thinking about mathematical ideas that will focus the whole-class discussion. For example, in their first draft of this section nearly half of PSTs listed general mathematical ideas such as "adding fractions" or content-neutral ideas such as "solving word problems." Through recorded feedback from the instructor and follow up conversations, more task-specific ideas such as "iterating fractions or repeated addition can be modeled with multiplication" were developed. This shift supported PSTs later when imagining what their discussion might look like and trying to make connections among representations. Other common feedback from the instructor was related to the variety of strategies that PSTs anticipated for the task. For those that only anticipated strategies with number sentences or standard algorithms, the instructor encouraged PSTs to consider including strategies and tools explored in class (e.g., fraction circles, number lines, tape diagrams).

Phase 2: Beginning the Formal Lesson Plan

The lesson plan template used for this assignment was broken into two sections to allow for another point of feedback and opportunity for revision (Table 2 in the Appendix). In this phase, students included their revised important mathematical ideas (which could be substituted with objectives, learning goals, etc. to meet the needs of the instructor). Specific planned questions to ask students while engaged with the task, anticipated student responses, and possible challenges were also included in this section.

During this part of the lesson planning process PSTs detailed how they would launch the task. Many PSTs considered ways to connect students to the context of the problem, connect to background knowledge, or provide some type of review. A lesson launch description is provided below as an example of a common way PSTs considered task entry. The task discussed is related to creating multiple servings of a fruit smoothie.

Before I have students work on the problem in their groups, I would ask how many of them drink or like to make smoothies, what their favorite types of smoothies are, and their favorite ingredients to put in smoothies. Asking students these fun questions help them stay engaged with the task and gets them thinking about the task before they start. I would then remind them of important concepts like the steps to adding fractions and how that plays into multiplication and also multiplication of a fraction and a whole number. For example, I would give them an example of adding fractions like ⅓ + ¼ *and go through the steps of finding a common denominator, and asking them what it means to multiply.*

This section of the lesson plan led to numerous meaningful conversations between PSTs and the instructor about cognitive demand, and questions such as, "If you review fraction addition before students start this task, in what ways do you think students are going to approach this?" More than half of the PSTs designed task introductions that explicitly reviewed a mathematical concept or procedure, oftentimes reducing the opportunities for students to make sense of the tasks.

Instructor Feedback on Phase 2

Instructor feedback was provided to encourage PSTs to reflect upon ways to allow students access to the task without providing guidance or suggestions as to *how* to solve the problem. Feedback in this portion of the lesson plan was provided in a variety of formats— informal conversations with PSTs during class, video recordings, and written comments in a rubric. In this phase, PSTs included revisions to the important mathematical ideas, the strategies included in their anticipated student responses, and how they envisioned launching the task. Attention was given to providing specific, actionable feedback that also included reflective questions. An example of recorded instructor feedback illustrates this approach.

I appreciate that you are thinking about eliciting prior knowledge and personal connections to the task. Instead of talking generally about this, what specifically would you ask students? What questions would allow students to engage with the task, but not lead students towards a particular approach to the task?

During the recorded feedback, the instructor was able to reference the PST's written work, navigated to other resources (e.g., the Common Core State Standards for Mathematics document), and showed the rubric with written comments to support PSTs with processing the feedback. Many ideas discussed in the recorded feedback would be challenging to provide at the same level of detail in a written format. Additionally, recording the feedback allowed PSTs to view it repeatedly.

Phase 3: Planning for the Mathematical Discussion

This phase of the assignment built upon the previous ones and required PSTs to revise their writing and thinking from the submission of the first half of their lesson plan. In this final phase, PSTs selected two to four anticipated student solutions and provided a rationale their inclusion in the class discussion and the order in which they would sequence the responses they had selected. PSTs' rationales included ideas such as moving from more visual or concrete strategies to more abstract strategies, comparing the same strategy that led to two different answers, and showing an incorrect response that may be common. Consider the following example of how a PST justified her sequencing decisions. She decided to share the visual fraction circles first, followed by an iterating on a number line strategy. An incorrect use of the standard algorithm was third, followed by a strategy which used the standard algorithm correctly. She justified these decisions:

Another reason I chose to share the incorrect solution third is so that as a class, we are able to analyze the incorrect strategy, and notice areas that we could "fix." We could find similarities and differences between the previous models and make connections between them which would then lead to sharing the correct [use of the standard algorithm] where we would be able to compare and contrast this strategy to the incorrect solution and think about why this strategy works. By comparing it to the other two models, students are really able to see how the standard algorithm makes sense/works because they are able to make the connection between repeated addition and multiplication.

This example illustrates the depth of analysis that is possible when PSTs are given opportunities *over time* to anticipate and plan responses to students' thinking.

Preservice teachers also wrote specific questions to correlate with anticipated student responses, along with examples of how students might respond to those questions. The questions were meant to focus on *making connections* among the chosen strategies as well as to the important mathematical ideas.

It is important to note that, when PSTs cannot implement a lesson plan with students, one of the Five Practices is difficult to include. *Monitoring* is the second step of the Five Practices. Monitoring requires the teacher to notice and pay close attention to the ways in which students are engaging with and solving the task. It involves circulating throughout the classroom as students are working, watching and listening to students' thinking, and keeping track of the variety of ideas in the room. While the instructor modeled monitoring for PSTs throughout the course as they engaged in problem solving, it was not possible for the PSTs to practice with elementary students. While there are creative ways for PSTs to practice (e.g., microteaching), they were not included in the lesson planning experience detailed here. However, monitoring was addressed during debriefs of lessons that PSTs experienced as learners. The instructor also shared their examples of monitoring documents used to record observed student thinking (Figure 8). Certainly, PSTs could prepare monitoring sheets that they *would* use in a classroom setting.

CONCLUSION

The Five Practices for Orchestrating Productive Mathematical Discussions, a framework developed by Stein et al. (2008), has been used widely in mathematics education. These practices support teachers in planning for the use of rich mathematical tasks and facilitating class discussions around carefully selected and sequenced student strategies. The framework was instrumental in the authors' own development as student-centered elementary mathematics teachers; the practices closely align with their development of and reflection upon their own Mathematical Knowledge for Teaching. This framework can also be used to help preservice teachers attend to students' mathematical thinking and develop habits of mind related to teaching student-centered mathematics as they experience mathematical discussions that utilize the Five Practices. These practices provide actionable steps for PSTs and operationalize many ideas of reform-based mathematics.

Supporting preservice teachers to make connections—to important mathematical ideas, to their own experiences, to other coursework—is important. Through this extended lesson planning activity that incorporates the Five Practices, PSTs demonstrated that they could successfully engage in mathematical discussions, plan for their own mathematical discussions, and revise their thinking over time. Facilitating mathematical discussions around students' thinking and strategies is a meaningful practice which the authors argue all mathematics teachers should be able to do because it supports a student-centered mathematics class. Preservice elementary teachers can practice and apply the skills needed to do this throughout their mathematics content and methods coursework with explicit modeling of mathematical discussions, carefully planned lesson planning activities, and detailed feedback from the instructor. Using the Five Practices as a guide provided specific and actionable steps that allowed PSTs to engage with lesson planning centered on students' thinking in a deep way.

The activities described in this chapter were developed in several iterations. Over several quarters, the authors learned from and refined several aspects of the lesson planning assignment. Engaging with rich tasks and class discussion that centered on PSTs' mathematical strategies was already a norm in the course. However, the authors found that PSTs benefitted from more explicit reflection on and analysis of the instructors' actions. Taking more time to debrief from the teacher perspective supported students in making connections to their own lesson planning. Questions such as *Why do you think these strategies were included in the discussion? What was the role of the teacher during this discussion? What important mathematical ideas do you think the teacher was focusing the discussion around?* supported this reflection. When PSTs engaged in mathematical tasks as learners, the instructor was able to model numerous practices that they hoped PSTs would include in their future classrooms. Additionally, the authors found that making their planning practices public (e.g., sharing lesson plans, discussing instructional decisions, and providing their reflection of how the lesson was similar and different to the planned lesson) allowed for rich planning discussions.

Planning for mathematical discussions that center students' thinking requires a variety of skills and knowledge. Structuring the lesson planning assignment in a way that allowed for multiple points of feedback and included an expectation of revision made the assignment more accessible for PSTs. One PST noted, "Video feedback helped me revise my thinking because you asked questions and gave more details than in the rubric." Each the Five Practices requires teachers to draw upon various domains of their MKT, which can naturally create some anxiety for PSTs at a time when they are learning so many new things. The activities described in this chapter that occurred before the formal lesson planning assignment allowed PSTs to strengthen their knowledge and identities as mathematics teachers.

When reflecting on how this work will continue to develop, several opportunities exist. The authors plan to reflect upon and make revisions to the other courses in the math education series. *In what ways can this work be developed in PSTs' first experiences in content and methods courses for elementary teachers? How can the learning from this lesson planning assignment be expanded upon in the following course?* The opportunity to emphasize the importance of and practice anticipating student responses can be developed early in PSTs math education experiences. Watching videos or interacting with students that are solving rich tasks can be used more purposefully to build PSTs knowledge of student strategies. In the course that follows, the authors are planning to incorporate new learning around the practice of *connecting*, which requires interrelated knowledge of content, the CCSS-M, common student strategies, and developmentally appropriate tools. The foundation for this practice has been laid by the Five Practices lesson planning assignment, but more support is needed in future courses.

In addition to ongoing revisions, formal research will be conducted to analyze the effects of this lesson planning experience on PSTs' ability to implement mathematical discussions. The authors want to learn more about how PSTs' thoughts about lesson planning in mathematics may have been shaped by these experiences. In what ways do the Five Practices and the resulting habits of mind persist in new teachers' lesson planning and math instruction when they begin teaching in their own classrooms? Continuing to use and refine this lesson planning experience, the authors will develop a robust collection of tasks for PSTs to use to anchor their lesson planning activities. Currently, the tasks focus on fractions, but the content will be expanded and organized for other MTEs to use in their courses. Collaboration among MTEs and universities can strengthen and expand this important work.

Preservice elementary teachers bring a wide variety of experiences and ideas about what effective mathematics instruction looks like. By carefully selecting learning experiences, teacher educators have the opportunity to model the practices they hope PSTs will implement in their future classrooms. Further, PSTs can then practice applying these strategies as they write lesson plans. Attending to the connections among their own mathematical experiences and the practices of teaching can lead to meaningful outcomes for PSTs. Applying the Five Practices in lesson planning prepares PSTs to develop successful student discussions even when they do not have access to field experiences. By practicing and reflecting on isolated skills first (e.g., anticipating student responses to several tasks over the course of an academic quarter, using a new math tool or strategy), PSTs gain confidence and develop their MKT. Carefully planned activities that build upon each other allow PSTs to experience, practice, and reflect upon the value of planning for and facilitating productive mathematical discussions.

REFERENCES

Adiredja, A. P. (2019). Anti-deficit narratives: Engaging the politics of research on mathematical sense making. *Journal for Research in Mathematics Education*, *50*(4), 401–435. doi:10.5951/jresematheduc.50.4.0401

Ball, D. L., Thames, M. H., & Phelps, G. (2008). Content knowledge for teaching: What makes it special? *Journal of Teacher Education*, *59*(5), 389–407. doi:10.1177/0022487108324554

Banilower, E. R., Smith, P. S., Weiss, I. R., Malzahn, K. A., Campbell, K. M., & Weis, A. M. (2013). *Report of the 2012 National Survey of Science and Mathematics Education*. Horizon Research.

Courtney, S. A., Eliustaoglu, E., & Crawford, A. (2015). Examining the role lesson plans play in mathematics education. In T. G. Bartell, K. N. Bieda, R. T. Putnam, K. Bradfield, & H. Dominguez (Eds.), *Proceedings of the 37th Annual Meeting of the North American Chapter of the International Group for the Psychology of Mathematics Education* (pp. 632-639). East Lansing, MI: Michigan State University.

Dobie, T., & Anderson, E. R. (2021). Noticing and wondering to guide professional conversations. *Mathematics Teacher: Learning and Teaching PK-12, 114*(2), 94-102. doi:10.5951/MTLT.2020.0210

Gutstein, E., Lipman, P., Hernandez, P., & de los Reyes, R. (1997). Culturally relevant teaching in a Mexican American context. *Journal for Research in Mathematics Education*, *28*(6), 709–733. doi:10.2307/749639

Illustrative Mathematics. (n.d.). *Half of a recipe*. http://tasks.illustrativemathematics.org/content-standards/5/NF/B/6/tasks/296

Lee, M. Y. (2019). The development of elementary pre-service teachers' professional noticing of students' thinking through adapted lesson study. *Asia-Pacific Journal of Teacher Education*, *47*(4), 383–398. doi:10.1080/1359866X.2019.1607253

Parrish, S. (2010). *Number talks: Helping children build mental math and computation strategies, grades K-5*. Math Solutions.

Rumack, A. M., & Huinker, D. (2019). Capturing mathematical curiosity with notice and wonder. *Mathematics Teaching in the Middle School*, *24*(7), 394–399. doi:10.5951/mathteacmiddscho.24.7.0394

Schoenfeld, A. S. (1998). Toward a theory of teaching-in-context. *Issues in Education*, *4*(1), 1–95. doi:10.1016/S1080-9724(99)80076-7

Shaughnessy, M., Garcia, N., Selling, S. K., & Ball, D. L. (2016). What knowledge and skill do mathematics teacher educators need and (how) can we support its development? In M. B. Wood, E. E. Turner, M. Civil, & J. A. Eli (Eds.), *Proceedings of the 38th Annual Meeting of the North American Chapter of the International Group for the Psychology of Mathematics Education* (pp. 813-820). Tucson, AZ: The University of Arizona.

Silver, E. A., Shapiro, L. J., & Deutsch, A. (1993). Sense making and the solution of division problems involving remainders: An examination of middle school students' solution processes and their interpretations of solutions. *Journal for Research in Mathematics Education*, *24*(2), 117–135. doi:10.2307/749216

Smith, J. P. (1996). Efficacy and teaching mathematics by telling: A challenge for reform. *Journal for Research in Mathematics Education*, *27*(4), 387–402. doi:10.5951/jresematheduc.27.4.0387

Smith, M. S., & Stein, M. K. (2018). *5 practices for orchestrating productive mathematics discussions* (2nd ed.). National Council of Teachers of Mathematics.

Son, J., & Lee, J. (2016). Pre-service teachers' understanding of fraction multiplication, representational knowledge, and computational skills. *Mathematics Teacher Education and Development*, *18*(2), 5–28.

Stein, M. K., Engle, R. A., Smith, M. S., & Hughes, E. K. (2008). Orchestrating productive mathematical discussions: Five practices for helping teachers move beyond show and tell. *Mathematical Thinking and Learning*, *10*(4), 313–340. doi:10.1080/10986060802229675

Thurtell, E., Forrester, T., & Chinnappan, M. (2019). Building conceptual knowledge of fraction operations among pre-service teachers: Effect of a representation-based approach within a teacher education program. *Mathematics Teacher Education and Development*, *1*, 100–124.

KEY TERMS AND DEFINITIONS

Anticipating: The first step in *Five Practices for Orchestrating Productive Mathematical Discussions* (Smith et al., 2008). Anticipating refers to considering the variety of strategies students will use to solve a rich task, including correct, partially correct, and incorrect solutions.

Connecting: The last step in *Five Practices* (Smith, et al., 2008). Teachers pose questions that will support students in making connections to important mathematical ideas during a class discussion.

Equipartitioning: Splitting a set or region into equal-sized groups or pieces.

Iterating: A fraction concept in which a fractional piece is repeated or copied.

Jigsaw: An instructional strategy that splits up a large text or group of texts among students. Students thoroughly read their assigned text then teach it to other students who have not read that text.

Mathematical Knowledge for Teaching: A framework for the domains of knowledge of mathematics teachers (Ball et al., 2008). It includes three subtypes of content knowledge: common content knowledge, horizon content knowledge, and specialized content knowledge. It also includes three subtypes of pedagogical content knowledge: knowledge of content and students, knowledge of content and teaching, and knowledge of content and curriculum.

Monitoring: The second step in *Five Practices* (Smith, et al., 2008), monitoring is a practice where teachers observe students' thinking and solutions as they are solving rich math tasks.

Selecting: The third step in *Five Practices* (Smith, et al., 2008), selecting refers to intentionally choosing students to share their thinking during a discussion to illustrate an important mathematical idea.

Sequencing: The fourth step in *Five Practices* (Smith, et al., 2008), sequencing is the practice of choosing the order in which selected students' solutions are shared during a class discussion with the purpose of illuminating important mathematical ideas.

APPENDIX

Table 1. Readings for developing knowledge of the Five Practices

Source	Summary	Practices addressed
Smith, M. S., & Stein, M. K. (2018). *5 practices for orchestrating productive mathematics discussions* (2nd ed.). National Council of Teachers of Mathematics.	Originally published in 2011, *5 practices* is a foundational work for the practices discussed in this writing. Chapter 1 gives a general overview of the Five Practices.	All
Ahrendt, S., Monson, D., & Cramer, K. (2021). Promoting discourse: Fractions on number lines. *Mathematics Teacher: Learning and Teaching PK-12, 114*(4), 284-289.	Ideas for using the Five Practices with a lesson on fractions and number lines.	All
Campbell, M. P. (2020). Not just number: Representation talks. *Mathematics Teacher: Learning and Teaching PK-12, 113*(11), 918-924.	Discusses representation talks, a variation of number talks focused on mathematical representations. Topics include anticipating student responses and connecting students' ideas.	Anticipating Connecting
Candela, A. G., Boston, M. D., & Dixon, J. K. (2020). Discourse actions to promote student access. *Mathematics Teacher: Learning and Teaching PK-12, 113*(4), 266-277.	Presents two discourse practices, linking and press, that can be used to improve the quality of mathematical discussions	Connecting
Cengiz, N. (2013). Facilitating productive discussions. *Teaching Children Mathematics, 19*(7), 450-456.	Demonstrates how a disagreement about a task solution was used to encourage productive discourses, as well as selecting work for sharing and helping students build connections across ideas.	Selecting Connecting
Ely, R. E., & Cohen, J. S. (2010). Put the right spin on student work. *Mathematics Teaching in the Middle School, 16*(4), 208-215.	An overview of how the Five Practices was used guide selection and planning to teach a probability task using spinners. Addresses specific student responses and how they were selected and sequenced for sharing during whole class discussion.	Selecting Sequencing
Ghousseini, H., Lord, S., & Cardon, A. (2021). Supporting Mathematics Talk in Kindergarten. *Mathematics Teacher: Learning and Teaching PK-12, 114*(5), 362-368.	While not explicitly aligned to the Five Practices, this article discusses three practices that can support mathematical talk with young children and integrated with connecting.	Connecting
Hallman-Thrasher, A., & Spangler, D. A. (2020). Purposeful questioning with high cognitive-demand tasks. *Mathematics Teacher: Learning and Teaching PK-12, 113*(6), 446-459.	Strategies for planning lessons that incorporate meaningful classroom discussions. Connections are made to all Five Practices.	All
Hicks, T., & Bostic, J. D. (2021). Formative assessment through think alouds. *Mathematics Teacher: Learning and Teaching PK-12, 114*(8), 598-606.	Whole-class think-alouds, similar to the practice of monitoring, can be used as a formative assessment strategy.	Monitoring
Kimani, P. M., Olanoff, D., & Masingila, J. O. (2016). The locker problem: An open and shut case. *Mathematics Teaching in the Middle School, 22*(3), 144-151.	Strategies for posing productive questions (through anticipating student responses) and ideas for building connections among ideas during discourse about the task, using the familiar locker problem.	Anticipating Connecting
Lomax, K., Alfonzo, K., Dietz, S., Kleyman, E., & Kazemi, E. (2017). Trying three-act tasks with primary students. *Teaching Children Mathematics, 24*(2), 112-119.	Three-act tasks can be incorporated into primary classrooms. Provides examples of monitoring and making connections during different phases.	Monitoring Connecting
Meikle, E. M. (2016). Selecting and sequencing students' solution strategies. *Teaching Children Mathematics, 23*(4), 226-234.	Research on PSTs' task solutions with practical suggestions for selecting and sequencing.	Selecting Sequencing

Continued on following page

Table 1. Continued

Source	Summary	Practices addressed
Smith, M. S., Hughes, E. K., Engle, R. A., & Stein, M. K. (2009). Orchestrating discussions. *Mathematics Teaching in the Middle School, 14*(9), 548-556.	Summarizes the Five Practices and gives an example of their use with a "bag of marbles" task.	All
Staples, M., & Colonis, M. M. (2007). Making the most of mathematical discussions. *The Mathematics Teacher, 101*(4), 257-261.	Distinguishes between sharing discussions and collaborative discussions. Proposes ideas to help students build connections across concepts.	Connecting
Thomas, J., Fisher, M. H., Jong, C., Schack, E. O., Krause, L. R., & Kasten, S. (2015). Professional noticing: Learning to teach responsively. *Mathematics Teaching in the Middle School, 21*(4), 238-243.	Discussion of using professional noticing to recognize important features of student thinking. Integrates the Five Practices into mathematical discourse moves.	All

Table 2. Lesson plan template

Phase 2: Selecting and Setting Up a Mathematical Task	
Your name:	
INTENDED GRADE: What grade level will you be using this task with?	
PREVIOUS EXPERIENCE and BACKGROUND KNOWLEDGE What will students have done prior to this task to prepare? What important things will students need to know to successfully complete this task? (Sample ex: Have students worked with number lines before or is today's task the first time they will use them?)	
TASK Include full details of the task. If the task involves pictures, an activity sheet, etc., you may submit them as a separate attachment, but make sure you list them here.	
IMPORTANT MATHEMATICAL IDEAS What are the important *mathematical concepts* students will learn or engage with by performing this task?	
CCSS Standard(s) What Common Core standard(s) will you address with this lesson? Give the standard number **AND** write it out.	**Connection to Standard** How does this task address that standard?
CLASS SETTING Describe the class setting while students work on the task. Will students work alone, in partners, small groups or a combination of these options? Why? What materials and manipulatives will they have access to?	
TASK INTRODUCTION How will you introduce the task? How will you maintain the cognitive demand of the task and avoid suggesting specific solution strategies, but still make sure that students understand what they are being asked to do?	
Phase 2 continued: Anticipating Solutions, Misconceptions, and Planning for Support and Extension	
SOLUTION STRATEGIES Identify the strategies that you expect students will use to solve this task. What strategies do you expect to see? What do you expect to be the most common approach? Here is where your own work solving the task and asking others to solve will come in handy. You should include at least three correct solution strategies and two partially correct or incorrect responses. ***You can include this as a separate attachment if you would like.***	
MISCONCEPTIONS What misconceptions do you believe students might have about this task? (Reference your incorrect or partially correct solutions above)	**ADDRESSING MISCONCEPTIONS** How will you help students to recognize or confront misconceptions without telling them what to do? Provide suggestions for each of the incorrect or partially correct strategies that you listed.
CHALLENGES What challenges do you expect students to face while working on this task? What are the most difficult aspects of the task? Where might students become frustrated or confused?	**SCAFFOLDING** What questions will you ask to help students understand the task, to observe their errors, to combat frustration, and to help students move forward on the task, without lowering the cognitive demand of the task?
EXTENSIONS What will you do for students who quickly complete the task? How will you keep students engaged in the important mathematical ideas of the task?	
Phase 3: Sharing and Discussing the Task	

Continued on following page

Table 2. Continued

<table>
<tr><td>SELECTING STUDENT WORK TO SHARE

What solution strategies will you plan to have students share in a whole class discussion after students have completed the task? (Choose at least 2, no more than 4 – this can include the incorrect or partially correct solutions)

Revisit the strategies you wrote about earlier. Which of those strategies (correct, partially correct, or incorrect) is it essential to include in your discussion for this task?</td><td>RATIONALE for each chosen strategy
● Why did you choose this strategy to share?
● How does this solution strategy expose important mathematical ideas or misconceptions?
● What will the rest of the class learn by seeing these solutions?</td></tr>
<tr><td>SEQUENCING SHARED STUDENT WORK

In what order will you have students share the solution strategies?</td><td>RATIONALE
● Why are you having students share in this order?
● How does sequencing solutions in this order help students better understand the important mathematical ideas in the task?</td></tr>
<tr><td colspan="2">CONNECTING

● How will you connect the shared strategies to tell the mathematical story of the task?
● What specific questions will you ask to help students compare, contrast, and connect solution strategies? (How do you think they will respond to these questions? What do you hope they might say?)
● What evidence will you look for to show that students understand the important mathematics of the task?
How does your planned discussion of student strategies move students towards meeting the mathematical goals of your lesson and connect to the major mathematical concepts?</td></tr>
</table>

Figure 8. Sample monitoring sheet

Monitoring Students' Thinking – Recording Sheet

Monitoring is the second of the *5 Practices for Orchestrating Productive Mathematical Discussions* (Stein et al., 2008). After a teacher anticipates likely student responses, they monitor students in the classroom while they are working on the task. Keeping notes about different student strategies can help a teacher make intentional choices about which students' ideas to share and in what order. Below is an example of a monitoring sheet I used during one of our classes.

> You are making lemonade. You have 2 ½ cups of sugar at home. If a batch of lemonade needs ¾ cup of sugar, how many batches of lemonade can you make? (Please include a number sentence)

Note to self: Look for different answers to the question. 3 batches w/ leftover sugar, 3 ¼ batches, 3 1/3 batches, other

Strategy	**Who and What**	**Order**
Uses an area model (e.g. fraction circles or fraction bars) and creates equal groups of ¾.	MH "2½ ÷ 3/4" made equal groups "3 groups" AB "3/4 + 3/4 + 3/4 + ¼ = 2½" fraction bars MM ← broke down 1/4 in 1/3's	1 (with answer of 3 batches)
Uses a number line to "jump forward" or "jump backward", using ideas of repeated addition or repeated subtraction.	AW partitions bars in 3/4 groups KP "3¼ batches" JC 3⅓ batches BR ST 3/4 + 3/4 + 3/4 = 2¼	2 AW + JC w/ different answers
Creates a division number sentence ($2\frac{1}{2} \div \frac{3}{4} = 3\frac{1}{3}$) and uses an algorithm (e.g. common denominator algorithm, multiply by the reciprocal)	EF Multiply By reciprocal KG CD Common denominator algorith CW	3
Creates a multiplication number sentence to model the problem ($2\frac{1}{2} \times \frac{3}{4} = 1\frac{7}{8}$)	AG "3/4 of 2½ is 1 7/8" YC "multiplying by fraction makes it smaller"	
Uses approximation or counting up (ex: "I can get 1 batch for sure, 2 batches is one and a half cups, so I can definitely get 3 batches"	TQ "I guess 3 and checked"	
Other ideas	RH Cuisenaire Rods "3/4 is like my whole"	

Adapted from Orchestrating Discussions by Smith, Hughes, Engle & Stein, 2009.

Chapter 8
Supporting Pre-Service Teachers' Argumentation-Focused Visions of Mathematics Teaching and Learning in a Teacher Preparation Program

Marta T. Magiera
https://orcid.org/0000-0003-2944-9787
Marquette University, USA

Vecihi S. Zambak
Monmouth University, USA

Hyejin Park
James Madison University, USA

ABSTRACT

This chapter describes the design of integrated learning spaces (ILS) for prospective teachers (PSTs) to support PSTs' learning about reform-based teaching by helping them experience, think about, understand, and position mathematical argumentation in their future classrooms. Drawing on a longitudinal research program that explored PSTs' learning about argumentation in elementary and middle school mathematics, the authors discuss the design of ILS in the context of mathematics and pedagogy with field experience courses in a teacher education program. The authors also share the visions of mathematical argumentation as a teaching-learning practice that PSTs developed in the ILS context. Finally, this chapter poses questions for future research that might explore design principles for creating effective ILS in teacher education programs and examine how PSTs' learning in the ILS context supports their content and pedagogical knowledge development. This chapter also poses questions about ongoing support that future teachers might need to implement and sustain instructional reform.

DOI: 10.4018/978-1-7998-9422-3.ch008

INTRODUCTION

In recent years, the preparation of mathematics teachers has become an increasingly important area of interest to mathematics educators and policymakers in the United States. They agree that reforming teacher preparation in postsecondary institutions is a central factor in enhancing the capacity to improve mathematics education for all students (e.g., the Association of Mathematics Teacher Educators [AMTE], 2017; Conference Board of Mathematical Sciences [CBMS], 2001; National Research Council [NRC], 1996). Teachers play a critical role in implementing ambitious reform-oriented teaching (e.g., Kinser-Traut & Turner, 2020; Schoen et al., 2003). Thus, mathematics teacher educators, researchers, and policymakers recognize that improving mathematics education for all students needs to begin in teacher preparation (e.g., NRC, 2000; Schmidt, Cogan, & Houang, 2011; Schmidt, Blömeke, et al., 2011). To prepare teachers for implementing the reformed vision of mathematics education, they need to experience reform-based learning and teaching in their teacher preparation programs and professional development. To do so, mathematics teacher educators need to identify the approaches and tasks that facilitate prospective teachers' (PSTs') learning, support their content and pedagogical knowledge development, and help them develop productive beliefs about reform-based instruction in mathematics (Chapman, 2012; Philipp, 2007).

Preparing mathematics teachers for implementing reform-based curricula and instructional approaches raises many questions. Central to this chapter are dispositions that PSTs should develop in teacher preparation programs to position themselves for facilitating reform-oriented instruction upon entering the teaching profession. Research shows that teachers' beliefs about mathematics and student learning greatly influence how they implement the curriculum and what opportunities for student learning they provide in their instruction (e.g., Akinsola, 2009; Richardson, 2003). Many PSTs in the United States enter teacher education programs with conceptions of mathematics teaching and learning based on their K-12 experiences with mathematics, often limited to traditional mathematics instruction (e.g., Handal & Herrington, 2003; Maasepp & Bobis, 2015; Valentine & Bolyard, 2019). PSTs with limited experiences in learning mathematics from a reform-based perspective need to experience and develop visions of reform-oriented mathematics classrooms as they transition in teacher education programs from being students to teachers.

In this chapter, we draw on studies conducted as part of a longitudinal research project that explored prospective K–8 teachers' learning about argumentation in elementary and middle school mathematics (Magiera, 2020; Magiera & Zambak, 2020, 2021; Park & Magiera, 2019, 2020; Zambak & Magiera, 2020). We focus our discussion on creating *Integrated Learning Spaces* (ILS) in teacher education programs to illustrate coherent efforts of supporting PSTs in developing visions of mathematics teaching and learning consistent with reform-oriented pedagogies. We discuss PSTs' experiences and the visions of mathematical argumentation as a teaching-learning practice PSTs developed in the context of their mathematics and pedagogy with field experience courses that formed the ILS.

BACKGROUND

Mathematical Argumentation and Reform-Based Mathematics Teaching

Mathematical argumentation is broadly defined as "the whole activity of making claims, challenging them, backing them up by producing reasons, criticizing those reasons, rebutting those criticisms, and so on" (Toulmin et al., 1984, p. 14). In the context of a mathematics classroom, the term *mathematical argument* is interpreted as a line of reasoning that shows or explains why a mathematical result is true and how one arrived at a given conclusion. The mathematical result might be a general statement about some class of mathematical objects, or it might be the solution to a mathematical problem.

Instruction that facilitates student engagement with mathematical argumentation aligns with the tenets of reform-based pedagogies (Board of Studies [BOSNSW], 2002; Common Core State Standards Initiative [CCSSM], 2010; Davis, 2014; Ontario Ministry of Education, 2005a, 2005b, 2007). In reform-oriented mathematics classrooms, the roles of teachers and students are redefined, and students are actively engaged in discussing, making sense of, and reasoning about mathematics concepts (BOSNSW, 2002; National Council of Teachers of Mathematics [NCTM], 2000; NCTM, 2014; Sherin, 2002). Reform-based mathematics teaching is characterized by designing and managing learning environments focused on developing communities of learners and instruction that position students to be active participants in their own learning and help them develop complex cognitive skills (e.g., Makar, 2012; NCTM, 2000; Sherin, 2002; Vi-Nhuan et al., 2006). Through argumentation, students engage in worthwhile and meaningful mathematics learning. They engage in mathematical inquiries while working on cognitively demanding tasks and explore important mathematical relationships and concepts. They detect mathematical regularities, describe them, make conjectures, provide evidence, give reasons to justify their thinking, ask questions, make sense of and represent mathematical ideas, take another's point of view, analyze, revise, and evaluate their own reasoning and the reasoning of others.

Teachers who foster student engagement in argumentation through reform-based pedagogies shift the instructional emphasis from focusing on generating answers to revealing students' intuitive and informal ways of mathematical thinking. They routinely pose questions to engage students in reasoning and pay attention to how their students think about the procedures and concepts being developed. They also use their understanding of student thinking to make instructional decisions. In reform-oriented mathematics classrooms, teachers pay attention to students' emerging ideas and build on them to facilitate students' mathematical understanding. Facilitating the practice of argumentation in mathematics classrooms fosters the development of students' mathematical skills, allows teachers access to and monitor students' thinking, nurtures students' lifelong skills and dispositions (e.g., critical thinking, perseverance, independence), supports diversity and inclusion, and empowers students in their own learning (Krummheuer, 2007; Nussbaum, 2008; Staples et al., 2012; Staples & Newton, 2016; Whitenack & Yackel, 2002; Wood, 1999; Yackel & Cobb, 1996).

Challenges in Implementing Argumentation in Mathematics Classrooms

Research shows that implementing reform-based instruction is challenging for practicing teachers and teacher candidates alike (e.g., Frid, 2000; Frykholm, 1999; Huffman et al., 2008; Jarvis, 2016; Kosko et al., 2014; Lewis, 2014; Park Rogers et al., 2011). Lewis (2014) identified that some challenges include responding to unexpected student ideas, managing classroom discussions, and planning lessons that build on student ideas. Jarvis (2016) reported that some teachers are hesitant to adopt reform-based practices due to the lack of confidence with the mathematics content and their perceived sense of lack of control and efficiency in the classroom. They might view traditional instruction as more "safe" and thus, especially in the early stages of their careers, they appear to be reluctant to use reform-oriented approaches. Other researchers identified lack of adequate resources, concerns about appropriate instructional pacing, or insufficient training and teacher professional development as factors that adversely impact the implementation of reform-oriented mathematics instruction (e.g., Anderson & Helms, 2001; Frykholm, 1999; Huffman et al., 2008; Jarvis, 2016; Jong, 2016; King, 2015; Washor & Mojkowski, 2006).

Many teachers find it challenging to make mathematical argumentation a regular aspect of their practice. The instructional focus on argumentation constitutes a significant shift away from traditional instructional models in *what* and *how* mathematics is taught (National Mathematics Advisory Panel, 2008). Researchers show that teachers' prior knowledge of argumentation and their beliefs about teaching argumentation impact how teachers facilitate argumentation in their mathematics classrooms (Kosko et al., 2014; Lin, 2018; Staples & Newton, 2016). Lin (2018) documented that teachers need strong pedagogical knowledge of argumentation to effectively facilitate argumentation in their work with students. Kosko et al. (2014) reported that teachers have difficulty planning how to engage students in argumentation without having a clear vision of scaffolding and facilitating mathematical argumentation. For example, they might not know what specific questions to ask or how to sequence them to engage students in argumentation. It is often difficult for teachers to envision how argumentation can support student learning of concepts. Many teachers also do not routinely recognize classroom situations for engaging students in argumentation and thus miss opportunities to capitalize on unexpected student solutions to facilitate argumentation in their mathematics classrooms. Some teachers believe that argumentation is not accessible to all students. They might struggle to understand the role of argumentation in mathematics teaching and learning (Staples & Newton, 2016). Past studies revealed that teachers hold a broad range of views on classroom tasks that engage students in mathematical argumentation and that how they think about mathematical argumentation might vary depending on the grade level they teach (Aylon & Hershkowitz, 2018; Rogers & Kosko, 2019). Researchers also bring attention to the lack of adequate curricular resources that support argumentation in mathematics classrooms (e.g., Bieda, 2010; Bieda et al., 2014; Dolev & Even; 2015; Stacey & Vincent, 2009).

Numerous researchers highlight the connections between teachers' beliefs about student learning, teachers' beliefs about mathematics teaching, and teachers' instructional practice while addressing the challenges related to implementing reform-based instruction and fostering mathematical argumentation (e.g., Handal & Herrington, 2003; Lewis, 2014; Park Rogers et al., 2011; Sawyer, 2018; Taylor, 2002). Handal and Herrington (2003), for example, noted that limited classroom implementation of reform-based pedagogies often stems from the lack of alignment between teachers' beliefs about mathematics teaching and learning, and the actual tenets of reform-based pedagogies. Park Rogers et al. (2011) observed that teachers who believe that reform-based instruction does not meet students' social and academic needs minimize students' agency in the learning process. In Park Rogers et al.'s study, many teachers believed that students must master basic concepts and skills through drill-type practice prior to implementing reform-oriented pedagogies.

Personal experiences with traditional instructional approaches frequently shape how teachers think about mathematics teaching and student learning (e.g., Banilower et al., 2006; Drake & Sherin 2006; Raymond, 1997; Weiss & Pasley, 2004). Drake and Sherin (2006) documented that elementary school teachers frequently articulate their visions of mathematics instruction drawing on their own early mathematics experiences, their experiences as adult mathematics learners, and their personal contexts and professional identities. Past research shows that many prospective teachers come to teacher preparation programs with already established visions of mathematics teaching and visions of teachers' and students' roles in the mathematics classroom. They often develop these visions at early ages, drawing on their own experiences as mathematics learners (Raymond, 1997). Even though the student-centered, reform-oriented view of mathematics teaching and learning is at the center of many teacher education programs, prospective teachers often consider reform-based instruction as an idealistic classroom practice (Barak & Shakhman, 2008). Barak and Shakhman (2008) pointed out that preparing future teachers for implementing educational change requires highly specific and systematic approaches in teacher education that positively influence prospective teachers' orientations towards reform-based instruction.

Mathematics teacher educators agree that teachers need a rich understanding of mathematics to provide high-quality reform-based instruction (e.g., Ball et al., 2008; Ma, 2020; Manouchehri & Goodman, 2000; NRC, 1996). Manouchehri and Goodman (2000) conducted classroom observations of teachers implementing reform-based mathematics curricula. They reported that teachers' own content knowledge played a significant role in how they dealt with student thinking during their mathematics lessons and how they interacted with curricular materials. Teachers with stronger mathematics content knowledge were more likely to respond to student ideas appropriately and adapt cognitively guided instructional approaches in their teaching. Conversely, teachers with weaker content knowledge emphasized procedural skills and provided limited opportunities for students to generate and reason about mathematical ideas.

To develop the knowledge and dispositions needed for designing transformational experiences for students consistent with reform-based pedagogies, PSTs need a rich understanding of content and how their content knowledge applies to their teaching (Ebby, 2000; Neubrand et al., 2009). Neubrand et al. (2009) stated that teacher education programs need to help teacher candidates develop a sound basis of content knowledge for teaching with strong ties to issues of practice. They also recognized challenges in accomplishing this goal of mathematics teacher preparation given the complexity of mathematics knowledge required for teaching, its interconnectedness, and the fact that teachers develop their knowledge for teaching within and across multiple learning spaces in their teacher education programs. Ebby (2000) argued that in teacher education programs, careful attention should be paid to helping PSTs connect the knowledge of mathematics they develop in the mathematics content courses and pedagogical content knowledge they develop in the methods courses and field experiences.

CREATING CONSISTENT ARGUMENTATION-FOCUSED INTEGRATED LEARNING SPACES WITHIN TEACHER EDUCATION PROGRAMS

One way of helping PSTs connect mathematics content and pedagogy across different learning contexts (e.g., mathematics courses, pedagogy courses, field experiences) and foster their understanding of the reformed vision of mathematics teaching and learning is to create Integrated Learning Spaces (ILS) focused on mathematical argumentation in teacher education programs. We use an example from the Learning about Mathematical Argumentation and Proving (L-MAP) research project (Magiera & Zambak, 2020; Magiera, 2020; Park & Magiera, 2020; Zambak & Magiera, 2020) to illustrate argumentation-focused ILS designed for PSTs preparing for grades 1–8 teaching license. We define an Integrated Learning Space in a teacher education program in terms of the curricular experiences that facilitate making significant connections to increase PSTs' understanding and thinking about learning and teaching school mathematics.

In the L-MAP project, the purpose of the ILS was to provide unified learning environments for PSTs in which they could experience mathematical argumentation as both learners and future teachers. The design of the ILS was informed by the framework of *opportunity to learn to teach* described by Schmidt, Blömeke, et al. (2011). Schmidt and colleagues coined the term *opportunity to learn to teach* to denote the content and experiences with which future teachers are provided in teacher preparation programs. The term *opportunity to learn to teach* embraces the knowledge and dispositions individual teachers develop within their teacher education programs. These include the knowledge of mathematics content, knowledge of student mathematical thinking, the ability to make productive use of student thinking, and content-specific instructional strategies that PSTs learn as they prepare to enact ambitious instruction in their future practice.

All courses that formed the ILS within the L-MAP project provided PSTs with an opportunity to learn mathematics by experiencing the practice of inquiry and sense-making through mathematical argumentation. In all courses within the ILS, PSTs were also engaged in reflective activities designed around their argumentation experiences as learners and future teachers to help them connect their learning across the learning spaces. Reflection is an essential element of teacher preparation because it helps consolidate teachers' knowledge (Chamberlin, 2009, 2013; Chapman, 2015; Garcia et al., 2007; Shulman, 1987). Shulman (1987) advocated for teacher education programs that support PSTs in connecting theory and practice through reflective opportunities. The educational literature documents that reflective practice supports teachers' development of more sophisticated conceptual structures and enhances their learning (e.g., Driscoll, 2015; Wheatley, 1992). Teachers who reflect on their knowledge and experiences are more effective in their instructional practice (Santagata & Angelici, 2010; Star & Strickland, 2008) and have the potential to sustain professional growth after leaving teacher preparation and professional development programs (Garcia et al., 2007).

The ILS consisted of a sequence of three mathematics content courses for prospective teachers preparing for grades 1–8 teaching license (taught in the Mathematics Department) and two mathematics methods with field experience courses (taught in the College of Education; Figure 1). The curricula of the concurrent mathematics content and methods courses were purposefully coordinated to maximize PSTs' opportunity to experience argumentation in mathematics teaching and learning.

Figure 1. Structure of the integrated learning space in the L-MAP project

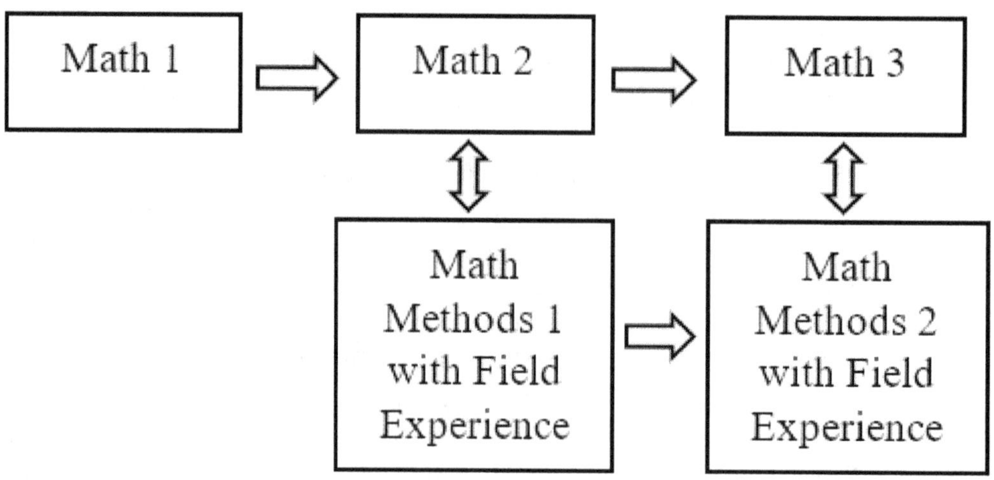

Each mathematics content course (Math 1—Problem Solving and Reasoning for Teachers, Math 2—Number Systems and Operations for Teachers, and Math 3—Algebra and Geometry for Teachers) provided space for prospective teachers to experience argumentation as learners of mathematics. Each pedagogy with field experience course (Math Methods 1—Teaching Elementary School Mathematics, and Math Methods 2—Teaching Middle School Mathematics) was designed to help prospective teachers explore argumentation as a pedagogical practice that supports the development of students' conceptual understanding and mathematical reasoning skills. Included in Table 1 is a summary of opportunities to learn about mathematical argumentation across all the courses that formed the ILS.

Table 1. PSTs' opportunities to learn about mathematical argumentation in the ILS

Course	Argumentation
Math 1. Problem Solving and Reasoning for Teachers	Prospective teachers: • were introduced to Toulmin's framework of argumentation to provide them with language for discussing mathematical arguments • generated, individually or collectively, and discussed different forms of mathematical arguments and ways of reasoning (e.g., inductive, deductive, by analyzing cases, by contradiction) • analyzed, individually or collectively, arguments generated in class and sample arguments presented by the course instructor. The focus of the analysis was on argument quality (i.e., chain of reasoning, validity of claims, clarity, precision sufficiency of supporting evidence, quality of justifications)
Math 2. Number Systems and Operations for Teachers	Prospective teachers: • constructed, individually or collectively, mathematical arguments about numbers and operations with a specific focus on the quality of included evidence and reasons • analyzed, individually or collectively, arguments generated in class and sample arguments presented by the course instructor. The focus of the analysis was on argument quality (i.e., chain of reasoning, validity of claims, clarity, precision, sufficiency of supporting evidence, quality of justifications, ways in which elementary school students might present mathematical arguments) • analyzed and discussed, individually or collectively, elementary students' arguments and student mathematical thinking about numbers and operations. Student arguments were presented in the form of student written work or video records of students sharing their arguments and reasoning
Math Methods 1. Teaching Elementary School Mathematics (with field experience)	Prospective teachers: • discussed ambitious teaching strategies that facilitate student engagement in argumentation (e.g., use of manipulative materials, selection of mathematical tasks, strategies for facilitating class discussions, questioning techniques, ways in which elementary school students might present mathematical arguments) • conducted observations of elementary mathematics classrooms with a focus on analyzing teacher and students' actions related to argumentation • posed and analyzed problems for their potential to engage elementary students in generating arguments about numbers and operations • designed and conducted problem-based interviews with elementary school students to facilitate student engagement in mathematical argumentation • analyzed student-generated mathematical arguments in their field classrooms • analyzed and reflected on their practice of engaging students in mathematical argumentation in the context of the interviews they conducted • reflected on their learning and experiences with mathematical argumentation

Continued on following page

Table 1. Continued

Course	Argumentation
Math 3. Algebra and Geometry for Teachers	Prospective teachers: • generated, individually or collectively, inductive arguments • generated, individually or collectively, deductive arguments • analyzed, individually or collectively, arguments generated in class and sample arguments presented by the course instructor. The focus of the analysis was on argument quality (i.e., chain of reasoning, validity of claims, clarity, precision, sufficiency of supporting evidence, quality of justifications) • analyzed and discussed middle school students' arguments and student thinking in the context of algebra and geometry. The focus of the analysis was on the argument quality
Math Methods 2. Teaching Middle School Mathematics (with field experience)	Prospective teachers: • discussed ambitious teaching strategies that facilitate middle school students' engagement in argumentation (e.g., characteristics of rich mathematical tasks, strategies for facilitating class discussions, questioning techniques, ways in which middle grades students might present mathematical arguments) • conducted observations of middle grades mathematics classrooms with a focus on analyzing teacher and students' actions related to argumentation • planned and conducted problem-based interviews with middle school students to engage students in argumentation • analyzed their practice of engaging students in mathematical argumentation and student-generated mathematical arguments • reflected on their learning and experiences with mathematical argumentation

The first mathematics content course, Problem Solving and Reasoning for Teachers, was designed to help PSTs develop flexibility in using a broad range of problem-solving strategies and understand the different structures and forms of mathematical arguments. The course activities were also planned to provide PSTs with language for discussing mathematical arguments. As one of the initial class activities, PSTs were introduced to Toulmin's (1958) framework for argumentation and engaged in analyzing a sample of mathematical arguments to identify claims, supporting evidence, and reasons. The instructor and PSTs negotiated social and sociomathematical norms (Cobb & Yackel, 1996; Lampert, 1990; Lakatos, 1976) and expectations about explaining and justifying problem solutions in class. The norms included acceptable ways of establishing the validity of generated claims, such as the sufficiency of supporting evidence or the clarity of reasoning. Throughout the semester, in their oral and written work, PSTs were expected to clearly articulate their reasoning by stating any assumptions, giving full explanations, and making logical inferences to justify their problem solutions. The established norms and expectations provided a framework for PSTs as they constructed, analyzed, interpreted, or evaluated the quality of mathematical arguments. Course activities engaged PSTs in generating arguments about complex, nonroutine problems. Class activities and discussions provided PSTs with an opportunity to develop an understanding of the process of constructing valid arguments and increase their awareness of the quality of evidence and reasons generated in support of mathematical claims.

In a typical class, in small groups, PSTs solved various problems for which they generated explanatory arguments. They also evaluated and critiqued mathematical arguments shared during class discussions and analyzed the instructor-provided samples of written arguments of elementary and middle school students. The problems addressed a wide range of mathematical topics providing PSTs with opportunities to explore various strategies, ways of mathematical reasoning, and ways in which K-8 students can present their arguments. Course activities also focused on helping PSTs understand classroom situations that could engage students in reasoning and generating mathematical arguments.

The second mathematics content course, Number Systems and Operations for Teachers, further built on these foundations by engaging PSTs as mathematics learners in reasoning about mathematics concepts and ideas fundamental to the elementary school mathematics curriculum (e.g., place value, operations with whole numbers and fractions, computational algorithms, and rational numbers). During a typical class session, PSTs solved and discussed problems in small groups and shared and critiqued the arguments presented in class. They were challenged to make sense of different computational algorithms and procedures, including nonstandard procedures, and test their understanding by developing mathematical arguments to validate, or refute, these procedures. They were encouraged to express their reasoning in formal and informal ways. They were also given different mathematical statements and asked to pose problems and reason about the given statements and the problem situations these statements represented. PSTs were frequently engaged in analyzing arguments and mathematical thinking of elementary school students. They discussed various ways of student reasoning, the merits of student solutions, and assertions. Systematically throughout the semester, PSTs were also asked to reflect on their knowledge development in the area of mathematical argumentation.

As shown in Figure 1, concurrently with their second mathematics content course, PSTs were enrolled in their first mathematics methods course with field experience, Teaching Elementary School Mathematics. Within the ILS, this methods course was designed to help PSTs think about mathematical argumentation as teachers. Assignments for the course were purposefully created to engage PSTs in discussions about curriculum materials and teacher actions that facilitate student engagement in argumentation. The assignments were scaffolded to provide numerous opportunities for prospective teachers to practice reform-oriented pedagogical skills. PSTs planned lessons and classroom activities, analyzed mathematical problems for their potential to engage students in argumentation, analyzed how different types of questions teachers ask can support (or hinder) students' reasoning skills, and anticipated ways to engage students in argumentation through mathematical discussions. PSTs also analyzed sample lessons that provided models for fostering mathematical argumentation in elementary school classrooms. In their field placements, PSTs conducted observations of elementary grade mathematics classrooms focusing on strategies classroom teachers use to engage students in mathematical argumentation and reasoning. Several times throughout the semester, PSTs were asked to respond to journal prompts that elicited their thinking about their ongoing learning experiences with argumentation.

Many of the activities in the methods course were coordinated with PSTs' learning in the mathematics content course. For example, when in the mathematics content course PSTs discussed the ideas related to place value, in their pedagogy course, they explored strategies and manipulative materials that a classroom teacher could use to help students make sense of place value and numbers. A joint project in the methods course and the concurrent mathematics content course asked PSTs to pose problems for elementary school students to engage students in argumentation, plan how to facilitate students' engagement in argumentation working with a student or a small group of students, and implement their problems with students in the field placement classroom. PSTs then analyzed the potential of the problems they posed and their practice of engaging students in argumentation and reflected on their experiences with mathematical argumentation within and across different contexts: their mathematics content courses, their methods course, and field experience.

The third mathematics course, Algebra and Geometry for Teachers, was also designed with a focus on inquiry-oriented pedagogies and argumentation to help PSTs explore foundational to middle grades mathematics concepts in algebra and geometry. The overarching goal of the course was to help prospective teachers experience argumentation as learners of mathematics, heighten their awareness of middle grades students' reasoning in algebraic and geometric contexts, and engage them in inductive and deductive reasoning. In addition to generating arguments for the algebraic and geometric problems they solved, prospective teachers frequently analyzed mathematical arguments (either those generated in the class or presented by the course instructor). Argument analysis assignments were scaffolded to draw prospective teachers' attention to the chain of reasoning; the validity of claims, clarity, and precision; the sufficiency of supporting evidence; and the validity of justifications. Like in the previously described mathematics content courses, PSTs were given different mathematical statements, asked to pose problems and reason about the given statements and the problem situations these statements represented, and reflect on the development of their knowledge in the area of mathematical argumentation.

Concurrently with their third mathematics course, prospective teachers were enrolled in their second mathematics methods course with field experience, Teaching Middle School Mathematics. This methods course was also designed to position prospective teachers in thinking about mathematical argumentation as teachers and further explore mathematics pedagogy and curricular materials that support middle school students' engagement in argumentation. The structure of the course was similar to that of the first mathematics methods course with a field experience component we described earlier. Class assignments provided further opportunities for PSTs to practice reform-oriented pedagogical skills and deepen their understanding of argumentation in the context of teaching middle grades mathematics. In addition to course activities and reflecting on their ongoing experiences with argumentation, PSTs conducted systematic observations of middle grades mathematics classrooms in their field placements. Their structured observations focused on strategies their mentor teachers use to engage students in mathematical argumentation and reasoning. A culminating project in the second methods course (a joint assignment with the concurrent mathematics content course) asked PSTs to design, implement in their field placement, and analyze their practice of engaging a student or small group of students in mathematical argumentation. PSTs analyzed what opportunities the problems they selected or modified provided to engage students in argumentation, the effectiveness of their questions, and student-generated arguments.

PSTs' Visions of Learning and Teaching Mathematics through Argumentation

The L-MAP project documented that ILS designed with a consistent focus on argumentation helped PSTs develop a wide range of positive views and dispositions about argumentation as a teaching/learning practice. To examine the PSTs' visions of argumentation in mathematics teaching and learning, we analyzed 52 PSTs' responses to journal prompts completed as part of their methods course journals and their end-of-semester papers in which they reflected on their learning about mathematical argumentation in the context of their coursework. The open-ended journal prompts were designed to elicit PSTs' reflections on their experiences with argumentation in the context of the ILS and their thinking about the work of a teacher who engages students in argumentation. The prompts, which we present below, were posed several times throughout the semester to encourage PSTs to build on their ongoing learning experiences and the development of their thinking about mathematical argumentation in mathematics teaching and learning.

1. Thinking about yourself as an elementary school teacher, define the term *mathematical argumentation*. How would you explain its meaning to a parent, for example?

2. Describe the practices that characterize an elementary/middle school mathematics classroom in which a teacher engages students in mathematical argumentation. What practices could a visitor (e.g., a parent) see observing that teacher? How do these practices support students' argumentation skills?
3. Describe the characteristics of mathematical tasks that have a high potential to engage students in mathematical argumentation. How do the tasks you described differ from those that do not encourage mathematical argumentation?
4. Describe how your experiences this semester influenced your ideas about teaching elementary/middle school mathematics with a focus on mathematical argumentation.
5. Drawing on your interview experiences, reflect on and compare your views about mathematical argumentation in teaching elementary/middle school mathematics.

PSTs' responses documented their strong awareness that teaching K–8 mathematics with an emphasis on mathematical argumentation and reasoning supports students' mathematical understanding. When reflecting on their experiences with argumentation across the learning spaces, PSTs recognized the positive value of instructional approaches that emphasize students' explanations and justifications. They shared observations and visions of classroom teachers who encouraged students to share their thinking and problem-solving strategies and engaged students in discussions about what strategies do or do not work in a given problem situation. One of the PSTs articulated that

The key aspect and benefit of mathematical argumentation is that students are encouraged to not just immediately accept an argument but to engage in the dialogue of reasoning and explanation. This method is a healthy way for students to understand the content and to further their understanding of the complexities of different approaches.

PSTs also frequently reflected on their own experiences with argumentation as students in their mathematics content courses while sharing their thinking about how engaging students in explaining and justifying contributes to developing students' mathematical understanding and reasoning skills. They saw argumentation as a tool for helping students gain confidence in their mathematical ability and build students' communication skills (Park & Magiera, 2019).

Many of the PSTs frequently linked their positive visions of mathematical argumentation as a teaching/learning practice to their own learning experiences with argumentation within the ILS. They shared how the emphasis on argumentation supported their mathematical content knowledge development. They also contrasted their prior learning experiences with "traditional" mathematics instruction with their learning experiences with mathematical argumentation in the ILS. One of the PSTs shared in her reflection:

My experiences have really reinforced for me that mathematical argumentation needs to be a part of elementary mathematics. While I am in the mathematics class, there are many justifications that I never knew because I was just taught arithmetic and never why it worked. From my mathematics class, I realized that I really need to foster argumentation because I did not know many of the things taught. I can see the major benefits of teaching kids to think deeper about math at an earlier age. If we, the teachers, help students learn math through this approach, it won't seem like such a mystery, and they will want to learn more about it.

In addition to recognizing that the instructional focus on argumentation positively impacts student learning, PSTs developed positive visions of how the instructional focus on argumentation benefits teachers' everyday practice and instructional planning. One prospective teacher shared,

My experiences this semester in class and at my field placement helped me realize the importance of mathematical argumentation and that it is key to knowing that you are teaching your students well and that they are grasping the concepts.... This course [pedagogy with field experience course] and my other class [mathematics content course] made me realize how very important justification and mathematical argumentation are, not only for yourself as a teacher but for the student as well. Both teachers and students benefit from justification and argumentation because it allows the students to dive deeper into a concept and understand it to the full extent. And, it allows the teacher to really know if how they are presenting material to students and how they are teaching is effective. This gives a teacher the chance to change the way they do certain things or how they introduce concepts to students to make sure they are doing it in the most effective way possible.

Another commented,

I've realized the importance of mathematical argumentation. Having a student talk out a problem and explain their reasoning and thinking behind solving it in a certain way is incredibly important for teachers because it shows us if a student is really getting a good grasp on the concepts discussed in class and reassures us [teachers] that our methods of teaching are working. Also, mathematical argumentation can clue teachers into the way that their students process concepts.

By coordinating the focus of instructional activities on argumentation across the learning spaces within the ILS, PSTs had the opportunity to experience the reform-based practices modeled in their university courses while concurrently observing classroom teachers enacting these practices in their field placement classrooms. One PST commented on their classroom observations and teacher actions for facilitating argumentation:

The teacher was continually prompting the students with, "How do you know?" I witness my cooperating teacher engaging in this sort of questioning continually with her students. This prompted students to explain and justify their answers rather than just regurgitate the correct answer.... Another characteristic was the challenge that the teacher often presented. Instead of having students give and justify their answer, the teacher tweaked an aspect of the problem and prompted students with the idea of, "What if we had this instead?" This forced students to reflect on their understanding of the problem and persevere in solving it. I believe that tweaking the problem challenges students to think on the spot, deeply and more abstractly about what makes something ALWAYS true and what makes something sometimes true.

PSTs frequently recognized the need to model good mathematical arguments and model the practice of argumentation in work with students. They articulated that teachers who model argumentation and launch classroom discussions about why a given justification works or does not work help students develop an understanding of what it means to argue and justify. Argumentation, said one of the PSTs, *also requires that the teacher sets an example of how they expect the students to explain and justify their work. A teacher should be explaining her work thoroughly, step by step, and using evidence just how their students would be expected to explain.*

PSTs also had the opportunity to reflect on curricular materials that can contribute to shifting students' mathematical activity from producing answers to engaging in deeper thinking and reasoning. They observed the nature of mathematical tasks that support reform-oriented teaching and student engagement in argumentation, describing tasks that have multiple entry points and are open to multiple solution strategies and interpretations, or tasks that are complex as the ones that contribute to productive student learning. One PST commented,

I think that a range of starting points is important because it allows the students to pick which starting point makes the most sense for them. Once they pick that starting point, they will and should be able to explain why they picked that point and why they thought it was the most beneficial place to start. If there are multiple strategies that the students can use to reach the solution, it allows them to reason through multiple ways of interpreting a problem, and the students must create a strong argument as to why they think their path and solution is the correct one.

The opportunity to observe reform-based student-centered mathematics teaching with a focus on argumentation helped many of the PSTs in the L-MAP project develop visions of classroom culture and students' actions in argumentation. In their journals, PSTs described that reform-based instructional pedagogies cultivate a sense of student responsibility in the classroom. They made observations about student actions, noticing students constructing arguments either individually or collectively, analyzing and critiquing their reasoning and reasoning of others (including the teacher), asking questions, or explaining and justifying their work to themselves and others. PSTs were thinking about how reform-based teaching and an emphasis on mathematical argumentation impact students' cognitive and affective traits: listening skills, the ability to be open to new ideas, the ability to be critical and reflective, a willingness to help others to understand mathematical concepts and problems, attentiveness, the ability to recognize mistakes or flaws in reasoning, confidence in one's reasoning skills, and a sense of responsibility for one's learning. Drawing on her field placement observations, one of the PSTs commented,

The students know that when they do not understand, they need to ask questions in order to gain a better understanding. For example, they may whisper to a friend or ask questions of the teacher. These students have a conditioned sense of responsibility for their learning and understanding. Responsibility fosters mathematical argumentation because it encourages students to ask "why" in order to grasp concepts that may be applied later in mathematical practices. It also compels students to independently reason and justify ideas to their peers. The students in my [field placement] class will often ask if they can work on their math sheets during free time, and the students are always excited to tell the teacher what the answer is and even more conditioned to explain how they got it. This sense of student independence and responsibility fosters mathematical argumentation and is an important environmental characteristic for the classroom.

Another PST articulated her learning about students while reflecting on their observations of students engaged with argumentation:

Students in an elementary classroom should be cognizant of other viable arguments that are being presented and critique one another on the strategies that are presented. ... They are able to express reasons and justifications for their decisions. They are able to help one another to recognize mistakes in reasoning when they are working together. They are able to self-correct and understand their mistakes. Then, they are able to explain the fix to their reasoning accurately.

FUTURE RESEARCH DIRECTIONS

Our work provides one example of an ILS designed to support PSTs' learning about mathematical argumentation. Future research into designing effective ILSs in teacher education programs is warranted. Given the limited research on the design of ILSs, future research needs to explore further the design principles for creating effective ILSs in teacher education programs.

While examining the effectiveness of the ILS described in this chapter, we focused on visions about argumentation in mathematics learning and teaching that our PSTs developed in the context of their learning experiences within the ILS. Future research needs to explore how teacher learning in the ILS context supports their content and pedagogical knowledge development. Future research is also needed to examine how PSTs transfer their learning within content-specific and reform-oriented ILSs to their own classroom practice upon completing their teacher education training.

CONCLUSION

This chapter illustrates how teacher educators can give conscious and overt attention to a range of activities that support reform-based teaching by helping PSTs experience, think about, understand, and position reformed-based instruction in their future classrooms. Raymond (1997) observed that helping PSTs develop beliefs and visions of mathematics learning and teaching consistent with curricular recommendations and a reformed vision of mathematics teaching constitutes a challenge for teacher educators. The professional preparation of mathematics teachers is influenced by many stakeholders. They include mathematics and mathematics education faculty, education faculty, supervisors in clinical experiences, and cooperating teachers in clinical settings. This chapter illustrates how the different components of mathematics teacher preparation can be conceptualized as the Integrated Learning Space to ensure a common voice across the different components and stakeholders in mathematics teacher preparation. We document how the ILS can be conceptualized and designed to provide PSTs with cohesive experiences on which they can build productive reform-oriented mathematics teaching and learning models.

Researchers have shown that helping teachers, including PSTs, implement and sustain student-centered instruction requires approaches that help teachers identify and understand the underlying philosophy of reform-based pedagogies (Loucks-Horsley et al., 2009; National Science Foundation [NSF], 2002). The L-MAP project illustrates that teacher education programs with carefully designed ILSs can help PSTs develop powerful personal theories about reform-based instruction. The ILS described in this chapter was intentionally designed to provide PSTs with consistent experiences focused on mathematical argumentation to help them think about argumentation from the perspective of learners and teachers. The analysis of PSTs' reflections on their learning and experiences shows that an argumentation-focused ILS helped PSTs develop productive visions of mathematics teaching and learning as they examined the interplay between teaching and learning elementary and middle school mathematics with an emphasis on the practice of argumentation. PSTs articulated the tenets of student-centered pedagogies: attending to student learning and dispositions, teacher actions, classroom norms, and student expectations for argumentation. Their personal theories about the nature and purpose of argumentation in the mathematics classroom were consistent with descriptions of purposes of argumentation in the mathematics classroom community reported in the existing literature (e.g., Staples et al., 2012; Staples & Newton, 2016).

Our work provides one example of creating interconnectedness across multiple learning spaces in teacher education programs. As the mathematics education community continues to advocate the use of reform-based pedagogies and emphasize mathematical argumentation, further professional development for teachers remains a central component of reforms in mathematics teaching. The positive personal theories about reform-based instruction that PSTs can develop in carefully designed teacher preparation programs provide only starting points from which PSTs can build their independent teaching practice upon completing their preparation programs. The research literature shows that teachers' beliefs and practices are mediated by the school contexts in which they are working (Skott, 2001; Sztajn, 2003). Researchers note that the productive beliefs about mathematics teaching and learning that PSTs develop during their teacher education are not necessarily stable. The research literature suggests that novice teachers often fall back on the beliefs and teaching models they experienced as K–12 learners (Banilower et al., 2006; Raymond, 1997; Weiss & Pasley, 2004). The critical question is, what professional development opportunities and further support do novice teachers need during their induction years to help them consistently position reform-based practices in their classrooms and school contexts? What further support would PSTs need upon completing their training, and how should it be structured to help PSTs maintain the positive dispositions and beliefs about mathematics teaching and learning they develop in teacher education programs?

Chapman (2012) brought attention to the sustainability of teacher professional development programs that focused on changing teacher beliefs and practices towards reform-oriented mathematics pedagogies after the initial learning experiences in teacher education. She described several challenging circumstances that might impact how teachers, including novice teachers, implement reform-based instruction (e.g., reform-based instruction is not implemented at all grade levels or by all teachers at a given school). The questions for future research and design of supportive learning environments in teacher education programs are, what opportunities can teacher educators provide, and how can they help novice teachers construct useful knowledge that transforms and sustains their practice?

ACKNOWLEDGMENT

This research was supported by the National Science Foundation's grant for the project "CAREER: L-MAP: Pre-service Middle School Teachers' Knowledge of Mathematical Argumentation and Proving," Grant No. DRL-1350802. Opinions, findings, conclusions, or recommendations are those of the authors and do not necessarily reflect the views of the National Science Foundation.

REFERENCES

Akinsola, M. K. (2009). Comparison of prospective and practicing teachers' mathematics efficacy beliefs regarding mathematics teaching. In J. Maass & W. Schloglmann (Eds.), *Beliefs and attitudes in mathematics education* (pp. 119–130). Sense Publishers.

Anderson, R. D., & Helms, J. V. (2001). The ideal of standards and the reality of schools: Needed research. *Journal of Research in Science Teaching*, *38*(1), 1–16. doi:10.1002/1098-2736(200101)38:1<3::AID-TEA2>3.0.CO;2-V

Association of Mathematics Teacher Educators (AMTE). (2017). *Standards for preparing teachers of mathematics*. amte.net/standards

Aylon, M., & Hershkowitz, R. (2018). Mathematics teachers' attention to potential classroom situations of argumentation. *The Journal of Mathematical Behavior*, *49*, 163–173. doi:10.1016/j.jmathb.2017.11.010

Ball, D. L., Thames, M. H., & Phelps, G. (2008). Content knowledge for teaching: What makes it special? *Journal of Teacher Education*, *59*(5), 389–407. doi:10.1177/0022487108324554

Banilower, E. R., Boyd, S. E., Pasley, J. D., & Weiss, I. R. (2006). *Lessons from a decade of mathematics and science reform: A capstone report for the local systemic change through teacher enhancement initiative*. Horizon Research Inc.

Banilower, E. R., Smith, P. S., Malzahn, K. A., Plumley, C. L., Gordon, E. M., & Hayes, M. L. (2018). *Report of the 2018 NSSME*. Horizon Research, Inc.

Barak, M., & Shakhman, L. (2008). Reform-based science teaching: Teachers' instructional practices and conceptions. *Eurasia Journal of Mathematics, Science and Technology Education*, *4*(1), 11–20. doi:10.12973/ejmste/75301

Bieda, K. (2010). Enacting proof-related tasks in middle school mathematics: Challenges and opportunities. *Journal for Research in Mathematics Education*, *41*(4), 351–382. doi:10.5951/jresematheduc.41.4.0351

Bieda, K., Ji, X., Drwencke, J., & Pickard, A. (2014). Reasoning-and-proving opportunities in elementary mathematics textbooks. *International Journal of Educational Research*, *64*, 71–80. doi:10.1016/j.ijer.2013.06.005

Board of Studies NSW (BOSNSW). (2002). K-6 *Mathematics Syllabus*. Sydney: BOSNSW.

Chamberlin, M. (2009). Teachers' reflections on their mathematical learning experiences in a professional development course. *Mathematics Teacher Education and Development*, *11*, 22–35.

Chamberlin, M. (2013). Prospective teachers' perspectives on mathematics teaching and learning: Lens for interpreting experiences in a standards-based mathematics course. *School Science and Mathematics*, *113*(8), 369–379. doi:10.1111sm.12042

Chapman, O. (2012). Challenges in mathematics teacher education. *Journal of Mathematics Teacher Education*, *15*(4), 263–270. doi:10.100710857-012-9223-2

Chapman, O. (2015). Reflective awareness in mathematics teachers' learning and teaching. *Eurasia Journal of Mathematics, Science and Technology Education, 11*(2), 313–324. doi:10.12973/eurasia.2015.1334a

Cobb, P., & Yackel, E. (1996). Constructivist, emergent, and sociocultural perspectives in the context of developmental research. *Educational Psychologist, 31*(3-4), 175–190. doi:10.1080/00461520.1996.9653265

Common Core State Standards Initiative. (2010). *Common Core State Standards for Mathematics.* Council of Chief State School Officers & National Governors Association Center for Best Practices.

Conference Board of the Mathematical Sciences (CBMS). (2001). *The mathematical education of teachers.* Washington, DC: Mathematical Association of America (MAA) in cooperation with American Mathematical Society (AMS).

Davis, J. (2014). Reasoning and proving within Ireland's reform-oriented national syllabi. *The Mathematics Enthusiast, 11*(3), 665–706. doi:10.54870/1551-3440.1321

Dolev, S., & Even, R. (2015). Justifications and explanations in Israeli 7th -grade mathematics textbooks. *International Journal of Science and Mathematics Education, 13*(S2), 309–327. doi:10.100710763-013-9488-7

Drake, C., & Sherin, M. G. (2006). Practicing change: Curriculum adaptation and teacher narrative in the context of mathematics education reform. *Curriculum Inquiry, 36*(2), 153–187. doi:10.1111/j.1467-873X.2006.00351.x

Driscoll, K. (2015). Improving the effectiveness of mathematics teaching through active reflection. In M. Marshman, V. Geiger, & A. Bennison (Eds.), *Mathematics education in the margins. Proceedings of the 38th annual conference of the Mathematics Education Research Group of Australasia* (pp. 205-212). Sunshine Coast: MERGA.

Ebby, C. B. (2000). Learning to teach mathematics differently: The interaction between coursework and fieldwork for preservice teachers. *Journal of Mathematics Teacher Education, 3*(1), 69–97. doi:10.1023/A:1009969527157

Frid, S. (2000). Constructivism and reflective practice in practice: Challenges and dilemmas of a mathematics teacher educator. *Mathematics Teacher Education and Development, 2*, 17–33.

Frykholm, J. A. (1999). The impact of reform: Challenges for mathematics teacher preparation. *Journal of Mathematics Teacher Education, 2*(1), 79–105. doi:10.1023/A:1009904604728

Garcia, M., Sanchez, V., & Escudero, I. (2007). Learning through reflection in mathematics teacher education. *Educational Studies in Mathematics, 64*(1), 1–17. doi:10.100710649-006-9021-9

Handal, B., & Herrington, A. (2003). Mathematics teachers' beliefs and curriculum reform. *Mathematics Education Research Journal, 15*(1), 59–69. doi:10.1007/BF03217369

Huffman, D., Thomas, K., & Lawrenz, F. (2008). Science and mathematics instruction in a reform-based teacher preparation program. *School Science and Mathematics, 108*(4), 137–145. doi:10.1111/j.1949-8594.2008.tb17820.x

Jarvis, D. (2016). Messy but meaningful: Exploring the transition to reform-oriented pedagogy with teachers of mathematics and coordinators in Ontario, Canada. *Teacher Development, 20*(1), 18–39. doi:10.1080/13664530.2015.1108928

Jong, C. (2016). Linking reform-oriented experiences to teacher identity: The case of an elementary mathematics teacher. *The Journal of Educational Research, 109*(3), 296–310. doi:10.1080/00220671.2014.947398

King, G. A. (2015). *A multiple case study examining secondary mathematics teachers' perceptions of reform-oriented instruction and obstacles to instructional change* (Publication No. 3719749) [Doctoral dissertation, Middle Tennessee State University]. ProQuest Dissertations Publishing.

Kinser-Traut, J. Y., & Turner, E. E. (2020). Shared authority in the mathematics classroom: Successes and challenges throughout one teacher's trajectory implementing ambitious practices. *Journal of Mathematics Teacher Education, 23*(1), 5–34. doi:10.100710857-018-9410-x

Kosko, K., Rougee, A., & Herbst, P. (2014). What actions do teachers envision when asked to facilitate mathematical argumentation in the classroom? *Mathematics Education Research Journal, 26*(3), 459–476. doi:10.100713394-013-0116-1

Krummheuer, G. (2007). Argumentation and participation in the primary mathematics classroom: Two episodes and related theoretical abductions. *The Journal of Mathematical Behavior, 26*(1), 60–82. doi:10.1016/j.jmathb.2007.02.001

Lakatos, I. (1976). A renaissance of empiricism in the recent philosophy of mathematics. *The British Journal for the Philosophy of Science, 27*(3), 201–223. doi:10.1093/bjps/27.3.201

Lampert, M. (1990). When the problem is not the question and the solution is not the answer: Mathematical knowing and teaching. *American Educational Research Journal, 27*(1), 29–63. doi:10.3102/00028312027001029

Lewis, G. M. (2014). Implementing a reform-oriented pedagogy: Challenges for novice secondary mathematics teachers. *Mathematics Education Research Journal, 26*(2), 399–419. doi:10.100713394-013-0092-5

Lin, P. (2018). Improving knowledge for teaching mathematical argumentation in primary classrooms. *Journal of Mathematics Education, 11*(1), 17–30. doi:10.26711/007577152790018

Loucks-Horsley, S., Love, N., Stiles, K. E., Mundry, S., & Hewson, P. W. (2009). *Designing professional development for teachers of science and mathematics* (3rd ed.). Corwin Press.

Ma, L. (2020). *Knowing and teaching elementary mathematics. Teacher understanding of fundamental mathematics in China and the United States*. Taylor & Francis: Routledge. doi:10.4324/9781003009443

Maasepp, B., & Bobis, J. (2015). Prospective primary teachers' beliefs about mathematics. *Mathematics Teacher Education and Development, 16*(2), 89–107.

Magiera, M. T. (2020). Prospective K-8 teachers' problem posing: Interpretations of tasks that promote mathematical argumentation. In A.I. Sacristán, J. C. Cortés-Zavala & P. M. Ruiz-Arias, (Eds.), *Mathematics education across cultures: Proceedings of the 42nd meeting of the North American Chapter of the International Group for the Psychology of Mathematics Education* (pp. 940-948). Cinvestav / AMIUTEM / PME-NA. 10.51272/pmena.42.2020-143

Magiera, M. T., & Zambak, V. S. (2020). Exploring prospective teachers' ability to generate and analyze evidence-based explanatory arguments. *International Journal of Research in Education and Science, 6*(2), 327–346. doi:10.46328/ijres.v6i2.765

Magiera, M. T., & Zambak, V. S. (2021). Prospective K-8 teachers' noticing of student justifications and generalizations in the context of analyzing written artifacts and video-records. *International Journal of STEM Education, 8*(1), 1–21. doi:10.118640594-020-00263-y

Makar, K. (2012). The pedagogy of mathematical inquiry. In R. Gillies (Ed.), *Pedagogy: New developments in the learning sciences* (pp. 371–397). Nova Science.

Manouchehri, A., & Goodman, T. (2000). Implementing mathematics reform: The challenge within. *Educational Studies in Mathematics, 42*(1), 1–34. doi:10.1023/A:1004011522216

National Council of Teachers of Mathematics. (2000). *Principles and standards for school mathematics*. Author.

National Council of Teachers of Mathematics. (2014). *Principles to Actions: Ensuring Mathematical Success for All*. Author.

National Mathematics Advisory Panel. (2008). *National Mathematics Advisory Panel Foundations for success: The final report of the National Mathematics Advisory Panel*. https://www.ed.gov/about/bdscomm/list/mathpanel/report/final-report.pdf

National Research Council. (1996). *The preparation of mathematics teachers: Considerations and challenges*. National Academy Press.

National Research Council. (2000). *Inquiry and the national science education standards*. National Academy Press.

National Science Foundation. (2002). Professional development that supports school mathematics reform. In *Foundations: A monograph for professionals in science, mathematics, and technology education* (Vol. 3, pp. 1-130). Arlington, VA: Author. Retrieved July 24, 2021, from https://www.nsf. gov/pubs/2002/nsf02084/start.htm

Neubrand, M., Seago, N., Agudelo-Valderrama, C., DeBlois, L., Leikin, R., & Wood, T. (2009) The balance of teacher knowledge: Mathematics and pedagogy. In RThe professional education and development of teachers of mathematics. New ICMI study series (Vol. 11, pp. 211-225). Springer. doi:10.1007/978-0-387-09601-8_21

Nussbaum, E. M. (2008). Collaborative discourse, argumentation, and learning: Preface and literature review. *Contemporary Educational Psychology, 33*(3), 345–359. doi:10.1016/j.cedpsych.2008.06.001

Ontario Ministry of Education. (2005a). *The Ontario Curriculum, Grades 1–8: Mathematics, Revised.* Queen's Printer for Ontario.

Ontario Ministry of Education. (2005b). *The Ontario Curriculum, Grades 9 and 10: Mathematics, Revised.* Queen's Printer for Ontario.

Ontario Ministry of Education. (2007). *The Ontario Curriculum, Grades 11 & 12: Mathematics, Revised.* Queen's Printer for Ontario.

Park, H., & Magiera, M. T. (2019). Pre-service teachers' conceptions of mathematical argumentation. In *Proceedings of the forty-first annual meeting of the North American Chapter of the International Group for the Psychology of Mathematics Education* (pp. 1264-1268). St Louis, MO: University of Missouri.

Park, H., & Magiera, M. T. (2020). Pre-service teachers' interpretations of mathematical reasoning. In *Mathematics Education Across Cultures: Proceedings of the 42nd Annual Meeting of the North American Chapter of the International Group for the Psychology of Mathematics Education* (pp. 1677-1681). Cinvestav/AMIUTEM/PME-NA. 10.51272/pmena.42.2020-265

Park Rogers, M., Cross, D. I., Sommerfield Gresalfi, M., Trauth-Nare, A., & Buck, G. (2011). First year implementation of a project-based learning approach: The need of addressing teachers orientations in the era of reform. *International Journal of Science and Mathematics Education, 9*(4), 893–917. doi:10.100710763-010-9248-x

Philipp, R. A. (2007). Mathematics teachers' beliefs and affect. In F. Lester (Ed.), *Second handbook of research on mathematics teaching and learning* (Vol. 1, pp. 257–315). National Council of Teachers of Mathematics and Information Age Publishing.

Raymond, A. M. (1997). Inconsistency between a beginning elementary school teacher's mathematics beliefs and teaching practice. *Journal for Research in Mathematics Education, 28*(5), 550–576. doi:10.2307/749691

Richardson, V. (2003). Preservice teachers' beliefs. In J. Raths (Ed.), *Teacher beliefs and classroom performance: The impact of teacher education* (pp. 1–22). Information Age Publishing.

Rogers, K. C., & Kosko, K. W. (2019). How elementary and collegiate instructors envision tasks as supportive of mathematical argumentation: A comparison of instructors' task constructions. *The Journal of Mathematical Behavior, 53*, 228–241. doi:10.1016/j.jmathb.2018.08.004

Santagata, R., & Angelici, G. (2010). Studying the impact of the lesson analysis framework on preservice teachers' abilities to reflect on videos of classroom teaching. *Journal of Teacher Education, 61*(4), 339–349. doi:10.1177/0022487110369555

Sawyer, A. (2018). Factors influencing elementary mathematics teachers' beliefs in reform-based teaching. *The Mathematics Educator, 26*(2), 26–53.

Schmidt, W. H., Blömeke, S., Tatto, M. T., Hsieh, F. J., Cogan, L., Houang, R. T., & Schwille, J. (2011). *Teacher education matters: A study of middle school mathematics teacher preparation in six countries.* Teachers College Press, Columbia University.

Schmidt, W. H., Cogan, L., & Houang, R. (2011). The role of opportunity to learn in teacher preparation: An international context. *Journal of Teacher Education, 62*(2), 138–153. doi:10.1177/0022487110391987

Schoen, H. L., Cebulla, K. J., Finn, K. F., & Fi, C. (2003). Teacher variables that relate to student achievement when using a standards-based curriculum. *Journal for Research in Mathematics Education, 34*(3), 228–259. doi:10.2307/30034779

Sherin, M. G. (2002). When teaching becomes learning. *Cognition and Instruction, 20*(2), 119–150. doi:10.1207/S1532690XCI2002_1

Shulman, L. (1987). Knowledge and teaching: Foundations of the new reform. *Harvard Educational Review, 57*(1), 1–23. doi:10.17763/haer.57.1.j463w79r56455411

Skott, J. (2001). The emerging practices of a novice teacher: The roles of his school mathematics images. *Journal of Mathematics Teacher Education, 4*(1), 3–28. doi:10.1023/A:1009978831627

Stacey, K., & Vincent, J. (2009). Modes of reasoning in explanations in Australian eight-grade mathematics textbooks. *Educational Studies in Mathematics, 72*(3), 271–288. doi:10.100710649-009-9193-1

Staples, M., & Newton, J. (2016). Teachers' contextualization of argumentation in the mathematics classroom. *Theory into Practice, 55*(4), 294–301. doi:10.1080/00405841.2016.1208070

Staples, M. E., Bartlo, J., & Thanheiser, E. (2012). Justification as a teaching and learning practice: Its (potential) multifaceted role in middle grades mathematics classrooms. *The Journal of Mathematical Behavior, 31*(4), 447–462. doi:10.1016/j.jmathb.2012.07.001

Star, J. R., & Strickland, S. K. (2008). Learning to observe: Using video to improve preservice mathematics teachers' ability to notice. *Journal of Mathematics Teacher Education, 11*(2), 107–125. doi:10.100710857-007-9063-7

Sztajn, P. (2003). Adapting reform ideas in different mathematics classrooms: Beliefs beyond mathematics. *Journal of Mathematics Teacher Education, 6*(1), 53–75. doi:10.1023/A:1022171531285

Taylor, M. P. (2002). Implementing the standards: Keys to establishing positive professional inertia in preservice mathematics teachers. *School Science and Mathematics, 102*(3), 137–142. doi:10.1111/j.1949-8594.2002.tb17907.x

Toulmin, S. (1958). *The uses of argument*. Cambridge University Press.

Toulmin, S., Rieke, R., & Janik, A. (1984). *An introduction to reasoning* (2nd ed.). Macmillan.

Valentine, K. D., & Bolyard, J. (2019). Lived moments of shifts in prospective elementary teachers' mathematical learning. *Journal for Research in Mathematics Education, 50*(4), 436–463. doi:10.5951/jresematheduc.50.4.0436

Vi-Nhuan, L., Stecher, B., Lockwood, J. R., Hamilton, L. S., Robyn, A., Williams, V., Ryan, G., Kerr, K. A., Mártinez, J., & Klein, S. (2006). *Improving mathematics and science education: A longitudinal investigation of the relationship between reform-oriented instruction and student achievement*. RAND.

Washor, E., & Mojkowski, C. (2006). Creating new steps: Innovating from the edge to the middle. *Phi Delta Kappan, 87*(10), 735–739. doi:10.1177/003172170608701006

Weiss, I. R., & Pasley, J. D. (2004). What is high-quality instruction? *Educational Leadership, 61*(5), 24.

Wheatley, G. H. (1992). The role of reflection in mathematics learning. *Educational Studies in Mathematics, 23*(5), 529–541. doi:10.1007/BF00571471

Whitenack, J., & Yackel, E. (2002). Making mathematical arguments in the primary grades: The importance of explaining and justifying ideas. *Teaching Children Mathematics, 8*(9), 524–528. doi:10.5951/TCM.8.9.0524

Wood, T. (1999). Creating a context for argument in mathematics class. *Journal for Research in Mathematics Education, 30*(2), 171–191. doi:10.2307/749609

Yackel, E., & Cobb, P. (1996). Sociomathematical norms, argumentation, and autonomy in mathematics. *Journal for Research in Mathematics Education, 27*(4), 458–477. doi:10.5951/jresematheduc.27.4.0458

Zambak, V. S., & Magiera, M. T. (2020). Supporting grades 1–8 pre-service teachers' argumentation skills: Constructing mathematical arguments in situations that facilitate analyzing cases. *International Journal of Mathematical Education in Science and Technology, 51*(8), 1196–1223. doi:10.1080/0020739X.2020.1762938

ADDITIONAL READING

Battista, M. T. (1994). Teacher beliefs and the reform movement in mathematics education. *Phi Delta Kappan, 75*(6), 462–470.

Conner, A., Edenfield, K. W., Gleason, B., & Ersoz, F. (2011). Impact of a content and methods course sequence on prospective secondary mathematics teachers' beliefs. *Journal of Mathematics Teacher Education, 14*(6), 483–504. doi:10.100710857-011-9186-8

Ensor, P. (2002). From preservice mathematics teacher education to beginning teaching: A study in recontextualizing. *Journal for Research in Mathematics Education, 32*(3), 296–320. doi:10.2307/749829

Graham, M., & Lesseig, K. (2018). Back-pocket strategies for argumentation. *Mathematics Teacher, 112*(3), 173–178. doi:10.5951/mathteacher.112.3.0172

Knudsen, J., Lara-Meloy, T., Stallworth, S., & Rutstein, D. W. (2014). Advice for mathematical argumentation. *Mathematics Teaching in the Middle School, 19*(8), 494–500. doi:10.5951/mathteacmiddscho.19.8.0494

Krummheuer, G. (2000). Mathematics learning in narrative classroom cultures: Studies of argumentation in primary mathematics education. *For the Learning of Mathematics, 20*(1), 22–32.

Melhuish, K., Thanheiser, E., Fasteen, J., & Fredericks, J. (2015). Teacher noticing of justification: Attending to the complexity of mathematical content and practice. In T. Bartell, K. Bieda, R. T. Putnam, K. Bradfield, & H. Dominquez (Eds.), *Proceedings of the 37th annual meeting of the North American Chapter of the International Group for the Psychology of Mathematics Education* (pp. 748-755). East Lansing, MI: Michigan State University.

Melhuish, K., Thanheiser, E., & Guyot, L. (2020). Elementary school teachers' noticing of essential mathematical reasoning forms: Justification and generalization. *Journal of Mathematics Teacher Education, 23*(1), 35–67. doi:10.100710857-018-9408-4

KEY TERMS AND DEFINITIONS

Integrated Learning Space in a Teacher Education Program: Curricular experiences that facilitate making significant connections to increase PSTs' understanding and their thinking about learning and teaching school mathematics.

Mathematical Argument: A line of reasoning that shows or explains why a mathematical result is true and shows how one arrived at a given conclusion.

Mathematical Argumentation: A sociomathematical practice that gives foundation to much of mathematics and mathematical development. Using definitions and previously established results to develop conjectures, explore, verify the truth, or refute.

Reform-Based Instruction: Instruction that positions students to be active participants in their learning and helps them develop complex cognitive skills by shifting focus from generating answers to revealing students' intuitive and informal ways of mathematical reasoning.

Visions of Mathematics Teaching and Learning: One's views and beliefs that provide frameworks through which an individual interprets mathematics teaching and learning.

Chapter 9
"Who Else Would I Even Talk To?": Supporting Prospective Teachers' Engagement With Student Thinking Through a Community of Practice

Katherine Baker
Elon University, USA

ABSTRACT

Engagement with student thinking involves eliciting and using evidence of student thinking in instruction. This case study explored how three prospective elementary teachers engaged with student thinking in mathematics and used this thinking to guide instruction during their student teaching experiences. The prospective teachers were situated at the same school and same grade level and were supported through team meetings facilitated by a mathematics teacher educator. The prospective teachers then voluntarily sustained a community of practice in which they unpacked the pedagogical and content needs of instructional units focused on fractions. Consideration is given to implementation of similar support structures within teacher preparation programs in order to encourage teachers' engagement with student thinking, especially in mathematics.

INTRODUCTION

Engaging with student thinking is a central mathematics teaching practice in the current vision of reform-based mathematics instruction (Huinker & Bill, 2017; NCTM, 2014). Engagement addresses how teachers elicit and use evidence of student thinking in instruction, and, specifically to this research, engagement means how prospective teachers (PTs) elicit and use student thinking in the context of their student teaching experience. When and how to elicit student thinking and how to integrate what is learned into instruction is demanding work. How to prepare PTs to enter their careers ready to do this work warrants exploration so that teacher educators can sustain awareness and thoughtfulness around support

DOI: 10.4018/978-1-7998-9422-3.ch009

approaches. Research indicates that the student teaching internship placement and experience impacts instructional beliefs and actions of PTs as they transition into future teaching (Borko & Mayfield, 1995; Borman et al., 2009; Slick, 1998) and into their mathematics teaching (Bieda et al., 2015; Philipp et al., 2007). The current study was grounded in the premise that what unfolds during the student teaching internship influences future practice. Specifically, this study unpacked how to support PTs' engagement with student thinking about fractions in order to equip PTs to continue engaging with student thinking beyond the student teaching experience. A case study was structured around the following question: in what ways does a Community of Practice (CoP) facilitate PTs' engagement with student thinking in the area of fractions?

This case study was designed to investigate PTs receiving contextualized support in learning how to engage with student thinking. The author reports on three PTs in the United States who were placed at the same elementary school and same 4th grade level during their student teaching experience. The PTs were required to plan and implement a six-week unit on fractions and were supported in their fraction instruction by a CoP (Wenger, 1998). Scheduled team meetings for the CoP occurred in which a mathematics teacher educator joined the PTs to share content insights and facilitate examination of student work. The PTs' lessons were then observed and debriefed to examine and support their engagement with student thinking. This study also followed the PTs as they moved from the scheduled team meetings facilitated by the mathematics teacher educator to their own meetings driven by their needs and goals. Findings are organized by themes of influences that the PTs expressed facilitated their engagement with student thinking. These findings provide insight into how teacher preparation programs might support PTs in the ambitious practice of engagement with student thinking, and how mathematics teacher educators could support engagement with student thinking in mathematics, especially in the area of fractions.

BACKGROUND

The National Council of Teachers of Mathematics (NCTM) focused attention on the characterization of effective mathematics instruction and highlighted what students need to know and do mathematically with a commitment to challenging and supporting students to learn mathematics conceptually (NCTM 1989, 1991, 1995, 2000). NCTM's (2014) *Principles to Actions: Ensuring Mathematical Success for All* defined eight effective mathematics teaching practices that align with the reform ideals. These eight practices provide the implementation roadmap for the organization's vision of effective mathematics teaching. This study moved from those eight practices, and specifically focused on the practice of *elicit and use evidence of student thinking* (Table 1). In implementing the mathematics teaching practice *elicit and use evidence of student thinking,* teachers need to know how to elicit student thinking, interpret student understandings and misunderstandings, and respond by adjusting instruction to support and extend learning (Chamberlin, 2005; Jacobs & Empson, 2015; Jacobs et al., 2019; NCTM, 2014). In preparing a PT to engage with student thinking in planning and instruction, attention to this teaching practice needs to be highlighted and thoughtful supports must be in place to help the PT learn to do the work of attending to, interpreting, and using student thinking regularly.

Table 1. Principles to actions mathematics teaching practice

Elicit and Use Evidence of Student Thinking
1. Identify what counts as evidence of student progress toward mathematics learning goals. 2. Elicit and gather evidence of student understanding at strategic points during instruction. 3. Interpret student thinking to assess mathematical understanding, reasoning, and methods. 4. Make in-the-moment decisions on how to respond to students with questions and prompts that probe, scaffold, and extend. 5. Reflect on evidence of student learning to inform the planning of next instructional steps.
Note. Adapted from *Principles to Actions: Ensuring Mathematical Success for All* (NCTM, 2014, p. 56)

In order to facilitate the PTs' learning and their development of eliciting and using evidence of student thinking, the study was heavily grounded in the work of the Cognitively Guided Instruction (Carpenter et al., 2014). In the 1980s, research and professional development around students' problem-solving strategies with whole number operations emerged as an important focus in Cognitively Guided Instruction (Carpenter, 1979; Carpenter et al., 1989; Carpenter & Moser, 1984). Cognitively Guided Instruction is grounded in how teachers use knowledge about children's thinking and problem solving to make decisions as they plan and implement instruction (Carpenter & Fennema, 1992). Cognitively Guided Instruction purports that listening to and attending to students' thinking should be at the forefront of instructional decision-making in order to support children's mathematical development (Carpenter et al., 2014). Teacher participants in Cognitively Guided Instruction professional development learn the research base around students' mathematical problem-solving strategies of various word problem structures, and then learn how to use this knowledge to make instructional decisions.

Empson and Levi's (2011) extended whole number Cognitively Guided Instruction strategies into fractional numbers. Through their book, *Extending Children Mathematics: Fractions and Decimals*, Empson and Levi (2011) affirmed past research that school-aged children can naturally partition objects and think about fraction concepts in the context of sharing items, e.g. *Four students share three candy bars. How much candy bar do they each get if they share equally?* (Charles & Nason, 2000; Confrey et al., 2009; Empson, 1995, 1999; Empson and Levi, 2011; Lamon, 1996; Pothier & Swada, 1983; Steffe, 2001; Streefland, 1993). Sharing equally, or equipartitioning, is the act of dividing into equal-sized groups or parts. Studies on equipartitioning focused on children's intuitive strategies and provided educators and researchers with sets of typical student understandings (Lamon, 2007). Acknowledging and building upon children's intuitive strategies about fractions supports their comprehension and conceptual understanding (Siegler et al., 2010). Additionally, providing children with conceptual and contextually-driven experiences with fractions, like in Cognitively Guided Instruction or the concrete-representational-abstract method to instruction (Flores et al., 2020) may increase students' understanding and performance in this content area of mathematics.

Empson and Levi (2011) presented partitioning through the context of equal sharing word problems and provided a construct of typical student strategies that result from solving equal sharing problems. Although the authors did not explicitly refer to their set of typical student partitioning strategies as a trajectory, they indicated that the evolution of these strategies follows a predictable pattern with children. The student partitioning strategies Empson and Levi (2011, p. 25) present within their equal sharing construct are: Non-Anticipatory Sharing (learner does not think in advance about how to coordinate sharing of objects and attempts partitions through trial-and-error or by ease of partition, like halves), Additive Coordination (learner either partitions and shares one object at a time or shares groups of items at a time), Ratio (learner applies knowledge of repeated halving or multiplication factors to create a simpler problem), and Multiplicative Coordination (learner understands that *a* objects shared by *b* people is a/b).

Empson and Levi's 2011 book became central to the work of the PTs because of its detailed instructional support for implementing fraction partitioning problems with children. Empson and Levi (2011) emphasized that the teacher plays a critical role in connecting and extending children's thinking through the number choices and discussions used in the equal sharing problems. The fact that children use different strategies at different times punctuated the need for PTs of this study to understand the typical strategies in order to look for and respond to them accordingly in the classroom environment. The PTs' understanding of a framework of typical partitioning strategies and how strategies look when produced by students could then support their anticipation and attention to students' strategies in real-time. A honed focus on one content area and within one mathematics domain, such as partitioning in fractions, could allow PTs to learn how to be more intentional in their noticing of strategies and their response to students, instead of having to split focus across multiple subjects and topics.

STUDY FRAMEWORK: COMMUNITY OF PRACTICE

A CoP is a group of people who share common concerns or passions for something they do, and learn how to do this thing better by interacting (Wenger, 1998; Wenger et al., 2002b). CoP as a term and construct stems from the work of situated learning theory (Lave, 1991; Lave & Wenger, 1991) in which learning is a situated activity and learners engage in sociocultural participation to gain knowledge. The collective learning that occurs within a CoP results in the development of shared practices that reflect the dynamics, pursuits, and social relations of the group (Wenger, 1998; Wenger et al., 2002a). While communities of practice include a network of people, not all networks of people constitute a CoP; to be a CoP there must exist "mutually negotiated competence around a domain of practice" (Farnsworth et al., 2016, p. 143).

Utilization of a CoP (Lave & Wenger, 1991; Wenger, 1998) structure in this study was as an innovation to the traditional PT preparation structure in which a clinical teacher and university supervisor serve as the assumed support personnel during the student teaching experience. A clinical teacher, university supervisor, and PT still interacted in a traditional triad during this study, addressing university expectations of observations and evaluation documentation, but additionally, a group of PTs engaged in a CoP that included facilitated team meetings and support related to engaging with student thinking as the focus of instruction. The intention was that this CoP structure would aid PTs to extend and reify the work of previous university coursework within their student teaching practices, and shape how they would then carry the work into future settings. This study furthers insight from Kirkby and colleagues (2019) around how CoPs might help shape teacher identity, and how they might support educator collaboration around innovative or new pedagogical practices (Aldana et al., 2018; Kelley et al., 2020).

The study investigated the CoP as a vehicle to help prepare PTs for the work of reform-based mathematics instruction, with specific attention to eliciting and using students' thinking in their instructional units on fractions. It was hypothesized that the PTs working together in the team meetings facilitated by a mathematics teacher educator would develop their craft, specifically with the focus on engaging with student thinking around fractions. The CoP lens and structure meant that the PTs could develop ownership of and goals for the study's team meetings to best support their efforts to problematize and address reformed-based mathematics pedagogy.

Structure of the Study's CoP Meetings

Throughout the study duration, three PTs participated in mathematics teacher educator-facilitated team meetings in which they were supported in their efforts to utilize student thinking in their instructional planning and instructional implementation. The community of PTs participated in these team meetings eight times over the span of a 12-week research project. Each meeting was scheduled for 60 minutes, but some lasted up to 180 minutes. Although the PTs participated in other faculty and grade-level meetings at their school site, the CoP team meetings were comprised solely of the PTs and the teacher educator. At these meetings, the mathematics teacher educator, who was also the researcher, acted as participant-researcher and facilitated professional learning and discussions around fraction instructional strategies. Appendix 1 provides the CoP's team meeting plans and shows how they were structured to prioritize engagement with student thinking and navigate cognitively demanding resources for instruction.

The team meetings also served to reorient the PTs to key pedagogical resources from their previous coursework and to support their prioritization of attending to student thinking when planning for instruction. These resources included a proposed learning pathway for 4th grade students that focused on developing conceptual understanding of fractions and featured the use of cognitively demanding tasks (pathway overview provided in Appendix 2). In addition, the team meetings reintroduced and reinforced the PTs' previous coursework initiative of facilitating productive classroom discussions from Chapin and colleagues (Chapin et al., 2003, 2013) and their talk tools for inviting students' thinking into the classroom space in real-time instruction.

Once the PTs were situated in their clinical teachers' classrooms and responsible for lead instruction, they collected student work to bring to team meetings. Their student work, along with their self-identified needs, then acted as the catalyst for the PTs' conversations and decision-making and took precedence over the researcher's suggested topics for team meetings. Empson and Levi's (2011) construct regarding the types of strategies children use to solve equal sharing problems was the basis for the PTs' initial analyses of their students' fraction understanding during the CoP team meetings.

CASE STUDY DESIGN AND METHODS

Case study methodology (Stake, 1995, 2005) was used to explore how the three PTs engaged with student thinking in mathematics and how they used this thinking to guide instruction during six weeks of their student teaching experience. Specifically, an instrumental case study design was used (Stake, 1995, 2005) in that the understanding of the case led to understanding the greater issue of PTs' engagement with student thinking and led to characterizing their engagement for the purposes of finetuning support for their engagement and reforming future teacher preparation around this needed support. The study addressed the following question: *In what ways does a CoP facilitate PTs' engagement with student thinking in the area of fractions?*

Participants and Context

Because of the use of a CoP, the study was bounded by three participants who were PTs placed at the same elementary school site and same grade level (4th grade) for their student teaching experience. They were enrolled in a two-year teacher preparation program embedded within their university's undergraduate experience. Kristi, Liz, and Nora (pseudonyms) self-identified as female and white or female and Latina (Nora). All three were seniors enrolled in an Elementary Education Program & pursuing K-6 Licensure. Although they were in a group placement at the 4th grade level in one elementary school, each was in a different classroom with a different clinical teacher at that grade level. The study followed their 12-week student teaching experience, and specifically their 6-week period of mathematics instruction in which they planned and executed a unit on fractions.

The placement school was Brookside Elementary (pseudonym), a Title I Elementary School in the United States serving 611 Prek-5th grade students with 64.2% of the student population considered economically disadvantaged. The overall student composition at Brookside Elementary was as follows (all demographic language reflects that of the public report): 46% Hispanic, 41.2% White, and 8.5% African American. Additionally, 23.5% of the students were classified as Limited English Proficiency and 13.3% of the school population was classified as Exceptional Children. The schoolwide composition represented the PTs' classroom compositions.

For three years, Brookside was a partner school with the local university and hosted six or more PTs during the student teaching experience. Each year approximately three PTs were placed in the K-2nd grade band, and three were placed in the 3rd-5th grade band. During this research, the three PTs in the 3rd-5th grade band were all placed in 4th grade for their student teaching experiences, thus facilitating the studying of their collaborative instructional design and implementation.

Because of its university partner school status, Brookside was the site of a previous two-year professional development book study for clinical teachers that used Empson and Levi's (2011) *Extending Children Mathematics: Fractions and Decimals.* The book study met once a month and supported teachers in unpacking the book's content and attempting the instructional implications in their own classrooms. Teachers brought student work to book study meetings to discuss with their colleagues, and to consider how to examine student thinking with the purpose of guiding their planning and instructional decisions. Therefore, Brookside Elementary provided a study setting in which the work of eliciting and integrating student thinking into instruction would not be unfamiliar to the clinical teachers that were hosting the PTs.

Data Sources

In order to capture the PTs' experiences within a CoP and their engagement with student thinking, the following sources of data were collected and later analyzed: field notes and recordings from CoP meetings, elementary student work artifacts and assessment data as provided by the PTs, field notes and recordings from PTs' instructional observations and debrief sessions, semi-structured qualitative interviews from the PTs and their corresponding clinical teachers. Field notes from the team meetings documented the ways that the PTs discussed student thinking and learning, and how they planned to use the thinking in their instruction. The team meetings were also recorded for reviewing to expand the field notes. The researcher then observed the PTs to see if proposed ideas were enacted in classroom practice, and in post-conferences discussed their implementation of lessons. Again, the process of recording and expanding field notes was used for the observations and their debriefs. Table 2 provides an overview of the study timeline over its 12-week duration and all of the study's data sources. The table is followed by further description of the data sources.

Table 2. Data collection timeline

	Months and Weeks of Research											
	1 Jan 11th*	**2 Jan 16th**	**3 Jan 23rd**	**4 Jan 30th**	**5 Feb 6th**	**6 Feb 13th**	**7 Feb 20th**	**8 Feb 27th**	**9 Mar 6th**	**10 Mar 13th**	**11 Mar 20th**	**12 Mar 27th***
Data Collection												
Participant 1 (Kristi)	c, a, i PT	-	-	-	o,v,v	d	v,o,v	d	v,v,o	d, a	i CT	iPT
Participant 2 (Nora)	c, a, i PT	-	-	-	v,o,v	d	v,v,o	d	o,v,v	d, a	i CT	iPT
Participant 3 (Liz)	c, a, i PT	-	-	-	v,v,o	d	o,v,v	d	v,o,v	d, a	i CT	iPT
Community of Practice Team Meetings												
All Participants		m1		m2	m3	m4	m5	m6	m7	m8		
Key: PT= prospective teacher, CT= clinical teacher c = consent, a= student fraction assessment (pre/post), i = interview (for PT or CT) m= community of practice team meetings scheduled, o = observation (in-person) & informal debrief v= video recording of mathematics lesson, d= data analysis of previous week's lessons												
Note. Table format adapted from Toles, Colón-Emeric, Naylor, Barroso, & Anderson, 2016 with permission from personal communication with Toles												

Interviews

Prior to beginning their mathematics instruction in the student teaching experience, each PT participated in an interview focused on her beliefs about the instruction and learning of mathematics. This interview was the initial data collected that focused on the PTs' engagement with student thinking. The interview protocol was based on questions from the Integrating Mathematics and Pedagogy (IMAP) Web-Based Beliefs Survey (Philipp & Sowder, 2003). The IMAP Survey is a mixture of open-ended and Likert scale responses that are evaluated on normed rubrics. IMAP items that addressed general mathematics teaching beliefs and those specific to the fraction domain were selected for use in the PTs' interviews. The IMAP was designed for administration through an online format with written response, but the questions were used in a face-to-face interview format in order to allow for additional probes into a participant's responses. A flexible interview style was used to clarify responses, probe for more detailed responses, and to follow the individual.

An interview was also completed with each corresponding clinical teacher in order to provide a historical context for the mathematics professional development done with Brookside Elementary School in previous years and to provide insight into the current mentorship of the PTs in the study.

Observations and Informal Debrief Sessions

The PTs engaged in teaching similar lessons and content that had been previously planned together in coursework or in CoP team meetings. Each PT had three scheduled in-person lesson observations over her six-week lead instructional period. Each observed lesson was also video-recorded for later analysis by the researcher. When the researcher was doing an in-person observation in one PT's classroom, the simultaneous lesson was video-recorded in the other two classrooms. This meant that a total of nine video-recorded lessons were collected for each PT. The lessons reflected instruction from the beginning (n=3), middle (n=3), and end (n=3) of the six-week sequence. This allowed the researcher to create a profile of instruction and instructional changes for each PT.

Observational field notes were collected during the in-person observations. As stated, special attention was given to watching if aspects discussed in the CoP team meetings were integrated into practice. The field notes first captured a description of physical setting and context. Then, a running record style of observation was completed to showcase the timing and duration of different portions of each mathematics lesson. Because of the study's emphasis on the use of student thinking, to the greatest extent possible, the field note running records captured direct quotations and included moments of teacher-to-student questioning, student-focused interactions, and student-to-student discussions. This same approach for the observational field notes was used upon viewing the video footage of the others' subsequent lessons. This allowed for the exposition of instructional similarities and differences within the same content.

The most prominent difference between in-person and video-recorded observations was that the in-person observation was followed by an informal debrief session with the observed PT. Although the observation itself was not a participant observation, the debrief time transitioned into a participant-researcher model. The debrief session was not evaluative, and instead served as a discussion time for the PT to reflect upon her teaching with supporting questions and suggestions from the researcher. It was also a time that the researcher was able to gain additional insight into the PT's beliefs about her instruction and engagement with student thinking.

Student Pre and Post Assessments

A formalized pre and post assessment on fractions was administered to the corresponding 4th grade students through the PTs. This was done prior to the fraction unit start date and at the end of the PT instructional segment. The assessment was administered to students as part of the natural classroom procedures and did not disrupt instruction. It was the expectation at Brookside Elementary that pre and post assessments be administered to students. Copies of the student assessments were shared with the researcher with identifying student information removed.

The assessment was designed around the constructivist view of student understanding. Rather than measure learning solely through a correct/incorrect item analysis, the assessment allowed for item analysis according to student learning progressions in fractions, and showcased the range in levels of conceptual sophistication for equal sharing problem types (more items than sharers, less items than sharers requiring halving, less items than sharers requiring fourths). The equal sharing items were tied to a review of research on how students learn fractions and innately work with fractional quantities (Petit et al., 2015).

Analysis

In order to answer the research question (*In what ways does a CoP facilitate PTs' engagement with student thinking in the area of fractions?*), first an understanding of PTs' engagement with student thinking during the study duration was needed. The initial round of data analysis focused on each PT individually, then on the CoP collectively, to examine if and how expressed beliefs and practices changed regarding engagement with student thinking. Analysis was done by coding across interviews, observations, debriefs, and team meetings according to the actions described in the teaching practice of *elicit and use evidence of student thinking* (see Table 1) from the *Principles to Actions: Ensuring Mathematical Success for All* (NCTM, 2014, p.10).

Then, to better understand if PT engagement with student thinking might influence student understanding, 4th grade students' pre and post assessment data was analyzed. Assessments were analyzed across classrooms for total accuracy change from pre to post assessment, and across each 4th grade student for growth in strategy use from pre to post assessment. Student strategies were classified according to Empson and Levi's (2011) strategies for equal sharing problems, in order of increasing sophistication: Non-Anticipatory Sharing (learner does not think in advance about how to coordinate sharing of objects and attempts partitions through trial-and-error or by ease of partition, like halves), Additive Coordination (learner either partitions and shares one object at a time or shares groups of items at a time), Ratio (learner applies knowledge of repeated halving or multiplication factors to create a simpler problem), and Multiplicative Coordination (learner understands that *a* objects shared by *b* people is a/b).

Finally, the research question necessitated that analysis examine the potential supports that facilitate PT engagement with student thinking. Qualitative data analysis in alignment with Stake (1995, 2005) was used and rounds of open coding were completed across all data sources. According to Saldaña (2015), a code is a "word or short phrase that symbolically assigns a summative, salient, essence-capturing, and/or evocative attribute for a portion of language-based or visual data" (p.292). Open coding allowed patterned responses within the data to emerge and then patterns were grouped as themes. All data sources were coded and re-coded to uncover the overarching themes of support for the PTs' engagement with student thinking. The themes allowed for insight into what aspects of the study, specifically related to the CoP structure, aided in PT engagement with student thinking in mathematics.

Statement of Positionality and Researcher Involvement

Central to the qualitative case study is the researcher as the instrument (Denzin & Lincoln, 2005), and because the instrument is human, this case study cannot be truly objective or transferable to other contexts (Merriam, 1998). The use of thick description (Geertz, 1994), or the detailed account of field experiences in which social and relational interactions are contextualized, lends to research credibility. The researcher continued to examine and re-examine biases, limitations, and assumptions about the space and participants, since the researcher and the researcher's positionality was a part of the space and each interaction. The researcher's positionality is described next to lend to study transparency and allow the reader to better interact with the findings and implications.

I am a white, female mathematics teacher educator, and specifically to this study's context, acted as the researcher and participant-researcher by facilitating the CoP team meetings. My values as teacher educator and researcher are grounded in my experience as an elementary teacher in Title I public schools. I am interested in student-centered elementary mathematics and this is the focus area and lens for my work with other educators. My personal history with learning mathematics was that of rote memorization without understanding. Exposure to reform-based practices during my undergraduate education experience urged me to unpack mathematics and make sense of it for myself in order to facilitate sense-making for my future PreK through higher education learners.

When the pronoun 'I' is used in this article, it is to reflect the researcher's role as participant-researcher as the teacher educator who facilitated the CoP team meetings and the PTs' instructional debrief sessions. This firsthand role provided insight into the PTs' expressed beliefs and practices that may not have been attained otherwise. The use of 'I' also acknowledges the researcher's potential influence in shaping some of these beliefs and practices. Claims of participants' beliefs made herein reflect participants' own statements made during study interactions.

FINDINGS

Initial analysis identified evidence that the PTs' expressed beliefs and practices were consistent with practices that engage student thinking in planning and instruction. Each PT was an individual, who held individual beliefs and experienced individual changes in practice that aligned with the *elicit and use evidence of student thinking* (see Table 1) practice. The PTs also began to identify as a community with common concerns, passions, and aim for competence around fraction instruction that elicited and used student thinking.

Analysis of the 4th grade assessment data then confirmed accuracy gains (Table 3) across all classrooms and strategy growth across 4th graders from pre to post (Table 4). Even if a student did not get an accurate answer, their solving strategy employed became more sophisticated, showing an advancement of approach and understanding towards fractions. While the PTs' engagement with their student thinking may not have been the cause or sole cause for these results, it is encouraging that their instructional practices did not negatively influence student scores or strategy use.

Once it was established that PTs were engaging with their students' thinking about fractions, and that there was improvement for 4th graders from pre to post assessments, data sources were open coded for the potential factors of influence that may lead to and support PTs' ability to elicit and integrate student thinking in instruction. These factors are shared next as the themes that emerged through coding.

Table 3. PTs' class assessment averages

	Pre	Post	Change
Classroom 1: Kristi, n=22	22%	40%	18%
Classroom 2: Nora, n=20	38%	58%	20%
Classroom 3: Liz, n=18	34%	62%	28%

Note. Averages reflect composite accuracy scores of students that took both pre and post assessments.

Table 4. Overall problem-solving strategy occurrences expressed as percentages

		Classroom 1: Kristi n= 22		Classroom 2: Nora n= 20		Classroom 3: Liz n=18	
		Pre	**Post**	**Pre**	**Post**	**Pre**	**Post**
Strategies for Equal Sharing Problems: least to most sophisticated	Non-Anticipatory	20%	14%*	53%	33%*	39%	15%*
	Additive Coordination by One	17%	35%	12%	12%	4%	19%
	Additive Coordination by Groups	20%	41%*	13%	38%*	19%	35%*
	Ratio	0%	0%	0%	0%	0%	0%
	Multiplicative Coordination	0%	4%	5%	7%	4%	11%

Note. Asterisks denote decrease in less sophisticated strategy with increase in more sophisticated strategy.

Themes of Influence

Open coding of data sources revealed two overarching themes of influence that arose for the PTs through the CoP structure that shaped how they planned for and enacted engagement with student thinking during their instruction. The themes reflected the types of support that the PTs expressed as critical to their development: (a) support of scheduled team meetings and (b) support of additional meetings (additional meaning meetings beyond those planned by the researcher for the study and organized by the PTs). Table 5 provides insight into key quotations and moments for each theme from each PT. The findings according to each theme are provided, followed by a deeper examination of a scenario from one CoP team meeting in order to highlight its influence on the PTs involved and its implications for the other teacher educators.

Table 5. Data examples by theme

	Team Meetings	Additional Meetings
Kristi	'But our CTs use that language. (Long pause) What do we use instead?' – Meeting 3, regarding their clinical teacher labeling students as 'low, medium, or high' achieving for planning purposes	'Whoa! I mean wait, when we flip the phrasing of 4 people sharing 3 items to 3 items being shared by 4 people then we get the fraction it's going to turn out to be at the end! That's crazy! I'm just making that connection now.' – Additional meeting, week 5
Nora	'Wait! I think I know what they're doing. Look at the way they're sectioning. They're not treating the whole as 0 to 1, they're treating the whole as the whole number line.' – Meeting 8, regarding student work being examined	Field note: Without prompting from researcher, Nora brings all her student work artifacts to additional meeting to analyze and ask the others questions about in order to guide next steps in instruction. – Additional meeting, week 5
Liz	'This is the best day ever; we're getting so much done.' – Meeting 7, describing curriculum suggestions that PTs provided to CT/PT team	Field Note: Raises question about a student's work sample and all members of team go through that student's work to discuss the student's strategy use with the fraction constructs. – Additional meeting, week 7

Theme One: Support of Team Meetings

The initial team meetings of the CoP were focused on reorientation to core reform-based mathematics practices originally addressed in the PTs' prior coursework. As the PTs became more involved with the classroom, they came to the team meetings with student work that intrigued them, their instructional dilemmas, and their moments of success to discuss and reflect upon. The focus on student work especially aligned with the mathematical teaching actions from Table 1: Identify what counts as evidence of student progress toward mathematics learning goals; Interpret student thinking to assess mathematical understanding, reasoning, and methods; Reflect on evidence of student learning to inform the planning of next instructional steps. Alignment of the team meetings around these teacher actions seemed to then facilitate the PTs' abilities to enact in-the-moment teacher actions focused on engaging with student thinking while teaching. As the study progressed, areas of shared needs and interests emerged for the PTs, and these began to give shape to the content of the team meetings. The PTs moved between figuring out students, and figuring out themselves.

Figuring Out Students

Although other aspects of students were discussed by the PTs (e.g. inter-personal interactions, management needs), the focus for the study precipitated analyzing for how the PTs discussed and determined the mathematical understandings and misunderstandings of their students. The first team meeting was launched with attention to student strategy fraction frameworks and discussion about the PTs' students' work on a fraction pretest. The intention was that this discussion of student work would establish an expectation that deep investigation of student learning was important, and analysis of student work should drive the mathematics planning and instruction. Previous to analyzing student work by strategies employed, the PTs had used raw accuracy data to group their students. The PTs were accustomed to using 'low, medium, or high' to describe the groups of learners in their classrooms, as this was the language they had heard within their placement school. Nora pondered, 'Correctness can't be the only thing we look for, right? There's more here.' Kristi troubled, 'But our CTs use that language. [Long pause] What do we use instead?' Kristi continued with an assertion that sparked a revelation for the group, 'The whole kid is not low; he's just missing a skill still.'

The focus on figuring out student understanding through a strategy construct like Empson and Levi's (2011) partitioning strategies, spurred the PTs to create student learning groups around similar skillsets for their classroom instructional time. Rather than use the grouping terms of 'low, medium, or high,' the PTs began to seek a descriptive language characterizing students around their learning and skillsets. The PTs began using the terms *approaching skill*, *on/with skill*, and *extending skill* to discuss students in relation to skills from their partitioning strategies work. The PTs agreed this seemed a better fit to label recognized skillsets around particular fraction topics. It allowed for flexibility of grouping students by skills and remedied the messaging of labeling a whole child as low, medium, or high performing. Admittedly, a change in language does not equal a change in belief or practice. However, even an acknowledgement of language dissonance and upholding the use of the new language and grouping strategies with one another in the community suggests that the supports helped the PTs to begin to problematize embedded schooling practices that do not reflect the ideals of the current vision of mathematics.

Figuring Out Selves

The PTs were not only reflecting around their pedagogical choices during team meetings, but also reflecting around their educational philosophies. Some of the team meeting time was devoted to discussion about their visions of teaching and how those visions could be enacted in real-time. Sometimes their discussions centered on what they learned about in their previous coursework and negotiating what class topics looked like in the context of their school and their individual classrooms. One of the assignments for their student teaching capstone course was writing an educational philosophy. Discussions around their philosophies surfaced during our team meetings and they shared their opening statements. Each had a differing 'I believe' lead statement that portrayed their personalities, but also their unification around engagement with student thinking being the focus of the classroom.

Kristi: I believe in the power of children, not of adults to empower children, but of the extreme power a child holds by simply being a child.

Nora: I believe that students all come from different backgrounds with their own life experiences and funds of knowledge. They can offer their knowledge and experiences to contribute to the learning of all students because each student has a topic that they are experts in.

Liz: I believe that the classroom should have a student-centered format, where the teacher acts more as a facilitator of discussion rather the sole discusser with all of the knowledge.

I maintained a safe space for the PTs' discussions around teaching philosophies and their embodiment of mathematics instruction and instruction holistically. In discussions around their beliefs, I listened but refrained from contribution unless I was asked a question. It was important that they had protected space to figure out who they were as practitioners in open and honest ways. Although I was at the team meetings as a facilitator for the mathematics at hand, ultimately this was their time to decide who they were as teachers and how they would move forward in the profession.

Theme Two: Support of Additional Meetings

Perhaps one of the most important aspects of support that shaped how the PTs engaged with student thinking were the additional meetings. These meetings were additional in that they were not designated by the research design nor were they part of the set schedule of team meetings established by the researcher. While I facilitated the scheduled team meetings that took place weekly, the additional meetings were established and orchestrated by the PTs. These meetings began within week 5 of the 12-week study (see Table 2). I was invited to two of these additional meetings and attended both. The first I attended occurred between team meetings 3 and 4 and it was in this meeting that the PTs expressed that the planning and instruction for mathematics in a manner that centered student thinking, coupled with planning for other subject matters, became an overwhelming reality. The second additional meeting I attended was at the end of their full-time student teaching instructional period, as the PTs were finalizing their mathematics instructional units.

The invitations to these meetings meant that I was viewed as a participating member of the CoP, but the fact that the meetings were established and took place regardless of my attendance revealed the voluntary nature of the CoP. The ownership of the CoP became that of the PTs. When I attended the additional meetings, I acknowledged that they were not mine to lead; rather, I was invited to follow their leads and support accomplishing their goals.

At the first additional meeting, I attended for three hours and an overarching agenda of topics seemed to be enacted, even if the agenda was not formalized. Hour one: Review their program-required mathematics portfolio expectations, Hour two: Unpack fractions content and consider student work in regard to types of partitioning strategies, Hour three: Revise future mathematics lesson plans based on student strategies. Even without the structure of the scheduled team meetings, the PTs adopted the practices from the NCTM *elicit and use evidence of student thinking* teacher actions (Table 1) that reinforced attention to student work and thinking in planning for instruction. The PTs informed me that they continued to meet for two more hours to discuss other content areas after the initial three-hour duration about mathematics. Individually, I asked each PT what caused her to start to meet outside our scheduled team meetings and each voiced a similar need.

Kristi: Fear; overwhelming stress.

Nora: To keep sane with each other together.

Liz: Stress; not knowing.

Their responses made obvious that outside of the team meetings, a voluntary CoP originated around shared concerns about instruction. Because I was not a PT, there were certain parameters of the community I no longer met. I did not have need for a space of intensive pacing and planning around the children of their classrooms as the three members of the community did. Liz later expressed that the group of PTs was viewed as her sole source of teaching companionship and support when she hypothetically asked me, 'I mean, who else would I even talk to?'

Findings revealed that the PTs aligned their expressed beliefs and practices with reform-based mathematics teaching practices. They engaged with student thinking in their fraction instructional units, even when outside influences made that ambitious work difficult. Although their CoP was initially facilitated by a mathematics teacher educator, the PTs voluntarily adopted and extended the community structure beyond the scheduled team meetings. At week 5 of the study (see Table 2), it seemed a true CoP structure was in place for the three PTs when they began to meet outside the official team meetings facilitated by the researcher. Although each in their own classroom, the PTs' shared experience of being in the same grade level at the same school spurred the CoP model's true intention of being a voluntary group of people working with shared concerns, passions, and goals (Wegner, 1998; Wenger et al., 2002b).

SOLUTIONS AND RECOMMENDATIONS: A LOOK INTO A TEAM MEETING

In order for other teacher educators to see how this structure may be beneficial within their programs, insights from one team meeting are shared in detail to highlight the PTs' ownership of mathematics content and deep reflection around it. A welcomed outcome from the PTs' focus on using student strategy constructs to analyze student work was their finetuned observations of student thinking and their push for a deeper understanding of content. At team meeting 5, a discussion around student thinking resulted from the PTs noticing different pictorial representations their students used within one particular equal sharing strategy category. The PTs were classifying student work by the equal sharing strategies that students used to solve an equal share problem (strategies: Non-Anticipatory, Additive Coordination, etc.), but they dissected the student strategies further within each classification. Kristi presented an example of three different students' pictorial representations of the same problem. All three students had the same answer and she classified their strategies as *Additive Coordination- sharing one item at a time*. Kristi noted, 'I mean, surely, these all show different levels of understanding within the same strategy, right?' Kristi alluded to the fact that the students seemed to exhibit different understandings even though they were all classified the same way according to the Empson and Levi (2011) partitioning strategies. Figure 1 shows a model of the three student drawings Kristi shared.

Once Kristi shared the three representations, the other PTs said that they too were wondering about the various student pictures and if they represented different levels of sophistication within a strategy. The PTs asked me for another classification system – one that would classify student pictorial representations within each of the original Empson and Levi partitioning strategy classifications. We consulted the book; although Empson and Levi's (2011) provided a 'progression of children's use of fraction word and symbols' (p. 27), there was not a specific breakdown of possible pictorial representations within each partitioning strategy. A group consensus was made among the three that since there was not a sophistication of pictorial representations provided within each equal sharing strategy, the PTs would use their own understanding of the student models to develop a pictorial progression. The order of Student 1 through Student 3 as presented in Figure 1, depicts the proposed hierarchy of student pictorial representations within an *Additive Coordination- sharing one item at a time* strategy, with Student 3 proposed as being the most sophisticated in their understanding of this strategy.

Figure 1. Additive coordination pictorial representations

Student Additive Coordination models to represent 3 objects shared equally by 4 people

Student 1: Depicts the 3 objects and 4 people. Draws lines to each portion of 4th to corresponding person.

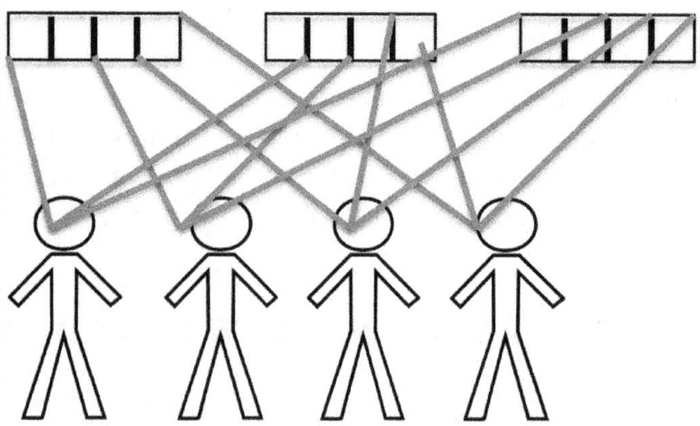

Student 2: Depicts 3 objects and 4 people. Organizes each partition by person with a notation system (letter or number).

Student 3: Depicts 3 objects, no people. Organizes each partition by person with a notation system.

Regarding the Student 1 pictorial strategy, the PTs felt students utilizing this pictorial strategy often lost track of how many lines went to each sharer. Although Student 1 used the Additive Coordination strategy, this student sometimes missed a direction line and had to recount the shares. The PTs likened this strategy to the Cognitively Guided Instruction whole number strategy of directly modeling all parts of a problem context. They referred to Student 1 strategy as 'the rainbow strategy' because it often resulted in many arched lines from object to sharer. Student 2 also directly modeled the people and objects, but students employing this strategy used an organization system of numbering or lettering the portions each sharer received. The PTs felt this was more advanced than the first strategy, but not as advanced as the Student 3 strategy. Student 3 used an organization system like Student 2, but did not need to represent the people. This student still drew the objects, but partitioned each for the correct number of sharers without physically representing the people. The PTs felt that they may be able to scaffold students utilizing the Student 3 strategy to a more sophisticated equal sharing strategy because these students no longer needed to directly model all parts of the problem. This detailed attention to student pictorial representations led to the PTs planning how to lead discussions that specifically questioned students about why they used the models they did. They also thought about creating classroom charts with student names attached to the different pictorial representations so that other classmates could consider various representations and their affordances.

The PTs' pictorial progression for the *Additive Coordination- sharing one item at a time* strategy resulted from a discussion of what each of Kristi's work samples, and any student who used a similar pictorial strategy, may be exhibiting conceptually. This discussion and targeted work around classification and scaffolding of their students' thinking, and the PTs' extension of published work to fit the needs of their contexts and students offers encouragement that a CoP model not only provides camaraderie for PTs, but a rich intellectual environment. The PTs recognized that the various dynamics of their classrooms raised unique challenges, and looked for trends across the grade level to consider how to best support one another and support the children they taught.

FUTURE RESEARCH DIRECTIONS

The study's findings highlight the importance of supporting PTs in the context of the student teaching experience in order to help transfer practices learned during teacher preparation coursework into teaching. The findings offer several contributions to the field, including a suggested CoP model of support to encourage PTs' engagement with student thinking, specifically in mathematics. The research findings and their interpretations suggest that certain nonnegotiable components should be included in a model of support. Two key components are: (1) A CoP structure in which the PTs can practice their instructional ideas, take risks, have disagreements, and receive feedback, both through a content expert's facilitation, and on their own. (2) Focused and substantive time for PTs to invest in the analysis of student work using strategy progressions and frameworks, like Empson and Levi's (2011) strategies for partitioning. Again, Appendix 1 provides the CoP's team meeting plans that emphasized engagement with student thinking, yet provided space for the PTs to take ownership of the meeting time based upon their needs and their students' needs. This ownership became evident when the PTs continued the CoP structure and meetings outside of the study's required team meetings. In the closing interviews, the PTs described how they believe this experience changed their teaching practices for life. Insomuch, Kristi and Liz changed their post-graduation plans and chose to move together to apply to the same school and same grade level so that they could continue teaching together with their already-established network. They expressed the value of working with someone who already knew them as a teacher and knew what they hoped to accomplish as they entered into their first year of teaching.

Worthwhile to note is that the separation of the PTs in their own CoP strengthened their bond and willingness to attempt new pedagogies and strategies; however, this separation without the inclusion of the clinical teachers could have also created an unattended barrier ('us versus them' mentality) between the PTs and their clinical teachers. Perhaps there is need for a concurrent PT and clinical teacher community around content and relevant educational topics in order foster the dyads' relationships. Even so, a separate PT time is still needed to provide space to work independently, explore, and have ownership of their development.

CONCLUSION

This study considered how the field might encourage and facilitate PTs' engagement with student thinking. The findings of this case study exposed the need for PTs to have a support system that facilitates their understanding and instruction of mathematics in order to sustain engagement with student thinking. However, within this system of support in learning the content and pedagogy of mathematics, PTs also required space to explore and question their surroundings, their teaching styles, and their teaching philosophies. They needed to do so in a space they considered their own and safe, in hopes that they would adopt the orientation of reform-based mathematics for themselves and not for the sake of coursework or mentor requirements. The voluntary supports that the PTs created outside of the study's structure indicate the need for PTs to shape their professional identities in a context-specific way, yet do so outside of the parameters of context.

The findings also reveal the opportunity for future research in how to utilize group placements of PTs to best support their navigation of reform-based mathematics ideals. Schools need teachers who engage with students' thinking in a way that constructively uses the mathematics content in order for students to grow in their conceptual understanding of the topics at hand. Therefore, teacher preparation programs need support models in place that encourage and facilitate prospective teachers' work in engagement with student thinking.

Engagement with student thinking is important to the advancement of equitable teaching because this practice puts students' mathematical ideas and contributions at the forefront of lesson planning and instruction (Chao et al., 2014). Engagement with student thinking means the teacher must elicit that student thinking. The hope is that through this eliciting, PTs will learn to value that thinking and position their students sharing of that thinking as valuable to the mathematics instruction. The PTs in the study learned that they could not simply implement stock lesson materials; rather, they had to adapt their means of instruction for the diverse student perspectives and learning needs of their classrooms. The support system of a CoP benefited the development of their craft, specifically with the focus on eliciting and utilizing student thinking, and thereby benefited the learners in their classroom spaces.

REFERENCES

Aldana, U. S., & Martinez, D. C. (2018). The development of a community of practice for educators working with newcomer, Spanish-speaking students. *Theory into Practice, 57*(2), 137–146. doi:10.1080/00405841.2018.1425813

Bieda, K. N., Sela, H., & Chazan, D. (2015). "You are learning well my dear": Shifts in novice teachers' talk about teaching during their internship. *Journal of Teacher Education, 66*(2), 150–169. doi:10.1177/0022487114560645

Borko, H., & Mayfield, V. (1995). The roles of the clinical teacher and university supervisor in learning to teach. *Teaching and Teacher Education, 11*(5), 501–518. doi:10.1016/0742-051X(95)00008-8

Borman, K. M., Mueninghoff, E., Cotner, B. A., & Frederick, P. B. (2009). Teacher preparation programs. In *International handbook of research on teachers and teaching* (pp. 123–140). Springer. doi:10.1007/978-0-387-73317-3_8

Burns, M. (2001). *Teaching arithmetic: Lessons for introducing fractions, grades 4-5.* Math Solutions.

Carpenter, T. P. (1979). *Cognitive Development Research and Mathematics Education.* Theoretical Paper No. 73. Conceptual Paper No. 2.Studies in Mathematics Series.

Carpenter, T. P., & Fennema, E. (1992). Cognitively guided instruction: Building on the knowledge of students and teachers. *International Journal of Educational Research, 17*(5), 457–470. doi:10.1016/S0883-0355(05)80005-9

Carpenter, T. P., Fennema, E., Franke, M. L., Levi, L., & Empson, S. B. (2014). *Children's mathematics: Cognitively guided instruction* (2nd ed.). Heinemann.

Carpenter, T. P., Fennema, E., Peterson, P. L., Chiang, C. P., & Loef, M. (1989). Using knowledge of children's mathematics thinking in classroom teaching: An experimental study. *American Educational Research Journal, 26*(4), 499–531. doi:10.3102/00028312026004499

Carpenter, T. P., & Moser, J. M. (1984). The acquisition of addition and subtraction concepts in grades one through three. *Journal for Research in Mathematics Education, 15*(3), 179–202. doi:10.2307/748348

Chamberlin, M. T. (2005). Teachers' discussions of students' thinking: Meeting the challenge of attending to students' thinking. *Journal of Mathematics Teacher Education, 8*(2), 141–170. doi:10.100710857-005-4770-4

Chao, T., Murray, E., & Gutiérrez, R. (2014). *What are classroom practices that support equity based mathematics teaching? A research brief.* National Council of Teachers of Mathematics.

Chapin, S. H., O'Connor, C., & Anderson, N. C. (2003). *Classroom discussions: Using math talk to help students learn, Grades K-6.* Math Solutions.

Chapin, S. H., O'Connor, C., & Anderson, N. C. (2013). *Classroom discussions in math: A teacher's guide for using talk moves to support the common core and more, Grades K-6.* Math Solutions.

Charles, K., & Nason, R. (2000). Young children's partitioning strategies. *Educational Studies in Mathematics*, *43*(2), 191–221. doi:10.1023/A:1017513716026

Confrey, J., Maloney, A., Nguyen, K., Mojica, G., & Myers, M. (2009). Equipartitioning/splitting as a foundation of rational number reasoning using learning trajectories. Presented at the *33rd Conference of the International Group for the Psychology of Mathematics Education*, Thessaloniki, Greece.

Denzin, N. K., & Lincoln, Y. S. (2005). Introduction: The discipline and practice of qualitative research. In N. K. Denzin & Y. S. Lincoln (Eds.), *The SAGE handbook of qualitative research* (3rd ed., pp. 1–32). Sage.

Empson, S. B. (1995). Using sharing situations to help children learn fractions. *Teaching Children Mathematics*, *2*(2), 110–115. doi:10.5951/TCM.2.2.0110

Empson, S. B. (1999). Equal sharing and shared meaning: The development of fraction concepts in a first-grade classroom. *Cognition and Instruction*, *17*(3), 283–342. doi:10.1207/S1532690XCI1703_3

Empson, S. B., & Levi, L. A. (2011). *Extending children's mathematics: Fractions and decimals.* Heinemann.

Farnsworth, V., Kleanthous, I., & Wenger-Trayner, E. (2016). Communities of practice as a social theory of learning: A conversation with Etienne Wenger. *British Journal of Educational Studies*, *64*(2), 139–160. doi:10.1080/00071005.2015.1133799

Flores, M. M., Hinton, V. M., & Meyer, J. M. (2020). Teaching fraction concepts using the concrete-representational-abstract sequence. *Remedial and Special Education*, *41*(3), 165–175. doi:10.1177/0741932518795477

Fosnot, C. T. (2007). *Field Trips and Fund-raisers: Introducing Fractions*. Heinemann.

Geertz, C. (1994). Thick description: Toward an interpretive theory of culture. *Readings in the philosophy of social science*, 213-231.

Huinker, D., & Bill, V. (2017). *Taking action: Implementing effective mathematics teaching practices in grades K-Grade 5*. National Council of Teachers of Mathematics.

Jacobs, V. R., & Empson, S. B. (2015). Responding to children's mathematical thinking in the moment: an emerging framework of teaching moves. *ZDM*, 1-13.

Jacobs, V. R., Empson, S. B., Pynes, D., Hewitt, A., Jessup, N., & Krause, G. (2019). The responsive teaching in elementary mathematics project. In P. Sztajn & P. H. Wilson (Eds.), *Designing professional development for mathematics learning trajectories* (pp. 75–103). Teachers College Press.

Kelley, T. R., Knowles, J. G., Holland, J. D., & Han, J. (2020). Increasing high school teachers self-efficacy for integrated STEM instruction through a collaborative community of practice. *International Journal of STEM Education*, *7*(1), 1–13. doi:10.118640594-020-00211-w

Kirkby, J., Walsh, L., & Keary, A. (2019). A case study of the generation and benefits of a community of practice and its impact on the professional identity of early childhood teachers. *Professional Development in Education*, *45*(2), 264–275. doi:10.1080/19415257.2018.1449003

Lamon, S. (1996). The development of unitizing: Its role in children's partitioning strategies. *Journal for Research in Mathematics Education, 271*(2), 170–193. doi:10.2307/749599

Lamon, S. J. (2007). Rational numbers and proportional reasoning. In F. K. Lester Jr., (Ed.), *Second handbook of research on mathematics teaching and learning* (pp. 629–667). National Council of Teachers of Mathematics.

Lave, J. (1991). Situating learning in communities of practice. *Perspectives on Socially Shared Cognition, 2*, 63-82.

Lave, J., & Wenger, E. (1991). *Situated Learning: Legitimate Peripheral Participation.* Cambridge University Press. doi:10.1017/CBO9780511815355

Merriam, S. B. (1998). *Qualitative research and case study applications in education. Revised and expanded from "Case study research in education"*. Jossey-Bass Publishers.

National Council of Teachers of Mathematics. (1989). *Curriculum and evaluation standards for school mathematics.* National Council of Teachers of Mathematics.

National Council of Teachers of Mathematics. (1991). *Professional standards for teaching mathematics.* National Council of Teachers of Mathematics.

National Council of Teachers of Mathematics. (1995). *Assessment Standards for School Mathematics.* National Council of Teachers of Mathematics.

National Council of Teachers of Mathematics. (2000). *Principles and Standards for School Mathematics.* National Council of Teachers of Mathematics.

National Council of Teachers of Mathematics. (2014). *Principles to actions: Ensuring mathematical success for all.* National Council of Teachers of Mathematics.

National Council of Teachers of Mathematics. (2016). *Illuminations playing fraction tracks.* Retrieved from https://illuminations.nctm.org/Activity.aspx?id=6637

Petit, M. M., Laird, R. E., Marsden, E. L., & Ebby, C. B. (2015). *A focus on fractions: Bringing research to the classroom* (2nd ed.). Routledge. doi:10.4324/9781315746098

Philipp, R., & Sowder, J. (2003). *Integrating mathematics and pedagogy (IMAP) web-based belief survey manual.* http://www.sci.sdsu.edu/CRMSE/IMAP/Beliefs-Survey_Manual.pdf

Philipp, R. A., Ambrose, R., Lamb, L. L., Sowder, J. T., Schappelle, B. P., Sowder, L., & Chauvot, J. (2007). Effects of early field experiences on the mathematical content knowledge and beliefs of prospective elementary school teachers: An experimental study. *Journal for Research in Mathematics Education*, 438–476.

Pothier, Y., & Sawada, D. (1983). Partitioning: The emergence of rational number ideas in young children. *Journal for Research in Mathematics Education, 14*(5), 307. doi:10.2307/748675

Saldaña, J. (2015). *The coding manual for qualitative researchers* (3rd ed.). Sage.

Siegler, R., Carpenter, T., Fennell, F., Geary, D., Lewis, J., Okamoto, Y., & Wray, J. (2010). *Developing Effective Fractions Instruction for Kindergarten through 8th Grade. IES Practice Guide. NCEE 2010-4039*. What Works Clearinghouse.

Slick, S. K. (1998). The university supervisor: A disenfranchised outsider. *Teaching and Teacher Education, 14*(8), 821–834. doi:10.1016/S0742-051X(98)00028-6

Stake, R. E. (1995). *The art of case study research*. SAGE Publications Ltd.

Stake, R. E. (2005). Qualitative case studies. In N. K. Denzin & Y. S. Lincoln (Eds.), *The SAGE handbook of qualitative research* (3rd ed., pp. 443–466). Sage.

Steffe, L. P. (2001). A new hypothesis concerning children's fractional knowledge. *The Journal of Mathematical Behavior, 20*(3), 267–307. doi:10.1016/S0732-3123(02)00075-5

Streefland, L. (1993). Fractions: A realistic approach. In T. P. Carpenter, E. Fennema, & T. A. Romberg (Eds.), *Rational numbers: An integration of research* (pp. 289–325). Routledge.

Toles, M., Colón-Emeric, C., Naylor, M. D., Barroso, J., & Anderson, R. A. (2016). Transitional care in skilled nursing facilities: A multiple case study. *BMC Health Services Research, 16*(1), 1–14. doi:10.118612913-016-1427-1 PMID:27184902

Wenger, E. (1998). *Communities of practice: Learning, meaning, and identity*. Cambridge University Press. doi:10.1017/CBO9780511803932

Wenger, E., McDermott, R., & Snyder, W. M. (2002b). Seven principles for cultivating communities of practice. *Cultivating Communities of Practice: A Guide to Managing Knowledge, 4*.

Wenger, E., McDermott, R. A., & Snyder, W. (2002a). *Cultivating communities of practice: A guide to managing knowledge*. Harvard Business Press.

KEY TERMS AND DEFINITIONS

Community of Practice: Defined as a group of people who share common concerns or passions for something they do, and learn how to do this thing better by interacting (Wenger, 1998; Wenger, McDermott, & Snyder, 2002b). This study used *team meetings* as required meeting time for the community of practice participants to unpack reform-based mathematics pedagogies. The voluntary aspect of a community of practice (Wenger, 1998) resulted from the participants' willingness to meet and extend their collaboration and cooperative learning outside of the required team meetings.

Context: Defined as the real-word situation of the case. When used in this case study, context refers to the situation of the participants' student teaching experience in both a university teacher preparation program and an elementary school site. The context introduced elements of potential influence that included participants' prior coursework, interpersonal relationships, and classroom dynamics.

Engagement With Student Thinking: Defined as the elicitation and use of evidence of student thinking in instruction (NCTM, 2014). This study's focus was prospective teachers' elicitation and use of students' mathematical ideas and strategies around fractions to inform instruction.

Fraction: Defined as a non-negative rational number that can be written in the form a/b where a and b are both integers and b does not equal zero (Behr, Lesh, Post, & Silver, 1983; Charalambous & Pitta-Pantazi, 2005; Kieran, 1976; Olanoff, Lo, & Tobias, 2014).

APPENDIX 1

COMMUNITY OF PRACTICE TEAM MEETING FACILIATION PLANS, MEETINGS 1-3

Note: All Plans in the style of the *Classroom Discussions in Math* template from Chapin, O'Connor, & Anderson (2013)

Title: Community of Practice Team Meeting 1
Time: 3:00-4:30
Identifying the Mathematical Goals:

- Prospective teachers (PTs) will review the content from *Classroom Discussions* (Chapin, O'Connor, & Anderson, 2013) and *Extending Children's Mathematics: Fractions and Decimals* (Empson & Levi, 2011) as introduced in their Math Methods course.
- PTs will be introduced to the Ongoing Assessment Project (OGAP) fraction progression of student solving strategies (Petit, Laird, Marsden, & Ebby, 2015)

Anticipating Confusion:

- PTs may need re-orientation to the talk moves and how to integrate them into the various instructional tools learned about in the fall semester: Marilyn Burns Fraction Kit (2001), Fosnot Fraction Mini-Lessons (2007), Illuminations/Math Investigations Fraction Tracks (2016)
- PTs may not remember the Types of Strategies – Equal Sharing Problems table that outlines the typical strategies students use to solve Equal Sharing problems.
- PTs may not see how Types of Strategies – Equal Sharing Problems and the OGAP Fraction Progression complement each other in regard to analyzing student work.

Asking Questions:

- What talk move(s) do you feel you're integrating into instruction already?
- What talk moves would you like to begin to use this semester? Why?
- Why did you classify student work that way? (Defend your choice)

Planning Implementation:

- PTs will be re-introduced to the Talk Moves and how they can be integrated into their pre-planned fraction units.
- These pre-planned units give structure to the instruction, but the PTs will also be reminded that they need to remain flexible to follow students' understanding.
- Units include lessons from Burns (2001), Fosnot (2007), and Illuminations/Math Investigations (2016)
- PTs will be re-introduced to the Types of Strategies – Equal Sharing Problems student strategy table and will practice sorting sample student work.
- PTs will be introduced to the OGAP Fraction Progression table and will practice sorting sample student work.
- Time for questions, needs, and instructional planning.
- Needs discussed here will be integrated into following meeting.

Title: Community of Practice Team Meeting 2
Time: 2:45-3:45
Identifying the Mathematical Goals:

- PTs will use Types of Strategies – Equal Sharing Problems and the OGAP Fraction Progression to analyze their pre-assessment student data.

Anticipating Confusion:

- PTs may have difficulty classifying student work with the OGAP Fraction Progression because it is the newer instrument to them.

Asking Questions:

- Why did you classify student work that way? (Defend your choice)
- What does this student seem to understand?

Planning Implementation:

- PTs will come to meeting having initially analyzed and classified their pre-assessments.
- PTs will re-analyze with researcher. They will defend their progression placements and adjust as necessary.
- PTs and researcher will analyze pre-assessments until all raters agree on student Equal Sharing strategies and OGAP Fraction Progression level classifications. This will account for inter-rater reliability.
- Address needs that arose at previous meeting.
- Time for questions, needs, and instructional planning.
- Needs discussed here will be integrated into following meeting.

Lesson Title: Community of Practice Team Meeting 3
Time: 2:45-3:45
Identifying the Mathematical Goals:

- PTs will reflect on day's instruction and set goals for remainder of unit.

Anticipating Confusion:

- PTs may be overwhelmed with their first day of full-time instruction of all subjects and first day of researcher videorecording/observing. They may need time to process full-time teaching responsibilities.
- PTs may be uncertain about how to adjust their pre-planned instructional unit based on student understanding.

Asking Questions:

- What was today's success? Today's stumble? Why? (Success/stumble is language used for a high point and point for improvement).
- What would you like to do moving forward? Why?
- How did you address student learning today?
- What student needs must still be met? How will you meet them?

Planning Implementation:

- PTs will discuss their self-perceptions and reflections about lessons and instruction today.
- PTs will set personal goals for moving forward.
- Researcher will address needs that arose at previous meeting.
- Researcher will address any emerging content and pedagogical needs/themes that arose from today's observations.

- Time for questions, needs, and instructional planning.

APPENDIX 2

PROSPECTIVE TEACHER INSTRUCTIONAL UNIT OVERVIEW CALENDAR

Note: Prior to teaching sequence, each elementary student made a Fraction Kit (Burns, 2001) for 1, ½, ¼, 1/8, and 1/16; corresponding Cover It and Uncover games were played. Then instructional sequence was used, as seen in Table 6.

Table 6. Unit overview calendar

DAY 1 Equal Sharing Problems	DAY 2 Equal Sharing Problems	DAY 3 Equal Sharing Problems	DAY 4 Review of Fraction Kit with ½, ¼, 1/8/ 1/16 & Cover Up Game	DAY 5 Cont. Cover Up Review
DAY 6 Equal Sharing Problems with Equivalent Fraction Situations	DAY 7 Construction of Fraction Kit 1/3, 1/6, 1/12 & Making Connections to Rest of Kit	DAY 8 Whole-Class *Fraction Tracks* (NCTM, 2016) Gameboard Introduction (Number-line model) & Comparison to Fraction Kits	DAY 9 Using Strategies to Compare Fractions (on *Fraction Tracks* Gameboard and with Fraction Kits)	DAY 10 Apply Equivalency and Comparison Ideas through *Fraction Tracks* as Whole Class
DAY 11 Intro *Fraction Tracks* as Partner Competition Game; Whole Group Debrief	DAY 12 Fraction Tracks as a Competition Continued; Discussion of Game Strategy Around Equivalency	DAY 13 Introduction of Submarine Sandwich Equal Sharing Task (Fosnot, 2007)	DAY 14 Work Day on Submarine Sandwich Task, with Challenge Component if Needed	DAY 15 Math Congress Set-up and Experience
DAY 16 Introduction to Common Denominators with Submarine Context; Share Out of Challenge Component (if applicable)	DAY 17 Readdressing the Submarine Sandwich Task with Common Denominators (Initial Student Exploration of Addition of Fractions)	DAY 18 Student Addition Exploration continued	DAY 19 Math Congress Around Addition Strategies	DAY 20 Formalize Class Addition Strategies
DAY 21 Student Exploration of Subtraction (Fraction Kit & Number Lines as Tools)	DAY 22 Student Exploration of Subtraction Continued	DAY 23 Formalize Class Subtraction Strategies	DAY 24 Practice with Subtraction Through Stations	DAY 25 Practice with Subtraction Through Stations
DAY 26 Mixed Operational Tasks and Stations	DAY 27 Mixed Operational Tasks and Stations	DAY 28 Mixed Operational Tasks and Stations	DAY 29 Mixed Operational Tasks and Stations	DAY 30 Posttest Administered

Chapter 10
Disrupting Their Histories:
Preservice Teachers' Mathematics Autobiographies as Catalysts for Change

Joan Gujarati
State University of New York at Oswego, USA

ABSTRACT

This qualitative study critically examines elementary preservice teachers' (PSTs) mathematics histories, predominantly negative relationships with mathematics, to see them through a more positive lens and the impact on their teaching futures. Data from 75 PSTs' mathematics autobiographies across three teacher education programs in the Northeast United States over a 10-year period reveal that prior teachers are the primary influence on the PSTs' mathematics identities. This chapter showcases specific teacher and their classroom characteristics which shaped these identities. Themes that are explored in this chapter include connections with the teacher, learning environment, understanding the purposes of mathematics, use of manipulatives and hands-on learning, mathematics and gender roles, and timed assessments. Implications for mathematics methods courses, teacher effectiveness and professional development, and research on autobiographical inquiry are discussed.

INTRODUCTION: FRAMING THE STUDY

Mathematics anxiety, a feeling of tension, apprehension, or fear that interferes with mathematics performance (Ashcraft, 2002), is a global phenomenon (OECD, 2015). In the United States, approximately 25% of four-year college students and up to 80% of community college students suffer from a moderate to a high degree of mathematics anxiety (Yeager, 2012). College students who major in early childhood and elementary education have among the highest levels of mathematics anxiety among college majors (Hembree, 1990; Kelly & Tomhave, 1985; Malinsky, Ross, Pannells, & McJunkin, 2006), especially the anxiety about teaching mathematics. Teacher mathematics anxiety can impact time spent on mathematics instruction, student mathematics achievement, and student mathematics dispositions (Ganley, Schoen, LaVenia, & Tazaz, 2019). This anxiety can shape their mathematics identities (their mathematics self-concepts) since those who think they are bad at mathematics are more likely to be anxious (Dowker,

DOI: 10.4018/978-1-7998-9422-3.ch010

Sarkar, & Looi, 2016). Mathematics reform in the United States (NCTM, 1989; 2000; 2014; CCSSO, 2010) calls for fundamental changes in content and pedagogy of K-12 mathematics; a considerable amount of teacher change and teacher learning must take place to teach toward reform expectations (Drake, Spillane, & Hufferd-Ackles, 2001), which can contribute to mathematics anxiety.

The publication of the *Curriculum and Evaluation Standards for School Mathematics* (NCTM, 1989) launched the standards-based movement in the United States. From then until the present day, with additional publications/guidelines from the National Council of Teachers of Mathematics (NCTM 1991; 1995; 2000; 2006; 2014) and the introduction of the Common Core State Standards (CCSSO, 2010), methods of teaching mathematics have been constantly evolving and transforming so that students can be competitive in a global market. There is a greater focus on problem-solving, discussing and explaining their mathematical thinking, and overall, students becoming more actively involved in their own mathematics learning (Drake, 2006). Teachers of mathematics have the job to prepare students not only with career readiness skills, but to simultaneously cultivate their positive dispositions toward the subject (Van de Walle, Karp, & Bay-Williams, 2019). However, standards alone will not realize the goal of high levels of mathematical understanding (NCTM, 2014). It is the teacher who is responsible for teaching toward those standards, helping students realize those standards, and fostering a learning environment which is conducive to doing so. In many instances, reform-oriented practices require teachers to adopt new strategies not just for teaching students mathematics, but for understanding mathematics themselves (Drake, 2006). Becoming a teacher of mathematics now involves a deep knowledge of mathematics, persistence, a productive disposition, readiness for change, willingness to be a team-player, and making the time to be self-aware and reflective (Van de Walle, Karp, & Bay-Williams, 2019). Teachers need to teach about the methods and strategies of doing mathematics, which involve five process standards: problem solving; reasoning and proof; communication; connections; and representation (NCTM, 2000). A considerable amount of teacher change must take place, often out of their comfort zones, to teach using the expectation of reform-oriented methods. Going beyond those comfort zones with all that reform-oriented methods encompass can play a role in their mathematics anxieties, which can shape their mathematics identities. Teachers must embark on a process of re-forming their mathematics identities (Drake, Spillane, & Hufferd-Ackles, 2001). Therefore, it is important to consider affective issues in addition to cognitive issues in mathematics teaching.

According to the Association of Mathematics Teacher Educators (AMTE, 2017), a leading voice in teacher mathematics preparation in the United States, within the social context of mathematics teaching and learning, one role of the teacher is to cultivate positive mathematical identities with their students. Therefore, it is important for teachers to face their own mathematics anxieties so they do not perpetuate cycles of negative affect based on those anxieties to their students. On an encouraging note, despite the high degree of mathematics anxiety among preservice elementary teacher candidates, teachers' mathematics anxiety appears to be malleable through participation in mathematics methods courses (Gresham, 2007; Johnson & vanderSandt, 2011; Swars, Smith, Smith, & Hart, 2009).

This chapter amplifies preservice elementary teachers' (PSTs) voices through their mathematics autobiographies written during mathematics methods courses by using them as a "disruptive strategy" (Genor & Goodwin, 2005) that is purposefully structured to enable PSTs to develop new ways of thinking about mathematics teaching. PSTs come into teaching with pre-conceived notions based on their own life histories, which is succinctly substantiated by Nieto (2003) as she writes:

Teachers do not leave their values at the door when they enter their classrooms. As much as they might want to hide or avoid them, their values and beliefs slip in the door with them. In fact, teachers bring their entire autobiographies with them: their experiences, identities, values, beliefs, attitudes, hang-ups, biases, wishes, dreams and hopes. (p. 24)

This "disruptive strategy" (Genor & Goodwin, 2005) can help initiate and support authentic discussions about controversial and difficult issues by enabling PSTs to articulate (in writing and orally) what is often unconscious and deeply personal for them by making it a priority to consciously examine these issues for their benefit and those of their students.

The study presented in this chapter takes a look at PSTs' mathematics histories, mainly negative relationships with mathematics, to see them through a more positive lens. Previous research has focused on the use of mathematical autobiographies more passively for a moment in time and not over an extended period of time. There are many oral accounts (Drake, 2006; Drake, Spillane, & Hufferd-Ackles, 2001; Sfard & Prusak, 2005), but limited accounts of written autobiographies in mathematics education research (McCulloch, Marshall, DeCuir-Gunby, & Caldwell, 2013) with limited time allotted for teacher candidates to write their autobiographies, with prompts more like a question and answer rather than a well-developed narrative, and limited follow-up once the autobiographies were written. This study takes an active approach to work to shift preservice elementary teachers' often negative mathematics identities by critically examining their pasts to look toward their teaching futures.

CONTEXT OF THE STUDY: METHODOLOGY

Participants

Data for this chapter come from 75 preservice elementary teachers' (PSTs) mathematics autobiographies that were written in mathematics methods courses all taught by the author between 2010 to 2020 (specifically fall 2010, Fall 2010, fall 2016, fall 2017, fall 2019, and spring 2020 semesters) across teacher education programs in three institutions in the Northeast United States: a small Liberal Arts College, an Ivy League University, and a medium-sized State University. Forty-two were undergraduate students and 33 were graduate students. Seventy were female and five were male. Ages of the participants ranged from 20 to mid-50s. Table 1 shows the breakdown of the participants by year, school setting, level, and gender. Throughout this study, PSTs may also be referred to as teacher candidates; both terms are synonymous.

Data Sources

For the mathematics autobiographies, the PSTs had to respond to five prompts in narrative form: their definition of mathematics and the evolution of that definition; their relationship with mathematics and experiences which led to that relationship; greatest influences on their definition of mathematics and their relationship with the subject; how they learn best and their perception of "best" practices for teaching mathematics; and provide a title they feel best captures their mathematics story. The Appendix provides specific guidelines given to the PSTs. It was the first assignment of the semester in the mathematics methods course and teacher candidates were given two to three weeks to write these autobiographies.

Table 1. Preservice teacher participants

Semester	Setting	Level	Female Preservice Teachers	Male Preservice Teachers
Fall 2010	Small Liberal Arts College	Undergraduate	4	0
Fall 2010	Small Liberal Arts College	Graduate	12	3
Fall 2010	Small Liberal Arts College	Undergraduate	9	1
Fall 2016	Ivy League University	Graduate	9	1
Fall 2017	Ivy League University	Graduate	8	0
Fall 2019	Medium-Sized State University	Undergraduate	10	0
Spring 2020	Medium-Sized State University	Undergraduate	18	0

Following the submission of their autobiographies each semester, an informal debriefing conversation, predominantly student-lead, of approximately 30 minutes was held so PSTs could discuss their thoughts on the process of writing this autobiography with their classmates in the mathematics methods course. The conversations varied each semester depending on what the teacher candidates chose to share. The author took some informal notes about themes which surfaced from the conversations.

Data Analysis

Although the mathematics autobiographies consisted of five prompts, analysis for this study primarily focused on the prompts about their relationship with mathematics, experiences which led to that relationship, and influences on that relationship. A qualitative research design (Creswell, 2013; Guba & Lincoln, 1989) was utilized where data analysis began as soon as data were collected. Data were analyzed within and across cases utilizing Grounded Theory (Strauss, 1987; Strauss & Corbin, 1998) where theorizing grows from the data rather than from a pre-existing framework used to confirm or disconfirm a theory. Analytic memos (Saldaña, 2016) were utilized in this process. Inductive coding methods were primarily utilized during the first and second cycles of the coding process. To ensure trustworthiness of the data, member-checking (Guba & Lincoln, 1989) was utilized. Participants were asked any clarifying questions, as needed, to ensure that the author understood the intention behind their quotes in their autobiographies. It is important to note that although the PSTs may have referenced specific teachers by name in their mathematics autobiographies, only the first letter of the last name is noted in their quotes throughout this chapter for the purpose of confidentiality.

FINDINGS: TEACHER IMPACT

Data reveal that regardless of undergraduate or graduate status or preparation program, these preservice elementary teachers' (PSTs) mathematics stories are similar. Although the stories are highly personal, there are common themes/trends over this ten year period. Only 24% of the PSTs reveal having positive mathematics identities with little to no mathematics anxiety whereas the other 76% were along the continuum from extremely mathematics anxious to having some degree of mathematics anxiety, but overall had negative mathematics identities.

Data further reveal that how these PSTs feel about themselves in relation to mathematics (their mathematics identity) can be pinpointed to the influence of specific teachers in their lives. The grade levels may vary, but the constant across all autobiographies is the teacher, as evidenced by the following quotes from their mathematics autobiographies:

I think math is a difficult area for many people, which I believe can be due in part with how it's introduced to people early on in their life...Teachers have a lot to do with how students feel about math. If a teacher rushes through the material, loses patience, or dismisses a child's question or concern, the child will struggle. I have had my fair share of teachers just like this, and it made learning the material unpleasant. (Graduate PST, Fall 2010)

I truly believe that my greatest influences for the way I look at math are my teachers. They are the ones that taught me the methods to understand the way the world works through numbers. (Graduate PST, Fall 2010)

I think a teacher/professor has the ability to show and give the support to make a great classroom and have a good experience on the way a student is taught...Teachers and professors have a large impact on their students and if I did not have some of the educators in my life, I do not know if this is the path [to become an elementary teacher] I would be on and I am very grateful. (Undergraduate PST, Fall 2019)

My relationship with mathematics has changed depending on the class and the teacher; it can either be a positive or negative relationship. The experiences that have led me to this relationship is that from first through sixth grade, I was horrible at math...However, in 8th grade, I had the best math teacher I have ever had in my life... I told him on day one that I was not good at math and he was determined to fix and change my attitude about math. And he did just that. I loved that year of math because of him and how he motivated me to do it by asking me 'fight or flight?' He asked me if I wanted to learn how to do math and feel confident while doing it (fight) or if I wanted to run away from it and never feel comfortable doing math (flight) and I chose to fight...It's unfortunate that the teachers I had after him stole my confidence and made me scared of math. They would never seem to want to help out. (Undergraduate PST, Spring 2020)

As just revealed in the PSTs' own words, teachers throughout their schooling have been shown to be the greatest influences on these PSTs' mathematics identities. However, what is it specifically about the teachers that shaped their relationships with mathematics and what can be learned from how they want or do not want to teach as a result? is the focus of this chapter. This chapter flips the script and uses often negative experiences to accentuate the positive, and what can be learned from these stories to help PSTs become the strongest mathematics teachers they can be in the mathematics reform-oriented teaching climate. Both in their mathematics autobiographies and in their debriefing conversations, many PSTs repeatedly noted that they do not want to teach in ways that their teachers did so this study sheds light on some characteristics of the teacher, even if negative, to showcase the positive that can come out of negative experiences and what lessons can be learned from those experiences.

This chapter analyzes the themes which surfaced in the PSTs' own words through detailed, rich quotes, which are excerpted primarily from their mathematics autobiographies, and the significance of them. All themes focus on teachers, including personal characteristics and the way they structured the teaching and learning environment. Although there are common threads woven throughout the themes, they are presented separately to highlight specific elements of each theme. The following themes are detailed in this chapter: connections with the teacher: care; learning environment; understanding the purposes of mathematics: the "whys"; use of manipulatives and hands-on learning; mathematics and gender roles, and timed assessments.

Connections with the Teacher: Care

Across all autobiographies, the most common theme among the preservice teachers (PSTs) which influenced the way they felt about mathematics is the connections or lack thereof with their teachers. Those connections took different forms, but the one constant was the idea of teachers caring (or lack of caring) about them, which is evident in the following quotes:

Mrs. B (grade 6) did much more for me than just helping me with my times tables. The way she taught made everything seem clear to me and the things that were not so clear she made me feel comfortable to ask for help. She really cared about math and her students and I think that rubs off on the student and makes them better...That is why I feel like the teacher has a major impact on a child and the way that child feels about a particular subject. (Undergraduate PST, Fall 2010)

I remember feeling like my teacher hated me and never wanted to help me in any subject [4th grade]... Once I moved to middle school, I regained my confidence through amazing helpful teachers in fifth and sixth grade...But in seventh grade, once again, I met another teacher who, despite my cries for help, would never take the time I needed or even bother to form a relationship with me. At this point, I started to dislike all math teachers and relied heavily on the help of my friends...Over time, I realized my main learning method is through my relationship with my professors. When I feel a connection with my teachers I am much more open and comfortable to try, fail, and learn. (Undergraduate PST, Fall 2019)

The PSTs noted that *caring* took on different forms. For some, it meant that the teacher was willing to expend the extra effort to make sure they were successful students since the teaching methods were tailored to their specific learning needs, as exemplified in the following quotes:

Having teachers who took the time to work with me also made a big difference. During my time in grade school I didn't always have teachers who helped me to understand as best I could. Finally, in seventh grade, my math teacher helped make everything click. She would work with me before school and during free periods and after school. Her extra effort helped me grow as a math student. She worked to break down the problems and taught me to look at one part of the equation at a time. She helped me slow down when working out a problem so I wouldn't make silly mistakes. (Undergraduate PST, Fall 2010)

When I reached high school, possibly the most positively influential teacher I ever had was my freshman math teacher, Mr. R. Not only did he know math better than anyone I had met before, he had a genuine gift for teaching it to his students. One way in which he was great was his ability to teach to all of the students, no matter their particular learning style. It was amazing that every student inside of his classroom, at least for that one year, enjoyed math. Another reason everyone felt comfortable in his class is because Mr. R was always there to offer extra help, always. He was the first teacher in the school, he ate lunch in his classroom while helping students, and he was the last teacher to leave at night. He would even help students who were not in his classes, and would help students who graduated from the high school and were having trouble in college. I knew that math would always be my favorite subject because I always knew there was someone to help if I ever got lost. (Graduate PST, Fall 2010)

My ninth grade math teacher truly helped me the most throughout high school. Even when I didn't have her as a teacher, I would go into her office during her free periods and ask her to explain something to me. She wouldn't care if it was 10th, 11th, or 12th grade math; she could explain it to me better than anyone else could. She took the time out of her busy schedule to help me, a previous student, understand something from a class she isn't even teaching to me. She always found a way to draw out or use manipulatives to show me what she was doing...She helped me believe in myself. (Undergraduate PST, Spring 2020)

Yet for other PSTs, *caring* took the form of teachers having high expectations for their students' successes which meant not giving up on them, as showcased in the following quotes:

When people expect less of you, often you only meet those lower expectations...I have had a mix of good and bad teachers that have, in a way, sent me on a roller coaster of successes and failures in mathematics. (Graduate PST, Fall 2010)

Once high school hit, I was getting A's and my view of math drastically changed. I was doing tremendously well, and I enjoyed what I was doing. I think my success was related to my teachers' ability to teach the subject, bring out my best potential in the subject, and not give up on me...My teachers are my greatest influences. They were the ones who made me despise math because I didn't understand what they were teaching or what I was learning and I never got the proper help or attention to understand. In addition, I felt as if they gave up on me and saw me as a lost cause. I wanted the chance to be able to succeed and I wanted them to think I could do it. The teachers that helped me succeed were also the ones who not only brought out the best in me, but challenged me and really made me think without frustration. These teachers were a positive influence on me and gave me the courage to not be completely afraid of math and taught me that if I applied myself I could succeed. (Undergraduate PST, Fall 2010)

Overall, the PSTs' mathematics autobiography excerpts in this section show that making connections with teachers and teachers who make connections with their students can have a tremendous impact on their mathematics identities. These connections do not happen automatically; they take work. However, having teachers who genuinely care about their students and go out of their way to tailor their instruction to their students' needs and have high, yet realistic, expectations is something which has lifelong impact.

Learning Environment

From reading the quotes in the previous section about caring and connections, it is evident that the teacher and classroom environment are intertwined since it is the teacher who sets the tone for the classroom environment. According to the preservice teachers (PSTs), the learning environment should be conducive for students learning about the mathematics content and forming positive dispositions toward the subject. Some characteristics of the learning environment which impacted the PSTs included structure, comfort, engagement, and variety as shown in the following excerpts:

I am the type of student that needs a serious and organized math teacher or else I am lost...The people who have helped me get through all the struggles in math are the ones I remember forever...having a good learning environment where students can be comfortable and focused is very important for me. I could become easily distracted and frustrated if I am not in a good learning environment. (Undergraduate PST, Fall 2019)

More than once, I have overheard elementary classroom teachers (when no students were present), declare that they never liked doing math when they were younger. So I can fully understand why it is important to emphasize to potential elementary classroom teachers why they need to improve their own attitudes towards math, for fear of passing on their negative attitudes to their impressionable students. Obviously, someone who has a limited math background will not be very creative when it comes time to begin crafting effective mathematics instructional activities...After casually observing how elementary teachers teach their students for the last four years, I also have a much greater realization now that the teacher's role is to make sure students understand the content, not just to 'cover the content and keep up with the calendar schedule.' I now understand that no matter how wonderfully you design a lesson plan, if you don't maintain the proper instructional setting that allows students to concentrate (management), and if you don't fully engage the minds of the students throughout the lesson (keeping them on-task), then you aren't really doing a very good job of teaching the students effectively. (Graduate PST, Fall 2010)

Though I enjoy math now, I did not always. I strongly believe that teachers scare children away from math by putting rules on them and presenting it in a very one-sided way. Students need variety when it comes to math and should be given all avenues when they are presented a unit or section. I think making math fun, interesting, and giving them variety will make it so children are not so scared anymore and will be able to enjoy math more. (Undergraduate PST, Fall 2010)

Although the types of learning environments that PSTs prefer varies based on their personal needs, these excerpts exemplify that a teacher needs to create an environment that is conducive to bringing out the best in their students. Some of those elements include the way the environment is structured, which relates to classroom management, how comfortable a student is made to feel, how engaged they feel, and how varied the teaching methods are. When students are in a learning environment which meets their needs, it can reflect their developing dispositions toward the subject.

Understanding the Purposes of Mathematics: The "Whys"

Extending what was noted about the learning environment in the previous section, teachers who structured their classrooms to teach the concepts for depth of understanding, as opposed to merely the procedures, positively influenced the preservice teachers (PSTs). Those teachers who explained the "why" behind the concepts and procedures deepened PSTs' knowledge and positively impacted their mathematics identities since they better understood the purpose(s) of what they were learning and why they were learning it, as showcased in the following quotes:

I did not really begin to find an enjoyment in math until I was in my senior year of high school. My teacher showed me that math had a purpose besides basic adding and subtracting and it could be fun. She was able to teach math that was complex and I became good at it. Eventually it changed from just answering the question to knowing why I was answering the question and how I got the answer. (Undergraduate PST, Fall 2010)

Freshman year of college, I had officially decided I wanted to become a teacher. I was required to take two courses pertaining to Mathematics of Elementary Education Majors. This is when I truly felt like everything began to click for me. My teacher for MAT 106 was amazing and for some reason everything just made sense…what I loved about this class was learning the 'why' behind everything so that when I was faced with a student who was like me, who didn't quite understand, I could answer them to the best of my ability and make sure they completely understood the problem they were faced with…I want my students in the future to not fear math the way I did. If math is something they are struggling with, I hope that I am the person that can instill a growth mindset in them. (Undergraduate PST, Spring 2020)

The best teacher, who probably changed my views the most, was Ms. G. She would sing formulas to help the class remember them. I once asked her, as I asked every other math teacher, 'When am I going to use this?' While most teachers would answer, 'Don't worry, just work on the problem,' she said, 'You need it for the test, and while you may not need the more difficult equations, you may need it in certain professions.' I thought that this was a good answer considering most teachers ignore that most-asked question. (Undergraduate PST, Fall 2010)

I have always learned mathematics best when it applies to something I might come across in real-life situations or when the lesson was truly engaging and didn't involve the students just sitting and listening for the entire class period…So often students ask 'When will I ever need to know this in real life?' and it is important to be able to have an answer when this questions does arise, otherwise it makes it much easier for students to give up. (Undergraduate PST, Spring 2020)

The latter three quotes reveal the larger perspective of "why" in terms of real-life applications whereas the first excerpt showcased the "why" of particular concepts more narrowly, more in the moment. Both types of "why", the more "practical why" and the more "fundamental why", are meaningful to formulate mathematical dispositions. Most importantly, however, is having a teacher who acknowledges students' questions and does not brush off or squelch their curiosity.

Use of Manipulatives and Hands-on Learning

Building upon the "whys" in the previous section, the use of manipulatives helped preservice teachers (PSTs) to get at the heart of a concept. They seemed to gain the most from teachers who employed interactive and kinesthetic methods as opposed to just lecturing and providing one-way/linear conversations (generally teacher to students), as illustrated in the following excerpts:

In kindergarten my teacher, Mrs. F, used the tactile kinesthetic approach. This is when teachers use actual manipulatives which students can touch and visually see to solve problems. Since I am a multisensory learner, this way of teaching was beneficial to me as I could solve the equations using my vision and sense of touch to add to or subtract objects. One lesson we used was actually going outside and picking leaves off of the trees, bringing them back to the classroom, and using them as our counters for addition or subtraction problems. (Undergraduate PST, Fall 2010)

My favorite math lesson was with Mrs. H in first grade. She knew how to make kids enjoy themselves while learning. The coolest lesson involved food. I'll never forget it. We had to complete a worksheet. If we got the answer correct we could take as many pieces of food as the number. For example, if you got number 1 right, you'd take one cookie. If you got number 10 right, you would take ten Skittles. The kids loved her. Our dentists and parents? Maybe not so much. What was awesome about Mrs. H was that we'd never forget the process. We'd work so diligently to get the answer. I suppose one could argue she was bribing us with food, but more importantly, she was teaching us how to group numbers together. (Graduate PST, Fall 2010)

Generally, manipulatives are thought of as tools to be used only in preschool and the early elementary grades. However, those teachers who employed them in older grades had a significant impact on their students' conceptual development, as evidenced by the following quotes:

Some of my fondest memories applying to math took place in my fifth and sixth grade classes with the same teacher, Mrs. C. From what I can recall, she was big on using manipulatives like blocks and geoboards. They were extremely helpful in keeping the subject alive, as opposed to the dryness most teachers unintentionally provide. (Undergraduate PST, Fall 2010)

I found that the most defining mathematical moments for me have been during the interactive math experiences and real-life applications of math concepts. In fifth and sixth grade, my math teachers had us participate in activities extended over multiple classes of a stock market game and a game that played out real-life financial and business situations to help us in seeing how being knowledgeable about math would help us in the future. I still remember certain parts of these activities, meaning that they were impactful to me and ultimately an efficient technique in teaching math in an easy-to-understand way. (Undergraduate PST, Fall 2019)

If anything, I learned the most out of trigonometry because of my senior year math teacher; she helped me love math and the fun project she had us do throughout the school year. My favorite thing we made was a hot dog cooker. We had to come up with the measurements then use the equations for the parabola to find the focal point. For this project, the focal point is the point that would cook the hot dog best when in the sun. We teamed up with the technology class and everyone got a partner. We worked together building the hot dog cooker. The tech students would show the math kids how to use the equipment and the math students would show the tech kids how we were figuring out the measurements. (Undergraduate PST, Spring 2020)

From reading these quotes, the PSTs explained the value of manipulatives to them. They could retain the mathematical concepts better since they had a deeper understanding of them because they were brought to life through real-world applications. The also noted the benefit of using them in higher grade levels too. Although it may take some creativity on the part of the teacher to match appropriate manipulatives with concepts, it is well worth the effort when students greater retain the material.

Mathematics and Gender Roles

At an early age, students pick up on who can and cannot do mathematics. Sometimes those messages are conveyed directly while other times they are indirect; messages are gained through actions and who students see in front of them teaching mathematics. From some of the preservice teachers' (PSTs) quotes, they have clearly internalized those messages, which have had an impact on their mathematics identities, as showcased through the following excerpts:

The summer between kindergarten and first grade, I went to a town camp. I vividly remember the girls talking about how hard first grade would be with 'real math'. However, first grade came and I had the most incredible teacher who redefined math for me. Looking back, I realize that she was accustomed to having female students come to her from the male kindergarten teacher with the predisposed notion that math was for boys. The first thing she did in first grade was mix the girls and the boys for math group. To this day, my family and I laugh about me learning that math wasn't only for boys in first grade... Throughout the rest of middle school, my math experience was less than positive. It clearly had much to do with my feelings for the teacher. I now know that this was related to my level of comfort as well as the tone set in the classroom with regard to math. (Graduate PST, Fall 2010)

Interestingly, the academic influences on my mathematical development have also been largely male. From sixth grade through my college graduation, I had just one female math teacher in seventh grade. All of my other teachers and professors were male. I took honors math courses as soon as they were available to me in eighth grade, and it just so happened that all of the teachers of honors math courses in my schools were men. I do wonder if the presence of so many male teachers in my education has influenced me to think that as a female, I am predisposed to do poorly in mathematics. As I continued in honors math classes, I did not necessarily have a female role model in the field. I wonder if I would have had greater confidence in my abilities if I had a female role model. Over the years, I never received an explicitly gendered approach to mathematics, but I have considered the unspoken truth that all the honors math teachers, who were theoretically the 'best' teachers of mathematics, happened to be men. Perhaps I internalized the message that the 'best' mathematicians were male, and thus I could never attain that level of competence and skill in mathematics. (Graduate PST, Fall 2017)

[Freshman Year College] At this point in my own math journey, I was starting to realize a trend. Although my late high school teachers were amazing in their own respects, I noticed that math was done by mostly males. I believe that this started to subconsciously manifest itself in me...At this point, my thoughts were confirmed: mathematics was a man's game. (Graduate PST, Fall 2017)

The aforementioned quotes reveal that this unspoken idea of who is capable of doing mathematics is internalized early on in one's education. It is linked to the learning environment a teacher creates, which sets the tone for the classroom, and who the student consistently sees in front of them teaching mathematics. However, on a larger scale, it is also linked to societal messages and stereotype threats (Steele & Aronson, 1995) as they relate to mathematics and gender.

Timed Assessments

A huge factor reported to cause mathematics anxiety by many preservice teachers (PSTs) is the use of timed assessments, often having to be completed within a minute or two. Students internalize the message that only speed matters in their mathematics abilities and that being faster equates with better or more competent/capable/advanced abilities, as demonstrated by the following quotes:

Second grade: Much of our math time is devoted to competition. The competition is concerned with the completion of math books. The color-coded math books make it clear who is advancing quickly and who is not. I couldn't help but notice that Alex is on Purple, while I'm still on Yellow. In a school founded on beliefs opposed to pointless competition between individuals, racing through math books provided an outlet for our competitive drive. It was a moment of pride and self-gratification when you announced to the teacher, and in actuality everyone else in the room, that you were ready for the next color of book. The math books often asked at the end of section, 'how are you feeling?' The book provides a blank face and instructed us students to draw our current emotional state. I often made the face into my favorite Pokémon. But often too, I made my face sad. The math was getting harder with each color. (Graduate PST, Fall 2016)

One thing I absolutely hated in elementary school was when we had to do the sixty second timed worksheets. I am the person that likes to take their time with math so I can actually understand and comprehend what I am doing. I felt anxious and rushed during these quizzes because I had very little time to actually do the worksheet. (Undergraduate PST, Spring 2020)

As a child in elementary school, I remember feeling stressed during math lessons because my teachers clearly were not comfortable with the content area either. They would make us do 'math minutes' and I honestly think that if a 6 year old could have a heart attack, I probably came close in those 60 seconds. Those do not teach us or teachers much. (Undergraduate PST, Spring 2020)

I remember having timed math quizzes that required mostly memorization and a calm composure. If you were like me and needed the time to count the numbers on your fingers underneath the desk, you ran out of time. These first memories of math shaped the way I viewed myself as a math student and hindered the way I looked at math as a whole. (Undergraduate PST, Spring 2020)

The last quote was particularly poignant because of the lasting impact that these seemingly benign short assessments can have on one's life. Although these sixty second assessments may seem innocuous, the accumulation of them and the message students come away with is significant, notably feelings of stress, anxiety, and being "less than".

DISRUPTING THEIR HISTORIES: TOWARD CULTIVATING FUTURE POSITIVE RELATIONSHIPS WITH MATHEMATICS FOR THEIR STUDENTS

Despite the anxiety and negative feelings toward mathematics that many preservice teachers (PSTs) formed as a result of their teachers and their classroom environments, as shown by the themes in the previous sections, something encouraging that came out of this study is that by having to write their mathematics autobiographies in mathematics methods courses, PSTs are becoming more self-aware and beginning to disrupt their history to now want to teach in ways many never experienced growing up. They are taking their often negative experiences to internalize how they want/do not want to teach their future students, which help them to see mathematics in a more positive light. They seem to aspire to teach using reform-oriented methods which focus more on process, depth of understanding, and students becoming more actively involved in their own mathematics learning so that their students form positive dispositions toward mathematics, which impact their mathematics identities. How the PSTs envision teaching mathematics is personal since it stems from their mathematics histories. Although they may have had some deficits, they do not want to transmit those to their future students and be an impediment. A representative quote from some of the themes which surfaced as the PSTs described potential relationships with their future students are presented to show the range of critical reflection which stemmed from looking inward at their mathematics histories.

Some PSTs want to make sure their future students *feel supported*:

I don't want math to be a scary subject for my students. I know that in my classroom there will be students like me struggling with math, but I want to make sure I take the extra time out of the period to help them better understand. Also, I feel that it is important to be supportive and positive by developing my students' confidence in math. I know when I had a teacher that was supportive it made me want to strive. (Graduate PST, Fall 2010)

Others want to make mathematics *enjoyable*:

I want my students to have a positive experience with math, not a negative and stressful one I had growing up. I never want my students to feel stupid or dumb like I did. So math, please be good to my future students. Help me make it enjoyable for them. (Undergraduate PST, Spring 2020)

Yet other PSTs are making it a priority to *not instill gender mathematics stereotypes*:

[Junior Year College] I was majoring in elementary education. In my math foundations course, I was asked to write a research paper that dealt with elementary mathematics. I chose to focus on the feelings of elementary girls in math. I found a book titled: The Truth About Girls and Boys: Challenging Toxic Stereotypes About Our Children by Caryl Rivers and Rosalind C. Barnett. In this book, I found a possible answer to this problem. The book discusses how girls are less likely to succeed in math if their elementary teachers mention 'I'm bad at math', or if the teacher treats the girls in math differently. At this, I decided to not be a teacher that pushed girls away from math. (Graduate PST, Fall 2017)

Still others want to *re-define failure*:

Not only did math become irrelevant and dull, it also became increasingly associated with speed and competition, with less emphasis on knowledge and understanding…It's surprisingly hard to write about this, even now. Mathematical failure – much like romantic failure – leaves us raw and vulnerable. It demands excuses. But I'm done making excuses for the sake of my future students. Failure isn't about a lack of 'natural intelligence,' whatever that is. Instead, failure is born from a messy combination of bad circumstances: high anxiety, low motivation, gaps in background knowledge. Most of all, we fail because, when the moment comes to confront our shortcomings and open ourselves up to teachers and peers, we panic and deploy our defenses instead. (Graduate PST, Fall 2017)

Furthermore, other PSTs want their students to take their time with mathematics to see it as a *learning process*:

I wish I had a good relationship with math. One of my main goals as a future elementary teacher is to make sure my students enjoy learning math. I don't want them to despise math as much as I did. I want my students to think of math as a slow learning process. Sometimes it is better to take time in completing problems then to rush and make little mistakes along the way. (Undergraduate PST, Spring 2020)

Finally, PSTs want students to feel *comfortable* with mathematics so that they *learn to love it*:

Although mathematics has been a subject that I have not always loved or found the most interesting, as a future teacher I would love to try and make it interesting for my students so that they can love math like how I did when I was a kid. When I become a teacher, my goal is going to be to make the kids that hate math feel more comfortable with it. (Undergraduate PST, Spring 2020)

As noted in the introduction of this chapter, becoming a teacher of mathematics in the climate of mathematics reform involves a deep knowledge of mathematics, persistence, a productive disposition, readiness for change, willingness to be a team-player, and making the time to be self-aware and reflective (Van de Walle, Karp, & Bay-Williams, 2019). The notions of support, enjoyment, combating gender stereotypes, re-defining failure, mathematics as a learning process, and comfort with the content area all have a place within mathematics reform. By writing their mathematics autobiographies, many PSTs have a new self-awareness and want to actively make a change in/disrupt how they were taught so they do not perpetuate negative cycles of mathematics affect in their future students.

EDUCATIONAL SIGNIFICANCE: IMPLICATIONS FOR PRACTICE AND RESEARCH

Teachers of all grade levels have an impact on student learning. Reading the rich quotes in the preservice teachers' (PSTs) own words throughout this chapter exemplifies this in addition to prior research and one's own anecdotal experiences. This chapter, through PSTs' own voices, showcases specific areas of impact. This study takes a comprehensive look into how teachers can shape dispositions toward mathematics, and therefore identities, in their students which can have lifelong impact. This shaping can be in large or small ways; some of which might not seem significant to an outside reader, but could mean the difference in how a person feels about themselves in relation to mathematics for their lifetime, as revealed in some of the quotes in earlier sections. The power of this chapter is that it puts PSTs at the heart of it since teachers' voices are often stifled or muted in educational conversations. This chapter is not from a deficit perspective, but uses some negative experiences to move in positive directions and understand what can be learned from them. Based on ten years' worth of data across three different teacher education programs, this study has tremendous educational significance for coursework, professional development, and the research base.

Mathematics Methods Courses

This study has implications for mathematics methods courses. The importance of including a mathematics autobiography assignment at the start of a semester is evident. The mathematics autobiographies can serve as baseline data each semester so that a professor can use that information to structure in-class activities/tasks, methods of instruction, and to select additional course readings. Knowing their teacher candidates' dispositions toward mathematics and their relationships with it can also help a professor connect on a deeper level with their PSTs so that meaningful experiences can be planned during the semester. The prompts for the mathematics autobiography can also be tailored to reflect specific information a professor wants to know about their teacher candidates. The work in the mathematics methods course can also inform student teaching and early career teaching since it has the potential to set teacher candidates on more positive mathematics teaching journeys if time is taken during the methods course to have PSTs confront their own mathematics identities.

Teacher Effectiveness and Professional Development

Stemming from the discussion in the earlier section, Disrupting their Histories: Toward Cultivating Future Positive Relationships with Mathematics for their Students, this study can also add to the base of teacher effectiveness in mathematics from which professional development can be designed. In beginning to disrupt their histories, PSTs noted their personal mathematics teaching aspirations based on how they were or were not taught. What they really described were varied characteristics of mathematics reform-oriented practices which they considered effective. But, teacher effectiveness is not as simple as formulating a list of "do's" and "don'ts" particularly since effectiveness is a relative term; what is effective for one student may not be effective for another. However, if a professor has teacher candidates engage in autobiographical inquiry each semester, and compiles and analyzes those data from one semester to the next, trends in effective mathematics teaching practices can be identified over time and linked to aspects of mathematics reform-oriented practices. Data from those trends can be the basis for both preservice and inservice teacher professional development.

Research Base on Mathematics Autobiographical Inquiry

This study can add to the base of knowledge on autobiographical inquiry as a "disruptive strategy" (Genor & Goodwin, 2005) not just for the moment, but for longer impact since it can be helpful for the preservice teachers (PSTs) and their students. During the debriefing conversations each semester, the PSTs noted that just being afforded the opportunity to write about their mathematics histories and then to debrief the experience with their peers already helped some PSTs to be less anxious about teaching mathematics. The process was "cathartic" for some and put some on the path of "self-discovery" since they could be truthful with themselves. They felt that being "given permission" to talk about the topic made mathematics "less daunting". They also appreciated not having to "stifle their emotions". This experience, in turn, can help them to find ways not to perpetuate negative cycles of mathematics affect. Additionally, the PSTs also appreciated being given two to three weeks to write their autobiographies since it gave them time to really think through the prompts to select vignettes which exemplify their relationship with mathematics and not do it on the fly. Since they did not have to answer the prompts in any specific order, it also allowed for some creativity in composing their narratives.

FUTURE RESEARCH DIRECTIONS

With data over a 10 year period of time, this study shows that still more work needs to be done to prepare teachers to teach in reform-oriented ways, and to work to understand and shift preservice elementary teachers' (PSTs) often negative mathematics identities. Future research building upon and extending this study will use mathematics autobiographies as a springboard to set individualized mathematics goals with PSTs in mathematics methods courses to support their teaching and learning in the course and beyond in clinical experiences. The longitudinal study will follow a sampling of PSTs through student teaching and into the first year or two of their teaching careers to learn about their progress in realizing those goals, and the impact that goal-setting can have on their students and their own mathematics identities. The author can then also see which reform-based teaching practices PSTs are applying in their teaching.

The author also plans to build in an additional component to the mathematics autobiographies by asking PSTs toward the end of the semester to write a continuing paragraph or two to their initial stories based on any shifts/new understandings they may have made over the course of the semester as a result of being in the mathematics methods course.

CONCLUSION

Mathematics anxiety is often a barrier to high quality teaching for preservice elementary teachers (PSTs). This qualitative study critically examined PSTs' mathematics histories, predominantly negative relationships with mathematics, to see them through a more positive lens and the impact on their teaching futures. This study focused on the affective side of mathematics teaching and learning, which is often given less attention than the cognitive side. At the heart of the study was a focus on mathematics identities through autobiographical inquiry, and bringing difficult issues to light in productive discourse to disrupt perpetuating often negative cycles of mathematics affect. This chapter showcased specific teacher and their classroom characteristics which shaped these identities. However, this chapter was not from a deficit perspective. By focusing on themselves, PSTs could focus more directly on their students and could confront tough questions about their mathematics identities and how they want or do not want to teach as a result of their mathematics histories.

REFERENCES

Ashcraft, M. H. (2002). Math anxiety: Personal, educational, and cognitive consequences. *Current Directions in Psychological Science, 11*(5), 181–185. doi:10.1111/1467-8721.00196

Association of Mathematics Teacher Educators (AMTE). (2017). *Standards for Preparing Teachers of Mathematics.* Retrieved from amte.net/standards

Council of Chief State School Officers (CCSSO). (2010). *Common Core State Standards for Mathematics.* Washington, DC: CCSSO.

Creswell, J. W. (2013). *Qualitative inquiry and research design: Choosing among five approaches* (3rd ed.). SAGE Publications.

Dowker, A., Sarkar, A., & Looi, C. Y. (2016). Mathematics anxiety: What have we learned in 60 years? *Frontiers in Psychology, 7*(508), 1–16. doi:10.3389/fpsyg.2016.00508 PMID:27199789

Drake, C. (2006). Turning points: Using teachers' mathematics life stories to understand the implementation of mathematics education reform. *Journal of Mathematics Teacher Education, 9*(6), 579–608. doi:10.1007/s10857-006-9021-9 doi:10.100710857-006-9021-9

Drake, C., Spillane, J. P., & Hufferd-Ackles, K. (2001). Storied identities: Teacher learning and subject matter context. *Journal of Curriculum Studies, 33*(1), 1–23. doi:10.1080/00220270119765

Ganley, C. M., Schoen, R. C., LaVenia, M., & Tazaz, A. M. (2019). Construct validation of the math anxiety scale for teachers. *AERA Open, 5*(1), 1–16. doi:10.1177/2332858419839702

Genor, M., & Goodwin, A. L. (2005). Confronting ourselves: Using autobiographical analysis in teacher education. The *New Educator, 1*(4), 311–331. doi:10.1080/15476880500283021

Gresham, G. (2007). A study of mathematics anxiety in pre-service teachers. *Early Childhood Education Journal, 35*(2), 181–188. doi:10.1007/s10643-007-0174-7 doi:10.100710643-007-0174-7

Guba, E. G., & Lincoln, Y. S. (1989). *Fourth generation evaluation.* Sage Publication, Inc.

Hembree, R. (1990). The nature, effects, and relief of mathematics anxiety. *Journal for Research in Mathematics Education, 21*(1), 33–46. doi:10.2307/749455

Johnson, B., & vanderSandt, S. (2011). "Math makes me sweat": The impact of pre-service courses on mathematics anxiety. *Issues in the Undergraduate Mathematics Preparation of School Teachers, 5*, 1–8.

Kelly, W. P., & Tomhave, W. K. (1985). A study of math anxiety/math avoidance in preservice elementary teachers. *The Arithmetic Teacher, 32*(5), 51–53. doi:10.5951/AT.32.5.0051

Malinsky, M., Ross, A., Pannells, T., & McJunkin, M. (2006). Math anxiety in pre-service elementary school teachers. *Education, 127*(2), 274–279.

McCulloch, A. W., Marshall, P. L., DeCuir-Gunby, J. T., & Caldwell, T. S. (2013). Math autobiographies: A window into teachers' identities as mathematics learners. *School Science and Mathematics, 113*(8), 380–389. doi:10.1111/ssm.12041 doi:10.1111sm.12041

National Council of Teachers of Mathematics (NCTM). (1989). *Curriculum and evaluation standards for school mathematics*. Reston, VA: NCTM.

National Council of Teachers of Mathematics (NCTM). (1991). *Professional standards for teaching mathematics*. Reston, VA: NCTM.

National Council of Teachers of Mathematics (NCTM). (1995). *Assessment standards for school mathematics*. Reston, VA: NCTM.

National Council of Teachers of Mathematics (NCTM). (2000). *Principles and standards for school mathematics*. Reston, VA: NCTM.

National Council of Teachers of Mathematics (NCTM). (2006). *Curriculum focal points for prekindergarten through grade 8 mathematics: A quest for coherence*. Reston, VA: NCTM.

National Council of Teachers of Mathematics (NCTM). (2014). *Principles to actions: Ensuring mathematical success for all*. Reston, VA: NCTM.

Nieto, S. (2003). Teaching as autobiography. In S. Nieto (Ed.), *What keeps teachers going?* (pp. 22–36). Teachers College Press.

OECD. (2015). *Does math make you anxious? PISA in Focus, 48*. doi:10.1787/5js6b2579tnx-

Saldaña, J. (2016). *The coding manual for qualitative researchers*. Sage Publications, Inc.

Sfard, A., & Prusak, A. (2005). Telling identities: In search of an analytic tool for investigating learning as a culturally shaped activity. *Educational Researcher, 34*(4), 14–22. doi:10.3102/0013189X034004014

Steele, C. M., & Aronson, J. (1995). Stereotype threat and the intellectual test performance of African Americans. *Journal of Personality and Social Psychology, 69*(5), 797–811. doi:10.1037/0022-3514.69.5.797 PMID:7473032

Strauss, A. L. (1987). *Qualitative analysis for social scientists*. Cambridge University Press. doi:10.1017/CBO9780511557842

Strauss, A. L., & Corbin, J. (1998). *Basics of qualitative research: Techniques and procedures for developing grounded theory* (2nd ed.). Sage Publications, Inc.

Swars, S. L., Smith, S. Z., Smith, M. E., & Hart, L. C. (2009). A longitudinal study of effects of a developmental teacher preparation program on elementary prospective teachers' mathematics beliefs. *Journal of Mathematics Teacher Education, 12*(1), 47–66. doi:10.1007/s10857-008-9092-x doi:10.100710857-008-9092-x

Van de Walle, J. A., Karp, K. S., & Bay-Williams, J. M. (2019). *Elementary and middle school mathematics: Teaching developmentally* (10th ed.). Pearson Education, Inc.

Yeager, D. S. (2012). *Productive persistence: A practical theory of community college student success*. Paper presented at the 2012 Annual Meeting for the American Educational Research Association, Vancouver, Canada.

ADDITIONAL READING

Gujarati, J. (2013). An "inverse" relationship between mathematics identities and classroom practices among early career elementary teachers: The impact of accountability. *The Journal of Mathematical Behavior*, *32*(3), 633–648. doi:10.1016/j.jmathb.2013.08.002

Saran, R., & Gujarati, J. (2013). Moving toward positive mathematics beliefs and developing socio-mathematical authority: Urban preservice teachers in mathematics methods courses. *Journal of Urban Learning. Teaching and Research*, *9*, 100–111.

KEY TERMS AND DEFINITIONS

Manipulatives: Hands-on objects used to aid students in conceptual mathematics development.

Mathematics Anxiety: Feelings of fear, apprehension, or discomfort with mathematics which can interfere with mathematics performance.

Mathematics Autobiography: Story that describes the evolution of who one is in relation to mathematics.

Mathematics Identity: One's sense of self in relation to mathematics.

Mathematics Reform: Standards-based approach to mathematics education in the United States.

Preservice Teachers: Students who are pursuing an initial teaching certification and have little to no prior classroom experience. Also synonymous with teacher candidates.

Stereotype Threat: Risk of conforming to stereotypes about a particular group to which one belongs.

Teacher Mathematics Anxiety: Feelings of fear, apprehension, or discomfort about having to teach mathematics.

Timed Assessments: Quizzes, often focusing on mathematics facts, which students have to complete within a short period of time, often sixty seconds.

APPENDIX

Mathematics Autobiography Guidelines

How we as teachers operate in the classroom—the activities we facilitate, the teaching methods we utilize, and our interactions with students—are often rooted in the ways in which we learned and felt about learning ourselves. The intent of this mathematics autobiography is for you to honestly recall and reflect on your prior mathematics experiences from your early childhood to the present. Use the following prompts to guide/frame your reflections:

- What is mathematics? In other words, if you were asked to define it, how would you respond? How did you come to believe mathematics as you have defined it? How might your definition have evolved over time?
- What is your relationship with mathematics? What experiences led you to this relationship?
- Who or what do you believe have been your greatest influences on how you define mathematics and how you feel about the content area?
- How do you learn mathematics best? Subsequently, what do you believe are "best practices" for teaching mathematics? How did you arrive at those beliefs?
- Add a title to your autobiography that you believe best captures your mathematics story.

It may help you to consider the success, or lack of success, that you felt learning mathematics as a child or adult. You might also describe specific activities or classroom events and reflect upon the impact they have had on your present ideas about mathematics and teaching and learning mathematics.

The style in which you choose to write this mathematics autobiography is up to you because I want it to reflect who you are. There is no page limit, although somewhere in the vicinity of 4-6 double-spaced pages is appropriate. However, I would like it to be comprehensive enough so that after reading it, I have a strong sense of who you are in relation to mathematics and what led you to your beliefs.

Section 3

Teacher Experiences With Reform-Based Mathematics Teaching

The chapters in this section present studies that explore teachers' experiences with and opportunities to learn reform-based mathematics teaching in various contexts. There are implications for teacher development and practice.

Chapter 11
Implementing Reform-Based Mathematics Teaching Practices to Address Student Errors

Susie Morrissey
Mercer University, USA

Gorjana Popovic
Illinois Institute of Technology, USA

ABSTRACT

The relationship between enactment of reform math (namely Common Core State Standards – Mathematics [CCSSM]), student misconceptions, and teacher variables was investigated. After providing professional development on enactment for the Standards of Mathematical Practice (SMP) of the CCSSM, observations were conducted to determine the degree of enactment of SMP in 22 classrooms of nine teachers in an urban charter school network consisting of three high schools. Students were all boys, 98% African American, and predominantly of low socio-economic status. Data included curricular documents, observation notes, quarterly assessments, and a teacher survey. Results indicated experienced teachers with high efficacy who expected students to reason and discuss their work were related to higher levels of SMP enactment and fewer misconceptions. Implications for teacher education and teacher practice are discussed.

INTRODUCTION

Based on the Third International Mathematics and Science Study (TIMSS) of 1995 (now called Trends in International Mathematics and Science Study) findings that examined what is taught in 37 countries, Schmidt et al. (2002) characterized mathematics content as seen in standards and textbooks in the United States as unfocused, highly repetitive, undemanding, and incoherent, compared to the curricula of higher-performing countries. In response, the Common Core State Standards - Mathematics (CCSSM) were created as narrower, yet deeper in focus than previous standards. The CCSSM were released in 2010,

DOI: 10.4018/978-1-7998-9422-3.ch011

and include the Standards for Mathematical Practice (SMP), which have eight standards that teachers are expected to develop in their students in order to build deeper understanding rather than a reliance on procedures. These are: SMP1: Make sense of problems and persevere in solving them, SMP2: Reason abstractly and quantitatively, SMP3: Construct viable arguments and critique the reasoning of others, SMP4: Model with mathematics, SMP5: Use appropriate tools strategically, SMP6: Attend to precision, SMP7: Look for and make use of structure, and SMP8: Look for and express regularity in repeated reasoning (National Governors Association Center for Best Practices, & Council of Chief State School Officers, 2010).

The narrower, deeper focus of the CCSSM content standards and the research-based SMP provide a response to student errors and misunderstandings created by what has been described as a national mathematics curriculum that is varied in content yet shallow in conceptual understanding. Specific causes of student errors have been found to be within the curriculum (Monaghan, 2000; Schmidt et al., 2002), and teaching methods that stress procedural knowledge over conceptual understanding of the material (Hiebert & Grouws, 2007; Kieran, 2007).

This study responds to the need to examine the relationship between classroom enactment of CCSSM (i.e., reform) and student achievement (Heck et al., 2011). Teachers received professional development (PD) on enacting CCSSM, specifically the SMP. Then, the current study examined enacted curricular documents, teaching practice, a teacher survey, and quarterly assessments to determine the correlation among enactment of SMP, alleviation of student errors, and teacher variables. Specifically, the following questions were asked:

1. What is the relationship between enactment of the SMP and persistence of the following student errors?
    - Negative sign – moving, deleting, or adding a negative sign, subtracting when addition is indicated or adding when subtraction is indicated (Booth et al., 2014).
    - Distribution – inappropriately applying the distributive property (Booth et al., 2014).
    - Diagram - making incorrect assumptions based on a given diagram, or drawing an incorrect diagram to represent a problem (Battista, 2007; Monaghan, 2000).
2. What are the various factors that relate to teachers' ability to enact SMP?

Review of the Literature

Teachers in this study were provided PD on enacting SMP, followed by observations to gauge their degree of enactment, while at the same time pre and post tests were administered to students to determine student errors present. Therefore, the literature review focuses on best practices for PD, teachers' struggles with enacting reform-based mathematics curricula, and students' algebra and geometry errors.

Loucks-Horsley et al. (1996) reviewed literature on effective PD for science and mathematics educators, and found common principles that included the beliefs that all students are capable, and learning should emphasize challenging problem-solving activities that include collaboration and develop deep conceptual understanding. Also, effective PD engages teachers in learning and developing their pedagogical content knowledge, builds on teachers' current knowledge, allows teachers to construct knowledge, and to collaborate. Teachers also need time to develop and reflect on their new knowledge, so PD should be presented over time with ongoing support. Confrey and Krupa (2010) recommend focusing on content, drawing on curricular materials, analyzing student work, and utilizing a long-term approach to PD. Hull et al. (2012) suggested that all teachers need PDs to acquire a rich mathematical content knowledge, and characterized useful PDs as the ones that always reflect back to the students and their achievements.

Research on PD for reform-based curricula emphasized the need for pedagogical strategies that include innovative methods for introducing content to students (Rimbey, 2013; Wilson & Downs, 2014). Teachers need time to gradually implement these strategies, along with positive reinforcement of their efforts (Wilson & Downs, 2014). PD should provide teachers with the tools they need to implement discussions within groups of students working on hands-on and/or open-ended questions that have several solutions (McGee et al. 2013; Patel et al., 2012). For example, Goodnough and Hung (2009) used a collaborative inquiry method of PD to enact problem-based learning (groups of students working on open-ended problems with multiple solutions), and found teachers had increased efficacy, increased belief in students' ability, and greater understanding of students' thinking processes following enactment of the problem-based learning activity. Nilsson (2014) engaged teachers in *learning study* PD such that teachers collaborated on lesson planning, identified critical aspects of the lesson, reflected on the lesson, analyzed student responses and teacher instruction, and revised the lesson, in a cyclical process. Teachers in Nilsson's study increased their knowledge of students' thinking and understanding of the material

Teachers can be reluctant to implement the changes required to transition to a reform-based curriculum for a variety of reasons. Heck et al. (2011) suggested research investigating how teacher factors such as interpretations of the CCSSM and expectations for student learning affect their enactment of the standards. Research has also identified teacher concerns about finding or creating appropriate materials as relevant to moving towards a more student-centered classroom as recommended in the SMP. For example, Borko et al., (2000), Christou et al. (2004), Gainsburg (2008), Johnson (2006), and Towers (2012) all found that a lack of materials were of concern in teachers' enactment of student-centered materials, and Supovitz and Turner (2000) found availability of materials to be positively correlated with enactment of investigative practices. Student ability was a particular concern in the study by Gainsburg (2008), while Borko et al. (2000), Bruce and Ross (2008) and Bruce et al. (2010) all found the change in pedagogy required, namely letting go and allowing students to struggle, was very difficult for teachers.

Peretin (2014) provided PD focused on CCSSM SMP, with a checklist for teachers to evaluate their own progress. The author found that while teachers' use of SMP improved slightly, student scores did not show improvement. Moschkovich (1998) demonstrated how use of a variety of problems with different properties, connections with other mathematical knowledge, and discussion could lessen student errors. Wilder (2015) used a research-based formative assessment to discover and dispel misunderstandings. The researcher found that discussion and the necessity to defend solutions aided students in alleviating student errors. Schmidt et al. (2002), in their comparison of mathematics curricula in the US to the curricula of higher-performing countries, considered the curriculum to be an overall cause of student errors. Monaghan (2000) concurred, in an exploration of students' geometry errors. A reliance on procedures was also related to student errors, as chapters by Hiebert and Grouws (2007) and Kieran (2007) attested to in the NCTM Second Handbook of Research on Mathematics Teaching and Learning.

Students' algebra and geometry errors identified in the literature may be placed into three broad categories: procedural, variable, and diagram errors. Procedural errors include negative sign errors, such as moving, deleting, or adding a negative sign, subtracting when addition is indicated or adding when subtraction is indicated, equality errors, misuse of the distributive, associative or commutative properties, or using the wrong operation (Booth & Koedinger, 2008; Booth et al., 2014). Variable errors include incorrect manipulations of an algebraic expression or equation such as combining unlike terms, moving, deleting, or adding a variable, believing that a variable has a specific value, seeing a variable as a label, and solving for only one variable in a system of equations (Booth et al., 2014; Knuth et al., 2005; Lucariello et al., 2014; Wilder, 2015). Errors in geometry primarily concern diagrams, involving students making incorrect assumptions based on a given diagram, or drawing an incorrect diagram to represent a problem (Battista, 2007; Monaghan, 2000).

Students' algebra errors have been seen to be related to student achievement (e.g., Booth & Koedinger, 2008; Booth et al., 2014; Knuth et al., 2005; McNeil & Alibali, 2005). Booth and Koedinger (2008) found students who began a lesson on solving algebraic problems with misunderstandings about the meaning of the equal sign or negative signs solved fewer equations correctly, and had difficulty learning how to solve the problems. Booth et al. (2014) assessed errors in Algebra I at different points throughout the school year in order to identify the errors that were most persistent and pernicious on standardized test items. They found that making more variable errors at the beginning of the year (on simpler content), making more mathematical property and wrong operation errors at the beginning and middle of the year (on simple and moderate content), making more equality/inequality errors in the middle and end of the year (on increasingly more difficult content), and making more negative sign and arithmetic errors at the end of the year (on difficult content) all predicted lower end-of-year test scores. Knuth et al. (2005) investigated middle school students' understanding of equality and variables, and the relationship of that understanding to performance in problem solving. Results showed that students who showed a relational view of the equal sign outperformed other students on the performance question, and students who indicated that a variable could represent more than one number performed better than other students on the variable performance problem. McNeil and Alibali (2005) examined whether previous arithmetic knowledge and practice interfered with successful equation solving. Researchers first found a correlation between an operational view of the equal sign and difficulty solving equations. Researchers then activated an operational view of the equal sign, and established causation between activating an operational view and difficulty solving equations.

METHODS

This study was intended to determine correlations between degree of enactment of SMP, student errors, and teacher variables, following PD on CCSSM content and SMP. To answer the question of what student errors in Algebra I and Geometry concerning negative signs, distribution and diagrams may be alleviated during enactment of the SMP, student errors were evaluated using quarterly assessments that were part of the curriculum. This study examined results for the third quarter, with the assessments administered at the beginning of the quarter serving as a pre-test and the assessments administered at the end of the quarter serving as the post test. Booth et al. (2014) asserted that observing the kinds of errors students produce in work they were already submitting revealed errors without requiring any work outside of regular curricular assignments. Fully answering the questions requires assurance that SMP have been enacted in classrooms; therefore, enacted curricular materials were collected and analyzed to determine their level of alignment with SMP. Additionally, targeted observations were made to determine the degree of enactment of SMP in each classroom during quarter three. To answer the question of what are the various factors that relate to teachers' ability to enact SMP, teachers completed a teacher survey measuring variables identified in the literature.

Sample

A total of nine teachers of Algebra I and Geometry from an urban charter school network[1] that consists of three high school campuses comprised the sample, with three teachers at each school participating. These were all-boys schools in an urban area and have an enrollment consisting of 98% African-American, 2% Hispanic and multiracial, with 84% low-income households (www.illinoisreportcard.com). The teachers were selected for the study because the chief academic officer of the charter school network requested PD to assist teachers in the network in enacting SMP. Although all mathematics teachers in the network participated in the PD, the chief academic officer requested content specifically targeted to Algebra I and Geometry teachers. Therefore, the teachers who participated in this study have had the benefit of targeted PD and have been supported in their efforts to transition to a reform-based curriculum. In addition, these teachers have students who took network-wide common quarterly assessments, allowing comparison of errors by classroom.

Professional Development and Observations

PD and other activities preceding the research period (i.e., third quarter) served to create an acceptable sample of teachers for answering the research questions. PD provided was based on recommendations for effective PD for mathematics and science educators, recommendation for PD for implementation of SMP, and previous research on PD for implementation of standards-based curricula. PD on enactment of the standards was provided to all mathematics teachers in the charter school network quarterly, followed by observations. Quarterly PD sessions were attended by all mathematics teachers from each of the three schools in the charter network. Observations of Algebra I and Geometry teachers followed each PD, and were conducted by the PD providers (the authors) and either the chief academic officer or the school principal. Additional PD in the form of targeted feedback was provided following each observation. Targeted feedback consisted of instances where SMP was observed in the classroom, as well as instances in the classroom where SMP could have been present if changes were made. Additional observations of Algebra I and Geometry teachers, solely for the purpose of gauging enactment of SMP, were conducted by one author during quarter three. During those observations the first author took observation notes regarding teacher and student actions to be coded for SMP, and collected classroom artifacts to be coded for SMP and content. She scheduled three observations with eight of the teachers and two observations with the ninth teacher. Quarter three was chosen because by this time, after the third quarterly PD, it was expected that teachers would display at least some, and perhaps a strong, enactment of SMP.

Quarter One Professional Development

Following recommendations as outlined, PD was presented to all levels of mathematics teachers from an urban charter school network of three high schools at the beginning of the school year. Eighteen teachers attended the PD, which was provided by the second author and observed by the first author, who also assisted in preparation of the PD structure and content. Observation notes concerning PD content, teacher actions and teacher comments were taken by the first author and discussed with the second author following the PD, to ensure 100% agreement between them concerning the description of the PD process and teacher responses. Content was specifically targeted to Algebra I and Geometry, at the request of the charter school chief academic officer, and was presented in a way that modeled student-centered instruction for the teachers. Challenging, engaging problems aligned with curriculum and assessments for quarter one were presented for teachers to work collaboratively. Student work was also presented for teachers to analyze. For Algebra I content, teachers were asked to work in four groups of six on a fraction division activity that required the division of 5¾ by 5/8 (see Appendix 1). After each group agreed on a solution, groups were given sample student solutions, including both correct and incorrect solutions, to discuss. This was followed by a whole-group discussion of the methods students used for their solutions, as well as why the prescribed procedure of "turn the second fraction over and multiply" produced a misleading answer (i.e., in the answer of 9 1/5, 1/5 indicated 1/5 of 5/8, not 1/5 of a whole unit). These components of PD provoked discussion concerning problem solving strategies and students' errors, as well as teachers' errors. Some teachers insisted on students' following the prescribed procedure, while other teachers argued that students should be able to solve the problem using the method of their choice.

This activity was followed by an activity that required ordering fractions from greatest to least, which teachers felt would be appropriate for their students. Third, an order of operations problem was given to each group of teachers. Teachers were given three student solutions; one incorrect and two correct. For each of the correct solutions, the operations were performed a different order, demonstrating to teachers that order of operations does not rely on mnemonics such as *pemdas (parenthesis, exponents, multiplication, division, addition, subtraction)*, but rather relies on the strength of each operation. This activity also sparked much discussion amongst the teachers, with one teacher stating that the answer for one solution was correct, but the student work was wrong because the student performed addition of unassociated numbers before multiplication (i.e., $2+2+2+2\times4=6+2\times4=6+8=14$).

The two authors briefly discussed with teachers an activity with multiple correct answers. In this activity, students were asked to determine which city had the highest buildings. The answer varied, depending on whether students used the number of tall buildings, the mean, median, mode, or absolute tallest building to make their determination. This discussion was followed by geometry content activities. The development of geometric theorems based on axioms was discussed and demonstrated, using parallel lines cut by a transversal to prove theorems about interior and exterior angles of a triangle. Teachers asked questions to clarify how these concepts could be introduced to their students.

Teachers then worked in groups once again, this time completing an activity that led to development of the Pythagorean Theorem and its converse. Each group was given a packet of squares sized from 1×1 to 15×15, and asked to design a rug of three squares such that the probability of a dart landing on the largest square is greater than, less than, or equal to the probability of a dart falling on the two smaller squares combined (Alper et al., 1998). Teachers commented that this activity would also be appropriate for their students. The concluding discussion focused on the common facets of all activities that were presented: they were student-centered, they challenged the notion that there is one correct way to find a solution or only one correct solution, they connected to previous material, they addressed grade-appropriate standards from CCSSM, and they incorporated SMP.

Quarter One Observations

Following the quarter one PD, three teachers were observed by the two authors and the chief academic officer; three teachers were observed by the two authors and the school principal; one teacher was observed by the second author and the school principal; and one teacher was observed by the first author. These teachers were selected for observation at the request of the chief academic officer, presumably based on the belief that these selected teachers were the most in need of additional PD. The two authors each took observation notes, specifically noting instances where SMP were observed in the lesson, and compared notes following the observation. In the case of the two teachers who were not observed by both authors, the recorded document was shared so the non-observer could identify instances of SMP. Instances of SMP were discussed until an agreement of 100% was reached.

Teachers displayed various degrees of enactment of SMP, the primary focus of those observations. For example, SMP3 *Construct viable arguments and critique the reasoning of others* was observed in four classrooms when discussing problem solutions written on the board by students. In one classroom, the discussion was almost completely student-led, whereas in another classroom, the teacher interrupted the student discussion quickly to provide the correct solution. This demonstrated how the use of mathematical practices could vary from one classroom to another. Other SMP observed included SMP1 *Make sense of problems and persevere in solving them*, seen in one classroom when students worked on classifying triangles; SMP2 *Reason abstractly and quantitatively*, seen in two classrooms when students factored quadratics and when students represented word problems in symbols; SMP4 *Model with mathematics*, observed in one classroom when students wrote inequalities to solve a problem in the context of the real world; SMP6 *Attend to precision*, seen when students used precise vocabulary to describe relationships in a triangle and in another classroom when students used precise vocabulary to discuss the coordinate plane; and SMP7, *Look for and make use of structure*, observed in one classroom when students used a shortcut for factoring quadratics.

Following each observation, teachers were provided with additional PD in the form of targeted feedback to aid in the enactment of SMP. Teachers were provided with feedback on SMP observed, and SMP that could have been present. Suggestions were included on how to increase SMP in the classroom. The most frequent recommendations made were to attend to precision, especially the use of mathematical vocabulary, and to have students explain and critique their work and the work of others. During these observations, authors noted both a lack of closure and a lack of planning, and so included a strong focus on planning a complete lesson in the PD for quarter two.

Quarter Two Professional Development

During the quarter two PD, two complete lesson plans on upcoming topics were presented for each subject, with teachers asked to complete introductory problems that connected to previous material, inductive lessons to introduce new material, application problems, and end of lesson discussion and questions to check for student understanding. One lesson plan included an inductive lesson on the distance formula, followed by application problems. The other lesson plan required students to translate patterns into expressions, recognize equivalent expressions, and create equivalent expressions. In this way, authors modeled CCSSM enactment, including content and SMP. Eighteen teachers again attended the PD, which was also provided by the second author and observed by the first author, who assisted in preparation of the PD content.

As before, observation notes concerning PD content, teacher actions and teacher comments were taken by the first author and discussed with the second author following the PD, to ensure accurate documentation. Discussion centered on components of the lesson plan and the importance of anticipating student errors. Teachers appreciated having written lesson plans for upcoming lessons, but were skeptical of having time to write such detailed lesson plans themselves for each day.

Quarter Two Observations

Observations of seven teachers were conducted by the two authors during quarter two. Teachers were asked in advance to provide complete lesson plans modeled on the lesson plans presented in the quarter two PD. During those observations, observers focused on materials and teacher actions that promoted SMP. Two geometry teachers were found to have an acceptable level of SMP. Two algebra teachers were lacking only in their requirements of students to evaluate each other's solutions. These teachers were given written feedback with suggestions to ask students if they agreed with solutions presented, and to explain why or why not. The remaining three teachers, one geometry teacher and two algebra teachers, had primarily teacher-centered classrooms. They received feedback on strategies to enact more SMP, such as having students present their work and explain their solutions, allowing students to discuss solution strategies, and allowing students to struggle to find a solution rather than providing the solution path for students.

Quarter Three Professional Development

Content of this PD, offered at the beginning of the third quarter, focused on connections between previous and upcoming material. Algebra content asked teachers to determine which of three given expressions were equivalent. An answer was given, and teachers had to decide whether the answer was correct and explain their reasoning. Evaluating a given answer was also required using geometry content, where two answers were given to a distance formula word problem. Again, teachers had to determine which answer was correct and explain their reasoning. This was followed by an inductive activity to determine surface area. An introductory assignment asked teachers to find the area of various polygons. Teachers then worked in pairs to create nets for solid shapes, find the surface area of the solid, and determine the amount of paint needed to cover the solid. A general formula for surface area was derived, followed by practice problems.

Finally, teachers were shown explicit connections between material covered in quarter two and upcoming material to be covered in quarter three. For example, solving equations built on students' previous work in simplifying expressions. In geometry, students would use their previous knowledge of the Pythagorean Theorem to determine ratios in special right triangles. Also, properties of parallel lines cut by a transversal were shown to be related to properties of squares, rectangles, parallelograms and trapezoids. Many of these connections were new to teachers, who commented that previously they would have presented these topics as unrelated to previous knowledge.

Quarter Three Observations

Additional, targeted observations were conducted after the third quarter PD. Seven of the nine teachers involved in this study were observed by both authors one time. These seven teachers were observed two additional times solely by the first author. The other two teachers involved in the study were observed solely by the first author, with one teacher observed two times and the other teacher observed three times.

The first observation made by both authors was for the purpose of providing feedback on enactment of CCSSM, specifically targeting SMP1 *Make sense of problems and persevere in solving them* and SMP3 *Construct viable arguments and critique the reasoning of others*. Of the seven teachers observed, four had an acceptable level, judged by the presence of students explaining their work and struggling to find solutions rather than the teacher providing solutions to the students. The other three teachers were advised to allow their students to explain their work, critique the solutions of others, and to struggle (with scaffolding as necessary) to solve problems themselves. Two of those teachers had been given the same feedback following quarter two observations; the third teacher was not observed during quarter two.

All other observations were made solely by the first author with the intention of collecting data to gauge the level of enactment of CCSSM. Teachers were still given feedback on ways to improve enactment of SMP. The first author collected relevant textbook pages, classroom tasks and assignments. Observation notes included teacher and student actions to be coded for instances of SMP.

Instruments and Data Collection

Instruments used in this study included alignment measurements, a teacher survey, an observation protocol, and quarterly assessments. Alignment of enacted curricular materials and student actions with SMP were determined by use of Levels for the Mathematical Practice Standards rubric used by Kartal (2015). This rubric was adapted from a rubric created by Hull et al. (2012), which has been widely used by teachers, leaders and leadership teams to transition to SMP (See Appendix 2). The rubric lists observed or expected student actions for each of the eight SMP, with initial, intermediate, and advanced levels of enactment described. Content validity for use of the instrument to gauge enactment of SMP was established by consultation with two experts in the field of mathematics education, who were provided with the rubric and the SMP for comparison. Level of agreement was 100%.

Teacher actions from observations were coded for SMP using an adaptation of the Rubric for Implementing Standards for Mathematical Practice (Park City Mathematics Institute, 2011) to determine the level of enactment of SMP (See Appendix 3). This instrument was intended for use by teachers when planning a lesson based on SMP. Since the instrument was not being used for the purpose for which it was created, content validity was established by consultation with two experts in the field of mathematics education. These experts were provided with the rubric and the SMP, and asked to determine whether the rubric adequately assessed teacher actions intended to enable students to engage in the SMP. Changes were made until 100% agreement was reached concerning validity of the instrument.

Consistency of coding for the Levels for the Mathematical Practice Standards rubric and the Rubric for Implementing Standards for Mathematical Practice was verified by asking another researcher to code random items for agreement. For example, an expert in mathematics education was asked to code classroom tasks and assignments using the Levels for the Mathematical Practice Standards rubric (for SMP), to code observation notes for student actions using the Levels for the Mathematical Practice Standards rubric, and to code observation notes using the Rubric for Implementing Standards for Mathematical Practice to measure teachers' actions that gave students the opportunity to engage in SMP during class. Initial level of agreement was 83%. Discrepancies in coding were resolved by discussion, until 100% agreement was reached.

The teacher survey, which was administered prior to quarter three observations, first asked how many years of teaching experience the teacher had; the survey was then broken into two parts, with the first part addressing a teacher's level of comfort with items from the Rubric for Implementing Standards of Mathematical Practice that relate to pedagogy changes, as well as comfort with the acquisition or availability of materials. The second part addressed a teacher's belief in their students' ability to perform actions described in the Levels for the Mathematical Practice Standards rubric. These sections of the survey are both 5-point Likert scales ranging from 1-not at all comfortable to 5-very comfortable for teachers' level of comfort, and 1- not able to 5-completely able for belief in students' ability; a selection for N/A was also included. Content validity for the survey was provided by two experts in the field of mathematics education, who were asked to compare each question on the survey with the intended goal of each survey item – namely, to determine the level of concern teachers have about student ability, changing their pedagogy, and availability of materials when enacting the SMP. Experts were provided with the table of specifications that was used to create the survey, as well as teacher and student actions from the enactment rubrics detailed above. Initial agreement was 90%; changes were made to the survey until all experts reach 100% agreement that the survey met the stated objectives. Reliability analysis was performed using Cronbach's alpha, with an acceptable value of 0.89.

The observation protocol agreed on by the two authors consisted of the identification of teacher and student behaviors regarding observed instances of SMP, as described in the CCSSM. Observations were recorded using observation notes. Following observations, researchers discussed observed instances of SMP until 100% inter-rater agreement was reached.

Students were assessed using quarter three assessments (Academic Approach, 2014, 2015). Quarter three assessments were provided by a test creation company and ordered by the charter school network, which specified the content desired for each quarter of assessments and each grade level. These tests were modeled on standardized tests and included sections on English, mathematics, reading, and science. Content validity for curricular alignment was assured by the test creation company. Content validity for the use of quarter three assessments as a measure of errors in student work was established by analyzing each item on each assessment to determine whether the opportunity existed for students to display the errors relevant to this study. The Algebra I assessment was coded by the author and by a mathematics education expert to identify items with the ability to showcase negative sign or distribution errors, and the Geometry assessment was coded by the author and by a mathematics education expert to identify items with the ability to showcase diagram errors. Discussion of items yielded an agreement of 100%. Reliability was determined by the use of the KR-20, with a value of .89 for the Algebra I quarter three assessment, and a value of .87 for the Geometry quarter three assessment, which indicated that the tests adequately discriminated between differing levels of student knowledge.

Data included PD materials, quarterly assessments, observation notes of teacher and student actions, relevant textbook pages, classroom tasks, and assignments from lessons observed during the third quarter. Pre and post test results of quarter three assessments, including answers by item, average scores by classroom, and student work were collected. The teacher survey was administered by the researcher before quarter three observations.

Data Analysis

Analysis of SMP Alignment

Alignment of enacted materials, including relevant textbook pages, classroom tasks and assignments, and student actions with SMP were determined by use of Levels for the Mathematical Practice Standards rubric. Observed instances of SMP at the Initial level in this rubric indicate progress in enactment of the standards, although Intermediate and Advanced levels are more desirable.

Student actions and classroom materials, including pages assigned from a textbook, problems from the board, and worksheets, were coded for instances of each SMP at each level, and recorded in a frequency chart. Because the same materials were used in each section of a class that was taught for multiple periods by a teacher, results were recorded by teacher, rather than classroom, with the exception of Teacher H who taught both Algebra and Geometry. As an example of coding, in one geometry class students worked on an activity that asked them to find the relationship between the area of a rectangle and the area of a kite inscribed in the rectangle, and derive a formula for the area of a kite. This classroom activity was intended to elicit MP8A, because students were required to discover an underlying relationship and derive a formula.

For an example of student actions, when solving for an angle measure, a student comments: "I'm looking at it from another way - we have adjacent so I'm looking at hypotenuse as x, adjacent is 36, opposite is 48," which was coded as MP3bI because the student was offering another approach to the problem being solved. In another class a student gave his answer to a multiple-choice question and explained "I put C because I did p for the price minus 25 percent for the discount price represents how much money you're paying," which was coded as MP3aA because the student justifies the correctness of his answer.

Teacher actions during observations were coded using the Rubric for Implementing Standards for Mathematical Practice (Park City Mathematics Institute, 2011) to determine the level of enactment of SMP. For this rubric, observed instances of SMP at the Emerging level indicated progress in enactment of the standards, but a level of at least Proficient is desired. Needs Improvement was not considered in determining the level of enactment, as it indicates a very low level and was observed infrequently. Again, instances of each of the SMP at each of the levels as described in the rubric were coded for each classroom using frequency tables. An example of MP2E comes from a discussion of surface area. The teacher stated "If you were given that a side was 12 and a side was 10 so you know that half of that side would be 5 we could create a right triangle right here using this," expecting students to use a single representation (that of the teacher) to interpret the diagram; additionally, the teacher explained the connections between the side of a pyramid and previous knowledge of a right triangle. A contrasting example was seen in another teacher's question: "Corrections, comments, who got something different?" which demonstrated MP3EX, because the teacher expected students to evaluate the arguments and work of their peers.

Teacher Survey Analysis

Responses to the teacher survey were analyzed independently for each teacher, and each teacher received a score on each of the three areas of concern being measured - lack of materials, student ability, and changes in pedagogy. Likert scale items ranged from 1 to 5, with some items having a reversed scale. After changing the reversed scales, each teacher's score on each area of concern was the average of Likert scales for the questions that addressed that area. This score, along with years of experience, was compared to the measurement for each teacher/classroom for enactment of SMP. The researcher searched for relationships and patterns qualitatively by comparing years of experience with the frequency of enactment of each of the SMP and content alignment index for enacted materials. The level of each concern was also compared qualitatively with the level of enactment of SMP in each corresponding category (materials, student actions and teacher actions).

Student Error Analysis

Both the pre and post administrations of quarter three assessment were coded for instances of errors related to negative signs, distribution or diagrams. Descriptions of negative sign errors and distribution errors from Booth et al. (2014) were used for determining those errors, and descriptions from Battista

(2007) were used for diagram errors. Specifically, Booth et al. described negative sign errors as moving, deleting, or adding a negative sign, subtracting when addition is indicated or adding when subtraction is indicated, and distribution errors as inappropriately applying the distributive property. An example of student work showing a negative sign error is seen in Figure 1, which shows the common error of subtracting a quantity from both sides of an equation, regardless of whether the quantity is positive or negative. Figure 2 shows student work with a distribution error, where the sign has not been distributed to each quantity inside the parentheses.

Figure 1. Example of negative sign error (© 2017, Copyright held by the author, S. Morrissey)

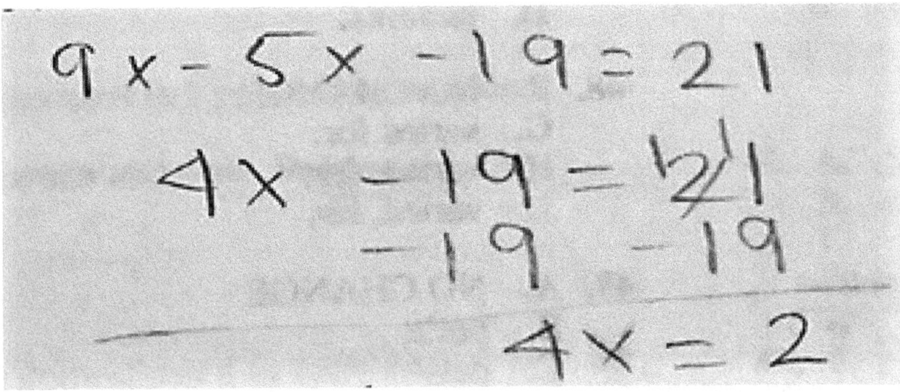

Figure 2. Example of distribution error (© 2017, Copyright held by the author, S. Morrissey)

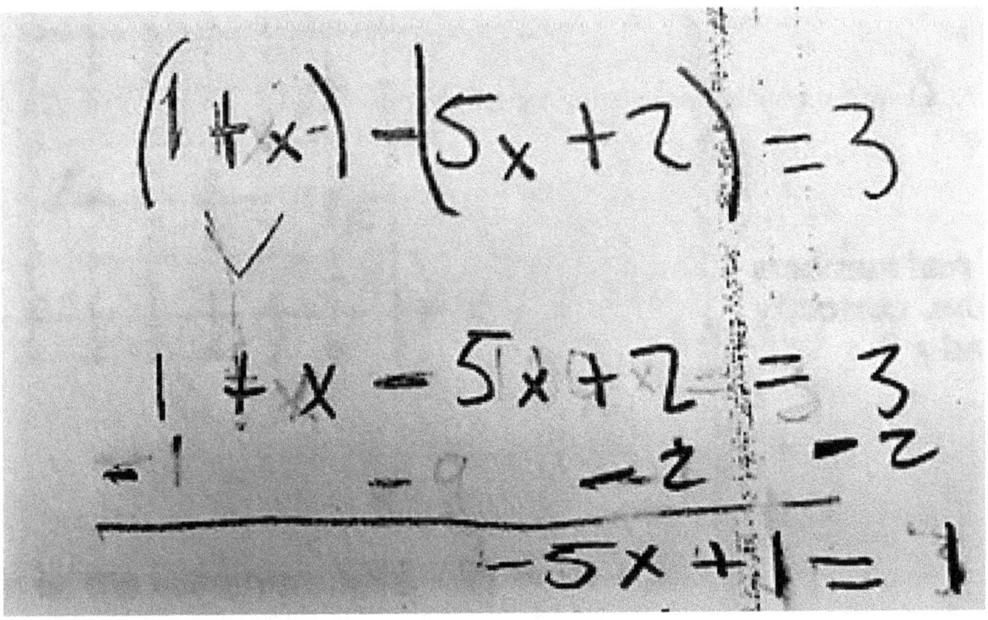

Battista described diagram errors as making incorrect assumptions based on a given diagram, or drawing an incorrect diagram to represent a problem. Two examples of diagram errors are shown in Figures 3 and 4. The first example evidences a student drawing an incorrect diagram for a problem that asks the student to find the value of x for the side of a rectangular prism with sides of 6 ft., 2 ft., and x ft., and a volume equal to that of a cube with side length of 6 ft. The second shows a student misinterpreting an existing diagram in order to find the area of a trapezoid with bases of length 9 and 19, height of 12 and slant height of 13.

Figure 3. Example of student drawing an incorrect diagram (© 2017, Copyright held by the author, S. Morrissey)

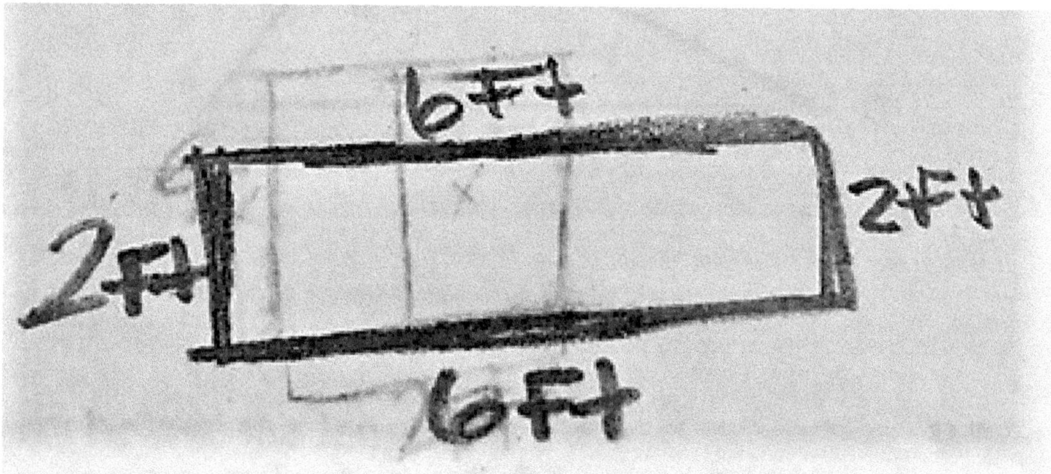

Figure 4. Example of student misinterpreting a given diagram (© 2017, Copyright held by the author, S. Morrissey)

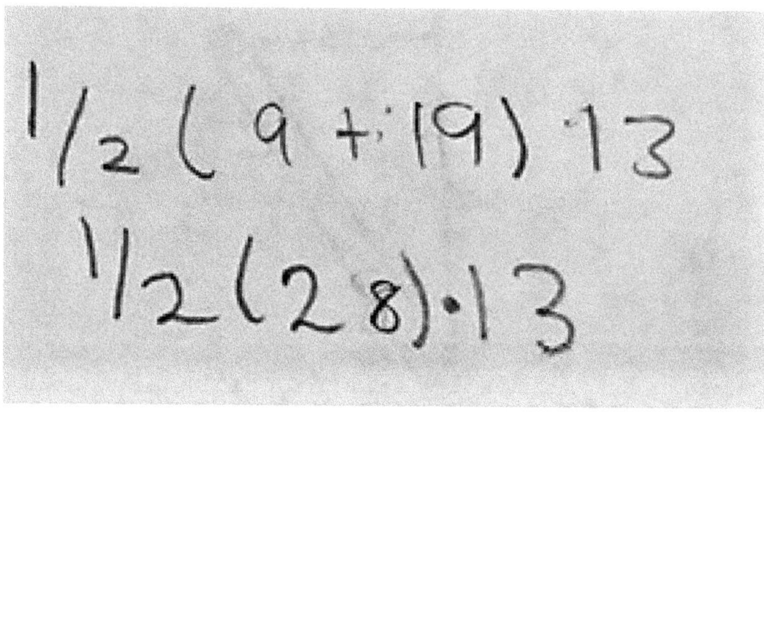

A chi-square test ($\alpha = .05$) was performed on instances of each type of student error in pre and post-tests by classroom to determine which student errors were alleviated through instruction during the quarter. Persistence of student errors was compared qualitatively to teacher's enactment of CCSSM, by classroom, to determine whether a relationship existed.

RESULTS

Enactment of SMP

Enactment of SMP by classroom was determined through frequency of SMP in enacted curricular materials, student actions and teacher actions. Ratings were assigned as low, medium and high, to correspond to ratings of initial, intermediate and advanced on the Levels for the Mathematical Practice Standards Rubric (Kartal, 2015) and emerging, proficient and exemplary on the Rubric for Implementing Standards for Mathematical Practice (Park City Mathematics Institute, 2011). Results showed that Algebra teachers most often implemented SMP2 at a low or medium level, and SMP3 at every level. Although materials did not often ask students to explain their reasoning, teachers did ask, and students responded with many instances of explaining their reasoning. However, most student enactments of SMP3 were at the lower level. Students seldom justified their reasoning or critiqued the reasoning of others. Teacher F, an Algebra teacher, had the most instances of materials that expected students to engage in SMP2, reasoning and problem solving, teacher expectations of SMP3, asking students to explain their reasoning, and students complying with the request by explaining their thought process.

Geometry teachers predominantly implemented SMP2 and SMP3, both at the low level. Geometry teachers and the materials they used had many instances of expecting students to display SMP2 at the low level, and students responded accordingly. Similar to Algebra classes, Geometry students showed many instances of explaining their reasoning (SMP3), again at the low level. Geometry teachers D and G had the highest instances of SMP2, expecting their students to engage in reasoning and problem solving, combined with students explaining their reasoning (SMP3). For both Algebra and Geometry classes, the other six SMP were seldom observed in materials, teacher actions, or student actions.

Student Errors

Evidence of student errors were based solely on instances where students displayed their work. Unfortunately, only the honors classroom of Teacher B displayed much work on these assessments, limiting the amount of information that could be gathered from student work. Regardless, student errors were analyzed for type and frequency for algebra and geometry classrooms. Four classrooms showed a significant decrease in instances of student errors from pre-test to post test. One Algebra class of Teacher F had a significant decrease in negative sign errors [$\chi 2(18, N=19)=30$, $\rho=0.037$] as well as distribution errors [$\chi 2(1^{8}, N=19)=26.44$, $\rho=2.04E-7$]. The honors class of Teacher B had a significant decrease in diagram errors [$\chi 2(25, N=26)=120$, $\rho=2.21E-14$], as did one regular geometry class of Teacher B [$\chi 2(16, N=17)=28$, $\rho=0.032$]. Additionally, one geometry class of Teacher C had a significant decrease in diagram errors [$\chi 2(10, N=1^{1})=19$, $\rho=0.04$].

Teacher Survey

The second research question addressed the various factors that impact teachers' ability to enact SMP. Teachers responded to a survey regarding years of experience and concerns about lack of materials aligned with SMP, changes in pedagogy required to implement the SMP, and belief in student ability to perform actions expected by the SMP. Results of the teacher survey are displayed in Table 1. All teachers who responded to the survey appear to be somewhat comfortable with changes in pedagogy required to implement the SMP (M=3.4075*)*. Two teachers do not feel comfortable with finding SMP materials, while two teachers feel very comfortable. Most teachers indicated a belief in their students' abilities to engage in the SMP, although one teacher had a low level of belief in the students' abilities. Teachers' years of experience ranged from three years to 19 years, providing a variety of capabilities.

Table 1. Teacher survey results

Teacher	Score on level of comfort with SMP teacher actions	Score on level of comfort with finding SMP materials	Score on belief in students' ability	Years of experience
A	3.31	4.25	3	10
B	3.56	3.88	4.75	7
C	3.63	3.56	3.97	19
D	3.25	4.13	5	6
E	3.38	2.25	2.56	3
F	3.44	2.38	4.31	15
G	3.38	3.69	3.19	12
H	3.31	3.19	4.44	3
J	No response	No response	No response	No response

Relationships Identified

For the first question, asking about the relationship between enactment of SMP and persistence of student errors, the most notable relationship found was between teacher and student actions for SMP3, construct viable arguments and critique the reasoning of others, at low, medium and high levels, and a decrease in both negative sign and distribution errors in the algebra classroom of Teacher F. This classroom also evidenced many instances of SMP2, reason abstractly and quantitatively, at a low level in materials. Two geometry classrooms of Teacher B and one geometry classroom of Teacher C showed a significant decrease in diagram errors. The geometry classroom of Teacher C, like Teacher F's classroom, had teacher and student actions for SMP3, construct viable arguments and critique the reasoning of others, at low, medium and high levels, as well as instances of SMP2, reason abstractly and quantitatively, at a low level in materials. Students in Teacher C's classroom also displayed SMP2, reason abstractly and quantitatively, at a low level. The classrooms of Teacher B displayed fewer instances of SMP3, and at only the low and medium levels, compared to Teachers C and F. However, many instances of SMP2 were displayed in teacher actions, student actions, and materials in Teacher B's classrooms. In each of these teachers' classrooms, students were expected to and did explain their reasoning.

The second question examined teacher factors, and this study determined that the three teachers with highest level of comfort concerning making pedagogical changes necessary in the classroom for enactment of SMP (Teachers F, C and B) were the three teachers whose students had significant decreases in student errors; two of those were the teachers with the most years of experience (15 and 19 years), while the third teacher had seven years of teaching experience. Teacher A, with ten years of experience, had the highest score on the teacher survey for comfort with finding SMP materials, and also had classroom materials with the most SMP. These materials, for Algebra, evidenced many instances of SMP2, SMP3, and SMP4, model with mathematics. Teacher G, with 12 years of experience, had many instances of teacher actions to elicit SMP2, as well as many instances of students' displaying SMP3.

DISCUSSION

Teachers' displays of SMP aligned with the feedback they had received following PD and observations. Feedback after quarter one observations included attending to precision, but this feedback was not re-enforced in the following quarters. The feedback most often provided after observations centered on having students explain and critique their work and the work of others, as well as allowing students to struggle to find a solution rather than providing the solution path for students. Therefore, it could be expected that teachers most often displayed SMP2 and SMP3, through their actions and the use of materials, and students responded with those same practices. These two SMP are related, in that SMP2 asks students to reason and SMP3 asks students to explain their reasoning. The other six SMP were seldom observed.

Teacher variables of increased comfort with enacting SMP, more years of experience, and expecting students to explain their reasoning were related to a decrease in student errors. Wilder (2015) had a similar finding in a group activity, where students were required to defend their categorizations. Places where the groups disagreed sparked deep discussion and illuminated student errors. Wilder concluded that discussions and defending their conjectures resulted in deeper student understanding. Moschkovich (1998) also demonstrated that discussion, seen in descriptive language, gestures and actions, contributed to the alleviation of student errors.

Student errors were alleviated significantly for the three teachers who showed the highest level of comfort with making pedagogical changes necessary to enact the SMP. Similarly, Nurlu (2015) administered a survey to teachers, and found a relationship between teachers with a high level of efficacy in teaching mathematics and openness to trying new methods and demonstrating a higher level of effort and persistence with their students. Harris (2010) found that higher teachers' beliefs that student ability was within the locus of control of the teachers correlated with higher student achievement in mathematics.

These findings further align with those of Moore and Esselman (1992), who found that higher teacher efficacy correlated with greater gains in mathematics knowledge for students. The studies of Harris and of Moore and Esselman both looked for relationships between teacher efficacy and students' standardized test scores. Bruce et al. (2010) conducted PD using a basis comparable to that used in this study - teachers collaborated in groups on problem solving activities and analyzed student work - and found that teacher efficacy on ability to use a range of mathematical practices affected student achievement indirectly, as increased teacher efficacy led to persistence using teaching strategies that benefit students.

Of the three teachers mentioned above, who showed the highest level of comfort with making pedagogical changes necessary to enact the SMP, two had the most years of experience in teaching (15 and 19 years), while the third teacher had seven years of teaching experience. The relationship between years of teaching and comfort with making pedagogical changes necessary to enact the SMP of common core standards is somewhat in line with conclusions by Christou et al. (2004), who found highly experienced teachers (more than 20 years) to be more confident in their abilities to deal with a change in curriculum. Christou et al. administered a survey to teachers who had very little PD, so PD was not a factor for consideration of teachers' confidence in dealing with curricular changes. The teachers in the present study had all received the same PD over the past year, and all but one teacher had received the same PD in the previous year, essentially removing PD as a mediating factor and leaving years of experience as a factor, as in the Christou et al. study.

A survey of the literature by Darling-Hammond (2000) found teacher experience to be related to teacher effectiveness, with new teachers found to be less effective than more senior teachers. However, Darling-Hammond cautions that effectiveness may wane for teachers who do not continue to learn and grow. Teachers in the present study have participated in PD and continued to expand their knowledge of effective teaching, and thus years of experience were positively related to student achievement.

IMPLICATIONS

This study revealed a relationship between teacher beliefs and actions, and student errors, for an underserved population of primarily African-American boys from low-income households. Finding that decreased student errors and students engaged in productive discourse were related to enactment of SMP2 and SMP3 underscores the importance of increased student reasoning and discourse.

Implications for Teaching Practice

Administrators who are considering PD, PD providers, and teachers should note that a combination of teachers' willingness to enact pedagogical changes, years of experience, and high levels of SMP in teacher actions, yield the most beneficial results for students, as detailed above. These stakeholders should also note that materials with a high level of SMP are not enough by themselves – teachers must make use of the materials in a way that elicits SMP from students.

In order to enact curricular changes in the classroom, administrators should provide effective PD as recommended by Loucks-Horsley et al. (1996) that engages teachers in learning and developing their pedagogical knowledge and abilities, builds on teachers' current knowledge, allows teachers to construct knowledge, and allows teachers to collaborate, as seen in Goodnough and Hung (2009) as well as Nilsson (2014).

Teachers must also believe in their ability to affect students' learning. Initial attempts to enact new strategies may lower efficacy; teachers need to be made aware of this and receive support and training during this time (Tschannen-Moran et al., 1998). Prelli (2016) found leader actions that improved culture, worked towards a common vision, brought the teachers to a consensus about goals, and supported individual teachers to be related to increased teacher efficacy. These actions are similar to findings by Hipp (1996) that indicated principal actions of modeling behavior, inspiring group purpose and providing contingent rewards were related to increased teacher efficacy. Hipp identified additional needs of teachers related to increased efficacy to be access to resources and updated materials, constructive feedback, accessible instructional leadership and support for parent and student issues.

Implications for Teacher Education

Researchers and teacher educators develop coursework for future teachers, and are guided by outcomes of empirical studies such as this. The results of this study are of interest to this group, as findings will aid in bringing research into practice. Specifically, in the classrooms where student errors were alleviated, teachers expected their students to display SMP2 and SMP3 by asking their students to engage in reasoning, explain their conjectures and evaluate the arguments of their peers. Those language functions were shown by Wilder (2015) and Moschkovich (1998) to aid in alleviating student errors.

Another important factor to be emphasized is ensuring preservice teachers are prepared to incorporate the pedagogy necessary to enact the SMP, and believe in their ability to affect student achievement. Several researchers (Bruce et al., 2010; Harris, 2010; Moore & Esselman, 1992; Nurlu 2015) reported finding that increased teacher efficacy led either directly or indirectly to increased student achievement. A belief in student ability should also be instilled in preservice teachers, as Muijs and Reynolds (2002) found that teacher beliefs indirectly affect student achievement through teacher behavior.

Shulman and Shulman (2004) recommended reflection on the learning process and learning as part of a community in an environment that supports, sustains and creates a collective vision and understanding in its members. Accordingly, McGinnis et al. (2005) surveyed preservice teachers who had taken content courses taught using the reform methods that were a focus of their education classes. Preservice teachers in that study indicated that they learned more than they had in other content classes and, after experiencing classes taught using reform methods, wanted to teach in that way.

Implications for Future Research

Heck et al. (2011) noted that a key priority in the research agenda concerning enactment of CCSSM is studying the relationship between the classroom enactment of CCSSM and students' achievement of the designated learning goals. Heck et al. specifically called for research on enactment of CCSSM for "students at different grade levels; for different standards for mathematical content and practice; with students of various backgrounds; and with students who experience different types of programs" (p. 33). A continuation of research that provides variations in enactment and related student outcomes, also recommended by Kartal (2015), is necessary to have results that can be generalized. This research should be conducted over time, so teachers have the opportunity to develop skills related to SMP.

Although all teachers in this study showed evidence of CCSSM enactment after receiving PD, level of enactment varied. Other factors that may be related to enactment of SMP should be explored, such as the learning environment in each classroom. Fraser (1998) defined learning environment as a social, psychological and pedagogical context that affects the attitude and success of students and where learning occurs. One social and psychological aspect of the environment in particular, rapport between teacher and student, was found to be positively correlated with final course grades in a study by Lammars et al. (2017). Further exploration of learning environment and other factors that may relate to enactment of SMP would help to highlight specific combinations of factors that relate to student achievement.

REFERENCES

Academic Approach. (2014). *Grade 9 quarter three assessment.* Author.

Academic Approach. (2015). *Grade 10 quarter three assessment*. Author.

Alper, L., Fendel, S., Fraser, S., & Resek, D. (1998). *Interactive Mathematics Program, Year Two*. Key Curriculum Press.

Battista, M. T. (2007). The development of geometric and spatial thinking. In F. K. Lester (Ed.), *Second handbook of research on mathematics teaching and learning* (pp. 843–908). National Council of Teachers of Mathematics.

Booth, J. L., Barbieri, C., Eyer, F., & Pare-Blagoev, E. J. (2014). Persistent and pernicious errors in algebraic problem solving. *The Journal of Problem Solving*, *7*(1), 10–23. doi:10.7771/1932-6246.1161

Booth, J. L., & Koedinger, K. R. (2008). Key misconceptions in algebraic problem solving. In *Proceedings of the 30th annual conference of the Cognitive Science Society* (pp. 571-576). Academic Press.

Borko, H., Davinroy, K. H., Bliem, C. L., & Cumbo, K. B. (2000). Exploring and supporting teacher change: Two third-grade teachers' experiences in a mathematics and literacy staff development project. *The Elementary School Journal*, *100*(4), 273–306. doi:10.1086/499643

Bruce, C. D., Esmonde, I., Ross, J., Dookie, L., & Beatty, R. (2010). The effects of sustained classroom-embedded teacher professional learning on teacher efficacy and related student achievement. *Teaching and Teacher Education*, *26*(8), 1598–1608. doi:10.1016/j.tate.2010.06.011

Bruce, C. D., & Ross, J. A. (2008). A model for increasing reform implementation and teacher efficacy: Teacher peer coaching in grades 3 and 6 mathematics. *Canadian Journal of Education/Revue canadienne de l'éducation, 31*(2), 346-370.

Christou, C., Eliophotou-Menon, M., & Philippou, G. (2004). Teachers' concerns regarding the adoption of a new mathematics curriculum: An application of CBAM. *Educational Studies in Mathematics*, *57*(2), 157–176. doi:10.1023/B:EDUC.0000049271.01649.dd

Cohen, D., & Ball, D. (1990). Relations between policy and practice: A commentary. *Educational Evaluation and Policy Analysis*, *12*(3), 331–338. doi:10.3102/01623737012003331

Confrey, J., & Krupa, E. (2010). *Curriculum design, development, and implementation in an era of common core state standards*. Summary Report of a Conference. Center for the Study of Mathematics Curriculum.

Darling-Hammond, L. (2000). Teacher quality and student achievement. *Education Policy Analysis Archives*, *8*, 1. doi:10.14507/epaa.v8n1.2000

Fraser, B. J. (1998). Classroom environment instruments: Development, validity and applications. *Learning Environments Research*, *1*(1), 7–33. doi:10.1023/A:1009932514731

Gainsburg, J. (2008). Real-world connections in secondary mathematics teaching. *Journal of Mathematics Teacher Education*, *11*(3), 199–219. doi:10.100710857-007-9070-8

Goodnough, K., & Hung, W. (2009). Enhancing pedagogical content knowledge in elementary science. *Teaching Education*, *20*(3), 229–242. doi:10.1080/10476210802578921

Harris, M. (2010). *Teacher efficacy beliefs: Understanding the relationship between efficacy and achievement in urban elementary schools*. UC Berkeley: Educational Leadership. https://escholarship.org/uc/item/44h29077

Heck, D. J., Weiss, I. R., & Pasley, J. D. (2011). *A Priority research agenda for understanding the influence of the common core state standards for mathematics*. Horizon Research, Inc.

Hiebert, J., & Grouws, D. A. (2007). The effects of classroom mathematics teaching on students' learning. In F. K. Lester (Ed.), *Second handbook of research on mathematics teaching and learning* (pp. 371–404). National Council of Teachers of Mathematics.

Hipp, K. (1996). *Teacher efficacy: Influence of the principal leadership, and student achievement.* Paper presented at the Annual Meeting of the American Educational Research Association, New York, NY.

Hull, T. H., Miles, R. H., & Balka, D. S. (2012). *The common core mathematics standards: Transforming practice through team leadership*. NCTM and Corwin. doi:10.4135/9781544308692

Johnson, C. C. (2006). Effective professional development and change in practice: Barriers science teachers encounter and implications for reform. *School Science and Mathematics*, *106*(3), 150–161. doi:10.1111/j.1949-8594.2006.tb18172.x

Kartal, O. (2015). *Enactment of common core state standards for mathematics: Relationship between teachers' choices of curriculum, teaching, and professional development* (Order No. 3733992). Available from Dissertations & Theses @ Illinois Institute of Technology; ProQuest Dissertations & Theses Global. (1732166593). http://216.47.136.160/docview/1732166593?accountid=28377

Kieran, C. (2007). Learning and teaching of algebra at the middle school through college levels: Building meaning for symbols and their manipulation. In F. K. Lester (Ed.), *Second handbook of research on mathematics teaching and learning* (pp. 707–762). NCTM.

Knuth, E. J., Alibali, M. W., McNeil, N. M., Weinberg, A., & Stephens, A. C. (2005). Middle school students' understanding of core algebraic concepts: Equivalence & variable. *Zentralblattfür Didaktik der Mathematik*, *37*(1), 68–76. doi:10.1007/BF02655899

Lammers, W. J., Gillaspy, J. A. Jr, & Hancock, F. (2017). Predicting academic success with early, middle, and late semester assessment of student–instructor rapport. *Teaching of Psychology*, *44*(2), 145–149. doi:10.1177/0098628317692618

Loucks-Horsley, S., Stiles, K., & Hewson, P. (1996). Principles of effective professional development for mathematics and science education: A synthesis of standards. *NISE Brief*, *1*(1), n1.

Lucariello, J., Tine, M. T., & Ganley, C. M. (2014). A formative assessment of students' algebraic variable misconceptions. *The Journal of Mathematical Behavior*, *33*, 30–41. doi:10.1016/j.jmathb.2013.09.001

Luft, J. A. (2001). Changing inquiry practices and beliefs: The impact of an inquiry-based professional development programme on beginning and experienced secondary science teachers. *International Journal of Science Education*, *23*(5), 517–534. doi:10.1080/09500690121307

McGee, J. R., Wang, C., & Polly, D. (2013). Guiding teachers in the use of a standards-based mathematics curriculum: Teacher perceptions and subsequent instructional practices after an intensive professional development program. *School Science and Mathematics, 113*(1), 16–28. doi:10.1111/j.1949-8594.2012.00172.x

McGinnis, J. R., Watanabe, T., & McDuffie, A. R. (2005). University mathematics and science faculty modeling their understanding of reform based instruction in a teacher preparation program: Voices of faculty and teacher candidates. *International Journal of Science and Mathematics Education, 3*(3), 407–428. doi:10.100710763-005-3591-3

McNeil, N. M., & Alibali, M. W. (2005). Why won't you change your mind? Knowledge of operational patterns hinders learning and performance on equations. *Child Development, 76*(4), 883–899. doi:10.1111/j.1467-8624.2005.00884.x PMID:16026503

Monaghan, F. (2000). What difference does it make? Children's views of the differences between some quadrilaterals. *Educational Studies in Mathematics, 42*(2), 179–196. doi:10.1023/A:1004175020394

Moore, W. P., & Esselman, M. E. (1992). *Teacher efficacy, empowerment, and a focused instructional climate: Does student achievement benefit?* Paper presented at the annual meeting of the American Educational Research Association, San Francisco, CA.

Moschkovich, J. N. (1998). Resources for refining mathematical conceptions: Case studies in learning about linear functions. *Journal of the Learning Sciences, 7*(2), 209–237. doi:10.120715327809jls0702_3

Muijs, D., & Reynolds, D. (2002). Teachers' beliefs and behaviors: What really matters? *Journal of Classroom Interaction, 37*(2), 3–15. https://www.jstor.org/stable/23870407

National Governors Association Center for Best Practices & Council of Chief State School Officers. (2010). *Common core state standards for mathematics*. Author.

Nilsson, P. (2014). When teaching makes a difference: Developing science teachers' pedagogical content knowledge through learning study. *International Journal of Science Education, 36*(11), 1794–1814. doi:10.1080/09500693.2013.879621

Nurlu, O. (2015). Investigation of teachers' mathematics teaching self-efficacy. *International Electronic Journal of Elementary Education, 8*(1), 21–40.

Park City Mathematics Institute. (2011). *Rubric for Implementing Standards for Mathematical Practice.* Retrieved March 15, 2016 from http://mathforum.org/pcmi/hstp/resources/rubric/rubric.pdf

Patel, N., Franco, S., Miura, Y., & Boyd, B. (2012). Including curriculum focus in mathematics professional development for middle-school mathematics teachers. *School Science and Mathematics, 112*(5), 300–309. doi:10.1111/j.1949-8594.2012.00146.x

Peretin, J. (2014). *The effectiveness of professional development using a checklist of common core state standards for mathematical practice for designing and delivering instruction on Algebra I achievement* (Doctoral dissertation). Robert Morris University.

Peterson, P. L., Fennema, E., Carpenter, T. P., & Loef, M. F. (1989). Teachers' pedagogical content beliefs in mathematics. *Cognition and Instruction, 6*(l), 1–40. doi:10.12071532690xci0601_1

Prelli, G. E. (2016). How school leaders might promote higher levels of collective teacher efficacy at the level of school and team. *English Language Teaching*, *9*(3), 174. doi:10.5539/elt.v9n3p174

Priestley, M. (2011). Schools, teachers, and curriculum change: A balancing act? *Journal of Educational Change*, *12*(1), 1–23. doi:10.100710833-010-9140-z

Prothero, A. (2018). Charter Schools. *Education Week*. https://www.edweek.org/policy-politics/what-are-charter-schools/2018/08

Rimbey, K. A. (2013). *From the common core to the classroom: A professional development efficacy study for the common core state standards for mathematics* (Doctoral dissertation). Arizona State University.

Schmidt, W., Houang, R., & Cogan, L. (2002). A coherent curriculum. *American Educator*, *26*(2), 1–18.

Shulman, L. S., & Shulman, J. H. (2004). How and what teachers learn: A shifting perspective. *Journal of Curriculum Studies*, *36*(2), 257–271. doi:10.1080/0022027032000148298

Simmons, P. E., Emory, A., Carter, T., Coker, T., Finnegan, B., Crockett, D., Richardson, L., Yager, R., Craven, J., Tillotson, J., Brunkhorst, H., Twiest, M., Hossain, K., Gallagher, J., Duggan-Haas, D., Parker, J., Cajas, F., Alshannag, Q., McGlamery, S., ... Labuda, K. (1999). Beginning teachers: Beliefs and classroom actions. *Journal of Research in Science Teaching*, *36*(8), 930–954. doi:10.1002/(SICI)1098-2736(199910)36:8<930::AID-TEA3>3.0.CO;2-N

Supovitz, J. A., & Turner, H. M. (2000). The effects of professional development on science teaching practices and classroom culture. *Journal of Research in Science Teaching*, *37*(9), 963–980. doi:10.1002/1098-2736(200011)37:9<963::AID-TEA6>3.0.CO;2-0

Towers, J. (2012). Administrative supports and curricular challenges: New teachers enacting and sustaining inquiry in schools. *Canadian Journal of Education/Revue canadienne de l'éducation, 35*(1), 259-278.

Tschannen-Moran, M., Hoy, A. W., & Hoy, W. K. (1998). Teacher efficacy: Its meaning and measure. *Review of Educational Research*, *68*(2), 202–248. doi:10.3102/00346543068002202

Wilder, S. (2015). Classroom challenge: A 3D snapshot of student learning in mathematics. *The Clearing House: A Journal of Educational Strategies, Issues and Ideas*, *88*(3), 77–84. doi:10.1080/00098655.2015.1023244

Wilson, P. H., & Downs, H. A. (2014). Supporting mathematics teachers in the common core implementation. *AASA Journal of Scholarship and Practice*, *11*(1), 38–47.

Wood, T., Cobb, P., & Yackel, E. (1991). Change in teaching mathematics: A case study. *American Educational Research Journal*, *28*(3), 587–616. doi:10.3102/00028312028003587

Zawojewski, J. (2008). The banner problem [Class handout]. Illinois Institute of Technology., MSED 501.

ENDNOTE

1 Charter Schools are tuition-free, government funded schools that are independently run. For more details, see Prothero (2018).

APPENDIX 1

The Banner problem, seen below, was provided by Zawojewski (2008). Sample student work (Figure 5) was created by the authors for teachers to evaluate.

The Banner problem:

Ms. Z has felt fabric that is 5 ¾ yards long, with a consistent width of 24". She is making banners that require 5/8 yards of fabric for each banner. She will make as many banners as possible out of the felt fabric, and then see if there is enough to cover the top of a trophy table. Ms. Z has asked your group to figure out how much fabric remains after making all of the banners.

Figure 5. Sample student work

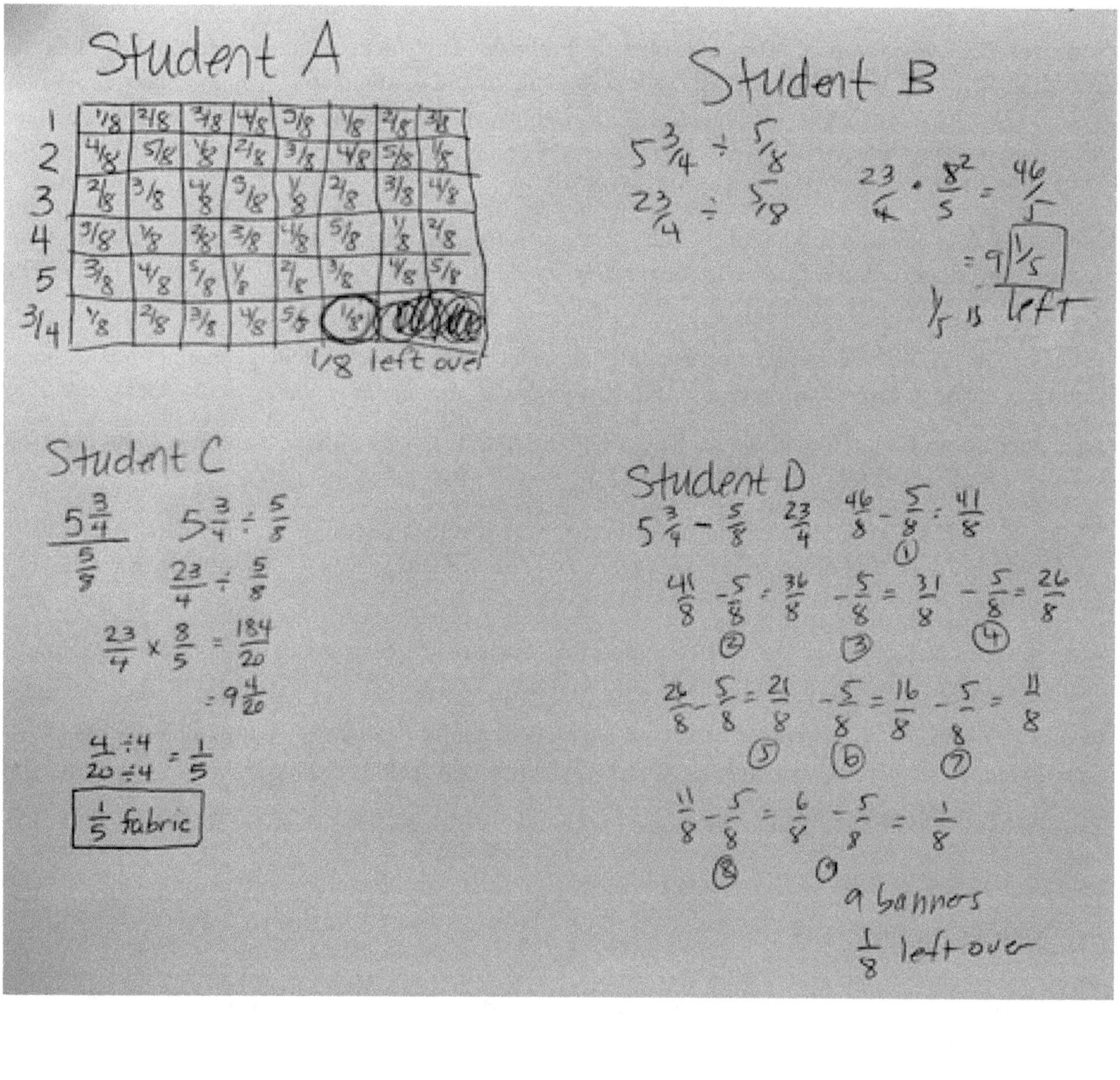

APPENDIX 2

Alignment of enacted curricular materials and student actions with SMP were determined by use of Levels for the Mathematical Practice Standards rubric used by Kartal (2015, p. 99), shown in Table 2.

Table 2. Levels for the mathematical practice standards rubric

		(I)=Initial	(IN)=Intermediate	(A)=Advanced
P1a	Make sense of problems.	Explain their thought processes in solving a problem one way.	Explain their thought processes in solving a problem and representing it in several ways.	Discuss, explain, and demonstrate solving a problem with multiple representations and in multiple ways.
P1b	Persevere in solving them.	Stay with a challenging problem that would need more than one attempt.	Try several approaches in finding a solution.	Struggle with various attempts over time and learn from previous solution attempts.
P2	Reason abstractly and quantitatively.	Reason with models and pictorial representations to solve problems.	Translate quantitative relations into symbols for solving problems.	Convert relations into symbols to solve problems as well as convert symbols into meaningful situations.
P3a	Construct viable arguments.	Explain own thinking and reasoning.	Explain own thinking and thinking of others.	Justify and explain why the solution is correct.
P3b	Critique the reasoning of others.	Understand and discuss other ideas and approaches.	Explain other students' solutions and identify strengths and weaknesses of the solutions.	Compare and contrast various solution strategies, and explain the reasoning of others.
P4	Model with mathematics.	Translate a mathematical situation into mathematical symbols.	Simplify a complicated mathematical situation by making assumptions and approximations, and translate into mathematical symbols.	Simplify and model a complicated mathematical situation or relationship using such tools as diagrams, two-way tables, graphs, flowcharts and formulas, and interpret and reflect on the results in the context of the situation to possibly improve the model.
P5	Use appropriate tools strategically	Use the appropriate tool to find a solution.	Select from a variety of tools the ones that can be used to solve a problem, and explain the reasoning for the selection.	Combine various tools, including technology, explore, and solve a problem as well as justify the tool selection and problem solution.
P6	Attend to precision.	Calculate accurately and efficiently, express numerical answers with a degree of precision.	Incorporate appropriate vocabulary and symbols in communicating own reasoning and procedure to others.	Use appropriate symbols, vocabulary, and labeling to effectively communicate and exchange ideas.
P7	Look for and make use of structure.	Look for structure within mathematics to solve problems efficiently.	Compose and decompose number situation and relationships through observed patterns in order to simplify solutions.	See complex and complicated mathematical expressions as component parts.
P8	Look for and express regularity in repeated reasoning.	Look for obvious patterns, and use if/then reasoning strategies for obvious patterns.	Find and explain subtle patterns.	Discover deep, underlying relationships (uncover a model or equation that unifies the various aspects of a problem such as discovering an underlying function).

APPENDIX 3

Teacher actions from observations were coded for SMP using an adaptation of the Rubric for Implementing Standards for Mathematical Practice (Park City Mathematics Institute, 2011), seen in Table 3.

Table 3. Adapted rubric for implementing standards for mathematical practice

PRACTICE	NEEDS IMPROVEMENT	EMERGING (teacher does thinking)	PROFICIENT (teacher mostly models)	EXEMPLARY (students take ownership)
MP1: Make sense of problems and persevere in solving them.	Teacher: ❑ Does not allow for wait time; asks leading questions to rush through task. ❑ Does not encourage students to individually process the tasks. ❑Is focused solely on answers rather than processes and reasoning.	Teacher: ❑ Allots too much or too little time to complete task. ❑ Encourages students to individually complete tasks, but does not ask them to evaluate the processes used. ❑ Explains the reasons behind procedural steps. ❑ Does not check errors publicly.	Teacher: ❑ Allows ample time for all students to struggle with task. ❑ Expects students to evaluate processes implicitly. ❑ Models making sense of the task (given situation) and the proposed solution.	Teacher: ❑ Differentiates to keep advanced students challenged during work time. ❑ Integrates time for explicit meta-cognition. ❑ Expects students to make sense of the task and the proposed solution.
MP2: Reason abstractly and quantitatively.	Teacher: ❑ Does not expect students to interpret representations. ❑ Expects students to memorize procedures with no connection to meaning.	Teacher: ❑ Expects students to model and interpret tasks using a single representation. ❑ Explains connections between procedures and meaning.	Teacher: ❑ Expects students to interpret and model using multiple representations. ❑ Provides structure for students to connect algebraic procedures to contextual meaning. ❑ Links mathematical solution with a question's answer.	Teacher: ❑ Expects students to interpret, model, and connect multiple representations. ❑ Prompts students to articulate connections between algebraic procedures and contextual meaning.
MP3: Construct viable arguments and critique the reasoning of others.	Teacher: ❑ Does not ask students to present arguments or solutions. ❑ Expects students to follow a given solution path without opportunities to make conjectures.	Teacher: ❑ Does not help students differentiate between assumptions and logical conjectures. ❑ Asks students to present arguments but not to evaluate them. ❑ Allows students to make conjectures without justification.	Teacher: ❑ Identifies students' assumptions. ❑ Models evaluation of student arguments. ❑ Asks students to explain their conjectures.	Teacher: ❑ Helps students differentiate between assumptions and logical conjectures. ❑ Prompts students to evaluate peer arguments. ❑ Expects students to formally justify the validity of their conjectures.
MP4: Model with mathematics.	Teacher: ❑ Identifies appropriate variables and procedures for students. ❑ Does not discuss appropriateness of model.	Teacher: ❑ Verifies that students have identified appropriate variables and procedures. ❑ Explains the appropriateness of model.	Teacher: ❑ Asks questions to help students identify appropriate variables and procedures. ❑ Facilitates discussions in evaluating the appropriateness of model.	Teacher: ❑ Expects students to justify their choice of variables and procedures. ❑ Gives students opportunity to evaluate the appropriateness of model.

Continued on following page

Table 3. Continued

PRACTICE	NEEDS IMPROVEMENT	EMERGING (teacher does thinking)	PROFICIENT (teacher mostly models)	EXEMPLARY (students take ownership)
MP5: Use appropriate tools strategically.	Teacher: ❑ Does not incorporate additional learning tools.	Teacher: ❑ Demonstrates use of appropriate learning tool.	Teacher: ❑ Chooses appropriate learning tools for student use. ❑ Models error checking by estimation.	Teacher: ❑ Allows students to choose appropriate learning tools. ❑ Creatively finds appropriate alternatives where tools are not available.
MP6: Attend to precision.	Teacher: ❑ Does not intervene when students are being imprecise. ❑ Does not point out instances when students fail to address the question completely or directly.	Teacher: ❑ Inconsistently intervenes when students are imprecise. ❑ Identifies incomplete responses but does not require student to formulate further response.	Teacher: ❑ Consistently demands precision in communication and in mathematical solutions. ❑ Identifies incomplete responses and asks student to revise their response.	Teacher: ❑ Demands and models precision in communication and in mathematical solutions. ❑ Encourages students to identify when others are not addressing the question completely.
MP7: Look for and make use of structure.	Teacher: ❑ Does not recognize students for developing efficient approaches to the task. ❑ Requires students to apply the same algorithm to a task although there may be other approaches.	Teacher: ❑ Identifies individual students' efficient approaches, but does not expand understanding to the rest of the class. ❑ Demonstrates the same algorithm to all related tasks although there may be other more effective approaches.	Teacher: ❑ Facilitates all students in developing reasonable and efficient ways to accurately perform basic operations. ❑ Continuously questions students about the reasonableness of their intermediate results.	Teacher: ❑ Prompts students to identify mathematical structure of the task in order to identify the most effective solution path. ❑ Encourages students to justify their choice of algorithm or solution path
MP8: Look for and express regularity in repeated reasoning.	Teacher: ❑ Does not show evidence of understanding the hierarchy within concepts. ❑ Presents or examines task in isolation.	Teacher: ❑ Hides or does not draw connections to prior or future concepts.	Teacher: ❑ Connects concept to prior and future concepts to help students develop an understanding of procedural shortcuts. ❑ Demonstrates connections between tasks.	Teacher: ❑ Encourages students to connect task to prior concepts and tasks. ❑ Prompts students to generate exploratory questions based on current task. ❑ Encourages students to monitor each other's intermediate results.

Chapter 12
Reform-Based Mathematics in Kenya: A Case Study

Yasmin Sitabkhan
RTI International, USA

Linda M. Platas
https://orcid.org/0000-0001-9253-3880
San Francisco State University, USA

Tabitha Nduku
https://orcid.org/0000-0001-7076-2244
RTI International, Kenya

Wendi Ralaingita
RTI International, USA

ABSTRACT

The focus of this chapter is to describe efforts to implement reform-based mathematics practices in Kenya preprimary classrooms. As much of the research behind reform-based practices comes from high-income contexts, the authors explore how practices can be adapted to middle income contexts through a case study with seven teachers. Using observations, surveys, and interviews, they present data around two key strategies common in reform-based math: questioning strategies and small group, hands-on work. They show that teachers struggled to use open ended questions, which may be related to their pedagogical knowledge. Teachers were successful with implementing small group work, especially after they were comfortable with the content. They discuss what types of research are needed to better understand how reform-based math practices can be adapted to be used across countries with varying levels of resources.

DOI: 10.4018/978-1-7998-9422-3.ch012

INTRODUCTION

The evidence behind reform-based mathematics teaching is vast, and most mathematics education professionals view reform-based mathematics teaching as best practice in classroom instruction (National Council of Teachers of Mathematics, 2014). However, much of the literature on the success of reform-based mathematics come from higher income countries, where teachers receive training in mathematics teaching, classrooms have resources, and class sizes are small. In low- and middle-income countries (LMIC), where primary school teachers are often not trained in mathematics teaching, classrooms frequently have few resources, and class sizes are large, there is instead a body of research on the challenges of implementing reform-based teaching methods and questions around their applicability (Komatsu et al., 2021; Schweisfurth, 2011; Tabulawa, 1997, 2003). These studies show how efforts to implement new practices such as group work, questioning strategies, and active learning have often failed to take hold because of resource constraints as well as cultural and political contexts. Teachers are typically not trained in reform-based math pedagogies, and these pedagogies are often not well adapted to suit the realities of the context.

More research is needed into how elements of reform-based teaching can be adapted to local contexts to increase uptake in LMIC. We define reform-based mathematics as classroom instruction that includes classroom discourse and an emphasis on conceptual understanding that is implemented through activities that connect with and build from students' informal mathematical knowledge. In this chapter, we describe a case study which allowed us to study a real-life phenomenon and test specific theories (Yin, 2017). The case study included classroom observations, teacher interviews, and a teacher knowledge, beliefs, and attitudes survey with seven teachers to examine elements of reform-based mathematics teaching introduced during an intervention in pre-primary classrooms in Kenya. We gathered data in classrooms of 4- to 6-year-old students in government schools located in Uasin Gishu County. We describe how teachers implemented two core elements of the intervention that were based on reform teaching practices: questioning strategies and small-group work. These two elements were integral to the intervention, and teachers were supported through the provision of teachers' guides, training, and ongoing classroom support to implement these elements. Questioning strategies included asking higher-order, open-ended questions that encouraged students to discuss mathematics, explain how they were solving problems, and move away from one-word responses. Small-group work created opportunities for students to construct their own understanding and practice new skills through hands-on practice with materials.

We present data around these two elements, describing what it looked like in classrooms, and pondering how these elements were adapted to the context. Specifically, our research seeks the following answers:

- How were questioning strategies and small-group work taken up by teachers?
- How were teachers' knowledge of mathematical development and their beliefs around teaching pre-primary mathematics related to their classroom practices?

BACKGROUND

Research has shown that children develop a large repertoire of informal mathematical competencies from birth through the pre-primary years, which form the foundation for learning primary school mathematics (Clements & Sarama, 2014; Hungi & Ngware, 2018). Toddlers are able to manipulate the number of objects, organize them into groups, and change amounts by adding or taking away objects (Baroody et al., 2008; Langer et al., 2003). Many of these informal competencies are developed through play and activities throughout children's everyday lives in daily routines and interactions with family members, friends, and teachers. Effective instruction during the pre-primary years focuses on providing opportunities to further develop these skills, while at the same time exposing children to the more formal mathematics they will encounter in their first year of primary school (Baroody & Wilkins, 1999; Carpenter et al., 1996; Cooke & Bruns, 2018; Ginsburg et al., 1998; Resnick, 1992; Van de Walle et al., 2014). This research on effective instruction has informed many reform-based methods in the United States and other high-income country contexts.

While the majority of the research on mathematics instruction and, in particular, reform-based instruction, has been undertaken in higher-income contexts, there is a growing body of literature on mathematics instruction in LMICs. There is some evidence on the potential positive impact of reform-based methods on learning outcomes in LMICs. Sitabkhan and Platas (2018) reviewed 24 studies of mathematics interventions in LMICs that had resulted in gains in learning outcomes to identify core instructional strategies and found four primary strategies, all in line with reform-based instruction: using multiple representations (Guajardo et al., 2013; Jonason et al., 2014; Näsland-Hadley et al., 2014); knowing and using developmental progressions (Gallego et al., 2017; King et al., 2015; Vaijayanti et al., 2016); supporting student explanation and justification (Education Development Center, 2017; Vula et al., 2017); and integrating formal and informal mathematics (Dillon et al., 2017; McEwan, 1998). Other reports discussing the status of mathematics education in LMICs provide examples of interventions that have shown positive results (Bethell, 2016; Bolton, 2019; Evans & Acquaye, 2019). While many of these efforts incorporated reform-based methods, they also included other inputs and did not systematically measure the impact attributable to the new teaching methods.

A larger body of literature, however, points to the many challenges involved in implementing reform-based instruction in these contexts. Curricula and standards in LMICs may not always reflect developmental progressions and there can be a heavy emphasis on procedural skill rather than conceptual understanding (Evans & Acquaye, 2018; Kusaka, 2019; Pritchett & Beatty, 2012). Textbooks and learning materials are frequently in short supply and children often have to share resources (Read, 2017; Yuan & Evans, 2016). For example, across sub-Saharan Africa, there is on average one mathematics book for every three students (United Nations Educational, Scientific and Cultural Organization [UNESCO], 2016). Access to teaching aids and manipulatives is rare, and when they are available, teachers often have little training or support in their effective use. Mathematics teachers face classrooms that are under-resourced and overcrowded. At the same time, these teachers may struggle with foundational math skills themselves due to their own educational experiences. A cross-national survey of African countries found that, while 90% of teachers could solve basic addition problems, 89% could not interpret data presented in a graph and 85% were unable to solve a word problem (Bold et al., 2017). A World Bank study showed similar problems in Latin America and the Caribbean. For example, 84% of teachers in Peru scored below level 2 as defined by the Ministry of Education, meaning they could not "adapt routine and simple mathematical procedures and strategies" (Metzler & Woessmann, 2012). Similarly, in Paraguay 90% of preschool teachers demonstrated a lack of understanding of the math content they were expected to teach (Inter-American Development Bank, 2015). Given their own struggles with mathematics, the difficult context in which they are teaching, and insufficient ongoing support, many teachers in LMICs will resort to traditional methods they themselves experienced as students: rote memorization and a reliance on procedural drills (Akyeampong, 2017; Hardman et al., 2009; Pryor et al., 2012).

Questioning Strategies and Groupwork

Given the many challenges experienced in math teaching and learning in LMIC contexts, researchers and practitioners have sought to identify entry points for improvement. Given that sweeping changes would likely be difficult to implement and overwhelming to teachers, these efforts have focused on trying to identify instructional strategies that would be contextually appropriate, relatively easy for teachers to begin learning and practicing, and would have impact on student learning, particularly for the development of conceptual understanding. Two key areas that have shown promise in this respect relate to the use of productive questioning and effective use of student grouping.

A core element of reform-based instruction, classroom "math talk" includes rich problems for children to explore and probing questions from the teacher and has been shown to support the development of conceptual understanding as well as procedural fluency (Ball, 1993; Hiebert & Wearne, 1993; Klibanoff et al., 2006; Lampert & Cobb, 2003). Use of productive questions supports the use of explanation and justification, a core focus of many reform efforts. In order to be productive, questions need to support higher order thinking, help teachers to identify student errors or misconceptions, and provide students opportunities to engage in explanation or justification (Lutfeali et al., 2021).

Facilitating small-group work within a math class can be supportive of multiple strategies (Sitabkhan et al., 2019; Lutfeali, et al., 2021). Particularly in the context of large class sizes, such grouping can facilitate explanation and justification, as it will allow more opportunities for children to discuss with each other and engage in math talk. Additionally, when large class sizes are coupled with insufficient resources, teachers can use small groups as a way for all children to engage with manipulatives, which supports understanding and use of representations.

While questioning and use of small groups represent potential entry points for helping teachers to incorporate reform-based methods intro their instructional practice, there remain a number of potential challenges. Teachers who struggle with foundational math concepts themselves, or lack confidence in their own math skills or their instruction, may be hesitant to ask open-ended questions or to facilitate math talk in the classroom that could expose their own knowledge gaps (Akyeampong et al., 2013; Schwille & Dembélé, 2007; UNESCO, 2012). Similarly, groupwork moves some of the control over lesson activities from the teacher to the students. For many teachers in LMIC contexts, this shift may threaten existing norms of teacher authority, as well as present unfamiliar challenges around classroom management (Bethell, 2016).

The Tayari Project

Tayari was a research-based program piloted in Kenya in the four counties of Laikipia, Nairobi, Siaya and Uasin Gishu from 2014-2019. The program was piloted in all public early childhood centers in the four counties. In Nairobi County the program was also piloted in Alternative Provision of Basic Education and Training centers (informal private schools located primarily in poor, peri-urban areas of the capital city). The program was funded by the Children's Investment Fund Foundation with an overarching goal of supporting the Kenyan government to develop a cost-effective and scalable early childhood model to enhance the proportion of pre-primary children (5 to 6-year-olds) transitioning to Class 1 with the requisite school readiness skills and competencies in literacy, numeracy, executive functioning and social-emotional skills.

Results from the Primary Math and Reading (PRIMR) Initiative implemented by RTI International (2011–2014) indicated that well-designed, syllabus-based teachers' guides; intensive teacher training; and targeted ongoing teacher support through coaches improved instruction and led to significant gains in lower primary pupils' literacy and numeracy performance (Piper & Mugenda, 2016; Piper et al., 2016). Building on the strength and success of PRIMR, the Tayari program was designed to test whether a program dependent on the existing Ministry of Education and county structures could also improve early childhood outcomes, as it did in primary grades. Thus, the project worked closely with the National and County governments to develop teacher guides and student books based on the Kenyan Early Childhood syllabus (Kenyan Institute of Curriculum Development, 2017), train and support teachers, and monitor implementation.

Prior to the implementation of the Tayari program, the status quo in most pre-primary classes in Kenya consisted of children sitting in desks with teachers delivering rote instruction, especially in early mathematics. Often, students spent time writing numbers for a large portion of the math instructional time. To improve on this, the Tayari mathematics materials covered a range of competencies including number sense, classification, measurement, geometry and spatial awareness. These activities help students build a strong conceptual base in these domains. Both the teacher's guide and student workbooks included explicit mathematical lessons that were meaningful, engaging, and enabled students to begin to make the necessary connections between informal and formal mathematical concepts.

To achieve changes in instruction, project staff and mathematics education experts, along with experts from the Kenyan Institute of Curriculum Development, created teacher guides and accompanying student books to support teachers to implement small-group work and questioning strategies that promoted mathematical discussions. These materials were based on a spiral approach whereby the program integrated several different topics into one core 45-minute lesson per day. These core lessons were then repeated throughout the year with increasing difficulty each week. The weekly structure was very predictable with Day 1–3 covering number concepts, Day 4 sorting and classifying, and Day 5 measurement. This weekly structure ensured that teachers and students could anticipate topics, making the flow of lessons, materials preparation and management easier. It also meant that students and teachers had multiple opportunities to delve more deeply into topics over time. In this way, the Tayari materials ensured teachers and students saw the connections among topics taught through the daily lessons.

Each lesson consisted of simple activities which were presented clearly and consistently and followed a predictable structure (lesson introduction, main (whole-class) activity, small-group activity, and conclusion). The majority of time in each lesson was dedicated to small-group work where children were actively engaged in a math activity modeled during the whole-group session. Teachers made most of the manipulatives using locally available materials for use during these small-group activity sessions. To support teachers with questioning strategies, "teachable moments" were included in which the teacher prompted the students to think critically around a key concept or skill.

METHODS

Participants

This case study, a detailed look at the implementation of specific practices within a larger study, took place in Uasin Gishu County, Kenya as part of the Tayari project. Participants in our research were seven teachers and their classrooms in four schools. Class size was between 20-40 in each classroom, which is typical for Uasin Gishu County classrooms. Schools and teachers were chosen by project staff using recommendations from County Early Childhood coordinators (government-employed coaches who provide support to public schools). Coordinators were asked to identify teachers who had consistently been delivering the lessons. We specifically asked coordinators not to identify their star teachers or the teachers that were struggling, but rather the teachers that would be representative of majority of their peers. As part of the Tayari program, all teachers received a 5-day training at the beginning of the first term and a 3-day training at the beginning of the second and third term each year. The training focused on supporting teachers to implement lessons, including structured discussion around how to manage group work and how to ask higher-order questions. Most of the time during the training was spent on teachers practicing lessons with colleagues and receiving feedback.

Instruments

The classroom observation focused on three core areas: instructional practices, use of manipulatives, and small-group or independent work. These three areas were chosen as they represent important elements of reform-based mathematics instruction in both the research literature as well as the design of the Tayari program.

The teacher survey asked teachers to fill in a validated survey measuring preschool teachers' mathematics knowledge and beliefs (Platas, 2014, 2015). The Knowledge of Mathematical Development (KMD) consisted of 20 multiple-choice items that measured respondents' knowledge of early mathematical development. The Mathematical Development Beliefs Survey (MDBS) was designed to measure beliefs concerning (a) age-appropriateness of mathematics instruction, (b) classroom locus of generation of mathematical knowledge (teacher vs child), (c) mathematical development as a primary goal of preschool education, and (d) confidence level in providing mathematics instruction. The MDBS used a six-point Likert scale.

Table 1. Exemplar items from the KMD and MDBS

<table>
<tr><th>Knowledge of Mathematical Development (KMD)</th><th>Mathematical Development Beliefs Survey (MDBS)</th></tr>
<tr><td rowspan="4">☐ Amari is presented with two groups of bottlecaps, one with five bottlecaps and one with two bottlecaps. When asked "How many altogether?" in the two groups, Amari counts all of the bottlecaps beginning with the group of five bottlecaps ("1, 2, 3, 4, 5, 6, 7").
☐ Amari is presented with two groups of bottlecaps, one with five bottlecaps and one with two bottlecaps. When asked "How many altogether?" in the two groups, Amari counts on from the first set ("5, 6, 7").
☐ Same
☐ Do not know</td><td>Math is an important part of the preprimary curriculum.</td></tr>
<tr><td>It is better to wait until Standard 1 for math activities</td></tr>
<tr><td>Mathematical activities are an inappropriate use of time for preprimary learners; because they aren't ready for them.</td></tr>
<tr><td>Preprimary learners are capable of learning math.</td></tr>
<tr><td>☐ Rispa answers the following addition question: "If you have these five biscuits and I give you two more, how many biscuits will you have altogether?" (all cookies are present).
☐ Rispa answers the following subtraction question: "If you have these three biscuits and you give me one, how many biscuits will you have left?" (all cookies are present).
☐ Same
☐ Do not know.</td><td></td></tr>
</table>

The teacher interview asked teachers to reflect on the lesson just taught (see Table 2 for exemplar questions). This semi-structured interview also included questions on the answers the teachers had given on the MDBS and about their perceptions of the four parts of the daily lessons (lesson introduction, main activity, small-group activity, and conclusion). Each interview was approximately 45 minutes long.

Table 2. Exemplar questions from teacher interview

Teacher Interview Questions
Imagine you are talking to a new teacher who is just about to use the Tayari programme for the first time. What advice would you give this new teacher?
Have you created any materials to use during your math lessons that are not provided by the program? Please describe.
Prior to teaching in Tayari project, please describe the materials you used to teach math (i.e. teacher's guide, lesson plans, etc.)

Taken together, these three data sources shed light on how the elements of reform-based mathematics teaching embedded in the program were being manifested in the classroom. This also allowed us to provide a better understanding of the classroom practices of the seven teachers.

Procedures

All seven classrooms were visited one time by the research team. All data collection took place during one week in April 2018. The research team consisted of two international researchers and a Tayari staff member who participated in developing the lessons. In each classroom, we observed one 45-minute math lesson. Each observer took their own notes during the lesson on the observation form, focusing on the three core areas- instructional practices, use of manipulatives, and small-group or independent work. At the end of each observation, the three members of the research team met, compared notes, and created one set of final notes that compiled all the observations into one form. After the observation, one member of the research team asked the teacher to fill out the teacher survey, and then conducted the interview. All notes were written by hand.

RESULTS AND DISCUSSION

In the following section, we interrogate two areas integral to reform mathematics, questioning strategies and active learning through small-group work. We draw on our three data sources to inform our investigation.

Questioning Strategies

Tayari lessons were designed with "teachable moments," which aimed to foster discussions that would promote mathematical thinking. An example is shown in Figure 1. This was done through higher-order questions without one correct answer, such as "How do you know?"

Figure 1. Teachable moments excerpt from Tayari materials

Whole class

- Draw the following table on the board or use number cards.

3	
15	
9	

- Point to the first number (3) and count an equal number of concrete objects with the learners.
- Draw the same number of objects next to the number. Write the number 3 next to the objects you drew, showing learners how to form the number. Guide learners to practise writing the number in the air.
- Repeat the activity for other numbers on the board.
- Ask:
  - ❖ *Which number is greater, 3 or 15? Why?*
  - ❖ *Which number is greater, 3 or 9? Why?*

In the observations, five of the seven lessons did not use the "teachable moments" despite their integration into the lessons. For example, Teacher ID01 was modeling a correct object counting strategy for students by counting five bottle tops. As she did this, she asked the class, "How many do I have?" to which there was a choral response of "five." She then asked a student to come up, and said "Can you confirm that there are five?" The student then mirrored the teacher's actions and counted the five bottle tops. In this way, the questions which the teacher asked were answerable by either one correct response or by repeating a previously-modeled action.

Teachers ID06 and ID07 were the only teachers who asked higher-order questions that did not have one correct answer. Both of these lessons were about buying and selling using money, and the small-group activity asked the students to role play buying items in a market. Teachers ID06 and ID07 were also in the same school, so it is possible that they shared strategies with each other. Teacher ID06, when introducing the play market and modeling it in front of the class, asked "Which coin cannot buy an orange?" After one student answered, she then asked, "Why?" She frequently followed up the questions she asked with "Why?" asking students to justify their answers. Teacher ID07 did not ask "why" questions, but she did ask questions that did not have only one right answer, and could therefore generate more discussion, such as "Can this coin buy a mandazi (Kenyan donut-style pastry)?" and "What do you want to buy?"

While close-ended questions are an essential part of any mathematics lessons, overusing them can lead to children developing a view that mathematics is only about the correct answer. Five of the seven classrooms used only close-ended questions. Questions with one correct answer do not promote the reform-based practices fostered by including the teachable moments in each lesson. However, often teachers do not ask open-ended questions because they may think students are not able to reason about mathematics, or they do not have enough knowledge themselves about the mathematics. To better understand this, we turn to the data from the KMD survey.

The KMD survey measures teachers' content knowledge of children's foundational mathematical development in numeracy. Teachers may miss opportunities to support children's learning if they themselves do not possess an understanding of children's mathematical development (Ginsburg et al., 2014; Pollitt et al., 2020). Most teachers in the study answered lower level KMD items such as whether it was easier for children to start counting from one or six and whether counting past ten was significantly more difficult than counting up to ten. Teachers ID06 and ID07 responded accurately to additional items that required more knowledge of children's numerical development. Only few teachers were able to answer questions regarding children's foundational development in early arithmetic. For instance, items asking about children's ability to share objects fairly between two or three children and the influence of the size of the set of objects on addition and subtraction problems (a subtraction problem involving three objects is easier than an addition problem involving seven objects) were rarely answered correctly. This data points to the importance of supporting the development of teachers' content and pedagogical knowledge during training.

Group Work

An essential component of reform-based math practices in the early years is hands-on engagement with materials by the students. The Tayari lessons incorporated students' use of materials in small groups. This was done for two reasons: first, most classrooms did not have enough available materials for all students in the class to use them at the same time during independent work. Thus, small groups provided a way for students to share the materials. Second, the small groups encouraged peer learning and collaboration, and students could observe and contribute to the activity. As discussed in the literature review earlier, there is a large body of research both on the importance of group work, and also how difficult it is for many teachers to implement, especially in LMIC.

In particular, pre-primary teachers in Kenya are frequently not used to relinquishing control of what was happening in the classroom. This form of control was mainly observed by teachers wanting to use more whole-class approach and not giving children opportunities to work with others in smaller groups during instruction time. As Teacher ID05 pointed out "*I like teaching whole-group because small groups cause distractions especially if there are not enough materials.*" This form of control was also noted with Teacher ID04 who said that she found the whole class easy to teach because "*I'm doing it, I could capture all student attention.*" Teacher ID07 thought that students could not work well with others in their small groups while sharing and role taking, because "*they're still egocentric.*" To allow teachers to relinquish control during instruction time, the Tayari lessons integrated reform-based mathematical practices by asking them to put students in small groups and allow them to work independently, providing support as needed. The lessons were structured as a mini-lesson where teachers and students did an activity together in a whole-class setting, and then students in small groups were asked to do that same exact activity with peers. Because this would be difficult, support was provided through classroom coaching and in-service training to enable teachers to successfully support small groups.

In the classroom observations, we found that six of the seven lessons (IDs 01, 04, 06, and 07) utilized small groups. In four out of these six lessons, students had opportunities to use the materials independently during the small-group time, which was the intended purpose of the small-group time. These four teachers were able to relinquish control of the materials and allow students to work with them independently or with peers. For example, Teacher ID01 taught a lesson about number sequencing, displayed in Figure 2. After demonstrating the task of matching number and object cards and then putting them in order, the teacher had students go into their groups. She then passed out a set of number and object cards to each small group. Students in the group each took a few number and object cards, and they talked to each other to match and order the cards. The teacher monitored the progress of the groups, walking around the room and checking each group.

Figure 2. Sequencing activity

Whole class

Demonstrate the following activity:

- Pick a number card and ask the learners to say the number you have picked. Show the learners an object card that matches the number card picked.
- Match the number cards and object cards, as in the photo below.

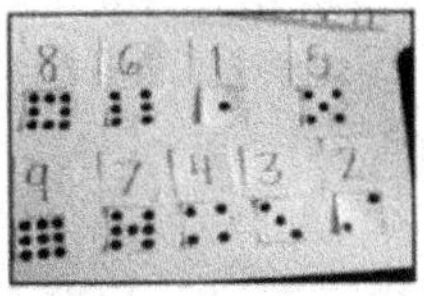

- Put the number and object cards in order with the learners, from as shown below.

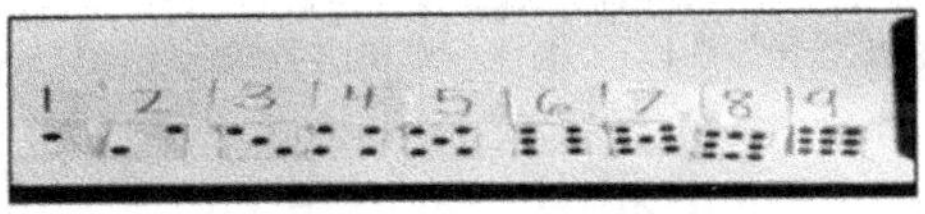

Small group

- Give groups of learners a set of number and object cards from 1–9 for them to match.
- Guide learners to put the cards in order from 1–9.

In contrast, Teachers ID02 and ID03 did not relinquish control to the students during small-group. Teacher ID02 was teaching a lesson where students would pick a number card, count out the appropriate number of bottle tops to match the number card, and then draw the bottle tops in their workbook. After demonstrating this game, the teacher gave each group a set of number cards and bottle tops. She then went to each group to model the activity once more, and then assigned students in a group to do the activity on their own while other students in the group watched. Each group waited until the teacher came around before touching the materials. Because of this, only one student at a time was engaged in the activity, with the rest of the class either watching that student or waiting for the teacher to come to their table. Although this teacher engaged students in small-group work, it was qualitatively different than the structure described above, where all students were actively engaged with materials and working independently.

We now turn to the MDBS survey to understand how teacher knowledge and beliefs played out with regard to group work. We focused on the locus of control, which describes the extent to which teachers believe learning happens through teacher or child agency. Some teachers may believe that they solely are responsible for children's learning and that children play no part in the construction of that knowledge. Other teachers may believe that children alone construct that knowledge. In reality, these beliefs occur on a spectrum between teachers' and children's role in the acquisition of knowledge (Ginsburg et al., 2006, LeFevre et al., 2009). Locus of control is relevant to the observations we made during group work, as we hypothesized that teachers who were more likely to let go of control during group work and allow children to work on their own might have stronger beliefs that children have agency in their own learning.

In examining teachers' beliefs about mathematical development with the MDBS, the seven teachers in the Tayari group included *both* children and teachers as a locus in the generation of mathematical knowledge. Researchers have found that teachers' beliefs around the learning of mathematics relate to their support of mathematical development in the classroom. The teachers we observed agreed with items in the MDBS like "In preschool children construct mathematical knowledge." Group work, a classroom practice where children actively engage in mathematical concepts outside of the more didactic chalkboard lessons, provide opportunities for children to play a larger role in the construction of their mathematical knowledge.

In addition to completing the MDBS survey, we interviewed teachers on why they marked specific answers on the survey. In terms of locus of control, we asked teachers to explain their answer to this statement: *Pre-primary learners learn mathematics without support from teachers.* All seven teachers disagreed with this statement, with two of them strongly disagreeing. When we asked them to explain, teachers offered justifications in which they emphasized that teachers have to demonstrate how to do something or a particular skill or concept before the student can learn it. Teacher ID02 said "[Students] can't learn without support. Teacher may not give objects, but need to show how/demo first," focusing on how the teacher should first model something before giving it to the student. This aligns with the observation, where the teacher in small groups modeled counting objects over and over at the beginning of the lesson. When one student tried to count seven objects but counted "1, 2, 3, 5," the teacher modeled how to count correctly to seven and asked the student to also count to seven.

We also asked teachers about their perceptions of the different parts of the lesson during the interviews, including small-group work and whole-group instruction. Three out of the seven teachers interviewed felt the whole-group instruction part of the lesson was easiest to teach, while two felt the introduction was the easiest. Most of these sections of the lesson involve the teacher addressing the whole group, which means that five teachers were more comfortable with direct instruction and thought it as the best way to teach math, as opposed to small-group or one-on-one instruction. The whole-group session was also considered an easy activity to teach because it was a demonstration session done by the teacher and during this session, the teacher could capture all students' attention. One teacher shared that she was confident that "*If I demonstrate well during whole-class, children will do it well during the small-group activity.*"

Results such as these shed light on how teachers are using new reform-based practices that were introduced to them in the classroom. While five teachers thought direct instruction was the easiest and best way to teach math, four teachers felt that small-group was easiest for students. Teachers said that small-group activity was the easiest for the students because everyone participates, and students can collaborate, share ideas, enjoy, and learn more. Many teachers reported that they previously had not conducted small groups within their lessons, and that the Tayari lessons pushed them outside of their comfort zone through the daily incorporation of small groups.

Many teachers also referred to the hands-on use of materials during the lessons, and especially during small-group work. When we asked the teachers how their teaching and confidence in teaching math had changed upon joining Tayari, four teachers commented on the increase in the amount of materials they had for teaching. Teacher ID01 acknowledged that no activity within the lesson plan was difficult for them to teach as long as materials are provided because students need them to understand the concepts taught. "*[Math is] not difficult.*" "*With activities, [the] children find it easy. They need to do it. [Teaching] 'take away' [e.g., subtraction] with concrete materials so it's easy for children. Before it was abstract, so [it was] hard for students.*" Teacher ID 6 also added that "*students can use tangible things*" focusing on how her approach to teaching math is now different.

CONCLUSIONS AND NEXT STEPS

This was a small case study, with clear limitations on generalization to teachers beyond the sample. However, the multiple sources of data, taken together, show the struggles and successes of reform-based mathematics teaching practice in new contexts that may have culturally-based ways of teaching and interacting. Even though open-ended questions were provided to teachers, a majority of teachers only asked questions with one correct answer. Asking open-ended questions can be difficult for teachers, especially teachers who are learning a new methodology, because teachers may not understand the student's answer. For example, if a teacher shows the numeral "4" and asks a student, "what number is this?" the answer is known, and if there is an incorrect response, the teacher knows what the correct one is and how to correct that student. However, if a teacher asks a child to explain how they know an answer is correct, or why they think a particular answer is correct, there are multiple ways a child can answer that question, with no clear correct/incorrect judgement to be made.

This situation can be uncomfortable and unfamiliar for teachers in general, and specifically in Kenya where direct instruction has dominated mathematics instruction in the past. Cultural factors come into play as well. A study conducted in Tanzania to understand teacher beliefs behind pedagogical decisions found that embarrassment is a key factor in the classroom (Jukes & Sitabkhan, 2021)—that is, teachers do not want to embarrass students, and teachers do not want to be embarrassed in front of their students. It might be that teachers in Kenya primarily ask questions that they know that students can answer, and that they themselves can respond to appropriately. What can be done to further support teachers and encourage them to ask students more open-ended questions? If further research shows that teachers are worried about embarrassing students, then it could be that adapting questioning strategies to reduce embarrassment but still encouraging mathematical discussion is needed. For example, teachers can begin by asking open-ended questions only for topics they feel equipped to answer, and this can be practiced during a training. Alternatively, teachers can model how to answer these types of questions for students so that students understand what is being asked of them. Finally, students can tell their peers in pairs or small groups how they solved a problem to reduce embarrassment in front of the entire classroom. More research is needed to try these and other methods in the classroom.

From the interviews, teachers indicated that they did not regularly use group work during lessons prior to the Tayari project. We found that teachers in our sample, with support from the project, regularly used group work during math lessons, pointing to the success of the intervention to introduce some reform-based practices to these teachers. We found that teachers mentioned how the materials and activities supported the learning of children, and praised the value of active, hands-on learning, especially when compared with their prior instruction. We also identified challenges. We saw that teachers continued to struggle with letting go of the control of the classroom; it may be that direct instruction is something that teachers are more comfortable with, so asking them to completely turn control of the learning over to students during small-group work may not be feasible initially or may take more time to scaffold teachers into this process. More follow-up studies, where teachers are directly asked about the control of the classroom, as well as studies which aim to understand teacher attitudes and beliefs around how best to teach, are needed to better understand teacher decision-making in the classroom.

One interesting insight from the observations was that, in general, students did not seem to struggle with the content; the majority of questions asked were answered correctly, and most small-group activities were done with ease. Although the content was aligned with the country-level curriculum, it may be that the activities were easy for students. In fact, in subsequent revisions of the materials, the project has increased the difficulty level of the activities. In some ways, though, it could be that the simplicity of the activities for the first years of implementation provided an entry point into teachers' ability to allow students to work in groups and begin to do activities somewhat independently, as we saw several, but not all, of the teachers do. Although more research is needed, simplicity of content may be one avenue of making teachers comfortable with reform-based practices that they may not be familiar with, where they can be confident that students will not struggle, and they can experiment with different strategies. Then, when content becomes more difficult, teachers will have prior experience to draw on in terms of implementing these elements of reform teaching.

In sum, this study provided insight into classrooms that are just being introduced to reform-based teaching practices in a LMIC. More research is needed across diverse contexts to adapt practices which achieve the same goal as those of reform-based mathematics, such as allowing students to develop rich and nuanced understanding of foundational mathematics skills, while at the same time ensuring that classroom culture is taken into account.

REFERENCES

Akyeampong, K. (2017). Teacher educators' practice and vision of good teaching in teacher education reform context in Ghana. *Educational Researcher*, *46*(4), 194–203. doi:10.3102/0013189X17711907

Akyeampong, K., Lussier, K., Pryor, J., & Westbrook, J. (2013). Improving teaching and learning of basic maths and reading in Africa: Does teacher preparation count? *International Journal of Educational Development*, *33*(3), 272–282. doi:10.1016/j.ijedudev.2012.09.006

Ball, D. (1993). With an eye on the mathematical horizon: Dilemmas of teaching elementary school mathematics. *The Elementary School Journal*, *93*(4), 373–397. doi:10.1086/461730

Baroody, A. J., Li, X., & Lai, M.-L. (2008). Toddlers' spontaneous attention to number. *Mathematical Thinking and Learning*, *10*(3), 240–270. doi:10.1080/10986060802216151

Baroody, A. J., & Wilkins, J. L. (1999). The development of informal counting, number, and arithmetic skills and concepts. In J. V. Copley (Ed.), Mathematics in the early years (pp. 48–65). National Council of Teachers of Mathematics & National Association for the Education of Young Children.

Bethell, G. (2016). *Mathematics education in Sub-Saharan Africa: Status, challenges, and opportunities*. World Bank. https://openknowledge.worldbank.org/handle/10986/25289

Bold, T., Filmer, D., Martin, G., Molina, E., Stacy, B., Rockmore, C., Svensson, J., & Wane, W. (2017). Enrollment without learning: Teacher effort, knowledge, and skill in primary schools in Africa. *The Journal of Economic Perspectives*, *31*(4), 185–204. doi:10.1257/jep.31.4.185

Bolton, L. (2019). *Foundational mathematics education in developing countries*. Institute of Development Studies. https://resourcecentre.savethechildren.net/node/16285/pdf/657_foundational_mathematics_education_in_developing_countries.pdf

Carpenter, T., Fennema, E., & Franke, M. L. (1996). Cognitively guided instruction: A knowledge base for reform in primary mathematics instruction. *The Elementary School Journal*, *97*(1), 3–20. doi:10.1086/461846

Clements, D. H., & Sarama, J. (2014). *Learning and teaching early math: The learning trajectories approach* (2nd ed.). Routlege. doi:10.4324/9780203520574

Cooke, A., & Bruns, J. (2018). Early childhood educators' issues and perspectives in mathematics education. In I. Elia, J. Mulligan, A. Anderson, A. Baccaglini-Frank, & C Benz (Eds.), Contemporary research and perspectives on early childhood mathematics education (pp. 267-289). Springer.

Dillon, M. R., Kannan, H., Dean, J. T., Spelke, E. S., & Duflo, E. (2017). Cognitive science in the field: A preschool intervention durably enhances intuitive but not formal mathematics. *Science, 357*(6346), 47–55. doi:10.1126cience.aal4724 PMID:28684518

Education Development Center. (2017). *Literacy, language, and learning initiative (L3).* Education Development Center. http://idd.edc.org/ projects/literacy-language-and-learning-l3-rwanda

Evans, N., & Acquaye, E. (2019, March 25-29). *Problematizing the familiar in Ghana: Can an in-service model based on authentic problem-solving help teachers construct deeper understandings of new mathematics instructional models?* [Paper presentation]. Comparative and International Education Society 2018 Annual Conference, Mexico City, Mexico.

Gallego, F. A., Näslund-Hadley, E., & Alfonso, M. (2017). *Tailoring instruction to improve mathematics skills in preschools: A randomized evaluation.* https://www.povertyactionlab.org/sites/default/ files/publications/613_1026_Tailoring-Instructions-to- Improve-Mathematics-Skills-in-PreSchool_June2017.pdf

Ginsburg, H. P., Hyson, M., & Woods, T. A. (2014). *Preparing early childhood educators to teach math: professional development that works.* Paul H. Brookes.

Ginsburg, H. P., Kaplan, R., & Cannon, J. (2006) Helping early childhood educators to teach mathematics. In Critical issues in early childhood professional development (pp. 171-202). Paul H. Brookes.

Ginsburg, H. P., Klein, A., & Starkey, P. (1998). The development of children's mathematical thinking: Connecting research with practice. In W. V. B. Damon (Ed.), Handbook of child psychology: Vol. 4. *Child psychology in practice* (5th ed., pp. 404–427). Wiley.

Guajardo, J., Mabeti, F., & Kelly, M. (2013). *Numeracy Boost Malawi: Endline report July 2013.* Save the Children. http://resource-centre-uploads.s3.amazonaws.com/uploads/nb_malawi_zomba_endline_report_final.pdf

Hardman, F., Abd-Kadir, J., Agg, C., Migwi, J., Ndambuku, J., & Smith, F. (2009). Changing pedagogical practice in Kenyan primary schools: The impact of school-based training. *Comparative Education, 45*(1), 65–86. doi:10.1080/03050060802661402

Hiebert, J., & Wearne, D. (1993). Instructional tasks, classroom discourse, and learners' learning in second-grade arithmetic. *American Educational Research Journal, 30*(2), 393–425. doi:10.3102/00028312030002393

Hungi, N., & Ngware, M. (2018). The effects of preschool participation on mathematics achievement of Grade 6 pupils in Uganda. *Educational Research for Policy and Practice, 17*(2), 105-126. https://link.springer.com/article/10.1007/s10671-017-9218-y

Inter-American Development Bank. (2015). *All children count.* Inter-American Development Bank. https://publications.iadb.org/en/publication/17038/all-children-count-early-mathematics-and-science-education-latin-america-and

Jonason, C., Guajardo, J., Nath, B. D., & Hossain, M. (2014). *Literacy & Numeracy Boost: Bangladesh endline.* Save the Children. https://resourcecentre.savethechildren.net/pdf/literacy_boost_and_numeracy_boost-_bangladesh-_endline_report_april_2014.pdf/

Jukes, M., & Sitabkhan, Y. (2021). *Understanding the social classroom: The basis of effective pedagogy?* RTI International SharEd. https://ierc-publicfiles.s3.amazonaws.com/public/resources/Jukes_Social_Classroom_Apr_24.pdf

Kenyan Institute for Curriculum Design. (2017). *Curriculum designs-preprimary PP1-Grade4.* https://kicd.ac.ke/cbc-materials/curriculum-design/

King, S., Korda, M., Nordstrum, L., & Edwards, S. (2015). *Liberia teacher training program: Endline assessment of the impact of early grade reading and mathematics interventions.* RTI International SharEd. https://shared.rti.org/content/liberia-teacher-training-program-endline-assessment-impact-early-grade-reading-and

Klibanoff, R., Levine, S., Huttenlocher, J., Vasilyeva, M., & Hedges, L. (2006). Preschool children's mathematical knowledge: The effect of teacher math talk. *Developmental Psychology, 42*(1), 59–69. doi:10.1037/0012-1649.42.1.59 PMID:16420118

Komatsu, H., Rappleye, J., & Silova, I. (2021). Student-centered learning and sustainability: Solution or problem? *Comparative Education Review, 65*(1), 6–33. doi:10.1086/711829

Kusaka, S. (2019). Issue analysis of competency-based mathematics curriculum design in African countries: A case study of Mozambique's primary mathematics education. *Journal of Education and Learning, 9*(1), 41–50. doi:10.5539/jel.v9n1p41

Lampert, M., & Cobb, P. (2003). Communication and language. In W. G. Kilpatrick & M. D. Schifter (Eds.), *A research companion to the principles and standards for school mathematics.* National Council of Teachers of Mathematics.

Langer, J., Gillette, P., & Arriaga, R. I. (2003). Toddlers' cognition of adding and subtracting objects in action and in perception. *Cognitive Development, 18*(2), 233–246. doi:10.1016/S0885-2014(03)00022-4

LeFevre, J., Fast, L., & Sowinski, C. (2009) *Who's counting? Numeracy and literacy practices of early learning and child care practitioners.* Final report. Canadian Council on Learning. https://www.researchconnections.org/childcare/resources/19205

Lutfeali, S., Sitabkhan, Y., Ralaingita, W., & Piper, B. (2021). *Designing effective numeracy programs in low- and middle-income countries.* Science of Teaching. https://scienceofteaching.site/wp-content/uploads/2021/11/Practical-Guide-for-Numeracy_converted.pdf

McEwan, P. J. (1998). The effectiveness of multigrade schools in Colombia. *International Journal of Educational Development, 18*(6), 435–452. doi:10.1016/S0738-0593(98)00023-6

Metzler, J., & Woessmann, L. (2012). The impact of teacher subject knowledge on student achievement: Evidence from within-teacher within-student variation. *Journal of Development Economics*, *99*(2), 486–496. doi:10.1016/j.jdeveco.2012.06.002

Näslund-Hadley, E., Parker, S. W., & Hernandez-Agramonte, J. M. (2014). Fostering early math comprehension: Experimental evidence from Paraguay. *Global Education Review*, *1*(4).

National Council of Teachers of Mathematics. (2014). *Principles to actions*. National Council of Teachers of Mathematics.

Piper, B., & Mugenda, A. (2016). *USAID/Kenya Primary Math and Reading (PRIMR) Initiative: Endline impact evaluation* (revised edition). RTI International. https://learningportal.iiep.unesco.org/en/library/the-primary-math-and-reading-primr-initiative-endline-impact-evaluation

Piper, B., Ralaingita, W., Akach, L., & King, S. (2016). Improving mathematics outcomes in Kenya: The impact of the Primary Math and Reading (PRIMR) Initiative. *Journal of Development Effectiveness*. Advance online publication. doi:10.1080/19439342.2016.1149502

Platas, L. M. (2014). Knowledge of Mathematical Development Survey: Testing the validity and reliability of the survey and interpreting its results. *NHSA Dialog*, *17*(1). https://journals.charlotte.edu/dialog/article/view/123

Platas, L. M. (2015). The Mathematical Development Beliefs Survey: Validity and reliability of a measure of preschool teachers' beliefs about the learning and teaching of early mathematics. *Journal of Early Childhood Research*, *13*(3), 295–310. doi:10.1177/1476718X14523746

Pollitt, R., Cohrssen, C., & Seah, W. T. (2020). Assessing spatial reasoning during play: Educator observations, assessment and curriculum planning. *Mathematics Education Research Journal*, *32*(2), 331–363. doi:10.100713394-020-00337-8

Pritchett, L., & Beatty, A. (2012). *The negative consequences of overambitious curricula in developing countries*. Center for Global Development Working Paper 293. https://www.cgdev.org/sites/default/files/1426129_file_Pritchett_Beatty_Overambitious_FINAL_0.pdf

Pryor, J., Akyeampong, K., Westbrook, J., & Lussier, K. (2012). Rethinking teacher preparation and professional development in Africa: An analysis of the curriculum of teacher education in the teaching of early reading and mathematics. *Curriculum Journal*, *23*(4), 409–502. doi:10.1080/09585176.2012.747725

Read, N. (2017). Measures of learning and teaching material availability and use in Sub-Saharan Africa and other low-income countries. *European Journal of Education: Research, Development and Policy*, *52*(4), 523–532. doi:10.1111/ejed.12242

Resnick, L. (1992). From protoquantities to operators: Building mathematical competence on a foundation of everyday knowledge. In G. Leinhardt, R. Putnam, & R. A. Hattrup (Eds.), *Analysis of arithmetic for mathematics teaching* (pp. 373–429). L. Erlbaum Associates.

Schweisfurth, M. (2011). Learner-centred education in developing country contexts: From solution to problem? *International Journal of Educational Development*, *31*(5), 425–432. doi:10.1016/j.ijedudev.2011.03.005

Schwille, J., & Dembélé, M. (2007). *Global perspectives on teacher learning: Improving policy and practice*. UNESCO International Institute for Educational Planning. https://files.eric.ed.gov/fulltext/ED496753.pdf

Sitabkhan, Y., Davis, J., Earnest, D., Evans, N., Ketterlin-Geller, L., Lutfeali, S., Ngware, M., Perry, L., Pinto, C., Platas, L., Ralaingita, W., Smith, K., & Srikantaiah, D. (2019). *Instructional strategies for mathematics in the early grades*. Mathematics Working Group Working Paper. USAID. https://shared.rti.org/content/instructional-strategies-mathematics-early-grades

Sitabkhan, Y., & Platas, L. M. (2018). *Early Mathematics Counts: Promising Instructional Strategies from Low- and Middle-Income Countries*. RTI Press Publication No. OP-0055-1807. RTI Press. doi:10.3768/rtipress.2018.op.0055.1807

Tabulawa, R. (1997). Pedagogical classroom practice and the social context: The case of Botswana. *International Journal of Educational Development, 17*(2), 189–204. doi:10.1016/S0738-0593(96)00049-1

Tabulawa, R. (2003). International aid agencies, learner-centred pedagogy and political democratisation: A critique. *Comparative Education, 39*(1), 7–26. doi:10.1080/03050060302559

UNESCO. (2012). *Challenges in basic mathematics education*. UNESCO. https://unesdoc.unesco.org/images/0019/001917/191776e.pdf

UNESCO. (2016). *School resources and learning environments in Africa: Key results from a regional survey on factors affecting quality of education*. UNESCO. http://uis.unesco.org/sites/default/files/school-resources-and-learning-environment-in-africa-2016-en/school-resources-and-learning-environment-in-africa-2016-en.pdf

Vaijayanti, K., Suma, M. N., & Mondal, A. (2016). *The impact of Akshara Ganitha—A longitudinal study 2012– 13 to 2014–15*. Akshara Foundation. https://akshara.org.in/en/resources-research/reports/

Van de Walle, J. A., Lovin, L. H., Karp, K. S., & Bay-Williams, J. M. (2014). Teaching student-centered mathematics: Developmentally appropriate instruction for grades pre-K–2 (Vol. 1). Teaching Student-Centered Mathematics Series. Pearson Higher Ed.

Vula, E., Avdyli, R., Berisha, V., Saqipi, B., & Elezi, S. (2017). The impact of metacognitive strategies and self-regulating processes of solving math word problems. *International Electronic Journal of Elementary Education, 10*(1), 49–59. doi:10.26822/iejee.2017131886

Yin, R. K. (2017). Case study research and applications: Design and methods. *Sage (Atlanta, Ga.).*

Yuan, F., & Evans, D. (2018) *The working conditions of teachers in low- and middle-income countries*. World Bank. https://riseprogramme.org/sites/default/files/inline-files/Yuan.pdf

Chapter 13
Co-Teaching to Develop Reform-Based Practices in Mathematics:
Classrooms as Shared Spaces for Teacher Learning

Raewyn Eden
Massey University, New Zealand

ABSTRACT

Improving opportunities for under-served (and all) students to learn mathematics requires shifts in what teachers know and can do. This chapter explores teachers' opportunities to learn: How might collaborative inquiry support them to make such shifts? In particular, the chapter explores co-teaching as a possible solution to the complex challenge of implementing reform-based mathematics pedagogy in primary (elementary) mathematics classrooms. It draws on the voices of teachers who worked together to develop ambitious, talk-based pedagogies in their classrooms where the use of "talk moves" was a particular focus for their shared work. It examines the transformative power of co-teaching as a feature of the teachers' collaborative activity, outlining three fields of practice that interacted to expand teachers' possibilities for action: their enacted practice, their conversations about practice, and their noticing of student thinking within practice. Working across difference served as catalyst and context for the transformation of teachers' practices.

INTRODUCTION

For educators in Aotearoa New Zealand, inequity in mathematics teaching | learning is an enduring challenge that has proven difficult to address. Some groups of learners experience significant and persistent underachievement. Improving opportunities for under-served (and all) students to learn mathematics requires shifts in what teachers know and can do. Teachers create the conditions within which mathemat-

DOI: 10.4018/978-1-7998-9422-3.ch013

ics can be learned, and so developing new, more equitable, teaching practices is central to addressing the persistent problem of underachievement for some. This raises the question of what opportunities teachers have, and how they can be supported, to make such shifts; specifically, how can teachers develop new, more equitable, teaching practices in the context of their day-to-day work?

This chapter examines how teachers' collaborative inquiry might address the complex challenge of implementing reform-based mathematics pedagogy in primary (elementary) mathematics classrooms. It draws on the voices of teachers who worked together to develop ambitious, discourse-based pedagogies in their classrooms where the use of "talk moves" (Chapin & O'Connor, 2007) was a particular focus for their shared work. The chapter specifically explores the transformative potential of co-teaching as a feature of the collaborative inquiry of groups of teachers who wanted to improve mathematics teaching I learning in their classrooms.

Background

Achieving equitable learning outcomes in mathematics for all learners in Aotearoa New Zealand is a significant ongoing challenge with systemic failures that appear to be "replicating and exacerbating social inequity, not reducing it" (Royal Society Te Apārangi, 2021, p. 20). Recommendations of the Royal Society's expert advisory panel, convened to provide advice on the mathematics and statistics learning area in the New Zealand Curriculum, highlight the importance of reform-based approaches to mathematics teaching and learning and the need for sustained professional learning to strengthen teachers' knowledge for teaching. Towards an imperative to provide experiences that help students to understand their worlds through "tasks that support learners as they investigate, represent and connect mathematical and statistical ideas through discussions in the contexts of problem-solving" (p. 44) the panel recommends that teachers be "supported to provide mathematical challenges that encourage students to engage in productive struggle and take risks in a trusting classroom community" (p. 48). Kazemi and colleagues (2009) describe such teaching as involving ambitious pedagogies and for many, if not most, teachers, it represents a profound shift from the status quo and the need for support is significant. This raises the question of how teachers might be supported to make such shifts, including what opportunities there might be to learn within the context of their day-to-day practice.

Collaboration and inquiry are fundamental to teaching and learning. Teacher collaboration is generally understood to be an element of teachers' community (e.g. Shulman & Shulman, 2004) and there is ongoing attention in policy, practice and research to the notion of community as a core element of teachers' professional learning (Horn & Little, 2010). Bobis et al. (2012) suggest that "understanding teachers' thinking about their own knowledge and its perceived impact on teaching practices is paramount to improving the professional learning of teachers" (p. 335). The notion of inquiry as stance distinguishes between teachers *doing* inquiry and *becoming* inquirers whereby inquiry is an identity that teachers adopt as they engage in practice (Cochran-Smith & Lytle, 2009). Teacher learning through inquiry and the role of teacher collaboration within community come together in the interrelated notions of teachers' collaborative inquiry within inquiry communities. Levine (2010) describes inquiry communities as foregrounding "the role of systematic inquiry conducted with the support of colleagues as a means of improving teaching and learning in schools" (p. 112). Highlighting mechanisms for learning he describes systematic teacher inquiry as characteristic of inquiry communities whereby teachers "articulate questions, problems and dilemmas" and knowledge is developed through the structured analysis of data "and/or conversations which generate new, revised, or more explicit understandings" (p. 122). Although this chapter focuses on the inquiry of teachers into their teaching, they were also engaging in inquiry in mathematics with their students. Thus, teaching *as* inquiry (the inquiring of teachers into their practice) and inquiry-based teaching and learning (the inquiry of students in the classroom) can be seen in parallel.

In Aotearoa New Zealand, collaboration and inquiry explicitly underpin current education policies aimed at addressing inequalities in primary schools. One of six professional standards for teachers requires teachers to "use inquiry, collaborative problem-solving and professional learning to improve professional capability to impact on the learning and achievement of all learners" (Education Council, 2017, p. 18). Elaborating on the standard, teachers are expected to "[i]nquire into and reflect on the effectiveness of practice in an ongoing way, using evidence from a range of sources . . . and engage in collaborative problem-solving and learning-focused collegial discussions" (p. 18). The Education Review Office [ERO] (2016) in Aotearoa New Zealand highlights the importance of *mahi tahi*, which is described as "the unity of a group of people working towards a specific goal or on a specific task, often in a hands-on fashion" and, in the school context, as "the business of working together collaboratively in the pursuit of learner-centered education goals" (p. 15). Mahi tahi is one of four concepts that provide the foundation for examining effective practice aimed at promoting equity for all learners. Accordingly, ERO identifies collaboration as essential for improvement in schooling and learner success, citing Fullan and colleagues (2015), who argue that "collaboration focused on the improvement of teaching and learning is one of the highest-yielding strategies to boost student, school and system performance" (p. 8).

An underlying assumption in this chapter is that the professional learning and growth of teachers occurs in the social and cultural contexts of their work. Teachers who are working to build new practices need time and space to talk about their work. Opportunities to articulate and evaluate their knowledge of mathematical content support teachers to shift from thinking about ideas to thinking about how they might be applied to enhance teaching (Bobis et al., 2012; McPhan et al., 2008). Teachers' conversations are a key resource for teacher learning and play an important role in creating opportunities for changed practice (Hunter, 2010). Talking with colleagues about what they do and why they do it creates space for teachers to reflect on those things they do in classrooms out of habit.

Reflective conversations can disrupt and transform what teachers believe and feel about teaching | learning. However, teachers' interactions can both support and constrain possibilities for their professional learning (Little, 2003). For instance, norms of collegiality where validation, rather than critical reflection, is the focus can shut down critical questions and thus constrain teacher learning (Allen, 2013). Thus, to be productive for teachers' ongoing learning, professional conversations need to surface and challenge problematic aspects of teachers' practice as well as affirming effective practice.

Teacher noticing is an important element of reflective practice, which in turn is key to strengthening teaching. As teachers make sense of teaching | learning experiences, they in turn use this to inform their future practice. Highlighting a connection between noticing and reflection, Jaworski (2003) argues that critical intelligence, a kind of "metaknowing," emerges from an increased awareness of classroom issues. Thus, she suggests, critical reflection outside the classroom promotes more explicit awareness of classroom conditions and events within practice, which in turn leads to possibilities for expanding action and shifting practice. Critical reflection requires challenging what is taken for granted in teachers' practice and involves "a deep exploration process that exposes unexamined beliefs, assumptions, and expectations . . . reflective practitioners challenge assumptions and question existing practices, thereby continuously accessing new lens to view their practice and alter their perspectives" (Larrivee, 2000, p. 296).

Sherin and colleagues (2011) point to the benefits of seeing noticing as an important focus for inquiry, suggesting that "noticing is a critical component of mathematics teaching expertise and thus a better understanding of noticing could become a tool for improving mathematics teaching and learning" (p. xxvi). Noticing is central to inquiry whereby teachers are required to notice, interpret, and respond to unpredictable classroom events as they occur (van Es & Sherin, 2021). Noticing generally refers to paying attention to and making sense of specific aspects of classroom events (Sherin et al., 2011). Teacher noticing in mathematics has been described as a "hidden practice of in-the-moment decision-making that is needed to respond to children's . . . strategy explanations" (Jacobs et al., 2010, p. 197), a conceptualization that is related to Schön's (1983) notion of reflection-in-action. Jacobs and colleagues conceptualize the professional noticing of children's mathematical thinking in terms of a set of interrelated skills comprising attending to the mathematical details in students' strategies; interpreting students' understandings as apparent in their strategies; and the reasoning teachers use in deciding how to respond to students' understandings. These authors found that noticing is challenging; it is not a routine skill. Noticing is learned through a combination of practice experience and professional learning. Perhaps counterintuitively, noticing children's thinking in a domain one knows well is a specialized and sometimes unnatural skill (Ball, 2011). Ball points out that a "paradox of expertise is that ideas that seem obvious are not so to the learner" (p xxi), perhaps suggesting that the more extensive the teacher's knowledge of the subject, the more difficult it can be to interpret students' understandings (and misunderstandings).

In the context of teachers' work, collaboration and inquiry are interconnected notions with nuanced meanings. Mason (2009) argues that collaborative critical reflection can lead to future noticing in the moment, thereby allowing for the possibility of alternative decisions and actions. This aligns with a complex view of teaching and suggests an increased capacity for teachers to notice both elements of the teaching activity itself and the impacts of their teaching (Jaworski, 2003). That is, an expanded capacity to notice both the activity itself and its shifting goals as student learning outcomes are realized. However, despite the substantial and growing interest in collaboration and inquiry for teachers within education settings, these notions are often underconceptualized and treated as largely independent of one another in research literature and policy documents. For instance, the teaching-as-inquiry model in the New Zealand Curriculum [NZC] (Ministry of Education, 2007) presupposes teachers' inquiry as an individual rather than collective process, as implied by the use of language indicating a first-person perspective (Benade, 2015). The teaching-as-inquiry cycle outlined in the NZC poses questions such as "Is there something *I* need to change?" and "What strategies are most likely to help *my* students learn this?" (p. 35, emphasis added). Critiquing the notion of teaching as inquiry, Benade argues that collaboration is a necessary condition of teachers' work if they are to make their theories of action explicit and open to challenge.

Teachers' learning is made apparent in their actions. Jaworski (2008) suggests that reflecting critically with colleagues on aspects of past practice can support the development of enhanced noticing within future practice, thus influencing classroom actions and potentially changing practice. Such a view provides a compelling rationale for a focus on teacher noticing in the context of teachers' learning through inquiry. This chapter adds to the growing body of work that answers Jaworski's (2003) call for further research to test related theoretical claims and gain improved insights into the associated processes of teacher noticing within inquiry activity. In particular, this chapter focuses on the ways in which teachers' collaborative inquiry can promote the understanding and enactment of reform-based practices in mathematics classrooms.

CO-TEACHING MATHEMATICS: TEACHERS LEARNING ACROSS DIFFERENCE

Collaborative Inquiry and the Emergence of Co-teaching: The Studies

This chapter draws on two related studies (Eden, 2017, 2019) that took place in four primary (elementary) schools in a large urban centre in Aotearoa New Zealand. They were all experienced teachers and included a range of ages and levels of confidence with teaching mathematics. In addition to their classroom teaching roles, many held additional leadership and specialised roles in a range of areas within their schools including one teacher from each school who was responsible for curriculum leadership in mathematics. The teachers predominantly taught in 'single-cell' classrooms and their class levels ranged from Year 1 (5-year-olds) to Year 8 (12-year-olds). Most were composite classes comprising two combined year groups and some classes were in open learning environments with more than one class sharing the space. The schools were all in multicultural communities with mid- to high-decile ratings for socio-economic status however the individual circumstances of students and their families were wide-ranging with some experiencing significant poverty while others had access to significant economic resources.

In each of the four schools, teachers worked together to learn about and implement ambitious, reform-based approaches to their mathematics teaching. They were aiming to develop and enact ways of working together that could expand what they knew, and what they could do in their mathematics classrooms. The overarching aim was to develop ways of working that could raise achievement for previously under-served students. In other words, the teachers aspired to learn about and implement reform-based teaching practices in mathematics. They were specifically aiming to develop discourse-based practices with a particular focus on using "talk moves" (Chapin & O'Connor, 2007) in their lessons. Although developing and enacting practices that could promote rich mathematical discussions were central to their professional learning goals, this was inextricably connected with other aspects of a complex system of pedagogy which included the nature of the mathematical tasks and range of mathematical representations being used, and the nature of the classroom culture including the ways in which mathematically productive norms were supported (Anthony & Walshaw, 2007).

The initial study (see Eden, 2019) involved the author and a group of four teachers in one primary school, working together over a six-month period to develop an approach to the teachers' collaborative inquiry. The second study (see Eden, 2017) involved approximately 30 teachers in three additional schools implementing and further refining the initial design over an 18-month period.

Both studies used a design-based approach that was iterative and cyclical and highlighted collaboration at multiple levels. In accordance with a view of professional learning as situated in the social and cultural contexts of teachers' work, the overarching aim of both studies was to explore the affordances and constraints of collaborative inquiry for promoting teacher learning. Design-based research focuses on connecting processes associated with enacting learning designs with outcomes of interest (Design-based Research Collective, 2003), and the design cycle broadly parallels cycles of professional inquiry and knowledge-building, such as the teaching-as-inquiry model in the NZC (Ministry of Education, 2007). The author was interested in how collaborative inquiry supported teachers to understand, develop and enact reform-based practices in their mathematics classrooms.

The co-designed inquiry approach involved pairs of teachers co-teaching one-off mathematics lessons and meeting as a larger group to discuss their experiences. During the first study, co-teaching emerged from the design process as a productive context for the teachers' professional learning and became a key characteristic of the designed inquiry approach. The idea of co-teaching surfaced quite serendipitously after one of the teachers (Kris) and the author co-taught a lesson. Co-teaching was initially proposed as a way of providing support for Kris to try something new in her practice. She had wanted to trial using decimats, a representation designed to support students' understanding of place value (see Ministry of Education, 2008, pp. 45–46), but had previously tried using decimats with limited success. The author had not used them before but believed they had merit and was keen to try. They decided to introduce the decimats together; that is, they decided to co-teach a lesson using decimats.

Subsequent to their initial co-taught lesson, Kris and the author reflected on the experience with the other teachers and it was decided to try co-teaching in pairs. This eventually led to the development of a model in which teachers in each of the schools engaged in iterative cycles of co-teaching in pairs and then came together as a larger school-based team to reflect on their co-teaching experiences. The co-teaching inquiry approach was designed to be flexible, and Figure 1 illustrates a typical in-school co-teaching inquiry cycle over one school term of approximately 10 weeks. Each co-teaching cycle consisted of two "rounds" of co-teaching during which teachers would typically have two experiences of co-teaching: one in their own classroom and one in a colleague's classroom. For instance, during the first "round" of co-teaching, Teacher A would go to Teacher B's classroom to co-teach a lesson, Teacher B would go to Teacher C and Teacher C would go to Teacher A's classroom. Likewise for Teachers D, E and F. For the subsequent round, the arrangement was reversed. That is, Teacher A would go to Teacher C, Teacher C would go to Teacher B and Teacher B would go to Teacher A's classroom to co-teach.

Figure 1. Illustration of a typical in-school co-teaching inquiry cycle

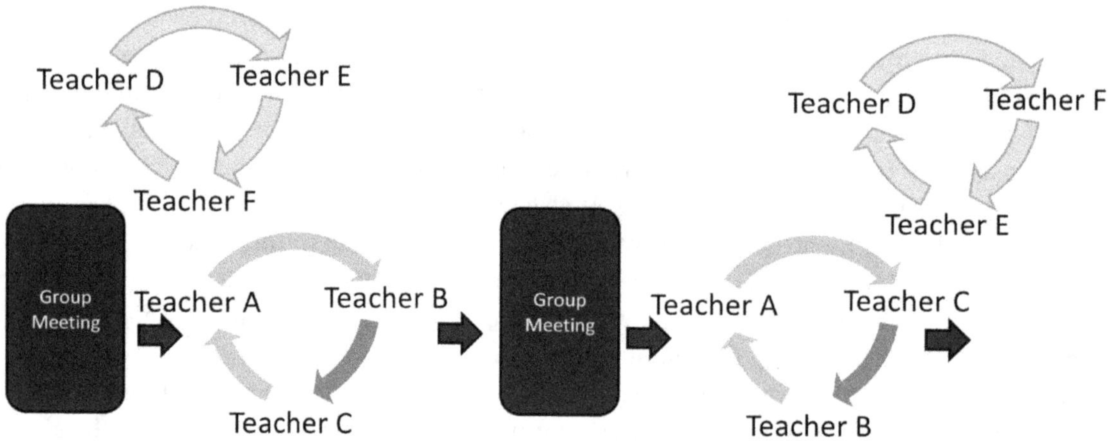

Co-teaching in the context of this work refers to pairs of teachers co-planning, co-instructing, and co-reflecting on "one-off" mathematics lessons in one teacher's classroom. Generally, the classroom teacher would plan a lesson as part of their usual mathematics program and share this with the teacher who was "coming in" to co-teach with them. The second teacher then contributed ideas and/or resources for the lesson. During the lesson, both teachers actively participated in instruction, taking roles that were more-or-less predetermined depending on the preferences of the two teachers involved. The research provided additional staffing resources for teachers to be released from their own classrooms to co-teach in a colleague's room. Beyond the research activity, however, some teachers created opportunities for additional co-teaching experiences, for example by combining their classes and co-teaching two classes together.

Following each co-taught lesson, co-teaching pairs reflected together on the experience. The teachers were highly motivated to review and reflect on co-taught lessons despite the challenge of scheduling opportunities for teachers to meet and talk about their lessons. One of the teachers explained how a reflective conversation was a priority:

We were so busy afterwards [but] we knew that was an important link to meet afterwards and so we saw each other in the morning and we both thought we'd just say hello but actually both came together cos we wanted to talk about [the lesson].

Teachers also noted the importance of the opportunity to talk about their shared practice as it was occurring:

We didn't just wait til the end to talk, we actually talked to each other all the time, every little piece we did. So, when I was modelling I was talking and saying "what do you think about this?" or she's giving me feedback, I'm giving feedback, so every little thing.... We were able to not just support but share our understanding as well.

Later, all participating teachers in the school came together for a group meeting to share and reflect on their co-teaching experiences. Sometimes, co-taught lessons were video recorded, and teachers shared and reflected on self-selected excerpts from the videos. Sharing planning to gain a sense of what to expect from the lesson appeared to be an important aspect of the teachers' reflections during and after the lesson. As such, at least some level of co-engagement in the three stages of co-teaching, that is co-planning, co-instruction and co-reflection, constituted the co-teaching activity. The interplay among the different stages of the co-teaching activity is illustrated by Kris' comment that:

Having that [co-planning] conversation allows you to process things that you can't process at the time when you're in there but you can prepare yourself for that interaction.

The teachers were able to reflect both during and after their co-taught lessons in light of the intent that had been made explicit before teaching.

As a key activity in the teachers' collaborative inquiry, co-teaching was a seemingly simple idea, yet teachers appeared to experience it as transformative in ways that were surprising for everyone involved. The remainder of this chapter draws on data from the recorded conversations of groups of teachers in each of the four schools. The interactions at the teachers' school-based group meetings as they reflected on shared classroom events, and the co-teaching activity itself, are the central focus. The chapter outlines how co-teaching supported the teachers to expand what they knew and the possibilities for what they could do when teaching mathematics.

Working Across Difference: Expansion within Three Fields of Practice

Co-teaching created new possibilities for mathematics teaching | learning within three fields of the teachers' collaborative practice: their enacted practice (what they did), their conversations about practice (what and how they talked about what they did), and their noticing of student thinking within practice (what they noticed about how what they did impacted on their students).

Enacting Practice Together

Co-teaching was a shared teaching experience that went beyond observations or representations of teaching removed from the actual practice. In various ways, each of the co-teaching partners brought something new and different to each co-taught lesson. The teachers had different levels of expertise and experience. Each time they co-taught, one partner was in their own classroom and the other was in an unfamiliar classroom teaching at an unfamiliar year level with students they did not know. In some cases, towards the end of the second project, teachers even traveled to co-teach with their colleagues in one of the other two participating schools.

Co-teaching allowed teachers to observe what happened in their own classrooms, with their own students, when another teacher introduced new ways of doing things. At the same time, their co-teaching partner experienced the impacts of their familiar and taken-for-granted practices in a new setting. For instance, a routine question in one class might prompt unanticipated responses in another. Reflecting on their co-taught lesson, two teachers (Kris and Pat) described how an unexpected question gave a student a new "way in" to understanding decimal fractions, and both teachers had to pay close attention to make sense of an unanticipated line of reasoning from the student.

What was great was that Pat hasn't been part of any of the prior teaching . . . but he [the student] had explained it in a way that Pat could follow it.

It was unanticipated because Kris could not predict the student's response to Pat's question (because she had never asked the students a question like that), and Pat did not know the students in Kris' class sufficiently well to anticipate their responses to the task. For Kris and Pat, co-teaching involved a dynamic sharing of the teaching role. They described how each teacher made a unique contribution to the lesson:

Kris: You came in on a part of it . . . you'd kind of contributed into the whole understanding but there was no sense during the whole thing of "oh I've stepped on her toes" or "she's stepped on mine".

Pat: What I felt was that you plugged in gaps that I may have missed out and I would have plugged in gaps that you would have missed out . . . because we listened to each other and because we felt that there was a particular thing or a point that we may have missed out and we would just quickly throw it in.

Kris: And you didn't have to do all of the thinking.

Sharing the teaching role opened space for teachers to access aspects of the other's practice as resources for their own teaching and, at the same time, gain feedback on their practice. Furthermore, working together as co-teachers created productive conditions for teachers to enact new and challenging pedagogies. For instance, one teacher supported her co-teaching partner and students to co-construct success criteria for the lesson. Incorporating this practice into her teaching had been an explicit, but until now elusive, learning goal for that teacher. Having the practice modeled by another teacher in her own classroom, with her own students, in the context of a co-taught lesson brought the practice to life.

It allowed her to see the value of the practice for her students and the relatively minor adjustment to her current practice that this new practice represented.

Observing and responding to differences in their teaching challenged each teacher's existing beliefs and practices:

Someone coming from Year 7–8 into the new entrant room . . . they're getting a view into my classroom and the language that I've got to choose. You know, scale it down for my new entrant kids and how far we have to break some of those tasks, and how far I have to speed up when I go to their rooms because everyone is a bit more worldly.

Teaching five-year-olds in a new entrant (kindergarten) class requires a repertoire of teaching practices that is very different to teaching students in Years 7 and 8 (Grades 6 and 7). When teachers were co-teaching at an unfamiliar level, the differences in approach catalyzed deeper levels of thinking about practice as teachers had to adjust their teaching in response to a different level of student thinking:

I quite enjoyed that cross-level actually because it made me think a bit more.

The visibility of one another's classroom practice was enhanced for both participants, including each teacher's actions and decisions, and students' responses. Co-teaching expanded the teachers' access to one another's practice beyond seeing or hearing about it; they were engaging in the practice together and could co-construct new practices in the moment. Furthermore, two teachers actively teaching a mathematics lesson together were more likely to bump up against practice ideas that were different to their own. Making differences in practice visible opened opportunities to learn about and from those differences (Tobin, 2014). As Katz and Dack (2014) argue, paying attention to data that interrupt the status quo "holds the potential to yield new professional learning" (p. 36).

Talking about Practice

The shared experience of co-teaching mathematics lessons in one another's classrooms supported shifts in both *what* teachers talked about and *how* they talked about mathematics teaching | learning. Teachers were able to talk about their teaching as they enacted it, and those conversations had student learning as the central focus, something Kris and the author noticed when they co-taught together:

Kris: We would have these side conversations as we went, and . . . the starting point was the kids . . . and there was that equality of roles.

Raewyn: We had those side conversations that made me think about the way the kids were responding in a way that I probably wouldn't have if I'd just been sitting there teaching it . . . I might have been thinking about what I was doing or—

Kris: The kids' needs were at the center, which allowed the adults to weave together because there's strengths in both.

Raewyn: Doing it together helped to really explicitly focus on what the kids were doing, and getting out of it, and connecting with, and not connecting with, and that's a real strength.

Kris: I totally agree . . . our kind of model puts the child at the center and what their need is and interestingly that's what's woven over our collaboration today.

Teachers noted that their discussions were increasingly focused on explicit details related to mathematics teaching | learning:

It [teaching] was siloed and our aim was to smack it open but the kinds of content of breaking that down is what's shifted it. It's not necessarily "Well I'm going to talk to you" . . . it's more a "Let's bat around mixed ability grouping and what that impact is, let's talk about rich tasks, what they actually are, and why that worked and that one didn't"—and because you are in a different whānau [teaching team], you've got different perspectives.

In particular, teachers found they needed to talk explicitly about the mathematics in a lesson they were planning to co-teach in order to coordinate their teaching:

If you're discussing a rich task that you're going to take with a group of children who are not in your class, for example, you actually need to sort of define the task just a bit more.

In addition to providing an important conceptual resource for teacher learning, talk that was explicitly focused on mathematics also supported productive working relationships:

You're sharing a task so the conversation is in front of you, you are not on opposite sides of the argument. You're a team . . . there's nothing to battle out. There's no inequality.

The shared experience of co-teaching meant that teachers had someone with whom they could reflect, and make sense of the events in a lesson and the impacts of teaching on students:

When you're working with someone else you always have someone . . . you go back and you sort of like reflect upon the day or the lesson that you've done and say "Oh I just noticed so-and-so . . . he's usually very shy but today he shared this with me," and it could be a moment of "ah-ha that's probably why he's not got it" but now he's got it because he's done it in a different way than you'd normally do.

Although finding time to reflect together after a co-taught lesson was challenging, teachers found that the shared experience of co-teaching prompted new opportunities to talk about their teaching:

[There are] conversations happening within the staff room, within each other's classrooms, because we are going in and out . . . those conversations are always happening.

Furthermore, teachers increasingly engaged in reflection with greater depth and rigor, and saw this as central to strengthening their practice:

We've all said that the reflection level, if you're reflecting together on a lesson, has been the most valuable part . . . all that noticing and that you can communicate with somebody else, people have just said that that's been the most beneficial thing for them and their teaching, because they know what they need to keep continuing with or try differently.

Noticing Student Thinking

Reflective conversations during and after a co-taught lesson appeared to support deeper noticing than when teachers worked alone. Reflecting on the first lesson they had taught together, Kris and the author were immediately struck by their heightened noticing of the children's thinking during the shared teaching episode, and they agreed that co-teaching was a "powerful" experience. Co-teaching appeared to make student learning the central focus of the lesson. Similarly, sharing a video excerpt of her first co-teaching experience with the author, Pat's reflection on the events of the lesson were focused on student talk, with particular attention to one student's actions:

We let them talk and let them share their ideas and just let them go . . . you can see that they're really starting to listen to each other and starting to talk to each other. One clicked on and he was sitting back and [student] . . . had his way of explaining like he didn't want to talk too much but he questioned the others about why they did these things . . . almost like, I've introduced it [talk moves] and then I sat back and just let the chaos run . . . it was really interesting just to see it just all come out and just the way that they talked to each other and I just sat back and then I did the repeat and I'd said "did you say?" or "you're saying this?" and they were "Ah, maybe that didn't sound right".

The detail in Pat's descriptions of the students' talk in the example above suggests greatly enhanced noticing compared to her more general comments at the previous group meeting; for instance:

They're doing a lot more talking in terms of the language as well with the use of the talk moves.

Previously, Pat had seen the role of talk moves differently: primarily as a tool for monitoring the students' attentiveness where the focus for both Pat and her students was on listening rather than talking. She had described using talk moves as a tool for

tracking to see whether they're being attentive or not and some just get caught out . . . I guess for me it's just the attentiveness of them being able to listen has been like tick number one first.

Co-teaching experiences appeared to extend the specificity of what teachers were noticing about students' actions and opened space for teachers to notice and respond to students' mathematical thinking. For instance, reflecting on a video excerpt from the lesson she co-taught with Kris, Pat described how she had supported students to make sense of adding 0.9 and 0.7 using decimats:

We were talking about the parts and he was going to just add on point seven and create more parts where I said "well what does point seven look like and shall we draw another bar and then you can show me that?" and so then they could see the connection of "oh actually I just need to make four bars to make one whole" and then they could see "oh okay well I'll just tag on that little bit there and I've got six left on here so it's one point six". So they can actually see that they're actually trying to make wholes from there ... I said "well let's make them point seven" so they could see that just like how we're adding our counters that they're just shifting over to make one whole.

For Kris, this was an important excerpt to share because this student would not previously have been expected to provide such an explanation. Pat's turn in the conversation illustrates how the student's explanation had opened space for Pat to engage in her own mathematical thinking and sense-making as she worked with the group to build on the student's explanation and apply that to solving the problem. Here, Pat was describing how students had used the decimats to make sense of the numbers and the addition operation, and in doing so she was illustrating how engaging with the student's explanation afforded new ways for her to understand the mathematics content. Responding to Pat's unfamiliar approach to questioning, the student extended his understanding of decimal place value by making an explicit connection between place value and decimal fractions.

Across the two projects, co-teaching was seen to expand what teachers were able to notice *in the moment* as they were teaching. Sharing responsibility for teaching | learning supported teachers to notice, in greater detail, the impacts of their teaching on students:

I had to follow everything you were doing so that at any point that I needed to support or take a different [angle] I had to be following.

They described enhanced opportunities to notice and respond to student learning as it was taking place:

When we co-teach together, some of the things that I didn't pick up from my own lessons, others were able to pick up. You notice different things and that's where the improvement comes from.

Through co-teaching, attention was increasingly directed towards the impacts of teaching on student learning:

I think it's the purposefulness of noticing. When you've got your own class and everything is going on, you kind of notice but you don't. When you're with someone else, though, you're really drilling in the "what I am looking and listening for?" and pulling some of those bits apart and having that conversation with someone else later . . . we've both seen very different things.

Attention, or noticing, appeared to be heightened because teachers needed to be engaged with what was happening in order to actively participate in co-teaching the lesson. The complexity of the co-teaching activity and the need to be responsive to unpredictable events meant that teachers needed to carefully observe one another's actions, and the impacts of those actions, in order to respond to events as they unfolded.

Two teachers actively involved in teaching a lesson together introduced new perspectives from which teachers could notice and make sense of what is "going on" for students during mathematics lessons:

That opportunity for noticing and hearing what's actually going on with the students from a different perspective . . . the idea of, you've got your own discourse, your own story going on and so you see and hear things you want to see and hear and you often miss things . . . it may be unimportant or you don't recognise whatever it is and another teacher coming in will have that opportunity to offer those [other perspectives].

Teachers found that what they noticed could be "pulled apart," and their reflections increased in depth and rigor.

We watched some of our staff members really struggling with [the co-teaching inquiry] for the last two years but watching their journey has been amazing because they've come out the other end and are really reflective practitioners now when it comes to their maths ability and what's going on in their classroom.

Teachers' enhanced noticing appeared to support a dynamic interrelationship amongst the co-planning, co-instruction and co-reflection aspects of their co-teaching.

Learning through Co-Teaching: Space to Engage across Difference and Learn

The co-teaching experience promoted teachers' reflexivity whereby they were able to reflect and respond "in the moment" and draw on diverse perspectives to make sense of what was noticed. The emergence of difference as the teachers enacted, talked about, and noticed the impacts of classroom practices was central to their ongoing learning through their co-teaching inquiry. Figure 2 illustrates a trialectical relationship among three fields of practice within which the co-teaching inquiry activity occurred.

Figure 2. Interconnected fields of teachers' co-teaching inquiry practice

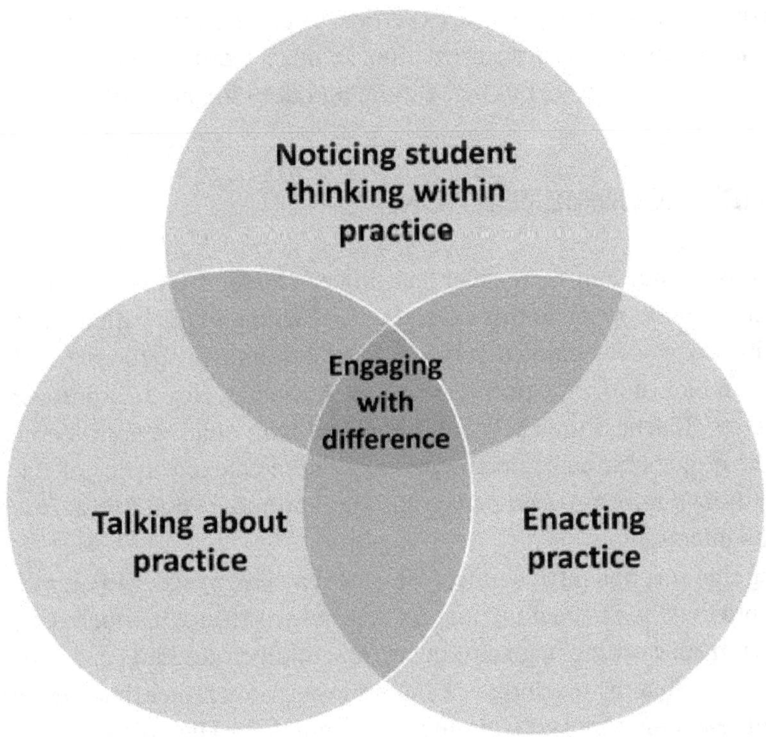

Through co-teaching, engaging with difference was central to teachers' opportunities to learn about and enact new, reform-based pedagogies. Towards the end of the first project, the author noted:

The conversation that sits around [co-teaching] is a tool but also the fact that people might have different approaches, so it might be "Oh, I was deliberately leaving space then and you kept talking" then gives a reason to have a conversation about that, that you wouldn't otherwise have.

As Skattebol and Arthur (2014) note, the need to explain thinking across different fields of practice and areas of expertise "spark[s] provocative lines of intellectual engagement" (p. 362). The co-teaching inquiry fostered conditions that allowed teachers to have expanded access to, and deeper engagement with, new, and often dissonant, practice ideas. Differences that emerged within the teachers' co-constructed and enacted practice, the conflicting ideas about practice that emerged in their conversations about practice, and what teachers noticed about students' sometimes unexpected responses to practice served to expand the teachers' conceptual and material resources for practice. Teachers described how co-teaching had expanded the boundaries and increased the specificity of their conversations about practice whereby they were explicitly focused on the effective teaching and learning of mathematics.

Opportunities to engage deeply with one another's practice opened space for an expanded set of actions for each of the teachers in their own practice. Furthermore, teachers experienced increasingly positive emotions, a greater sense of belonging, more of a shared responsibility for everybody's learning, and increased collective agency to act in the interests of their students. Co-teaching allowed teachers to reconstruct their identities from the lone operator whose professional reputation needs protection from exposure of any weaknesses in their mathematics knowledge or practice, to a learner whose naïve questions and gaps in practice serve as a resource for all in their learning of teaching.

Working across differences in perspective, expertise, roles, and contexts of teachers' day-to-day work acted as both catalyst and context for transformation. In other words, what each teacher encountered as different sparked new ways of seeing, talking, acting, and thinking about their own classroom practice.

SOLUTIONS AND RECOMMENDATIONS

Interestingly, productive conditions for teacher learning can be seen to parallel those for diverse (and all) learners in mathematics classrooms. In this case, the co-teaching inquiry model parallels what we know to be effective conditions for occasioning diverse students' learning in mathematics. Co-teaching engaged teachers in solving complex, authentic problems of practice using flexible approaches that needed to be explained and justified. Teachers took risks and engaged with challenge in a context that highlighted collaboration and the co-generation of understanding. These characteristics of the co-teaching experience mirror key features of reform-based pedagogies that promote equitable access to rich, challenging mathematics for all students.

It is worth noting that it is not sufficient to co-locate teachers in order to opportune teacher learning of the nature and depth that the co-teaching inquiry was able to afford, in much the same way as students are unlikely to benefit from working in groups but on essentially individual tasks. A key characteristic of the co-teaching inquiry is that the teaching is a shared experience rather than centering on observations or representations of teaching removed from the actual practice. The co-teaching experience promotes teachers' reflexivity whereby they are able to reflect and respond "in the moment" to what is noticed during the lesson and draw on diverse perspectives to make sense of what is noticed. The flexible and negotiated nature of the design meant individual schools and teachers had agency to construct co-teaching experiences that worked for their unique contexts and needs while at the same time having the support of the collective in doing so.

In Aotearoa New Zealand, and internationally, there is increasing interest in collaboration, and new possibilities for arrangements for mathematics teaching | learning in schools raise implications for resource allocation decisions including scheduling and staffing, and the design and use of teaching spaces. In Aotearoa New Zealand, for example, there is increasing across-school collaboration amongst communities of learning [CoLs] and a growing number of schools being built or remodeled with shared teaching | learning spaces. A primary challenge for teachers and their schools is how teachers' work can be restructured to support their ongoing collaborative inquiry, and to include sufficient time for teachers to co-plan and co-reflect. As one of the teachers noted,

it was the balance of trying to find time to sit down together and actually think about it carefully, find a good way of sharing the role, and then acting on that and then reflecting and saying OK so what do we need to do again. Every time I reflected I felt like it was . . . we need more time to sit down together and talk about it before, during and after. Could we do this better?

During the second study in particular, teachers found ways to restructure aspects of their work in order to mitigate this challenge; for example, one co-teaching pair co-planned lessons a week ahead, sometimes by email, and then spent the first 15 minutes of a co-teaching session talking together while the students in the class worked independently. Teachers saw that a school-wide restructuring of their work was needed to sustain co-teaching beyond the projects. For instance, one teacher suggested that co-planning and co-reflection meetings might be included in schools' regular staff meeting schedules. Despite being reluctant to add to teachers' workloads, teachers suggested that time scheduled for teachers to collaborate should be a priority:

That's what co-teaching is about . . . making the time to talk to the other people you are working with . . . you're prioritising it.

Despite the challenge of resourcing, the co-teaching inquiry model was seen as having the potential to be self-sustaining and to create opportunities for strengthening teaching beyond mathematics:

There will be more discussion how we can work together better across the schools and within our teams because of this . . . it might not come forward in a maths setting . . . but you will see the collaborative side of teaching popping up in different places because people have been exposed to it so they're more likely to want to do it again or pull other people in.

Overall, the co-teaching inquiry model represents a promising approach to restructuring teachers' work and harnessing opportunities for sustained improvements in mathematics teaching | learning.

FUTURE RESEARCH DIRECTIONS

Further research into the co-teaching inquiry approach is warranted, including studies aimed at i) further refining the co-teaching inquiry design and ii) exploring its implementation in different contexts.

The application of co-teaching structures in the context of learning areas other than mathematics is a potential area for further study. Investigating the implications of scaling up the approach to work with larger groups of teachers across multiple school sites is another possibility for future research. In order to address the challenge of connecting teacher learning with improved student learning, research that explicitly examines the impacts of teachers' co-teaching inquiry on both teachers and students over an extended period of time would be worthy of consideration. The teachers in the study described their co-teaching experiences as transformative and future studies should draw on data from within mathematics classrooms to investigate this claim empirically. Such research might include extended time in the classroom and the in-depth analysis of classroom events.

CONCLUSION

The research that informs this chapter builds on a long-held interest in learning through collaboration and the role of purposefully designed contexts that maximize the potential for productive interaction. This research supports a view that what is productive for teacher learning parallels what promotes learning in our primary mathematics classrooms. Specifically, the co-teaching inquiry approach shares many of the characteristics of the mathematical inquiry communities that are being developed within primary classrooms in Aotearoa New Zealand (e.g., Hunter & Anthony, 2010). For instance, the promotion of mathematical argumentation might be seen to mirror these teachers' experiences of engaging in increasingly robust conversations involving articulating and justifying aspects of their classroom practice.

Teachers' collaborative inquiry featuring co-teaching appears to be an adaptive and productive model of teachers' professional learning that can support the development and enactment of reform-based practices aimed at raising diverse (and all) students' achievement in mathematics. Co-teaching sharpened teachers' noticing of students' thinking and supported them to deepen their understanding of the impacts of their practice. Working across difference within three interconnected fields of practice, teachers increasingly developed an inquiry orientation towards their ongoing teaching. Enacting practice together, talking about practice, and noticing students' responses to their practice interacted to create spaces in which teachers could engage productively across difference and expand the repertoire of practice ideas they had access to. Working across difference appears to serve as both a catalyst and context for the transformation of teachers' practices.

A key characteristic of the co-teaching inquiry approach was that the teaching itself was a jointly enacted and shared experience whereby the co-teaching inquiry experience involved active participation in joint classroom practice. This contrasts with other, similar models that primarily involve teachers' engaging with observations or representations of one another's practice. Co-teaching simultaneously afforded opportunities for teachers to reflect and respond "in the moment" to events as they unfolded in the lesson, and an expanding range of resources to draw on for their responses. As such, teachers' collaborative inquiry involving co-teaching offers a promising contribution towards the complex and challenging goal of addressing the enduring and important problem of inequity in mathematics achievement.

ACKNOWLEDGMENT

This research was supported by a Victoria University of Wellington Doctoral Scholarship and the Ministry of Education's Teacher-led Innovation Fund [TLIF 1-126].

REFERENCES

Allen, D. (2013). Reconstructing professional learning community as collective creation. *Improving Schools, 16*(3), 191–208. doi:10.1177/1365480213501056

Anthony, G., & Walshaw, M. (2007). *Effective pedagogy in mathematics/pāngarau: Best evidence synthesis iteration (BES)*. Ministry of Education.

Ball, D. L. (2011). Foreword. In M. G. Sherin, V. R. Jacobs, & R. A. Philipp (Eds.), *Mathematics teacher noticing: Seeing through teachers' eyes* (pp. xx–xxiv). Routledge.

Benade, L. (2015). Teaching as inquiry: Well intentioned, but fundamentally flawed. *New Zealand Journal of Educational Studies, 50*(1), 107–120. doi:10.100740841-015-0005-0

Bobis, J., Higgins, J., Cavanagh, M., & Roche, A. (2012). Professional knowledge of practising teachers of mathematics. In B. Perry, T. Lowrie, T. Logan, A. MacDonald, & J. Greenlees (Eds.), *Research in mathematics education in Australasia 2008–2011* (pp. 313–341). Sense Publishers. https://link.springer.com/chapter/10.1007/978-94-6091-970-1_15

Chapin, S. H., & O'Connor, C. (2007). Academically productive talk: Supporting students' learning in mathematics. In W. G. Martin & M. E. Strutchens (Eds.), *The learning of mathematics* (pp. 113–128). National Council of Teachers of Mathematics.

Cochran-Smith, M., & Lytle, S. L. (2009). *Inquiry as stance: Practitioner research for the next generation*. Teachers College Press.

Design-based Research Collective. (2003). Design-based research: An emerging paradigm for educational inquiry. *Educational Researcher, 32*(1), 5–8. doi:10.3102/0013189X032001005

Eden, R. (2017). *Co-teaching to support pedagogical shifts to raise achievement in mathematics—Teacher-Led Innovation Fund Final Project Report* (Final). Ministry of Education. https://www.researchgate.net/publication/349026301_Co-teaching_to_support_pedagogical_shifts_to_raise_achievement_in_mathematics_-_Teacher-Led_Innovation_Fund_Final_Project_Report

Eden, R. (2019). *Restructuring mathematics teaching | learning: Co-teaching as a designed approach to teachers' collaborative inquiry* [Unpublished doctoral dissertation]. Victoria University of Wellington, Wellington, New Zealand.

Education Council. (2017). *Our code our standards: Code of professional responsibility and standards for the teaching profession*. Education Council, New Zealand. https://teachingcouncil.nz/professional-practice/our-code-our-standards/

Education Review Office. (2016). *School evaluation indicators*. New Zealand Government. https://ero.govt.nz/how-ero-reviews/schoolskura-english-medium/school-evaluation-indicators

Fullan, M., Rincón-Gallardo, S., & Hargreaves, A. (2015). Professional capital as accountability. *Education Policy Analysis Archives*, *23*(15), 1–18. doi:10.14507/epaa.v23.1998

Horn, I. S., & Little, J. W. (2010). Attending to problems of practice: Routines and resources for professional learning in teachers' workplace interactions. *American Educational Research Journal*, *47*(1), 181–217. doi:10.3102/0002831209345158

Hunter, R. (2010). Changing roles and identities in the construction of a community of mathematical inquiry. *Journal of Mathematics Teacher Education*, *13*(5), 397–409. doi:10.100710857-010-9152-x

Hunter, R., & Anthony, G. (2010). Developing mathematical inquiry and argumentation. In R. Averill & R. Harvey (Eds.), *Teaching primary school mathematics and statistics: Evidence-based practice* (pp. 197–206). NZCER Press.

Jacobs, V. R., Lamb, L. L., & Philipp, R. A. (2010). Professional noticing of children's mathematical thinking. *Journal for Research in Mathematics Education*, *41*(2), 169–202. doi:10.5951/jresemathe-duc.41.2.0169

Jaworski, B. (2003). Inquiry as a pervasive pedagogic process in mathematics education development. *Proceedings of the Third Conference of the European Society for Research in Mathematics Education*. http://erme.site/cerme-conferences/cerme3/cerme-3-proceedings/

Jaworski, B. (2008). Building and sustaining inquiry communities in mathematics teaching development. In K. Krainer & T. Wood (Eds.), *Participants in mathematics teacher education: Individuals, teams, communities and networks* (Vol. 3, pp. 309–330). Sense Publishers.

Katz, S., & Dack, L. A. (2014). Towards a culture of inquiry for data use in schools: Breaking down professional learning barriers through intentional interruption. *Studies in Educational Evaluation*, *42*, 35–40. doi:10.1016/j.stueduc.2013.10.006

Kazemi, E., Franke, M., & Lampert, M. (2009). Developing pedagogies in teacher education to support novice teachers' ability to enact ambitious instruction. In R. Hunter, B. Bicknell, & T. Burgess (Eds.) *Crossing divides: Proceedings of the 32nd annual conference of the Mathematics Education Research Group of Australasia* (*Vol. 1*, pp. 12–30). Academic Press.

Larrivee, B. (2000). Transforming teaching practice: Becoming the critically reflective teacher. *Reflective Practice*, *1*(3), 293–307. doi:10.1080/713693162

Levine, T. H. (2010). Tools for the study and design of collaborative teacher learning: The affordances of different conceptions of teacher community and activity theory. *Teacher Education Quarterly*, *37*(1), 109–130.

Little, J. W. (2003). Inside teacher community: Representations of classroom practice. *Teachers College Record*, *105*(6), 913–945. doi:10.1177/016146810310500601

Mason, J. (2009). Teaching as disciplined enquiry. *Teachers and Teaching*, *15*(2), 205–223. doi:10.1080/13540600902875308

McPhan, G., Pegg, J., & Horarik, S. (2008). Feedback about professional growth for teachers of mathematics: A developmental perspective. In M. Goos, R. Brown, & K. Makar (Eds.), *Proceedings of the 31st Annual Conference of the Mathematics Education Research Group of Australasia* (pp. 345–352). Mathematics Education Research Group of Australasia Inc. https://www.merga.net.au/Public/Publications/Annual_Conference_Proceedings/2008_MERGA_CP.aspx

Ministry of Education. (2007). *The New Zealand Curriculum*. Learning Media.

Ministry of Education. (2008). *Book 7: Teaching fractions, decimals, and percentages*. Ministry of Education.

Royal Society Te Apārangi. (2021). *Pāngarau mathematics and tuanga statistics in Aotearoa New Zealand: Advice on refreshing the English-medium Mathematics and Statistics learning area of the New Zealand Curriculum*. Royal Society Te Apārangi.

Schön, D. A. (1983). *The reflective practitioner: How professionals think in action*. Temple Smith.

Sherin, M. G., Jacobs, V. R., & Philipp, R. A. (Eds.). (2011). *Mathematics teacher noticing: Seeing through teachers' eyes*. Routledge. doi:10.4324/9780203832714

Shulman, L. S., & Shulman, J. H. (2004). How and what teachers learn: A shifting perspective. *Journal of Curriculum Studies*, *36*(2), 257–271. doi:10.1080/0022027032000148298

Skattebol, J., & Arthur, L. M. (2014). Collaborative practitioner research: Opening a third space for local knowledge production. *Asia Pacific Journal of Education*, *34*(3), 351–365. doi:10.1080/02188791.2013.871690

Tobin, K. (2014). Transforming science education by expanding teacher and student collaboration. In A. L. Tan, C. L. Poon, & S. S. L. Lim (Eds.), *Inquiry into the Singapore science classroom* (pp. 47–66). Springer Singapore. doi:10.1007/978-981-4585-78-1_3

van Es, E. A., & Sherin, M. G. (2021). Expanding on prior conceptualizations of teacher noticing. *ZDM Mathematics Education*, *53*(1), 17–27. doi:10.100711858-020-01211-4

ADDITIONAL READING

Hargreaves, A. (2019). Teacher collaboration: 30 years of research on its nature, forms, limitations and effects. *Teachers and Teaching*, *25*(5), 603–621. doi:10.1080/13540602.2019.1639499

Jacobs, V. R., & Spangler, D. A. (2017). Research on core practices in K-12 mathematics teaching. *Compendium for research in mathematics education*, 766–792.

Jaworski, B. (2014). Mathematics education development: Research in teaching ¬® learning in practice. In J. Anderson, M. Cavanagh, & A. Prescott (Eds.), Curriculum in focus: Research guided practice (pp. 2–23). Mathematics Education Research Group of Australasia Inc.

Murphy, C., & Scantlebury, K. (Eds.). (2010). *Coteaching in international contexts: Research and practice*. Springer., doi:10.1007/978-90-481-3707-7

Roth, W.-M., & Tobin, K. (2002). *At the elbow of another: Learning to teach by coteaching*. Peter Lang.

Roth, W.-M., & Tobin, K. (2004). Coteaching: From praxis to theory. *Teachers and Teaching, 10*(2), 161–180. doi:10.1080/0954025032000188017

KEY TERMS AND DEFINITIONS

Aotearoa New Zealand: Includes both the English and Māori (Indigenous) names for the country, which are used either interchangeably or together in a wide range of official contexts.

Co-Teaching: Two teachers co-planning, co-instructing, and co-reflecting on lessons together, both as active participants in the teaching activity.

Collaborative Inquiry: Teaching as an inquiry-focused activity enacted in collaboration with colleagues.

Noticing: The practice of paying attention to and making sense of specific aspects of classroom events.

Reflection: Examining and making sense of teaching events for the purpose of informing future teaching practice.

Talk Moves: A set of explicit instructional teacher moves used to orchestrate academically productive discussions, for example "revoicing" students' explanations, or pressing students to justify their thinking.

Teacher Learning: The multiplicity of events that might constitute learning opportunities for teachers, including both the deliberately planned-for events that are often described as professional development, and the wide-ranging situations, experiences, and opportunities from which teachers may learn.

Teaching | Learning: Emphasizes the reciprocal nature of teaching and learning which are assumed to constitute one whole. That is, without learning, teaching cannot be said to have occurred, and, similarly, teaching can be said to fundamentally involve learning.

Chapter 14
Using the Five Practices to Promote Active and Equitable Learning Environments

Keith A. Nabb
Piedmont Virginia Community College, USA

ABSTRACT

Research has shown that active learning can result in knowledge gains for mathematics students at all levels. Even so, establishing and maintaining active learning environments can be challenging for teachers. This chapter focuses on the planning, teaching, and learning of a specific student-centered framework—the five practices for orchestrating productive mathematics discussions. Classroom examples will be selected from a variety of different mathematics courses, highlighting the implementation of the five practices. Foci will include the monitoring, selecting, sequencing, and connecting of student work and how teacher intentionality—through the facilitation of meaningful dialogue—can result in more equitable learning environments for all students.

INTRODUCTION

Research on active learning in STEM fields has found that student-centered instruction can result in deeper learning and higher achievement (Freeman et al., 2014), as well as greater persistence and self-accountability on the part of students (Braun et al., 2017; Bressoud, 2015; Laursen, 2013). Active learning (AL) may take on many different forms, appear under different labels, embrace a variety of philosophies, and/or be grounded in different design theories. Some examples include:

1. **Process Oriented Guided Inquiry Learning (POGIL) and Inquiry-Based Learning (IBL).** POGIL and IBL are well-known in the mathematics community as evidenced by readily available curricular materials (e.g., Rockhurst University, 2018; Straumanis et al., 2013). For a discussion on what makes inquiry particularly unique, see Laursen and Rasmussen (2019).

DOI: 10.4018/978-1-7998-9422-3.ch014

2. **Thinking Classrooms.** Thinking Classrooms (Liljedahl, 2021) enhance mathematics learning by physically transforming the classroom space. They also reformulate the ways students work and communicate, and place students at the center of the learning process.
3. **Realistic Mathematics Education (RME).** RME's foundational roots (Freudenthal, 1991; Gravemeijer, 1994) have inspired a multitude of materials in post-secondary education, ranging from topic-specific lessons (Larsen & Lockwood, 2013; Wawro et al., 2012) to curricular units designed for guided reinvention across entire subject areas (e.g., for differential equations, see Rasmussen & King (2000) or Rasmussen et al. (2018)).
4. **3-Act Tasks.** Three-Act Tasks have earned a dedicated following of educators at all levels, PK-12 (Fletcher, n.d.; Meyer, n.d.). Three-Act Tasks harness the power of video to perplex students, engage them in deep thinking, and eventually allow students to see for themselves (by video) if they produced a reasonable solution.
5. **The Five Practices for Orchestrating Productive Mathematics Discussions.** The Five Practices (Smith & Stein, 2011, 2018) offer a framework for how to activate K-12 classrooms using carefully selected student work. Teacher decision-making and student explanations together help achieve the intended goals of the lesson.

The above list[1] provides evidence of the mathematics community's movement toward and commitment to student-centered learning. Regardless of the form of AL a teacher may choose to adopt, all share the following two characteristics: (1) students are the producers of the mathematics, and (2) teachers trust that the work of students will be the centerpiece of meaningful classroom discussions. Even while every major mathematics and mathematics education professional organization has endorsed this change (CBMS, 2016), one cannot overlook the fact that embracing a shift to AL is enormously challenging for teachers.

This article will cast light specifically on the five practices as one of the many ways to activate one's mathematics classroom. The five practices were initially conceived in the K-8 arena but they are effective in secondary settings (Smith, Steele, & Sherin, 2020), community college/university settings (Nabb, 2018b, 2020; Nabb et al., 2018), and graduate-level mathematics (Reinholz, 2020). The purpose of this article is to provide a detailed description of the practices, including how teachers prepare for instruction, monitor student work, and orchestrate in-the-moment classroom discussions. Examples will be drawn from the areas of decimals/percent, algebra, and calculus. The reader will see that the spirit of the discussion applies to grades 8-12, but extends to post-secondary settings as well.

RESEARCH ON THE FIVE PRACTICES

The value of communicating one's thinking in mathematics classrooms is evident at all levels—in the early/primary grades (Clements et al., 2011; Humphreys & Parker, 2015; Parrish, 2010, 2011; Stein et al., 2008; Smith & Stein, 2011, 2018; Smith, Bill, & Sherin, 2020), middle/secondary levels (Sun et al., 2018; Schoenfeld, 2016; Smith & Sherin, 2019; Smith, Steele, & Sherin, 2020), and post-secondary settings (Kogan & Laursen, 2013; Laursen, 2013; Laursen & Rasmussen, 2019; Larson, 2013; Rasmussen & Marrongelle, 2006; Wawro et al., 2012). Teachers learn more about their students when thinking is made public (Boaler, 2015). Teachers start with what students know, honor their strategies and voices as they explain their mathematics to others, and amplify the message that all students can do mathematics. When teachers see mathematical learning as a fundamental right for all, there is a natural tendency to shift away from teacher-centered instruction and toward student-centered learning. Using what students currently know about mathematics to facilitate new learning is inherently more equitable (Boaler, 2015; Kalinec-Craig, 2017).

The five practices—there are actually *six* (Smith, Steele, & Sherin, 2020)—are these: (0) setting goals/developing tasks, (1) anticipating, (2) monitoring, (3) selecting, (4) sequencing, and (5) connecting. They provide a way for teachers to consider students' ideas, highlight them to meet instructional goals, facilitate meaningful and relevant discussions around the mathematics, and ultimately shape the identities of students as doers of mathematics. These practices and their benefits are discussed below.

Practice 0: Setting Goals and Choosing Tasks

Ensuring successful implementation of the five practices includes (a) establishing learning goals with clear mathematical expectations of what students should learn, and (b) using cognitively-demanding, rich mathematical tasks. With respect to learning goals and objectives, it is important to provide a level of detail that can guide teachers in identifying student work to be highlighted in classroom discussion. For example, consider these three similar instructional goals:

Goal A: Students will recognize and use proportional reasoning.
Goal B: Students will be able to solve "missing information" problems by setting up and using a proportion.
Goal C: Students will be able to recognize and solve problems with proportional reasoning using scale factor and multiplicative reasoning.

It is clear that not all of the above goals help the teacher in the physical act of teaching. Students are wise to recognize the current instructional unit of learning so Goal A holds little promise in guiding instruction and decision making. Goal B provides more detail on what students might *do* but very little with respect to what we want students to *know*. Only Goal C guides the teacher with respect to students' mathematical understanding and further, it suggests what teachers might look for in student work and/or group discussion in order to achieve this goal (e.g., ratio tables and double number lines provide clear evidence of this type of proportional reasoning). When an instructional goal is sufficiently detailed with respect to the *mathematics* we want students to know and use, it assists the teacher in choosing artifacts to select/sequence during whole class mathematical discussion. It is well-known that clear learning goals are foundational to student learning and teacher decision-making (Hiebert et al., 2007; Mills, 2014).

With respect to the cognitive demand of the tasks, this is a delicate terrain to navigate. Two extremes are possible here: if the task is too demanding, students may disengage, and if the task is too routine, little discussion will take place. Neither outcome is desirable as they run counter to a successful active classroom. Finding (or developing) tasks deemed "just right" takes careful planning, repeated practice, and teacher reflection. Smith, Steele, and Sherin (2020) offer advice on how to use rich tasks in mathematics classrooms, how to find tasks that are readily available, how to create tasks from scratch, and how to ensure alignment between the task and the learning goals. Above all, rich mathematical tasks with clear learning goals allow for deeper learning and stronger connections between what is new to students and what is already known (Hiebert & Grouws, 2007). In particular, Black and Wiliam's (1998) work shows that well-designed tasks followed by rich discussions are key pillars of equitable teaching. These experiences lift the achievement of *all* learners, and most importantly, those who traditionally don't experience high achievement in mathematics.

Practice 1: Anticipation

Once a task is chosen for classroom use, the teacher can draft a few probable solutions—correct and incorrect—anticipating what they believe students will do with the task. With continued use of the same task, teachers will eventually acquire a body of work—a written record of *anticipation*—thus making future implementations of the task much more tenable. Teachers will amass information on what students find challenging, common pitfalls, and where questions are likely to emerge. This is also the stage where teachers begin (a) drafting questions to help student groups who may be having difficulty, and (b) reflecting on how to respond to student thinking (Smith, Steele, & Sherin, 2020).

When a task is sufficiently rich, it allows students to represent their ideas in a variety of ways. Depending on the specifics of the task, different groups may approach the problem with a variety of tools—drawing a picture, guessing and checking, reasoning inductively (e.g., determining an explicit formula or recurrence relation), and/or reasoning deductively. A teacher familiar with the task and what students are likely to do will be able to recognize the mathematics and ask thoughtful questions along the arc of the learning goal set for the task (Chapin et al., 2009; Smith & Stein, 2011). Moreover, teachers will recognize small errors in student thinking or instances where the group might have misunderstood or misread the task. Anticipating all that could occur prior to using the task "live" prepares the teacher to be a competent facilitator in the monitoring stage. The ties to mathematical knowledge for teaching (MKT) are as evident as they are important (Ball et al., 2001; Hill et al., 2007).

Practice 2: Monitoring

Monitoring is the first classroom practice that is visible to an outside observer. At this stage, students begin working on the task, often spending a significant amount of time understanding the problem, discussing the problem with other group members, deciding on a viable plan, and formulating a solution. The teacher has two main jobs at this stage—to assess student thinking (e.g., What are the students doing? How are they doing it? Is this what they should be learning?) and to organize/document who is doing what, in alignment with what was anticipated. The backbone of the monitoring stage is listening carefully to student groups and moving student work in the direction of the learning goals via probing questions. Wood et al. (2019) propose the teaching move of "observing unobtrusively" (p. 221)—giving groups the space so they eventually rely on each other, not the teacher. Sometimes listening to groups will cue the teacher as to what the group is doing but assessing student work is challenging and time-consuming (e.g, Ball et al., 2001), especially when the teacher drops in to a group discussion mid-flow. To add to this challenge, teachers should ensure they visit *every* group in an effort to be equitable and fair. This is a nontrivial matter if the teacher stumbles upon a particularly interesting discussion or assists a group that may be struggling with the task. Regardless, teacher questions should focus on what students are saying (verbal) and doing (written). Bearing this in mind, the future discussion that builds on *student* thinking (i.e., not what teachers necessarily want students to produce) will be inherently richer and more equitable (Jacobs & Philipp, 2010). Additionally, it explicitly demonstrates teacher trust in student work and gives the students agency and autonomy as doers of mathematics (Schoenfeld, 2016).

In order to manage the task of recording student strategies, it is prudent to have a monitoring chart in which information can be easily transcribed for future reference (Smith & Stein, 2011). Figure 1 illustrates the main idea; teachers can identify the tools and the thinking observed, decide whether these strategies get airtime in the class discussion, and forecast a suitable order for whole class discussion of student work (selecting/sequencing). This monitoring document is critical in transferring what is observed during the problem-solving stage into the decision(s) of who should be contributing to the discussion (Lampert, 2001). The business of recording student work is challenging, whether this occurs for teacher use only (Smith & Stein, 2011) or for public consumption (Garcia et al., 2021; Shaughnessy et al., 2021).

With continued use of the same task, teachers become more skilled in the monitoring process and increase their awareness of the type(s) of mathematical thinking that students are likely to produce with the task. In turn, the experienced teacher can channel attention away from interpreting the mathematics—because this will be easily recognizable—and more toward facilitating a meaningful discussion around such mathematics. Regardless, assessing what students are doing and finding opportunities for learning within them are nontrivial components of this work (Ball et al., 2001; Leatham et al., 2015).

Practices 3 and 4: Selecting and Sequencing

Selecting and sequencing—the heart of the five practices—are often discussed as a cohesive unit, and they carry the greatest weight in what/how the students experience the lesson content (Smith, Steele, & Sherin, 2020). When selecting student work, the teacher should consider two main questions—*What mathematical ideas should be made public?* and *Which student(s) will be sharing their work?* With regard to mathematics, it is important that teachers focus on choosing strategies that align with the goals of the lesson so that the discussion is productive and relevant. In rare instances, teachers can ask for volunteers to share their reasoning but these discussions can quickly backfire. For example, a student might end up dwelling on a piece of mathematics that does not align with the lesson goal; this may not be a problem but it uses valuable class time. Second, an unsolicited volunteer may offer a solution that neither the teacher nor the students fully understand. This can cause unnecessary confusion and result in a loss of continuity and momentum.

In addition to focusing on the mathematics, it is important to reflect on the choice of individual(s) asked to share/explain their work. While monitoring student strategies, it is possible for the teacher to come across three or four solutions that are similar or even identical. Can the teacher highlight the work of a student who has not presented recently? Can teachers consider covert mechanisms such as student status (e.g., how a student is viewed in the eyes of peers, social standing, academic standing) and how this creates participation barriers? Can teachers move to equalize this imbalance—so quiet students aren't continually ignored, so that active participants see value in other's work, or for "vocal" contributors to take a turn to listen to their classmates' contributions? Wood et al. (2019) discuss the teacher move of "highlighting student strengths," (p. 221) in which teachers can elevate the status of students in significant ways. For example, by publicly acknowledging the work of students who are not usually noticed by others—this can be either by addressing a specific person's work in a group setting or by asking the student to present their idea to the class—this increases the student's status and participation. It may result in a student of elevated status (e.g., a frequent presenter or a vocal student) calling attention to how they approached the problem similarly.

Figure 1. Monitoring chart for the organization of student work (a variation of Smith & Stein, 2011, p. 35).

Task: ____________________

Goal: ____________________

Who presented last time? ____________________

The time before? ____________________

Anticipate (Strategy/Ideas)	Who and What Group & Strategy **Monitor & Select**	**Sequence**	Comments **Connect**
Errors:	Not anticipated:		
Errors:	Not anticipated:		

Questions (Monitoring):

Questions (Connecting)

When sequencing student work, teachers should ask, *What order makes pedagogical sense?* For example, if student solutions range from informal to formal—using counting techniques or a picture on one end of the spectrum all the way to formal derivations and algebraic manipulation on the other end—this could support sequencing the solutions from "most accessible" to "least accessible." By choosing to begin the discussion with a concrete approach, every student will have access to the discussion and the solution methods can gradually progress to be more symbolic and abstract, if this is an intended goal of the task. This "informal to formal" continuum is far from the only way that makes sense, as this is greatly dependent on the work that students produce and the goals that are set for the task. Sometimes, the following extremes are observed in student work: incorrect to partially correct to correct, common to unusual, simple to complex, and inefficient to efficient. Moreover, it is common for these categories to intersect (see the *Bargain Task* in the next section for which the most common solution was, in fact, an incorrect one). The benefits of sharing a common approach are evident as are the benefits of sharing incorrect approaches (Boaler 2015; Kazemi & Hintz, 2014), especially if these latter conceptions are held by many individuals. Teachers' continued focus on communication, participation, and explanation has demonstrated learning gains for young children (Clements et al., 2011; Clements et al., 2013) and early work with the five practices in the upper grades shows similar promise (Nabb et al., 2018; Reinholz, 2020).

Practice 5: Connecting

Once definitive choices have been made for what work to share, the stage is set for the connection of mathematical ideas. Students will share their solutions with their classmates (e.g., using a classroom document camera or displaying a personal whiteboard for others to see) and offer explanations about how they solved the problem. At this point, it is common for several group members to contribute to the explanation and sometimes other classmates (outside the presentation group) will briefly comment on how their solutions are similar or different. Critical to this stage are the teacher-led questions that often have two distinct purposes—(1) clarification and meaning-making, and (2) building connections to the core mathematical ideas in the task (these questions often come from the teacher but sometimes from students) (Boaler & Brodie, 2004; Kazemi & Hintz, 2014; Leatham et al., 2015). Depending on the solutions selected, there are typically 3-4 group presentations in all (a few minutes each) in which the teacher can facilitate brief question/answer exchanges and/or inquire into whether classmates perceive the relationships between different solutions. In some cases, the final solution may be a capstone of sorts—something that alludes to previously shared work or extends beyond the mathematics that was initially discussed (see the *Counting Squares Task* and the *Elementary Integration Task* in the next section).

In order to ensure the connection stage goes beyond a mere display gallery of student work—a "show-and-tell discussion" (p. 5) as Smith and Stein (2011) put it—the teacher must be attune to where students are mathematically and what students find difficult. Facilitation of explicit connections between informal ways of thinking, symbolic/visual connections, nonstandard approaches, and common student errors are keys to advancing students' mathematical knowledge and pointing students to the deeper mathematical ideas set within the task. Teachers need to embrace the fact that students do much learning in the connecting stage. There will be questions of all types, (hopefully temporary) confusion as students grapple with new ideas, and eventual affirmation. Early in the connection stage, students should be encouraged to be active listeners while their classmates are presenting work. Since other students are expected to add to or disagree with previously discussed solutions, students must be held accountable to what is happening (Michaels et al., 2010). Students who are new to the norms of the five practices will likely need assistance on how to contribute to a productive discussion (Horn, 2012; Kazemi & Hintz, 2014; Smith & Stein, 2011, 2018) and this can begin early on with the understanding that this skill is acquired over time.

THE THREE TASKS

In this section, I provide summaries of three teaching episodes with the five practices in undergraduate mathematics classrooms. I offer information on why I chose the tasks, what I anticipated from my students, what unfolded during the monitoring stage (including notes transcribed for later reference), the details of selecting and sequencing, and the questions and conversations held in whole-group discussion. By doing so, the reader is provided a glimpse of how to facilitate classroom learning with the five practices and the nature of what students learned. Additionally, I draw attention to some of the teaching practices that support equity in classrooms. These include revoicing student work in the whole class environment, juxtaposing standard approaches with nonstandard strategies, positioning mistakes as areas to grow, elevating the status of nonparticipators, and allowing students to amplify another student's work in light of new information. These practices redefine the student role as facilitators in everyone's mathematical learning.

Task 1: The Bargain Task

Figure 2. The Bargain Task

Bargain

A shop advertised "Everything half price in our sale", but also now advertises that there is "An additional 15% off sale prices." Overall, this is equivalent to what reduction on the original prices?

Instructional Goal: Students will be able to solve decimal/percent problems with specific mathematical attention to recalibrating the whole.

The bargain task (NRICH, 2018) is suitable for a learning unit involving decimals and percent (see Figure 2). These topics arise in many different mathematics courses, including Math for Elementary Teacher Preparation, Liberal Arts Mathematics, and Quantitative Reasoning.[2] The bargain task is ideal for students with a wide variety of experiences because the problem can be solved using mathematical reasoning, models/manipulatives, algebra, or computation. Figure 3 shows the completed monitoring chart used to track student work and link them to the instructional goals. The reader will also see informal notes leading to the selection/sequencing of student work.[3]

Figure 3. Monitoring log of student work on the bargain task.

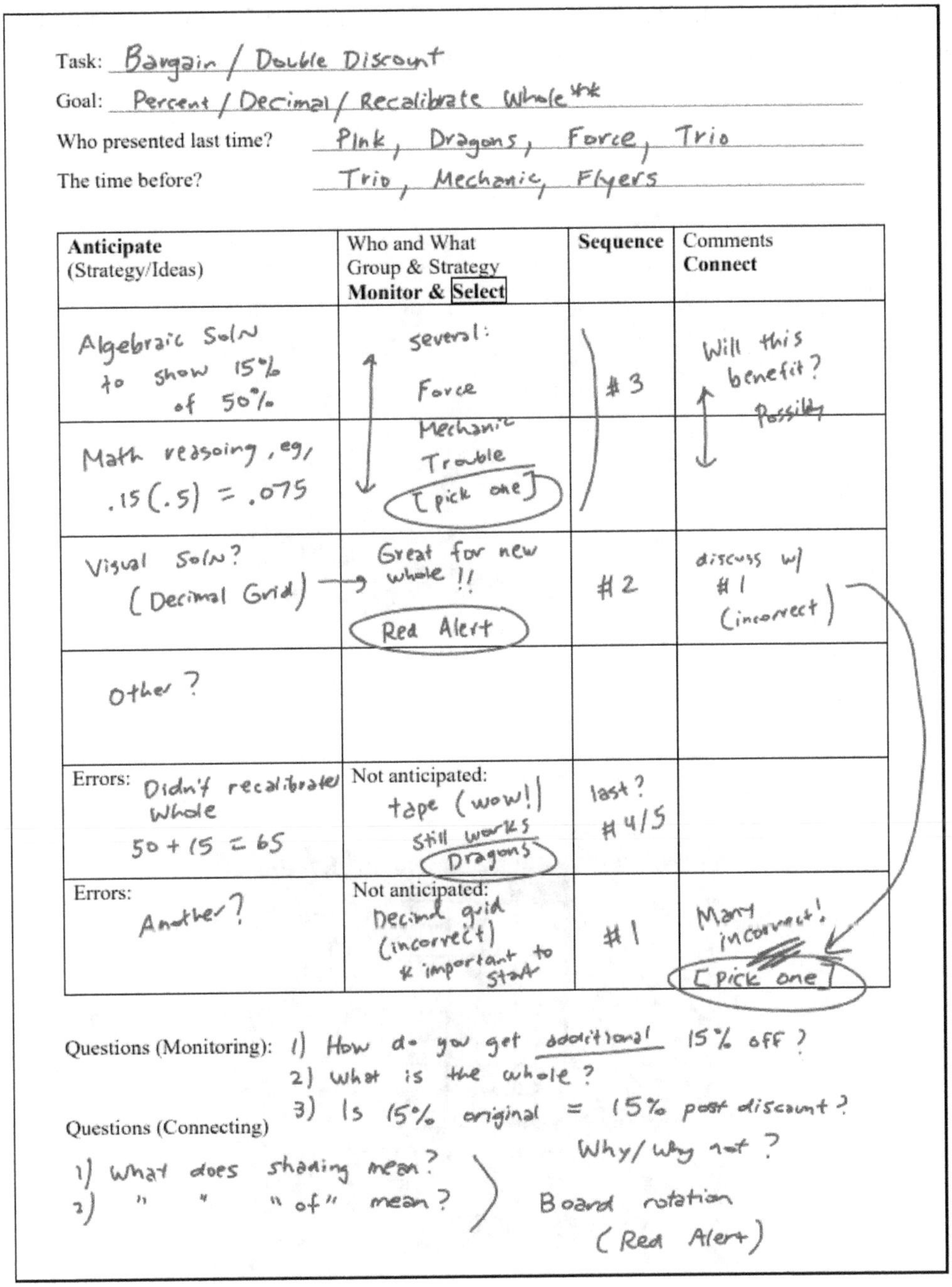

Task: Bargain / Double Discount

Goal: Percent / Decimal / Recalibrate Whole ***

Who presented last time? Pink, Dragons, Force, Trio

The time before? Trio, Mechanic, Flyers

Anticipate (Strategy/Ideas)	Who and What Group & Strategy **Monitor & Select**	**Sequence**	Comments **Connect**
Algebraic Soln to show 15% of 50%	several: Force	#3	Will this benefit?
Math reasoning, eg, .15(.5) = .075	Mechanic Trouble [pick one]		Possibly
Visual Soln? (Decimal Grid) →	Great for new whole!! Red Alert	#2	discuss w/ #1 (incorrect)
other?			
Errors: Didn't recalibrate whole 50 + 15 = 65	Not anticipated: tape (wow!) still works Dragons	last? #4/5	
Errors: Another?	Not anticipated: Decimal grid (incorrect) * important to start	#1	Many incorrect! [Pick one]

Questions (Monitoring): 1) How do you get additional 15% off? 2) what is the whole? 3) Is 15% original = 15% post discount? Why/why not?

Questions (Connecting)

1) What does shading mean?
2) " " "of" mean?

Board rotation (Red Alert)

The reader will notice several idiosyncrasies in Figure 3 including (a) references to group names established earlier in the semester (Pink, Dragons, Force, Trio, Mechanic, Flyers, Trouble, and Red Alert), (b) summaries of anticipated work, (c) attention to equity (documentation of groups that presented previously in an effort to hear different voices on this task), (d) specificity of novelties in solutions (e.g., "tape" = "tape diagram"), (e) miscellaneous notes (e.g., the importance of connecting invalid reasoning with sound reasoning via the same tool), and (f) questions to guide the monitoring and connection stages.

With a detailed log such as Figure 3, the teacher is freed from having to remember who did what so that facilitation in leading a mathematical discussion can become the focal point. Based on the monitoring log, I chose to select and sequence student solutions in the following manner[4]:

Figure 4.

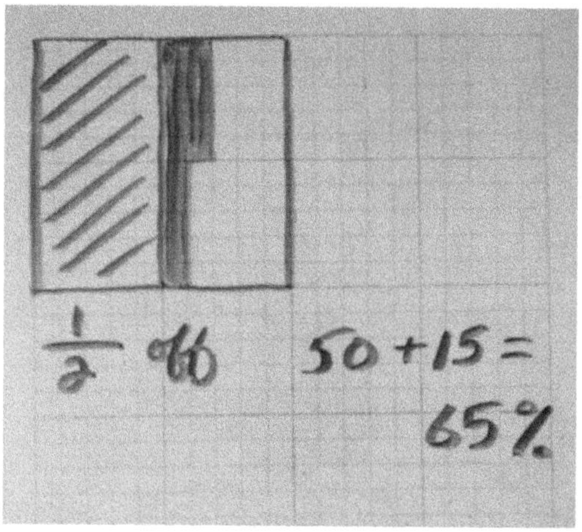

Figure 5.

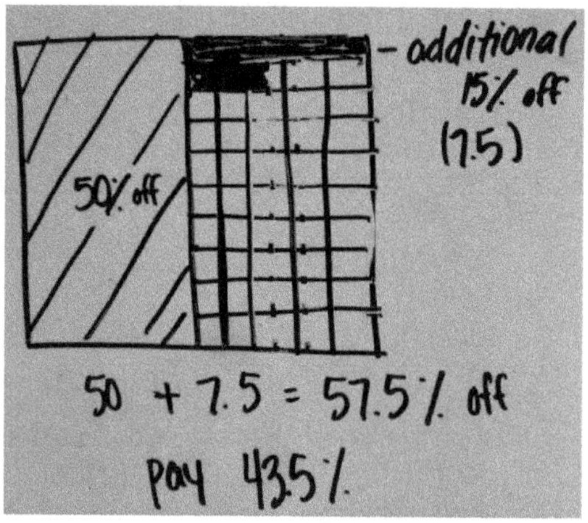

Figure 6.

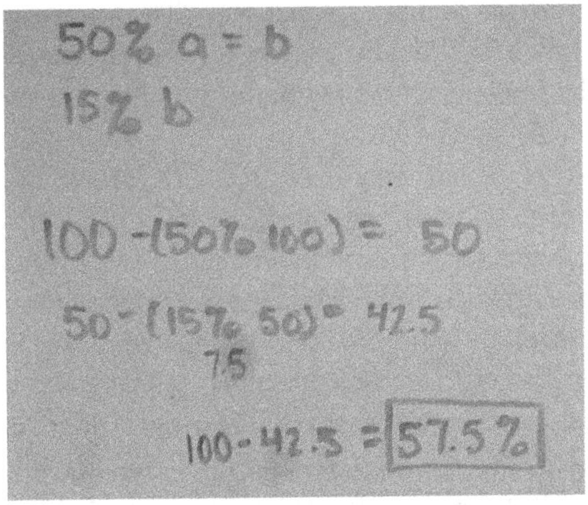

Figure 7.

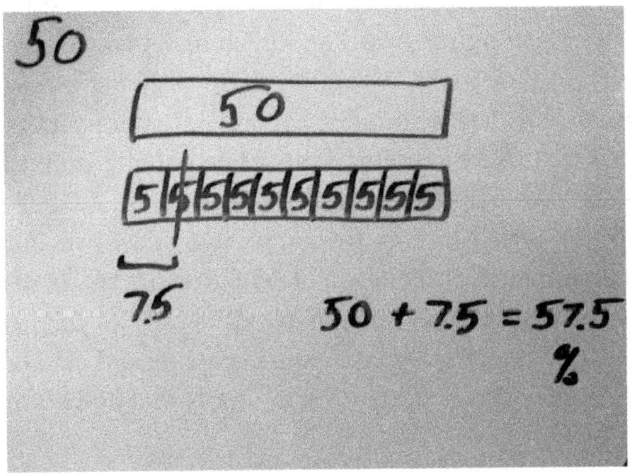

Figure 4, while incorrect, was important to open our classroom discussion. My rationale was grounded in the fact that nearly half of the groups in the room had obtained this answer and felt confident with the result. The source of the error is that students failed to recalibrate the "whole" once the initial 50% discount was applied. After Figure 4 was shared and I inquired about reactions, a classmate mentioned how "the dark shading is actually more than 15% of the 50% left." This presented an ideal segue into Figure 5. The authors of Figure 5 explained how it was difficult to find 15% of the 50 remaining squares "…so we rotated the board and we could see 10% as being one of these mini horizontal bars." They explained how they shaded one and a half of these bars (representing 15%) giving them the desired 7.5 boxes (representing 7.5% of the whole). This led to an immediate affirmation of understanding from one of the group members of Figure 4: "Oh…I can see 15 percent of 50% now. Ours is actually more than 20% of 50. I get it!" This student remark was in direct alignment with the instructional goal. Figure 6 was presented next. While the students explained they took a computational route, they referenced an important piece of language from the earlier explanations. "The toughest part of this was the 15% *of* 50 so we thought of this as 0.15 times 50." Finally, Figure 7 was offered as another way of reimagining the recalibration of the whole. The students worked with the 50 remaining pieces, made 10 equal parts in a tape diagram, and then easily calculated 15%.

While there is no rulebook on the best way to sequence the above solutions, my choice was to start with the incorrect solution since (a) many students were convinced of its validity, and (b) it was the centerpiece of the instructional goal (to identify/understand the new whole in the problem). Beginning with incorrect mathematics is a provocative move but it was successful in motivating other students to contribute to the discussion, especially in the case of a decimal grid. The discussion around Figures 4 and 5 was powerful to witness because we were able to see both where and why the error occurred. While teachers may hesitate to use student errors as discussion points, there are collective benefits to doing so. Above all, it tells our students that we value risk-taking in mathematical thinking and that mistakes augment everyone's learning.[5] If a positive atmosphere is cultivated from the start, students will see errors as learning opportunities—places to grow (Boaler, 2013). Teachers can reposition mistakes as snapshots of in-the-moment thinking and blueprints for future instructional decisions (Kalinec-Craig, 2017; Kazemi & Hintz, 2014; Lim et al., 2015).

The above exchange heightened everyone's awareness that the task was more difficult than they had initially thought. Once this was established, I felt it important to show other solutions that did not use the decimal grid. Figure 6 was a straightforward approach and I was thankful of the careful emphasis the group made in referencing "15% of 50." That is, these students knew exactly *why* many of their classmates initially made a miscalculation. Finally, Figure 7 supports a personal philosophy of mine in that students should use tools that make sense to them when solving problems. When this solution was being shared, the authors said, "We didn't even think to use the grid…" but I shot back, "Isn't this [tape diagram] just as good?" In summary, I was motivated by two factors in sharing student solutions—(1) to move from the common (and incorrect) solution of 65% to more mathematically valid solutions (i.e., proving that wrong can lead to right), and (2) to show a diversity of thinking processes with different mathematical tools (decimal grids, tape diagrams, and computation). With respect to the instructional goal, all of the solutions were focused on students using tools that helped them identify the new whole after the first discount was applied.

Task 2: Counting Squares Task

Figure 8. Counting Squares Task

MORE ON PATTERNS

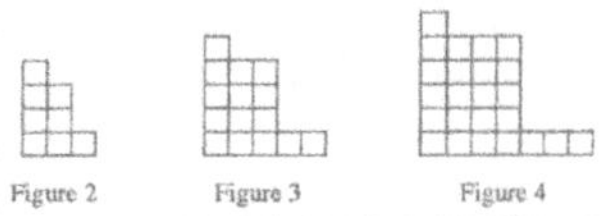

(a) Draw Figure 5. How many squares make the figure?

(b) Describe in words how the pattern is changing.

(c) Find a formula that gives the number of squares based on the figure number. Be sure to define your variables clearly.

(d) Based on part (c), how many squares would be in the 100th figure? How do you know?

Instructional Goal: (Part (c)) Students will be able to derive and use a mathematical formula for the number of squares based on the geometric qualities of the figure and the input value (figure number), and explain the visual connection between the algebra and geometry. Further, students will be able to recognize algebraic equivalence across various formulas via algebraic manipulations.

The counting squares task in Figure 8 works well in any math class where visual and algebraic connections are a central goal. The mainstream popularity of websites such as youcubed.org and visualpatterns.org make achieving this objective much more plausible. This task and similar ones are a good fit for classes such as Liberal Arts Mathematics and Algebra I/II—any class for which there is a unit on algebra with a focus on geometric meaning. While students discussed all components of this problem at length, the most challenging part of the problem was determining a formula for the number of squares based on the figure number (part c). I was expecting some groups to tackle this challenge recursively (from one figure to the next), and others to approach it explicitly (from figure number to formula, as instructed). The monitoring log can be seen in Figure 9.[6]

Figure 9. Monitoring log of student work on the counting squares task

Task: Counting Squares / Growth

Goal: Formula (w/ geom), fig # → # squares, algebraic equivalence

Who presented last time? Joe C, Emily, Carl, Micah

The time before? Micah, Jamison, Rachel, Jones

Anticipate (Strategy/Ideas)	Who and What Group & Strategy **Monitor & Select**	**Sequence**	Comments **Connect**
Numerical / Table 2 → 8 (2×4) 3 → 15 (3×5) 4 → 24 (4×6)			
As above but recursive 8 → 15 → 24 +7 +9 etc.	7, 9, 11, 13, etc. (focus on change) Rachel, Carl		linking figure # to next figure #
→ $n(n+2)$ (Algebra)	rearrangement ⇒ $n(n+2)$ see Jones	#3	hmmm.... do they know this algebra?
n^2+2n (Algebra	take apart method (square/ geometry is visible) Joe C	#2	→ good progression?
Errors:	Not anticipated: left in form: $n(n+1)+n$ Miles	#1? nice + color coded	What can we do w/ this formula? Equivalent version?
Errors:	Not anticipated: Stuck w/ recursion: 8+7=15, 15+9=24, 24+11=35	#4?	15 = 16 − 1 24 = 25 − 1 35 = 36 − 1 Discuss?

Questions (Monitoring):
1. Where is the formula visible in your work?
2. Other ways to write this? (Algebra)
3. Link figure → # squares (challenging)

Questions (Connecting):
1. Different ways to write/express this formula? (Algebraic equivalence)

I chose the following student work to select and sequence:

Figure 10.

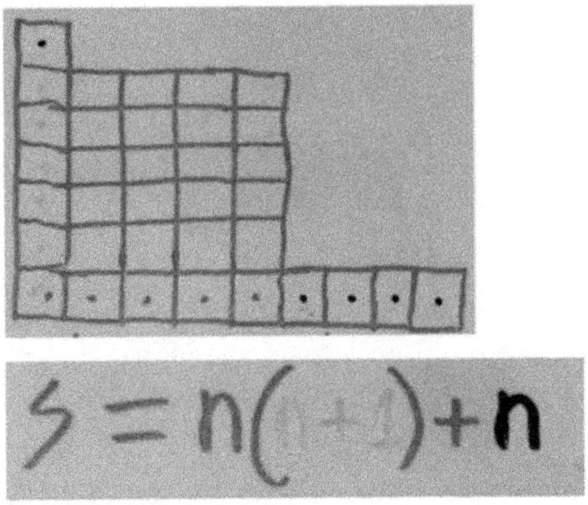

Figure 11.

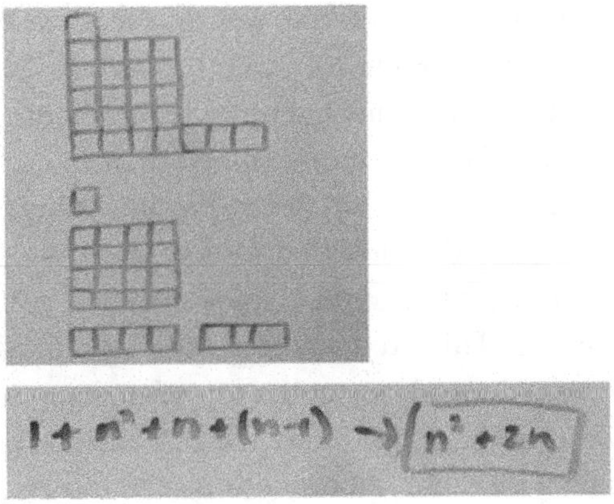

Figure 12.

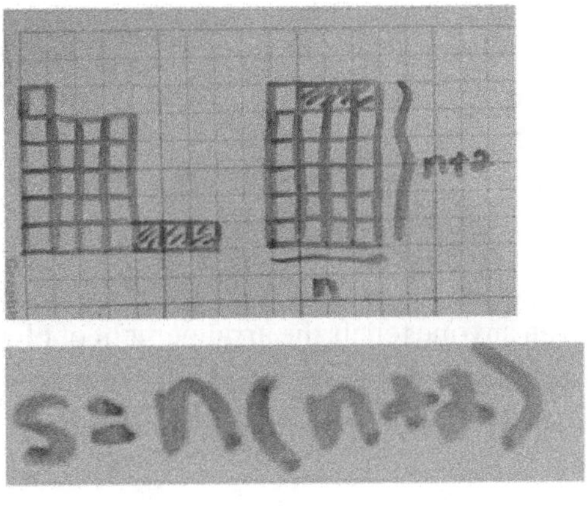

Figure 13.

We began our discussion with Figure 10 because of the detail and clarity displayed on the group's board. This solution was accessible to everyone in the room and it was annotated beautifully. My hope was that by sharing this solution first, anything that followed would possess at least this level of mathematical detail and clarity in communication. As the reader can see, these students color-coded their diagram and formula as a way of explaining their thinking. In line with the instructional goal, my expectation was that student derivations would include elementary counting procedures, rearrangement of pieces in the figure, and/or similar reasoning strategies. Thus, this was an excellent starting point.

Once we understood the construction of the formula, I asked students what they noticed about the formula. One asked, "Isn't that just n^2+2n?" This naturally led to Figure 11 as the formula n^2+2n stands out as the culmination of their work. When this group shared their thought process, it was clear to everyone they thought about the problem differently but the end result simplified to the equivalent expression n^2+2n. One classmate enthusiastically said, "This is not how we did it—that breaking apart—but I like it!" Continuing, I asked, "Are there different ways to write this formula?" Another group replied with $n(n+2)$ and we were on our way to Figure 12. This group treated the diagram like a game of Tetris—interlocking the tail of the figure into the remaining space at the top of the diagram (see Figure 12 image). From here, it was clear they had constructed an array where the dimensions are 4 ($n = 4$) and 6 ($n + 2 = 6$).

Given the diversity of thinking with these three solutions and the evidence that elementary algebraic manipulations were accessible to this audience, I thought to share one more solution as a whole class discussion. Moreover, I hoped we could indulge in a problem of reversal—if I shared with them a different looking formula, could they then figure out the accompanying grid arrangement? [7] Thus, I wrote the following on the side board

$$\begin{aligned} n^2+2n &= n^2+2n+1-1 \\ &= \left(n^2+2n+1\right)-1 \\ &= (n+1)(n+1)-1 \\ &= (n+1)^2-1 \end{aligned}$$

followed up with the question, "Can anyone tell us the arrangement of blocks that would fit this formula?" As I fielded contributions from around the room, we eventually figured it out with much fanfare from the class (see a graphical reenactment in Figure 13). In addition to "seeing" this was nearly a 5 by 5 square, a quiet student made the most profound statement in our discussion: "I actually think this [solution] is the best because you can predict all the numbers. It's always going to be things like 9 minus 1, 25 minus 1, 81 minus 1. The pattern is really obvious now."

The above episode shows the teacher's central role in connecting student solutions. Since Figure 10 was so well-prepared and the expression was in a less simplified form, this allowed me to work with their formula and connect this to other formulas that I had documented in the monitoring stage. From here, my vision was to connect the classroom discussion via the formulas that I knew were found by other groups:

$$\underbrace{n(n+1)+n}_{\text{first solution}} \rightarrow \underbrace{n^2+2n}_{\text{second solution}} \rightarrow \underbrace{n(n+2)}_{\text{third solution}}$$

The connections were gradual and natural, allowing students the space to appreciate the different ways their classmates viewed the problem. Based on the relatively smooth transition we experienced and the fact that there was clear enthusiasm for the solutions, I only then decided to take a risk and show the $(n+1)^2-1$ solution. This led to a mostly non-vocal student making a startling mathematical observation that the number of squares was always one shy of the perfect squares.

What is clear in this episode is that student work and student explanations are the seeds of all mathematical discussion; my job was to gently ask questions that could organically point to different student work. Using this task in a different class would likely result in a different classroom experience; however, this experience should still meet the intended goals of the lesson (aside: I have never had two identical discussions with students while using this task). Each time the teacher uses the five practices with a group of students, more knowledge is gained about how students think about the task. The $(n+1)^2-1$ solution was one I had "tucked away" from previous experiences and I thought this class would benefit from its discussion. This was my motivation for sharing Figure 13, which resulted in a wonderful closure of the task.

Task 3: Elementary Integration Task

Evaluate, by any means: $\int \cos^3 x \sin x \, dx$.

Instructional Goal**:** Students will be able to evaluate integrals involving various trigonometric powers and honor/respect/understand the natural choices that may lead to alternate but equivalent solutions.

The problem above was used in a Calculus II lesson involving integration of trigonometric functions with various powers—e.g., $\int \sin^4 x \, dx$, $\int \tan^3 x \, dx$, and $\int \sqrt{\tan x} \sec^2 x \, dx$. This material is generally considered a precursor to trigonometric substitution—a topic students usually find very challenging. The reader will notice this particular integral is routine; a simple substitution of $u=\cos x$ results in the integral transforming to the form $-\int u^3 du$, leading immediately to an answer. However, my experience with calculus students suggests that their understanding of trigonometry is often incomplete and checkered with self-imposed roadblocks (e.g., "I hate trigonometry"). What is interesting about this problem is that students were new to this process so not all students who solved this problem chose the most direct pathway. Later admissions revealed that some students didn't even *see* this path, even after it was explicitly pointed out. This is the inherent beauty of active learning and using student work to teach. See the monitoring log in Figure 14.

Figure 14. Monitoring log of student work on the integration task.

Task: $\int \cos^3 x \sin x \, dx$

Goal: antiderivative, * flexibility

Who presented last time? Samantha, Corey, James

The time before? Corey, Brianna, Melinda

Anticipate (Strategy/Ideas)	Who and What Group & Strategy **Monitor & Select**	**Sequence**	Comments **Connect**
$u = \cos x$, $du = -\sin x \, dx$, w/ u integral	Bryanne (good opener)	#1	
same but with x integral (mental math)	Preston, same, different notation	#2	How similar to #1? More efficient?
$u = \sin x$, $du = \cos x \, dx$, pathway → stuck?	another? See #4		
Errors:	Not anticipated: Save $\sin x \, dx$ after $\cos^2 x = 1 - \sin^2 x$ (Ian)	#3	interesting!! efficient?
Errors:	Not anticipated: $\cos^2 x ((\cos x \sin x) dx)$ Trig identity (James)	#4	matches #1, 2!! any way to tie to #3?

Questions (Monitoring): ① Isolate sine or cosine? ② Integrate w/ u or x?

Questions (Connecting): ① For overall approach: would work on $\int \cos^n x \sin x \, dx$ ← important ② other ways to approach this integral?

As evident in Figure 14, the tools I was provided to facilitate this solution were far richer than if I had simply demonstrated a solution for the students. The fact that I had anticipated three of the solutions allowed me to focus on the other (unexpected) solution and think about how it could be included in our class discussion. The selection/sequence process was as follows:

Figure 15.

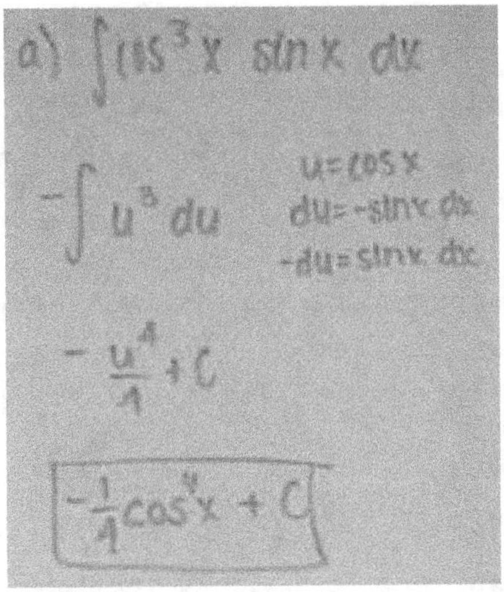

Figure 16.

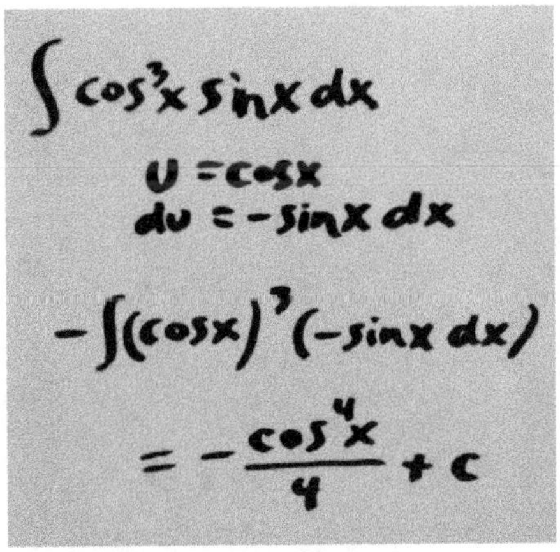

Figure 17.

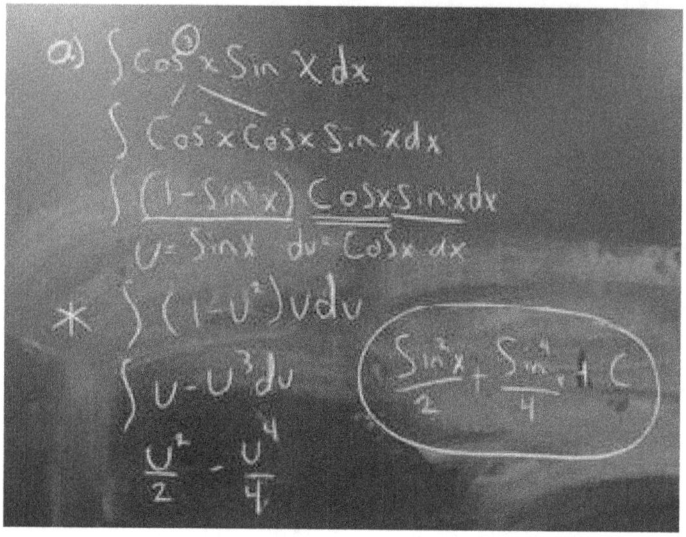

Figure 18.

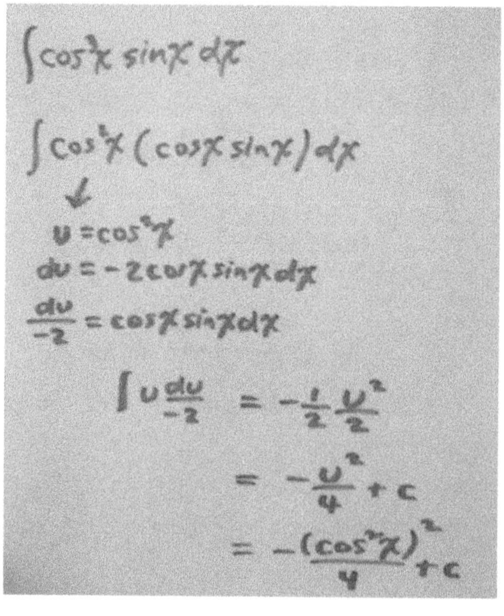

I shared Figures 15 and 16 to start because they showed the most direct pathway to a solution. As one student said, "It doesn't matter what power is on the cosine. Sine is right there for the *du*." This solution shows students' awareness of "sin *x dx*" as serving as the derivative (absent the sign) and is directly supported by a connecting question (see Figure 14, bottom). I displayed Figure 15 first because this highlights the substitution and transformation explicitly, followed by Figure 16, where the transformation is done mentally. In other words, Figure 16 shows a greater comfort level with substitution by way of leaving the original variable of x in the problem.

In a traditional lecture setting, this problem would be resolved and we would move on to different problems. However, there were two additional solutions which were quite special and these were in alignment with the instructional goal. The solution in Figure 17 has a focal point of "$\cos^3 x$" rather than the "$\sin x\,dx$." By way of this shift, this student saw the odd power of cosine as a way to save a cosine term and to use "$\cos x\,dx$" as the differential. Naturally, the answer looks different but we agreed this would be equivalent. The conversation around this solution was an interesting one. One student claimed this was "totally not necessary" (which is true) but a different student immediately replied by saying, "…but it's not wrong. It's just different." On that note, I shared one more solution, one that I had not anticipated before giving this task (Figure 18). This student began in a similar manner to Figure 17 but then noticed that the unconventional choice $u=\cos^2 x$ would lead to a solution. This solution was given the moniker "absolutely amazing" by a classmate and one student marveled at how Figure 18 connected both of the previously discussed solutions: "It starts out like that second way [Figure 17] but the answer matches the first one! [Figure 15, 16]" With respect to the neutral/negative remark made previously about Figure 17, this new comment accomplished two things: (1) it elevated Figure 17 as a legitimate strategy, and (2) it showed the power of revoicing (one student referencing and amplifying another student's work). While this is commonly a teacher-led instructional move—e.g., highlighting the work of a quiet student, or a student perceived to have low mathematical status, or for an emerging bilingual (Chapin et al., 2009; Wood et al., 2019)—the student comment here allowed the class to reevaluate a solution method.

So what are the takeaways from this experience? As a teacher, I learned that sometimes what we perceive to be routine is viewed quite differently by students. Additionally, active learning allows teachers to see and share mathematics that would otherwise remain hidden in student's notebooks. (Students might even label such work as a false start.) What was most interesting in this discussion was the shift in opinion I sensed in Figure 17. While this solution was initially greeted with skepticism—*Why would you do that?*—it was the even more unusual approach in Figure 18 that allowed the class to certify Figure 17 as "just as correct" as Figures 15 and 16. This conversation is a clear example of how the rewards of active learning are often worth the risks. A student communicated to me toward the end of our session that, "…it is nice we have lots of ways to do these problems. I read the book before class and I got the impression that you had to follow very strict guidelines for these problems."

EQUITY ACROSS THE THREE TASKS

Implementing equitable teaching practices in mathematics classrooms is one of the key recommendations in *Catalyzing Change* (NCTM, 2018):

…students are active participants, engaging in reasoning and sense making, striving to make their mathematical thinking visible and intelligible to others…equitable mathematics teaching practices support identity and agency by creating structures for having students' mathematical ideas considered during instruction, supporting students in viewing themselves as having ownership of mathematical meaning… (p. 28-29)

How can practitioners interpret the above statements and translate them into action? Perhaps the most important component of this recommendation is placing students at the center of the learning process by supporting their thinking as they learn mathematics (Schoenfeld, 2014, 2016). The experiences shared in this article provide concrete examples of equity-based teaching practices that consider student ideas and, consequently, help develop students' sense of mathematical agency. The three classroom episodes highlight many of the components of active and equitable instruction, as summarized in Table 1.

Table 1. Equity-based teaching practices in mathematical learning

Examples of Equitable Teaching Practices (Task)	Key Question(s)	Supporting Research
1. Welcoming mistakes and integrating them into the learning experience (Bargain)	How can mistakes be showcased in productive and nonthreatening ways to advance everyone's learning?	Boaler, 2013, 2015; Kalinec-Craig, 2017; Kazemi & Hintz, 2014; Lim et al., 2015; Vanoli & Luebeck, 2021
2. Juxtaposing traditional approaches (e.g., computation) with more personalized, novel approaches (Bargain, Integration)	How can teachers use what students know rather than what we hope they know?	Schoenfeld (2014) calls this "meeting students where they are." (p. 407); Star et al., 2021
3. Facilitating rich discussions that build upon student work (Counting Squares, Integration)	How can student work be organized to maximize learning?	Black & Wiliam, 1998; Kalinec-Craig, 2017; Shaughnessy et al., 2021; Smith & Stein, 2011, 2018; Smith, Bill, & Sherin, 2020; Smith, Steele, & Sherin, 2020
4. Hearing a significant piece of mathematics from an individual who rarely contributes (Counting Squares).	Who has access to mathematical discussions and who is contributing? Can teachers elevate the status of those students who are rarely seen or heard?	Nabb, 2020; Boaler, 2016b; Chapin et al., 2009; Wood et al., 2019
5. Facilitating a discussion that pushes the limits of what students know (Counting Squares, Integration).	Can students be empowered to think beyond what they initially conceived and participate accordingly?	Black & Wiliam, 1998; Boaler, 2002, 2015, 2016b; Cohen & Lotan, 2014; Lim et al., 2015; Smith, Bill, & Sherin, 2020; Smith, Steele, & Sherin, 2020; Wood et al., 2019
6. Reevaluating student work in light of new information (Integration)	How can teachers facilitate and encourage students to make connections?	Nabb et al., 2018; Garcia et al., 2021; Lim et al., 2015; Reinholz, 2020; Shaughnessy et al., 2021; Wood et al., 2019

For many of my years in mathematics classrooms, I have cohabitated a space with students where the following dialogue is common:

- *When are we ever going to use this?*
- *I'm terrible at math.*
- *Can you redo that? I don't get it.*
- *All I want is a C.*

The above statements and similar ones paint a painful picture of the subject that we cherish. Many students have low confidence in their abilities, a fixed mindset, and have accepted that mathematics has no relevance to their lives. With a shift to active learning, many students gain confidence as participants in and doers of mathematics. Building discussions around student work can have a lasting impact on how students view themselves in the learning of mathematics. While the evidence below is self-reported, initial student feedback across different content areas has been positive and encouraging:

[Calculus I, using the Five Practices] I liked the style the class was taught with. The emphasis on group work and small-group discussion helped me understand the material better than a straight lecture (Nabb et al., 2018).

[Mathematics for Liberal Arts, using the Five Practices] I've never been allowed to do the problems my way. I feel heard in this class (Nabb, 2020).

[Differential Equations, using the Inquiry-Oriented Differential Equations curriculum (Rasmussen, Keene, Dunmyre, & Fortune, 2018)] I like this class because it makes me think more. It's different from other math classes that I took before, and I believe your way of teaching is the best way because you make information[s] stick in our minds (Nabb, 2018a).

Every student deserves to have their voice heard in the classroom. We know who our "strong" students are after about two weeks of instruction. We also know who the active contributors are—whether they contribute mathematics that teachers want to incorporate into instruction or something less desirable to the plan for the day. The equity challenge lies in taking these contributions as *relevant* to our teaching, and facilitating a quality mathematical discussion around these ideas. This is attuned to redefining what mathematical competence actually is—i.e., recasting what students say and do as resources to be incorporated into mathematical learning (Celedón-Pattichis et al., 2018). As Boaler (2016b) asserts, "mathematics classrooms that have been shown to create more equitable opportunities for learning are those in which a range of student participation is supported, students are encouraged to ask questions and share ideas, and to feel safe in doing so, and teachers express a commitment to students' academic and social development" (p. 175).

Equity in mathematics classrooms can be tricky to define and sometimes, even a contentious topic to discuss (Boaler, 2002). For too long, mathematics has been a tool of oppression—a gatekeeper to college success and a barrier to degree completion (AMATYC, 2018). But on the flipside, mathematics is a byproduct of humanity—a subject where we can experience all that is beautiful about being human (Boaler, 2016a; Su, 2017). The question we should be asking is, *What specific teaching practices can be implemented to mitigate the negative experiences that typically shape student's mathematical identities?* Active learning, while not a silver bullet, shows much promise in this regard (Aguirre et al., 2017; Boaler, 2002; 2016b; Braun et al., 2017; Celedón-Pattichis et al., 2018; Chapin et al., 2009; Cohen & Lotan, 2014; Freeman et al., 2014; Kalinec-Craig, 2017; Laursen, 2013; Schoenfeld, 2014, 2016; Smith & Stein, 2018; Tang et al., 2017; Wood et al., 2019).

As a way of situating active learning within equity initiatives, Tang et al. (2017) mapped student-centered pedagogy to Gutiérrez's (2012) four dimensions of equity—access, achievement, identity, and power. Active learning strategies are inherently asset-based since teachers focus on what students know instead of dwelling on what they do not know. This increases accessibility to mathematics discussions and reframes who students view as content experts. This also has the potential to impact how students identify as participants in mathematical discussions—from passive listener to active contributor. While Gutiérrez's framework goes well beyond what happens in classrooms—e.g., opening the STEM pipeline to all individuals (achievement), using culture and language to empower individuals (identity), and disrupting the social systems that continue to oppress (power)—one can see that active learning sets in motion some of the fundamental principles to address inequities in mathematics classrooms.

A final caveat is that mathematics and equity are not separate entities within the work of mathematics teaching—e.g., see the "false dichotomy" between mathematics and equity work (Aguirre et al., 2017, p. 130). Teachers should not view equity as external to mathematics or something to be "added" to the subject we teach. One way to make equity intentional is to use active learning in classrooms; even starting small is an encouraging first step (e.g., one lesson or a small unit of instruction). Highlighting the knowledge of students with equitable teaching practices is a collective responsibility, and these actions should be sincere and intentional across the entire field of mathematics education (Aguirre et al., 2017).

CONCLUSION

The focus of this article has been student-centered learning in mathematics classrooms—specifically, the five practices for orchestrating mathematics discussions (Smith & Stein, 2011, 2018) and how they redefine the roles of teachers and students. When a group of students produces a solution, should the teacher incorporate this work into whole-class discussion? Does the solution highlight the mathematics that aligns with the instructional goal? Where does this solution fit into the flow of the discussion? What presentation order makes pedagogical sense to connect solutions and why? Teacher decision-making is the most impactful factor here and it is driven by student accessibility, equity in participation, and continuously fostering an environment where students can thrive mathematically and socially.

Equity should be organically built into everything we do. Students who are normally disengaged or feel "unseen" in traditional classroom environments should be provided a space where their thinking can be seen and heard. The classroom examples shared in this article demonstrate how *students'* mathematics—whether written or verbal—can be catalysts to beginning fruitful discussions or making important connections. In short, the end result is an equitable balance of contributions from students of various readiness levels. The five practices provide an effective framework for embracing the unique assets that students bring to mathematics classrooms. With continued use of student-centered instruction, teachers can develop their praxis of cultivating environments where every student feels welcome to contribute to the mathematical discussions of the day.

REFERENCES

Aguirre, J., Herbel-Eisenmann, B., Celedon-Pattichis, S., Civil, M., Wilkerson, T., Stephan, M., Pape, S., & Clements, D. H. (2017). Equity within mathematics education research as a political act: Moving from choice to intentional collective professional responsibility. *Journal for Research in Mathematics Education*, *48*(2), 124–147. doi:10.5951/jresematheduc.48.2.0124

American Mathematical Association of Two-Year Colleges (AMATYC). (2018). *IMPACT: Improving mathematical prowess and college teaching*. AMATYC.

Ball, D. L., Lubienski, S. T., & Mewborn, D. S. (2001). Research on teaching mathematics: The unsolved problem of teachers' mathematical knowledge. In V. Richardson (Ed.), *Handbook of research on teaching* (4th ed., pp. 433–456). American Educational Research Association.

Black, P., & Wiliam, D. (1998). Inside the black box: Raising standards through classroom assessment. *Phi Delta Kappan*, *80*(2), 139–144, 146–148.

Boaler, J. (2002). Learning from teaching: Exploring the relationship between reform curriculum and equity. *Journal for Research in Mathematics Education*, *33*(4), 239–258. doi:10.2307/749740

Boaler, J. (2013, March). Ability and mathematics: The mindset revolution that is reshaping education. *Forum*, *55*(1), 143–152. doi:10.2304/forum.2013.55.1.143

Boaler, J. (2015). *Mathematical mindsets: Unleashing students' potential through creative math, inspiring messages and innovative teaching*. John Wiley & Sons.

Boaler, J. (2016a). The math-class paradox. *Atlantic (Boston, Mass.)*, (Feb), 4.

Boaler, J. (2016b). Designing mathematics classes to promote equity and engagement. *The Journal of Mathematical Behavior*, *100*(41), 172–178. doi:10.1016/j.jmathb.2015.01.002

Boaler, J., & Brodie, K. (2004). The importance, nature, and impact of teacher questions. In D. E. McDougall & J. A. Ross (Eds.), *Proceedings of the twenty-sixth annual meeting of the North American Chapter of the International Group for the Psychology of Mathematics Education* (*Vol. 2*, pp. 774-782). Toronto: OISE/University of Toronto.

Braun, B., Bremser, P., Duval, A. M., Lockwood, E., & White, D. (2017). What does active learning mean for mathematicians? *Notices of the American Mathematical Society*, *64*(02), 124–129. doi:10.1090/noti1472

Bressoud, D. (2015). Insights from the MAA national study of college calculus. *Mathematics Teacher*, *109*(3), 179–185. doi:10.5951/mathteacher.109.3.0178

Celedón-Pattichis, S., Borden, L. L., Pape, S. J., Clements, D. H., Peters, S. A., Males, J. R., Chapman, O., & Leonard, J. (2018). Asset-based approaches to equitable mathematics education research and practice. *Journal for Research in Mathematics Education*, *49*(4), 373–389. doi:10.5951/jresematheduc.49.4.0373

Chapin, S. H., O'Connor, C., & Anderson, N. C. (2009). *Classroom discussions using math talk to help student learn*. Math Solutions.

Clements, D. H., Sarama, J., Spitler, M. E., Lange, A. A., & Wolfe, C. B. (2011). Mathematics learned by young children in an intervention based on learning trajectories: A large-scale cluster randomized trial. *Journal for Research in Mathematics Education*, *42*(2), 127–166. doi:10.5951/jresematheduc.42.2.0127

Clements, D. H., Sarama, J., Wolfe, C. B., & Spitler, M. E. (2013). Longitudinal evaluation of a scale-up model for teaching mathematics with trajectories and technologies: Persistence of effects in the third year. *American Educational Research Journal*, *50*(4), 812–850. doi:10.3102/0002831212469270

Cohen, E. G., & Lotan, R. A. (2014). *Designing groupwork: strategies for the heterogeneous classroom* (3rd ed.). Teachers College Press.

Conference Board of the Mathematical Sciences (CBMS). (2016, July 15). *Active learning in post-secondary mathematics education* (Meeting Report). Retrieved from http://www.cbmsweb.org/Statements/Active_Learning_Statement.pdf

Fletcher, G. (n.d.). *3 act task file cabinet*. Retrieved from https://gfletchy.com/3-act-lessons/

Freeman, S., Eddy, S. L., McDonough, M., Smith, M. K., Okoroafor, N., Jordt, H., & Wenderoth, M. P. (2014). Active learning increases student performance in science, engineering, and mathematics. *Proceedings of the National Academy of Sciences of the United States of America*, *111*(23), 8410–8415. doi:10.1073/pnas.1319030111 PMID:24821756

Freudenthal, H. (1991). *Revisiting mathematics education: China lectures*. Kluwer Academic Publishers.

Garcia, N., Shaughnessy, M., & Pynes, D. (2021). Recording student thinking in a mathematics discussion. *Mathematics Teacher: Learning & Teaching PK-12, 114*(12), 926-932.

Gravemeijer, K. P. E. (1994). *Developing realistic mathematics education*. Utrecht, The Netherlands: CD-β Press.

Gutiérrez, R. (2012). Context matters: How should we conceptualize equity in mathematics education? In B. Herbel-Eisenmann, J. Choppin, D. Wagner, & D. Pimm (Eds.), *Equity in Discourse for Mathematics Education* (pp. 17–33). Springer. doi:10.1007/978-94-007-2813-4_2

Hiebert, J., & Grouws, D. A. (2007). The effects of classroom mathematics teaching on students' learning. In F. K. Lester (Ed.), *Second handbook of research on mathematics teaching and learning* (pp. 371–404). Information Age Publishing.

Hiebert, J., Morris, A. K., Berk, D., & Jansen, A. (2007). Preparing teachers to learn from teaching. *Journal of Teacher Education*, *58*(1), 47–61. doi:10.1177/0022487106295726

Hill, H. C., Sleep, L., Lewis, J. M., & Ball, D. L. (2007). Assessing teachers' mathematical knowledge: What knowledge matters and what evidence counts? In F. K. Lester (Ed.), *Second handbook of research on mathematics teaching and learning* (pp. 111–155). Information Age Publishing.

Horn, I. S. (2012). *Strength in numbers: Collaborative learning in secondary mathematics*. National Council of Teachers of Mathematics.

Humphreys, C., & Parker, R. (2015). *Making number talks matter: Developing mathematical practices and deepening understanding, Grades 4-10*. Stenhouse Publishers.

Jacobs, V. R., & Philipp, R. A. (2010). Supporting children's problem solving. *Teaching Children Mathematics*, *17*(2), 98–105. doi:10.5951/TCM.17.2.0098

Kalinec-Craig, C. A. (2017). The rights of the learner: A framework for promoting equity through formative assessment in mathematics education. *Democracy & Education*, *25*(2), 1–11.

Kazemi, E., & Hintz, A. (2014). *Intentional talk: How to structure and lead productive mathematical discussions*. Stenhouse Publishers.

Kogan, M., & Laursen, S. L. (2013). Assessing long-term effects of inquiry-based learning: A case study from college mathematics. *Innovative Higher Education*, *39*(3), 183–199. doi:10.100710755-013-9269-9

Lampert, M. (2001). *Teaching problems and the problems of teaching*. Yale University Press.

Larsen, S., & Lockwood, E. (2013). A local instructional theory for the guided reinvention of the quotient group concept. *The Journal of Mathematical Behavior, 32*(4), 726–742. doi:10.1016/j.jmathb.2013.02.010

Larsen, S. P. (2013). A local instructional theory for the guided reinvention of the group and isomorphism concepts. *The Journal of Mathematical Behavior, 32*(4), 712–725. doi:10.1016/j.jmathb.2013.04.006

Laursen, S. L. (2013, November). From innovation to implementation: Multi-institution pedagogical reform in undergraduate mathematics. In D. King, B. Loch, & L. Rylands (Eds.), *Proceedings of Lighthouse Delta 2013: The 9th DELTA conference on the teaching and learning of undergraduate mathematics and statistics* (pp. 102-112). Retrieved from http://www.deltaconference.org/documents/program/Keynote-Monday-Laursen2013.pdf

Laursen, S. L., & Rasmussen, C. (2019). I on the prize: Inquiry approaches in undergraduate mathematics. *International Journal of Research in Undergraduate Mathematics Education, 5*(1), 129–146. doi:10.100740753-019-00085-6

Leatham, K. R., Peterson, B. E., Stockero, S. L., & Van Zoest, L. R. (2015). Conceptualizing mathematically significant pedagogical opportunities to build on student thinking. *Journal for Research in Mathematics Education, 46*(1), 88–124. doi:10.5951/jresematheduc.46.1.0088

Liljedahl, P. (2021). *Building thinking classrooms in mathematics*. Corwin.

Lim, W., Kim, H., Stallings, L., & Son, J. W. (2015). Celebrating diversity by sharing multiple solution methods. *Mathematics Teacher, 109*(4), 290–297. doi:10.5951/mathteacher.109.4.0290

Meyer, D. (n.d.). *Dan Meyer's three-act math tasks*. Google Docs. Retrieved from https://docs.google.com/spreadsheets/d/1jXSt_CoDzyDFeJimZxnhgwOVsWkTQEsfqouLWNNC6Z4/edit#gid=0

Michaels, S., O'Connor, M. C., Hall, M. W., & Resnick, L. B. (2010). Accountable talk® sourcebook. Pittsburg, PA: Institute for Learning University of Pittsburgh.

Mills, V. L. (2014). Mathematical goals: The alpha and omega of effective practice. *NCSM Summer Newsletter, 44*(4), 2–3.

Nabb, K. (2018a). Experiences using inquiry-oriented differential equations. *CODEE: Community of Ordinary Differential Equations Educators Journal, 11*(1), 1-10. Available: https://scholarship.claremont.edu/codee/vol11/iss1/3

Nabb, K. (2018b). Reflecting on the fruits of active learning. *MathAMATYC Educator, 9*(3), 9–11, 52.

Nabb, K. (2020). Active learning in undergraduate mathematics: The five practices and the five dimensions of TRU math. *MathAMATYC Educator, 12*, 30–35, 68–69.

Nabb, K., Hofacker, E. B., Ernie, K. T., & Ahrendt, S. (2018). Using the 5 practices in mathematics teaching. *Mathematics Teacher, 111*(5), 366–373. doi:10.5951/mathteacher.111.5.0366

National Council of Teachers of Mathematics. (2018). *Catalyzing change in high school mathematics: Initiating critical conversations*. National Council of Teachers of Mathematics.

NRICH. (2018). *The Nrich Maths Project*. https://nrich.maths.org/

Parrish, S. (2010). *Number talks: Helping children build mental math and computation strategies, Grades K-5*. Math Solutions.

Parrish, S. D. (2011). Number talks build numerical reasoning. *Teaching Children Mathematics, 18*(3), 198–206. doi:10.5951/teacchilmath.18.3.0198

Rasmussen, C., Keene, K. A., Dunmyre, J., & Fortune, N. (2018). *Inquiry oriented differential equations*. Course materials. Available at https://iode.wordpress.ncsu.edu

Rasmussen, C., & Marrongelle, K. (2006). Pedagogical content tools: Integrating student reasoning and mathematics in instruction. *Journal for Research in Mathematics Education, 37*, 388–420. doi:10.2307/30034860

Rasmussen, C. L., & King, K. D. (2000). Locating starting points in differential equations: A realistic mathematics education approach. *International Journal of Mathematical Education in Science and Technology, 31*(2), 161–172. doi:10.1080/002073900287219

Reinholz, D. L. (2020). Five practices for supporting inquiry in analysis. *PRIMUS: Problems, Resources, and Issues in Mathematics Undergraduate Studies, 30*(1), 19–35. doi:10.1080/10511970.2018.1500955

Rockhurst University. (2018). *IBL calculus: Concepts and applications*. Retrieved from https://www.iblcalculus.com/

Schoenfeld, A. H. (2014). What makes for powerful classrooms, and how can we support teachers in creating them? A story of research and practice, productively intertwined. *Educational Researcher, 43*(8), 404–412. doi:10.3102/0013189X14554450

Schoenfeld, A. H. (2016, April/May). What makes for powerful classrooms—And what can we do, now that we know? *MAA Focus, 36*(2), 11–14.

Shaughnessy, M., Garcia, N. M., O'Neill, M. K., Selling, S. K., Willis, A. T., Wilkes, C. E. II, Salazar, S. B., & Ball, D. L. (2021). Formatively assessing prospective teachers' skills in leading mathematics discussions. *Educational Studies in Mathematics, 108*(3), 451–472. doi:10.100710649-021-10070-z PMID:34966189

Smith, M., Bill, V., & Sherin, M. G. (2020). *The five practices in practice: Successfully orchestrating mathematics discussions in your elementary school classroom*. Corwin.

Smith, M., & Sherin, M. G. (2019). *The five practices in practice: Successfully orchestrating mathematics discussions in your middle school classroom*. Corwin.

Smith, M., Steele, M. D., & Sherin, M. G. (2020). *The five practices in practice: Successfully orchestrating mathematics discussions in your high school classroom*. Corwin.

Smith, M. S., & Stein, M. K. (2011). *Five practices for orchestrating productive mathematics discussions*. National Council of Teachers of Mathematics.

Smith, M. S., & Stein, M. K. (2018). *Five practices for orchestrating productive mathematics discussions* (2nd ed.). National Council of Teachers of Mathematics.

Star, J. R., Jeon, S., Comeford, R., Clark, P., Rittle-Johnson, B., & Durkin, K. (2021). Compare and discuss multiple strategies. *Mathematics Teacher: Learning and Teaching PK-12, 114*(11), 853-859.

Stein, M. K., Engle, R. A., Smith, M. S., & Hughes, E. K. (2008). Orchestrating productive mathematical discussions: Five practices for helping teachers move beyond show and tell. *Mathematical Thinking and Learning, 10*(4), 313–340. doi:10.1080/10986060802229675

Straumanis, A., Bénéteau, C., Guadarrama, Z., Guerra, J. E., & Lenz, L. (2013). *Calculus I: A guided inquiry*. Wiley.

Su, F. (2017). Mathematics for human flourishing. *The American Mathematical Monthly, 124*(6), 483–493. doi:10.4169/amer.math.monthly.124.6.483

Sun, K. L., Baldinger, E. E., & Humphreys, C. (2018). Number talks: Gateway to sense making. *Mathematics Teacher, 112*(1), 48–54. doi:10.5951/mathteacher.112.1.0048

Tang, G., El Turkey, H., Cilli-Turner, E., Savic, M., Karakok, G., & Plaxco, D. (2017). Inquiry as an entry point to equity in the classroom. *International Journal of Mathematical Education in Science and Technology, 48*(sup1), S4-S15.

Vanoli, L., & Luebeck, J. (2021). Examining errors and framing feedback. *Mathematics Teacher: Learning and Teaching PK-12, 114*(8), 616-623.

Wawro, M., Rasmussen, C., Zandieh, M., Sweeney, G. F., & Larson, C. (2012). An inquiry-oriented approach to span and linear independence: The case of the magic carpet ride sequence. *PRIMUS: Problems, Resources, and Issues in Mathematics Undergraduate Studies, 22*(8), 577–599. doi:10.1080/10511970.2012.667516

Wood, M. B., Sheldon, J., Felton-Koestler, M. D., Oslund, J., Parks, A. N., Crespo, S., & Featherstone, H. (2019). 8 teaching moves supporting equitable participation. *Teaching Children Mathematics, 25*(4), 218 223. doi:10.5951/teacchilmath.25.4.0218

ENDNOTES

1 This list is not exhaustive.

2 The latter two courses are typically offered at U.S. community colleges, four-year colleges, and universities as pathways for non-STEM majors to meet their mathematics requirement.

3 In the anticipation stage, I devise full solutions to a variety of different approaches (both correct and incorrect) to prepare myself for monitoring student work the following day. However, due to physical space limitations on the document, I only summarize the general method on the monitoring log (see Anticipate column in Figure 3) so I have a reference point with which to think about the different strategies. It is likely that teachers will develop their own methods for what works best in their classrooms (Smith & Stein, 2018).

4 As is readily evident from the solutions, students were familiar with decimal grids and tape diagrams.

5 Anecdotally, I find it interesting that students are consistently hyper-vigilant when "wrong" mathematics is being presented or discussed, as if to ask, "Why is this wrong?" Any situation that puts my students on high alert to an important piece of mathematics is welcome in my classroom.

6 Since I originally annotated the monitoring log electronically, I was able to change the names of students in my class. The names that appear in this article are pseudonyms.

7 Notice I contemplated this decision earlier in the monitoring stage (see Figure 9, Connect column, bottom cell).

Compilation of References

Abu-Elwan, R. (2016). Mathematics problem posing skills in supporting problem solving skills of prospective teachers. In C. Csikos, A. Rausch, & J. Szitanyi, (Eds.), *Proceedings of the 40th Conference of the International Group for the Psychology of Mathematics Education* (Vol. 2, pp 3-10). PME.

Academic Approach. (2014). *Grade 9 quarter three assessment*. Author.

Academic Approach. (2015). *Grade 10 quarter three assessment*. Author.

Adiredja, A. P. (2019). Anti-deficit narratives: Engaging the politics of research on mathematical sense making. *Journal for Research in Mathematics Education, 50*(4), 401–435. doi:10.5951/jresematheduc.50.4.0401

Aguirre, J., Herbel-Eisenmann, B., Celedon-Pattichis, S., Civil, M., Wilkerson, T., Stephan, M., Pape, S., & Clements, D. H. (2017). Equity within mathematics education research as a political act: Moving from choice to intentional collective professional responsibility. *Journal for Research in Mathematics Education, 48*(2), 124–147. doi:10.5951/jresematheduc.48.2.0124

Ahrendt, S., Monson, D., & Cramer, K. (2021). Promoting discourse: Fractions on number lines. *Mathematics Teacher: Learning and Teaching PK-12, 114*(4), 284-289.

Ainsworth, S. (1999). The functions of multiple representations. *Computers & Education, 33*(2-3), 131–152. doi:10.1016/S0360-1315(99)00029-9

Ainsworth, S. (2006). DeFT: A conceptual framework for considering learning with multiple representations. *Learning and Instruction, 16*(3), 183–198. doi:10.1016/j.learninstruc.2006.03.001

Akinsola, M. K. (2009). Comparison of prospective and practicing teachers' mathematics efficacy beliefs regarding mathematics teaching. In J. Maass & W. Schloglmann (Eds.), *Beliefs and attitudes in mathematics education* (pp. 119–130). Sense Publishers.

Akyeampong, K. (2017). Teacher educators' practice and vision of good teaching in teacher education reform context in Ghana. *Educational Researcher, 46*(4), 194–203. doi:10.3102/0013189X17711907

Akyeampong, K., Lussier, K., Pryor, J., & Westbrook, J. (2013). Improving teaching and learning of basic maths and reading in Africa: Does teacher preparation count? *International Journal of Educational Development, 33*(3), 272–282. doi:10.1016/j.ijedudev.2012.09.006

Aldana, U. S., & Martinez, D. C. (2018). The development of a community of practice for educators working with newcomer, Spanish-speaking students. *Theory into Practice, 57*(2), 137–146. doi:10.1080/00405841.2018.1425813

Alfieri, L., Brooks, P. J., Aldrich, N. J., & Tenenbaum, H. R. (2011). Does discovery-based instruction enhance learning? *Journal of Educational Psychology, 103*(1), 1–18. doi:10.1037/a0021017

Allen, D. (2013). Reconstructing professional learning community as collective creation. *Improving Schools, 16*(3), 191–208. doi:10.1177/1365480213501056

Alper, L., Fendel, S., Fraser, S., & Resek, D. (1998). *Interactive Mathematics Program, Year Two.* Key Curriculum Press.

Althauser, K. L. (2018). The emphasis of inquiry instructional strategies: Impact on preservice teachers' mathematics efficacy. *Journal of Education and Learning, 7*(1), 53-70.

Amador, J., & Earnest, D. (2018). Preservice mathematics teachers' curriculum visualization. In E. Bergqvist, M. Osterholm, C. Granberg, & L. Sumpter, (Eds.), *Proceedings of the 42nd Conference of the International Group for the Psychology of Mathematics Education* (Vol 2, pp. 19-26). PME.

Amato, S. A. (2011). Ratio: A neglected division at schools. In B. Ubuz, (Ed.), *Proceedings of the 35th Conference of the International Group for the Psychology of Mathematics Education* (Vol. 2, pp 25-32). PME.

American Mathematical Association of Two-Year Colleges (AMATYC). (2018). *IMPACT: Improving mathematical prowess and college teaching*. AMATYC.

Anderson, N. (2021). Keep calm and press for reasoning. *Mathematics Teacher: Learning and Teaching PK-12, 114*(8), 591-597.

Anderson, J., & Bobis, J. (2005). Reform-oriented teaching practices: A survey of primary school teachers. In H. L. Chick, & J. L. Vincent, (Eds.), *Proceedings of the 29th Conference of the International Group for the Psychology of Mathematics Education* (Vol. 2, pp. 65–72). PME.

Anderson, R. D., & Helms, J. V. (2001). The ideal of standards and the reality of schools: Needed research. *Journal of Research in Science Teaching*, *38*(1), 1–16. doi:10.1002/1098-2736(200101)38:1<3::AID-TEA2>3.0.CO;2-V

Anderson, S., & Gurnee, A. (2016, March). Home-grown citizens. *Educational Leadership*, *73*(6), 72–75.

An, S. A., & Tillman, D. A. (2015). Music activities as a meaningful context for teaching elementary students mathematics: A quasi-experiment time-series design with a randomly assigned control group. *European Journal of Science and Mathematics Education*, *3*(1), 45–60. doi:10.30935cimath/9420

Anthony, G., & Walshaw, M. (2007). *Effective pedagogy in mathematics/pāngarau: Best evidence synthesis iteration (BES).* Ministry of Education.

Apel, T. (2012). *Differential Planning.* CimeH Institute. https://members.tripod.com/math_human/id1.html

Apel, T. (2022). *The Relational Theory of Mathematics Teaching.* Las Huellas Matemáticas del Éxito, Cimeh Institute. https://www.amazon.com/LAS-HUELLAS-MATEM%C3%81TICAS-DEL-%C3%89XITO-ebook/dp/B09RSFWXM1

Apel, T., & Rave, A. (1979). *Pupil's Workshop: A learning approach to an interdisciplinary unit in environmental education.* National Congress of Education.

Apel, T. (1993). *Opening an account, Tools for Arithmetic teaching using an integrative approach.* Reches Publishers.

Apel, T. (2001). *Mathematics in Preschool.* CimeH Institute, Biniamina.

Ashcraft, M. H. (2002). Math anxiety: Personal, educational, and cognitive consequences. *Current Directions in Psychological Science*, *11*(5), 181–185. doi:10.1111/1467-8721.00196

Association of Mathematics Teacher Educators (AMTE). (2017). *Standards for preparing teachers of mathematics.* amte.net/standards

Association of Mathematics Teacher Educators (AMTE). (2017). *Standards for Preparing Teachers of Mathematics.* Retrieved from amte.net/standards

Audi, R. (2011). *Epistemology: A Contemporary Introduction to the Theory of Knowledge. Taylor & Francis Group.* Routledge. doi:10.4324/9780203846469

Ayalon, M., & Wilkie, K. (2018). Learning to assess: Exploring changes to pre-service teachers' criteria for a quadratics task. In E. Bergqvist, M. Österholm, C. Granberg, & L. Sumpter (Eds.), *Proceedings of the 42nd Conference of the International Group for the Psychology of Mathematics Education* (Vol. 2, pp. 91-98). PME.

Aylon, M., & Hershkowitz, R. (2018). Mathematics teachers' attention to potential classroom situations of argumentation. *The Journal of Mathematical Behavior, 49*, 163–173. doi:10.1016/j.jmathb.2017.11.010

Baki, M., Baki, A., Celik, D., Guler, M., & Sonmez, N. (2016). Improving prospective mathematics teachers' knowledge of student through lesson analysis. In C. Csikos, A. Rausch, & J. Szitanyi, (Eds.), *Proceedings of the 40th Conference of the International Group for the Psychology of Mathematics Education* (Vol. 2, pp 51-58). PME.

Bakker, A., Smit, J., & Wegerif, R. (2015). Scaffolding and dialogic teaching in mathematics education: Introduction and review. *ZDM, 47*(7), 1047–1065. doi:10.100711858-015-0738-8

Bakó, M. (2003). Differing projecting methods in teaching spatial geometry. *Proceedings of the Third European Research in Mathematics Education,* 1–9. http://www.mathematik.tu-dortmund.de/~erme/CERME3/Groups/TG7/TG7_Bako_cerme3.pdf

Ball, D. L. (2017). Uncovering the Special Mathematical Work of Teaching. In *Proceedings of the 13th International Congress on Mathematical Education* (pp. 11-34). Springer Open. 10.1007/978-3-319-62597-3_2

Ball, D. L. (1992). Magical hopes: Manipulatives and the reform of math education. American Educator. *The Professional Journal of the American Federation of Teachers, 16*(2), 14–18.

Ball, D. L. (1993). With an eye on the mathematical horizon: Dilemmas of teaching elementary school mathematics. *The Elementary School Journal, 93*(4), 373–397. doi:10.1086/461730

Ball, D. L. (2011). Foreword. In M. G. Sherin, V. R. Jacobs, & R. A. Philipp (Eds.), *Mathematics teacher noticing: Seeing through teachers' eyes* (pp. xx–xxiv). Routledge.

Ball, D. L., Lubienski, S. T., & Mewborn, D. S. (2001). Research on teaching mathematics: The unsolved problem of teachers' mathematical knowledge. In V. Richardson (Ed.), *Handbook of research on teaching* (4th ed., pp. 433–456). American Educational Research Association.

Ball, D. L., Thames, M. H., & Phelps, G. (2008). Content knowledge for teaching: What makes it special? *Journal of Teacher Education, 59*(5), 389–407. doi:10.1177/0022487108324554

Bandura, A. (1977). Self-efficacy: Toward a unifying theory of behavioral change. *Psychological Review, 84*(2), 191–215. doi:10.1037/0033-295X.84.2.191 PMID:847061

Banilower, E. R., Boyd, S. E., Pasley, J. D., & Weiss, I. R. (2006). *Lessons from a decade of mathematics and science reform: A capstone report for the local systemic change through teacher enhancement initiative.* Horizon Research, Inc.

Banilower, E. R., Boyd, S. E., Pasley, J. D., & Weiss, I. R. (2006). *Lessons from a decade of mathematics and science reform: A capstone report for the local systemic change through teacher enhancement initiative.* Horizon Research Inc.

Banilower, E. R., Smith, P. S., Malzahn, K. A., Plumley, C. L., Gordon, E. M., & Hayes, M. L. (2018). *Report of the 2018 NSSME.* Horizon Research, Inc.

Banilower, E. R., Smith, P. S., Weiss, I. R., Malzahn, K. A., Campbell, K. M., & Weis, A. M. (2013). *Report of the 2012 National Survey of Science and Mathematics Education*. Horizon Research.

Barak, M., & Shakhman, L. (2008). Reform-based science teaching: Teachers' instructional practices and conceptions. *Eurasia Journal of Mathematics, Science and Technology Education, 4*(1), 11–20. doi:10.12973/ejmste/75301

Baranes, R., Perry, M., & Stigler, J. W. (1989). Activation of real-world knowledge in the solution of word problems. *Cognition and Instruction, 6*(4), 287–318. doi:10.12071532690xci0604_1

Barke, H.-D., & Engida, T. (2001). Structural chemistry and spatial ability in different cultures. *Chemistry Education Research and Practice, 2*(3), 227–239. doi:10.1039/B1RP90025K

Barlow, A. T. (2020). Well played. *Mathematics Teacher: Learning and Teaching PK-12, 113*(1), 2-3.

Barlow, A. T., Edwards, C. M., Robichaux-Davis, R., & Sears, R. (2020). Enhancing and transforming virtual instruction. *Mathematics Teacher: Learning and Teaching PK-12, 113*(12), 972-982.

Barmby, P., & Milinkovic, J. (2011). Pre-service teachers' use of visual representations of multiplication. In B. Ubuz, (Ed.), *Proceedings of the 35th Conference of the International Group for the Psychology of Mathematics Education* (Vol. 2, pp 105-112). PME.

Baroody, A. J., & Wilkins, J. L. (1999). The development of informal counting, number, and arithmetic skills and concepts. In J. V. Copley (Ed.), Mathematics in the early years (pp. 48–65). National Council of Teachers of Mathematics & National Association for the Education of Young Children.

Baroody, A. J. (1989). Manipulatives don't come with guarantees. *The Arithmetic Teacher, 37*(2), 4–5. https://eric.ed.gov/?id=EJ405968. doi:10.5951/AT.37.2.0004

Baroody, A. J., Li, X., & Lai, M.-L. (2008). Toddlers' spontaneous attention to number. *Mathematical Thinking and Learning, 10*(3), 240–270. doi:10.1080/10986060802216151

Bartell, T. G. (2020). Reimagining homework to connect to home and community knowledge. *Mathematics Teacher: Learning and Teaching PK-12, 113*(12), 1023-1028.

Bartolini-Bussi, M. G., & Mariotti, M. A. (2002). Semiotic mediation in the mathematics classroom: Artifacts and signs after a Vygotskian perspective. In L. English, M. G. Bartolini-Bussi, R. Jones, & D. Tirosh (Eds.), *Handbook of international research in mathematics education* (2nd ed., pp. 746–783). Lawrence Erlbaum. http://math.unipa.it/~grim/YESS-5/Semiotic%20mediation%20Ch%2028%20Mariotti.pdf

Battista, M. T. (2007). The development of geometric and spatial thinking. In F. K. Lester (Ed.), *Second handbook of research on mathematics teaching and learning* (pp. 843–908). National Council of Teachers of Mathematics.

Bausch, I., Bruder, R., & Prescott, A. (2011). Personal constructs of planning mathematics lessons. In B. Ubuz, (Ed.), *Proceedings of the 35th Conference of the International Group for the Psychology of Mathematics Education* (Vol. 2, pp 113-120). PME.

Bayart, C., Gos, C., Hindelang, C., & Keyling, M. A. (2000). Voir et raisoner: À laconquête de l'espace au collège [See and reason: The survey of space at the college]. *Repères IREM, 33*(1), 51–73. https://publimath.univ-irem.fr/numerisation/PX/IGR14026/IGR14026.pdf

Bayrakdar-Çiftçi, Z., Akgün, L., & Deniz, D. (2013). Teachers' opinions and solution suggestions regarding encountered issues on the ninth-grade mathematics curriculum. *Anadolu University Educational Sciences Institute Journal, 3*(1), 2–21.

Benade, L. (2015). Teaching as inquiry: Well intentioned, but fundamentally flawed. *New Zealand Journal of Educational Studies*, *50*(1), 107–120. doi:10.100740841-015-0005-0

Ben-Haim, D., Lappan, G., & Houang, R. T. (1985). Visualizing rectangular solids made of small cubes: Analyzing and affecting student performance. *Educational Sciences in Mathematics*, *16*(4), 389–409. doi:10.1007/BF00417194

Bennett, C. A., & Morgan, M. J. (2020). Developing norms with silent discussions. *Mathematics Teacher: Learning and Teaching PK-12, 113*(10), 841-844.

Bennett, C. A. (2010). It's hard getting kids to talk about math: Helping new teachers improve mathematical discourse. *Action in Teacher Education*, *32*(3), 79–89. doi:10.1080/01626620.2010.10463561

Bernstein, B. (2011). Elaborated and Restricted Codes: Their Social Origins and Some Consequences. *The Ethnography of Communication, 66*(6), 55–69. https://www.jstor.org/stable/668161

Bernstein, B. (2003). *Class, Codes and Control*. Routledge.

Bethell, G. (2016). *Mathematics education in Sub-Saharan Africa: Status, challenges, and opportunities*. World Bank. https://openknowledge.worldbank.org/handle/10986/25289

Biazak, J. E., Marley, S. C., & Levin, J. R. (2010). Does an activity-based learning strategy improve preschool children's memory for narrative passages? *Early Childhood Research Quarterly*, *25*(4), 515–526. doi:10.1016/j.ecresq.2010.03.006

Bieda, K. (2010). Enacting proof-related tasks in middle school mathematics: Challenges and opportunities. *Journal for Research in Mathematics Education*, *41*(4), 351–382. doi:10.5951/jresematheduc.41.4.0351

Bieda, K. N., Sela, H., & Chazan, D. (2015). "You are learning well my dear": Shifts in novice teachers' talk about teaching during their internship. *Journal of Teacher Education*, *66*(2), 150–169. doi:10.1177/0022487114560645

Bieda, K., Ji, X., Drwencke, J., & Pickard, A. (2014). Reasoning-and-proving opportunities in elementary mathematics textbooks. *International Journal of Educational Research*, *64*, 71–80. doi:10.1016/j.ijer.2013.06.005

Birgin, O., & Acar, H. (2020). The effect of computer-supported collaborative learning using GeoGebra software on 11th-grade students' mathematics achievement in exponential and logarithmic functions. *International Journal of Mathematical Education in Science and Technology*, 1–18. doi:10.1080/0020739X.2020.1788186

Black, P., & Wiliam, D. (1998). Inside the black box: Raising standards through classroom assessment. *Phi Delta Kappan, 80*(2), 139–144, 146–148.

Blanton, M. L., & Kaput, J. J. (2005). Characterizing a classroom practice that promotes algebraic reasoning. *Journal for Research in Mathematics Education*, *36*(5), 412–446. https://mathed.byu.edu/kleatham/Classes/Fall2010/MthEd590Library.enlp/MthEd590Library.Data/PDF/BlantonKaput2005CharacterizingAClassroomPracticeThatPromotesAlgebraicReasoning-1974150144/BlantonKaput2005CharacterizingAClassroomPracticeThatPromotesAlgebraicReasoning.pdf

Blum, W., Artigue, M., Sträßer, R., & Van den Heuvel-Panhuizen, M. (2019). European didactic traditions in mathematics: Introduction and overview. In W. Blum, M. Artigue, M. A. Mariotti, R. Sträßer, & M. Van den Heuvel-Panhuizen (Eds.), *European Traditions in Didactics of Mathematics* (pp. 1–11). Springer. doi:10.1007/978-3-030-05514-1_1

Boaler, J., & Brodie, K. (2004). The importance, nature, and impact of teacher questions. In D. E. McDougall & J. A. Ross (Eds.), *Proceedings of the twenty-sixth annual meeting of the North American Chapter of the International Group for the Psychology of Mathematics Education* (*Vol. 2*, pp. 774-782). Toronto: OISE/University of Toronto.

Boaler, J. (2002). Learning from teaching: Exploring the relationship between reform curriculum and equity. *Journal for Research in Mathematics Education*, *33*(4), 239–258. doi:10.2307/749740

Boaler, J. (2013, March). Ability and mathematics: The mindset revolution that is reshaping education. *Forum, 55*(1), 143–152. doi:10.2304/forum.2013.55.1.143

Boaler, J. (2015). *Mathematical mindsets: Unleashing students' potential through creative math, inspiring messages and innovative teaching*. John Wiley & Sons.

Boaler, J. (2016a). The math-class paradox. *Atlantic (Boston, Mass.)*, (Feb), 4.

Boaler, J. (2016b). Designing mathematics classes to promote equity and engagement. *The Journal of Mathematical Behavior, 100*(41), 172–178. doi:10.1016/j.jmathb.2015.01.002

Board of Studies NSW (BOSNSW). (2002). K-6 *Mathematics Syllabus*. Sydney: BOSNSW.

Bobis, J., Higgins, J., Cavanagh, M., & Roche, A. (2012). Professional knowledge of practising teachers of mathematics. In B. Perry, T. Lowrie, T. Logan, A. MacDonald, & J. Greenlees (Eds.), *Research in mathematics education in Australasia 2008–2011* (pp. 313–341). Sense Publishers. https://link.springer.com/chapter/10.1007/978-94-6091-970-1_15

Bold, T., Filmer, D., Martin, G., Molina, E., Stacy, B., Rockmore, C., Svensson, J., & Wane, W. (2017). Enrollment without learning: Teacher effort, knowledge, and skill in primary schools in Africa. *The Journal of Economic Perspectives, 31*(4), 185–204. doi:10.1257/jep.31.4.185

Bolt, B. (1991). *Mathematics meets technology*. Cambridge University Press.

Bolton, L. (2019). *Foundational mathematics education in developing countries*. Institute of Development Studies. https://resourcecentre.savethechildren.net/node/16285/pdf/657_foundational_mathematics_education_in_developing_countries.pdf

Booth, J. L., & Koedinger, K. R. (2008). Key misconceptions in algebraic problem solving. In *Proceedings of the 30th annual conference of the Cognitive Science Society* (pp. 571-576). Academic Press.

Booth, J. L., Barbieri, C., Eyer, F., & Pare-Blagoev, E. J. (2014). Persistent and pernicious errors in algebraic problem solving. *The Journal of Problem Solving, 7*(1), 10–23. doi:10.7771/1932-6246.1161

Borko, H., Davinroy, K. H., Bliem, C. L., & Cumbo, K. B. (2000). Exploring and supporting teacher change: Two third-grade teachers' experiences in a mathematics and literacy staff development project. *The Elementary School Journal, 100*(4), 273–306. doi:10.1086/499643

Borko, H., & Mayfield, V. (1995). The roles of the clinical teacher and university supervisor in learning to teach. *Teaching and Teacher Education, 11*(5), 501–518. doi:10.1016/0742-051X(95)00008-8

Borman, K. M., Mueninghoff, E., Cotner, B. A., & Frederick, P. B. (2009). Teacher preparation programs. In *International handbook of research on teachers and teaching* (pp. 123–140). Springer. doi:10.1007/978-0-387-73317-3_8

Bostic, J. D., Vostal, B., & Folger, T. (2021). Growing TTULPs through your lessons. *Mathematics Teacher: Learning and Teaching PK-12, 114*(7), 498-507.

Boston, M., Candela, A. G., & Dixon, J. K. (2019). *Making sense of mathematics for teaching to inform instructional quality*. Solution Tree Press.

Bottge, B. (2001). Using intriguing problems to improve math skills. *Educational Leadership, 58*(6), 68–72.

Bowen, G. A. (2009). Document analysis as a qualitative research method. *Qualitative Research Journal, 9*(2), 27–40. doi:10.3316/QRJ0902027

Brakoniecki, A., Amador, J. M., & Glassmeyer, D. M. (2021). One task, multiple proportional reasoning strategies. *Mathematics Teacher: Learning and Teaching PK-12, 114*(1), 33-40.

Brand, S., Jung, H., Dorlack, A., & Gailliot, S. (2021). Launching a mathematical modeling lesson. *Mathematics Teacher: Learning and Teaching PK-12, 114*(4), 298-305.

Braun, B., Bremser, P., Duval, A. M., Lockwood, E., & White, D. (2017). What does active learning mean for mathematicians? *Notices of the American Mathematical Society, 64*(02), 124–129. doi:10.1090/noti1472

Bressoud, D. (2015). Insights from the MAA national study of college calculus. *Mathematics Teacher, 109*(3), 179–185. doi:10.5951/mathteacher.109.3.0178

Briley, J. S. (2012). The relationships among mathematics teaching efficacy, mathematics self-efficacy, and mathematical beliefs for elementary pre-service teachers. *Issues in the Undergraduate Mathematics Preparation of School Teachers, 5*, 1–13.

Brooks, J., & Brooks, M. (1993). *In search of understanding: The case for constructivist classrooms.* Association for Supervision and Curriculum Development.

Brown, J. P. (2021). Understanding the real world: Sweaty jo. *Mathematics Teacher: Learning and Teaching PK-12, 114*(7), 528-535.

Broza, O., & Lifshit, A. (2021). The influence of analytic model on critical reflective thought of pre-service mathematics teachers for elementary school. In M. Inprasitha, N. Changsri, & N. Boonsena (Eds.). *Proceedings of the 44th Conference of the International Group for the Psychology of Mathematics Education*, (Vol. 2, pp. 128-136). PME.

Bruce, C. D., & Ross, J. A. (2008). A model for increasing reform implementation and teacher efficacy: Teacher peer coaching in grades 3 and 6 mathematics. *Canadian Journal of Education/Revue canadienne de l'éducation, 31*(2), 346-370.

Bruce, C. D., Esmonde, I., Ross, J., Dookie, L., & Beatty, R. (2010). The effects of sustained classroom-embedded teacher professional learning on teacher efficacy and related student achievement. *Teaching and Teacher Education, 26*(8), 1598–1608. doi:10.1016/j.tate.2010.06.011

Bruner, J. S. (1964). The course of cognitive growth. *The American Psychologist, 19*(1), 1–15. doi:10.1037/h0044160

Bruner, J. S. (2017). *A study of thinking.* Routledge., doi:10.4324/9781315083223

Buforn, A., Fernandez, C., Coles, A., & Brown, L. (2015). The meaning of ratio: Prospective mathematics teachers' knowledge about the teaching and learning of proportional reasoning. In K. Beswick, T. Muir, & J. Fielding-Wells (Eds.). *Proceedings of 39th Conference of the International Group for the Psychology of Mathematics Education*, (Vol. 2, pp. 129-136). PME.

Buforn, A., Fernandez, C., & Llinares, S. (2018). Decision-making in noticing students' proportional reasoning. In E. Bergqvist, M. Österholm, C. Granberg, & L. Sumpter (Eds.), *Proceedings of the 42nd Conference of the International Group for the Psychology of Mathematics Education* (Vol. 2, pp. 187-194). PME.

Buforn, A., Fernandez, C., Llinares, S., & Badillo, E. (2017). Pre-service primary teachers' profiles of noticing students' proportional reasoning. In B. Kaur, W.K. Ho, T.L. Toh, & B.H. Choy (Eds.), *Proceedings of the 41st Conference of the International Group for the Psychology of Mathematics Education* (Vol. 2, pp. 193-200). PME.

Burns, M. (2001). *Teaching arithmetic: Lessons for introducing fractions, grades 4-5.* Math Solutions.

Bussi, M. G. B., & Funghi, S. (2019). Lesson study in primary pre-service teachers' education: Influences on beliefs about lesson planning and conduction. In M. Graven, H. Venkat, A. Eddien, & P. Vale, (Eds.), *Proceedings of the 43rd Conference of the International Group for the Psychology of Mathematics Education* (Vol 2, pp. 81-88). PME.

Cain-Caston, M. (1996). Manipulative queen. *Journal of Instructional Psychology*, *23*(4), 270.

Caine, G., Caine, R., & Crowell, S. (1994). *Mindshifts: A brain-based process for restructuring schools and renewing education*. Zephyr Press.

Candela, A. G., Boston, M. D., & Dixon, J. K. (2020). Discourse actions to promote student access. *Mathematics Teacher: Learning and Teaching PK-12, 113*(4), 266-277.

Canny, M. E. (1984). The relationship of manipulative materials to achievement in three areas of fourth-grade mathematics: Computation, concept development and problem-solving. *Dissertation Abstracts International*, *45*, 775–776. http://hdl.handle.net/10919/106313

Carbonneau, K. J., Marley, S. C., & Selig, J. P. (2013). A meta-analysis of the efficacy of teaching mathematics with concrete manipulatives. *Journal of Educational Psychology*, *105*(2), 380–400. doi:10.1037/a0031084

Carpenter, T. P. (1979). *Cognitive Development Research and Mathematics Education*. Theoretical Paper No. 73. Conceptual Paper No. 2.Studies in Mathematics Series.

Carpenter, T. P., & Fennema, E. (1992). Cognitively guided instruction: Building on the knowledge of students and teachers. *International Journal of Educational Research*, *17*(5), 457–470. doi:10.1016/S0883-0355(05)80005-9

Carpenter, T. P., Fennema, E., Franke, M. L., Levi, L., & Empson, S. B. (2014). *Children's mathematics: Cognitively guided instruction* (2nd ed.). Heinemann.

Carpenter, T. P., Fennema, E., Peterson, P. L., Chiang, C. P., & Loef, M. (1989). Using knowledge of children's mathematics thinking in classroom teaching: An experimental study. *American Educational Research Journal*, *26*(4), 499–531. doi:10.3102/00028312026004499

Carpenter, T. P., & Moser, J. M. (1984). The acquisition of addition and subtraction concepts in grades one through three. *Journal for Research in Mathematics Education*, *15*(3), 179–202. doi:10.2307/748348

Carpenter, T., Fennema, E., & Franke, M. L. (1996). Cognitively guided instruction: A knowledge base for reform in primary mathematics instruction. *The Elementary School Journal*, *97*(1), 3–20. doi:10.1086/461846

Carreira, S., & Jacinto, H. (2019). A model of mathematical problem solving with technology: The case of Marco solving-and-expressing two geometry problems. In *Mathematical problem solving* (pp. 41–62). Springer. doi:10.1007/978-3-030-10472-6_3

Carroll, W. M. (1996). Use of invented algorithms by second graders in a reform mathematics curriculum. *The Journal of Mathematical Behavior*, *15*(2), 137–150. doi:10.1016/S0732-3123(96)90011-5

Carroll, W. M., & Porter, D. (1997). Invented strategies can develop meaningful mathematical procedures. *Teaching Children Mathematics*, *3*(7), 370–374. doi:10.5951/TCM.3.7.0370

Celedón-Pattichis, S., Borden, L. L., Pape, S. J., Clements, D. H., Peters, S. A., Males, J. R., Chapman, O., & Leonard, J. (2018). Asset-based approaches to equitable mathematics education research and practice. *Journal for Research in Mathematics Education*, *49*(4), 373–389. doi:10.5951/jresematheduc.49.4.0373

Chamberlin, M. (2009). Teachers' reflections on their mathematical learning experiences in a professional development course. *Mathematics Teacher Education and Development*, *11*, 22–35.

Chamberlin, M. (2013). Prospective teachers' perspectives on mathematics teaching and learning: Lens for interpreting experiences in a standards-based mathematics course. *School Science and Mathematics, 113*(8), 369–379. doi:10.1111sm.12042

Chamberlin, M. T. (2005). Teachers' discussions of students' thinking: Meeting the challenge of attending to students' thinking. *Journal of Mathematics Teacher Education, 8*(2), 141–170. doi:10.100710857-005-4770-4

Chamoso, J. M., Cáceres, M. J., & Azcárate, P. (2012). Reflection on the teaching-learning process in the initial training of teachers. Characterization of the issues on which pre-service mathematics teachers reflect. *Teaching and Teacher Education, 28*(2), 154–164. doi:10.1016/j.tate.2011.08.003

Chandler, C. C., & Kamii, C. (2009). Giving change when payment is made with a dime: The difficulty of tens and ones. *Journal for Research in Mathematics Education, 40*(2), 97–118. doi:10.2307/40539328

Chao, T., Murray, E., & Gutiérrez, R. (2014). *What are classroom practices that support equity based mathematics teaching? A research brief.* National Council of Teachers of Mathematics.

Chapin, S. H., & O'Connor, C. (2007). Academically productive talk: Supporting students' learning in mathematics. In W. G. Martin & M. E. Strutchens (Eds.), *The learning of mathematics* (pp. 113–128). National Council of Teachers of Mathematics.

Chapin, S. H., O'Connor, C., & Anderson, N. C. (2003). *Classroom discussions: Using math talk to help students learn, Grades K-6.* Math Solutions.

Chapin, S. H., O'Connor, C., & Anderson, N. C. (2009). *Classroom discussions using math talk to help student learn.* Math Solutions.

Chapin, S. H., O'Connor, C., & Anderson, N. C. (2013). *Classroom discussions in math: A teacher's guide for using talk moves to support the common core and more, Grades K-6.* Math Solutions.

Chapman, O. (2013). Facilitating prospective secondary mathematics teachers' learning of problem solving. In A. M. Lindmeier & A. Heinze, (Eds.), *Proceedings of the 37th Conference of the International Group for the Psychology of Mathematics Education* (*Vol. 2*, pp. 153-160). Academic Press.

Chapman, O. (2004). Facilitating peer interactions in learning mathematics: Teachers' practical knowledge. *Proceedings of the 28th Conference of the International Group for the Psychology of Mathematics Education* (Vol. 2, pp. 191-198). PME.

Chapman, O. (2011). Prospective teachers' ways of making sense of mathematical problem posing. In B. Ubuz, (Ed.), *Proceedings of the 35th Conference of the International Group for the Psychology of Mathematics Education* (Vol. 2, pp 209-216). PME.

Chapman, O. (2012). Challenges in mathematics teacher education. *Journal of Mathematics Teacher Education, 15*(4), 263–270. doi:10.100710857-012-9223-2

Chapman, O. (2015). Reflective awareness in mathematics teachers' learning and teaching. *Eurasia Journal of Mathematics, Science and Technology Education, 11*(2), 313–324. doi:10.12973/eurasia.2015.1334a

Charles, K., & Nason, R. (2000). Young children's partitioning strategies. *Educational Studies in Mathematics, 43*(2), 191–221. doi:10.1023/A:1017513716026

Cheng, D., Feldman, Z., & Chapin, S. (2012). Mathematical discussions in preservice elementary courses. In T. Tso, (Ed.), *Proceedings of the 36th Conference of the International Group for the Psychology of Mathematics Education* (Vol. 2, pp 139-146). PME.

Chi, M. T. H. (2009). Active-constructive-interactive: A conceptual framework for differentiating learning activities. *Topics in Cognitive Science, 1*(1), 73–105. doi:10.1111/j.1756-8765.2008.01005.x PMID:25164801

Chissick, N. (2002). Factors affecting the implementation of reforms in school mathematics. In A. D. Cockburn, & E. Nardi, (Eds.), *Proceedings of the 26th Conference of the International Group for the Psychology of Mathematics Education* (Vol. 2, pp. 249-256). PME.

Christou, C., Eliophotou-Menon, M., & Philippou, G. (2004). Teachers' concerns regarding the adoption of a new mathematics curriculum: An application of CBAM. *Educational Studies in Mathematics, 57*(2), 157–176. doi:10.1023/B:EDUC.0000049271.01649.dd

Christou, C., Jones, K., Mousoulides, N., & Pittalis, M. (2006). Developing the 3DMath dynamic geometry software: Theoretical perspectives on design. *The International Journal for Technology in Mathematics Education, 13*, 168–174. https://www.learntechlib.org/p/107486/

Christou, C., Philippou, G., Pitta-Pantazi, D., & Menn-Eliophotou, M. (2002). The effect of efficacy on teachers' concerns with regard to the implementation of a new mathematics curriculum. In A. D. Cockburn, & E. Nardi, (Eds.), *Proceedings of the 26th Conference of the International Group for the Psychology of Mathematics Education* (Vol. 2, pp. 257-264). PME.

Ciftci, O., & Tatar, E. (2015). Teacher opinions about the updated secondary mathematics curriculum. *Turkish Journal of Computer and Mathematics Education, 6*(2), 285–298. doi:10.16949/turcomat.15375

Clements, D. H. (2003). Teaching and learning geometry. In J. Kilpatrick, W. G. Martin, & D. Schifter (Eds.), *A research companion to principles and standards for school mathematics* (pp. 151–178). National Council of Teachers of Mathematics.

Clements, D. H., & Battista, M. T. (1990). Constructive learning and teaching. *The Arithmetic Teacher, 38*(1), 34–35. doi:10.5951/AT.38.1.0034

Clements, D. H., & Battista, M. T. (1992). Geometry and spatial reasoning. In D. A. Grouws (Ed.), *Handbook of research on mathematics teaching and learning* (1st ed., pp. 420–464). Macmillan.

Clements, D. H., & Sarama, J. (2014). *Learning and teaching early math: The learning trajectories approach* (2nd ed.). Routlege. doi:10.4324/9780203520574

Clements, D. H., Sarama, J., Spitler, M. E., Lange, A. A., & Wolfe, C. B. (2011). Mathematics learned by young children in an intervention based on learning trajectories: A large-scale cluster randomized trial. *Journal for Research in Mathematics Education, 42*(2), 127–166. doi:10.5951/jresematheduc.42.2.0127

Clements, D. H., Sarama, J., Wolfe, C. B., & Spitler, M. E. (2013). Longitudinal evaluation of a scale-up model for teaching mathematics with trajectories and technologies: Persistence of effects in the third year. *American Educational Research Journal, 50*(4), 812–850. doi:10.3102/0002831212469270

Cobb, P., & Yackel, E. (1995). Constructivist, emergent, and socio-cultural perspective in the context of development research. In *Proceedings of the Annual Conference of North American Chapter of the International Group for Psychology of Mathematics Education.* Annual Conference of North American Chapter of the International Group for Psychology of Mathematics Education. https://eric.ed.gov/?id=ED389535

Cobb, P. (1994). An exchange: Constructivism in mathematics and science education. *Educational Researcher, 23*(7), 4. doi:10.2307/1176932

Cobb, P., & Yackel, E. (1996). Constructivist, emergent, and sociocultural perspectives in the context of developmental research. *Educational Psychologist, 31*(3-4), 175–190. doi:10.1080/00461520.1996.9653265

Cochran-Smith, M., & Lytle, S. L. (2009). *Inquiry as stance: Practitioner research for the next generation*. Teachers College Press.

Coffey, J. E., & Edwards, A. R. (2015). The role subject matter plays in prospective teachers' responsive teaching practices in elementary math and science. *Responsive Teaching in Science and Mathematics*, 163-179.

Cohen, D., & Ball, D. (1990). Relations between policy and practice: A commentary. *Educational Evaluation and Policy Analysis, 12*(3), 331–338. doi:10.3102/01623737012003331

Cohen, E. G., & Lotan, R. A. (2014). *Designing groupwork: strategies for the heterogeneous classroom* (3rd ed.). Teachers College Press.

Coleman, L. K. (2020). Build a bridge: Provide access to grade-level content for all students. *Mathematics Teacher: Learning and Teaching PK-12, 113*(7), 590-593.

Common Core State Standards Initiative (CCSSI). (2010). *Common core state standards*. National Governors Association for Best Practices and Council of Chief State School Officers.

Common Core State Standards Initiative. (2010). *Common Core State Standards for Mathematics*. Council of Chief State School Officers & National Governors Association Center for Best Practices.

Common Core State Standards Initiative. (2010). *Common core state standards for mathematics*. http://www.corestandards.org/assets/CCSSI_Math%20Standards.pdf

Common Core State Standards Initiative. (2010). *National governors association center for best practices*. Council of Chief State School Officers.

Conference Board of the Mathematical Sciences (CBMS). (2001). *The mathematical education of teachers*. Washington, DC: Mathematical Association of America (MAA) in cooperation with American Mathematical Society (AMS).

Conference Board of the Mathematical Sciences (CBMS). (2016, July 15). *Active learning in post-secondary mathematics education* (Meeting Report). Retrieved from http://www.cbmsweb.org/Statements/Active_Learning_Statement.pdf

Confrey, J., & Krupa, E. (2010). *Curriculum design, development, and implementation in an era of common core state standards*. Summary Report of a Conference. Center for the Study of Mathematics Curriculum.

Confrey, J., Maloney, A., Nguyen, K., Mojica, G., & Myers, M. (2009). Equipartitioning/splitting as a foundation of rational number reasoning using learning trajectories. Presented at the *33rd Conference of the International Group for the Psychology of Mathematics Education*, Thessaloniki, Greece.

Cooke, A., & Bruns, J. (2018). Early childhood educators' issues and perspectives in mathematics education. In I. Elia, J. Mulligan, A. Anderson, A. Baccaglini-Frank, & C Benz (Eds.), Contemporary research and perspectives on early childhood mathematics education (pp. 267-289). Springer.

Cooper, B., & Harries, T. (2009). Realistic contexts, mathematics assessment, and social class: Lessons for assessment policy from an English research programme. In Words and worlds (pp. 93–110). doi:10.1163/9789087909383_007

Cooper, M., & Sweller, J. (1989). School students' representations of solids. *Journal for Research in Mathematics Education, 20*(2), 202–212. doi:10.2307/749283

Corbin, J., & Strauss, A. (2014). *Basics of qualitative research: Techniques and procedures for developing grounded theory* (4th ed.). Sage publications.

Cornell, C. (1999). I hate math! I couldn't learn it, and I can't teach it! *Childhood Education, 75*(4), 225–230. doi:10.1080/00094056.1999.10522022

Council of Chief State School Officers (CCSSO). (2010). *Common Core State Standards for Mathematics.* Washington, DC: CCSSO.

Courtney, S. A., Eliustaoglu, E., & Crawford, A. (2015). Examining the role lesson plans play in mathematics education. In T. G. Bartell, K. N. Bieda, R. T. Putnam, K. Bradfield, & H. Dominguez (Eds.), *Proceedings of the 37th Annual Meeting of the North American Chapter of the International Group for the Psychology of Mathematics Education* (pp. 632-639). East Lansing, MI: Michigan State University.

Covell, K., & Howe, R. B. (2001). Moral education through the 3 Rs: Rights, respect and responsibility. *Journal of Moral Education, 30*(1), 29–41. doi:10.1080/03057240120033794

Creswell, J. W. (2013). *Qualitative inquiry and research design: Choosing among five approaches* (3rd ed.). SAGE Publications.

Cropley, D. H. (2015). *Teaching engineers to think creatively: Barriers and challenges in STEM disciplines. In Handbook of Research on Teaching Thinking. Routledge.*

Csikszentmihalyi, M. (2009). The promise of positive psychology. *Psihologijske Teme, 18*(2), 203–211. https://www.researchgate.net/publication/43170663_The_Promise_of_Positive_Psychology

Csikszentmihalyi, M. (2014). *CREATIVITY Flow and the psychology of discovery and invention.* Harper Perennial.

Cunningham, E. P. (2021). Infusing the day with mathematical moments. *Mathematics Teacher: Learning and Teaching PK-12, 114(*1), 70-73.

Curtis, D. (2002). The power of projects. *Educational Leadership, 60*(1), 50–53.

Darling-Hammond, L. (2000). Teacher quality and student achievement. *Education Policy Analysis Archives, 8*, 1. doi:10.14507/epaa.v8n1.2000

Davis, J. D. (2021). Constructing and adapting technology-infused lessons. *Mathematics Teacher: Learning and Teaching PK-12, 114*(4), 290-297.

Davis, J. (2014). Reasoning and proving within Ireland's reform-oriented national syllabi. *The Mathematics Enthusiast, 11*(3), 665–706. doi:10.54870/1551-3440.1321

De Lange, J. (1996). Using and applying mathematics in education. In A. J. Bishop, K. Clements, C. Keitel, J. Kilpatrick, & C. Laborde (Eds.), *International handbook of mathematics education: Part 1* (pp. 49–97). Springer Netherlands. doi:10.1007/978-94-009-1465-0_4

Dean, P. G. (2019). *Teaching and Learning Mathematics*. Routledge. doi:10.4324/9781315788753

DeLeon, A. (2003). *A curriculum of nonroutine problems in the middle school* [Unpublished masters project]. California State University.

Deleuze, G., & Guattari, F. (2006). *A Thousand Plateaus: Capitalism and Schizophrenia*. University of Minnesota Press. https://libcom.org/files/A%20Thousand%20Plateaus.pdf

Denzin, N. K., & Lincoln, Y. S. (2005). Introduction: The discipline and practice of qualitative research. In N. K. Denzin & Y. S. Lincoln (Eds.), *The SAGE handbook of qualitative research* (3rd ed., pp. 1–32). Sage.

Department for Education. (2013). *National curriculum in England. Mathematics programmes of study* (Publication No. DFE-00180-2013, Stages 1 and 2). https://www.gov.uk/government/publications/national-curriculum-in-england-mathematics-programmes-of-study/national-curriculum-in-england-mathematics-programmes-of-study

Design-based Research Collective. (2003). Design-based research: An emerging paradigm for educational inquiry. *Educational Researcher*, *32*(1), 5–8. doi:10.3102/0013189X032001005

DeVinne, P. B. (Ed.). (1985). The American heritage dictionary, Second college edition. Houghton Mifflin Co.

Dewey, J. (1933). *How we think: A restatement of the relation of reflective thinking to the educative process* (Vol. 8). DC Heath and Company.

Dickinson, P., & Hough, S. (2012). Using realistic mathematics education in UK classrooms. In P. Nickolson (Ed.), *Realistic mathematics education impact booklet.* https://www.hoddereducation.co.uk/media/Documents/Maths/Using-Realistic-Maths-Education-in-UK-classrooms.pdf

Dickinson, P., & Eade, F. (2005). Trialling realistic mathematics education (RME) in English secondary schools. *Proceedings of the British Society for Research into Learning Mathematics*, *25*(3), 1–14. http://www.bsrlm.org.uk/wp-content/uploads/2016/02/BSRLM-IP-25-3-1.pdf

Dickinson, P., Eade, F., Gough, S., & Hough, S. (2010). Using realistic mathematics education with low and high attaining pupils in secondary schools. In M. Joubert & P. Andrews (Eds.), *Proceedings of the British Congress for Mathematics Education* (pp. 73–80). https://www.researchgate.net/publication/266487229_Using_Realistic_Mathematics_Education_with_low_to_middle_attaining_pupils_in_secondary_schools

Dillon, M. R., Kannan, H., Dean, J. T., Spelke, E. S., & Duflo, E. (2017). Cognitive science in the field: A preschool intervention durably enhances intuitive but not formal mathematics. *Science*, *357*(6346), 47–55. doi:10.1126cience.aal4724 PMID:28684518

Dobie, T., & Anderson, E. R. (2021). Noticing and wondering to guide professional conversations. *Mathematics Teacher: Learning and Teaching PK-12, 114*(2), 94-102. doi:10.5951/MTLT.2020.0210

Dolev, S., & Even, R. (2015). Justifications and explanations in Israeli 7th -grade mathematics textbooks. *International Journal of Science and Mathematics Education*, *13*(S2), 309–327. doi:10.100710763-013-9488-7

Doruk, B. K. (2014). The educational approaches of Turkish pre-service elementary mathematics teachers in their first teaching practices: Traditional or constructivist? *The Australian Journal of Teacher Education*, *39*(10), 113–134. doi:10.14221/ajte.2014v39n10.8

Dowker, A., Sarkar, A., & Looi, C. Y. (2016). Mathematics anxiety: What have we learned in 60 years? *Frontiers in Psychology*, *7*(508), 1–16. doi:10.3389/fpsyg.2016.00508 PMID:27199789

Drake, C. (2006). Turning points: Using teachers' mathematics life stories to understand the implementation of mathematics education reform. *Journal of Mathematics Teacher Education*, *9*(6), 579–608. doi:10.1007/s10857-006-9021-9 doi:10.100710857-006-9021-9

Drake, C., & Sherin, M. G. (2006). Practicing change: Curriculum adaptation and teacher narrative in the context of mathematics education reform. *Curriculum Inquiry*, *36*(2), 153–187. doi:10.1111/j.1467-873X.2006.00351.x

Drake, C., Spillane, J. P., & Hufferd-Ackles, K. (2001). Storied identities: Teacher learning and subject matter context. *Journal of Curriculum Studies*, *33*(1), 1–23. doi:10.1080/00220270119765

Dreher, A., Kuntze, S., & Lerman, S. (2012). Pre-service teachers' views on using multiple representations in mathematics classrooms - an inter-cultural study. In T. Tso, (Ed.), *Proceedings of the 36th conference of the international group for the psychology of mathematics education* (Vol. 2, pp 211-218). PME.

Driscoll, K. (2015). Improving the effectiveness of mathematics teaching through active reflection. In M. Marshman, V. Geiger, & A. Bennison (Eds.), *Mathematics education in the margins. Proceedings of the 38th annual conference of the Mathematics Education Research Group of Australasia* (pp. 205-212). Sunshine Coast: MERGA.

Driscoll, M. J. (1983). *Research within reach: Secondary school mathematics.* National Council of Teachers of Mathematics & CEMREL, Inc. https://eric.ed.gov/?id=ED225842

Durkin, K., & Rittle-Johnson, B. (2012). The effectiveness of using incorrect examples to support learning about decimal magnitude. *Learning and Instruction*, *22*(1), 206–214. doi:10.1016/j.learninstruc.2011.11.001

Duval, R. (1998). Geometry from a cognitive point of view. In *Reasoning in geometry* (pp. 37–52). https://ci.nii.ac.jp/naid/10020806198/

Eade, F., & Dickinson, P. (2006). Exploring realistic mathematics education in English schools. In J. Novotná, H. Moraová, M. Krátká, & N. Stehlíková (Eds.), *Proceedings of the 30th conference of the international group for the psychology of mathematics education* (*Vol. 3*, pp. 1–8). Academic Press.

Easterby-Smith, M. (1997). Disciplines of Organizational Learning: Contributions and Critiques. *Human Relations, 50*(9), 1085–1113. https://www.researchgate.net/publication/241002680_Disciplines_of_Organizational_Learning_Contributions_and_Critiques/link/56a7f9f908ae0fd8b3fe44a0/download

Ebby, C. B. (2000). Learning to teach mathematics differently: The interaction between coursework and fieldwork for preservice teachers. *Journal of Mathematics Teacher Education*, *3*(1), 69–97. doi:10.1023/A:1009969527157

Eden, R. (2017). *Co-teaching to support pedagogical shifts to raise achievement in mathematics—Teacher-Led Innovation Fund Final Project Report* (Final). Ministry of Education. https://www.researchgate.net/publication/349026301_Co-teaching_to_support_pedagogical_shifts_to_raise_achievement_in_mathematics_-_Teacher-Led_Innovation_Fund_Final_Project_Report

Eden, R. (2019). *Restructuring mathematics teaching | learning: Co-teaching as a designed approach to teachers' collaborative inquiry* [Unpublished doctoral dissertation]. Victoria University of Wellington, Wellington, New Zealand.

Education Council. (2017). *Our code our standards: Code of professional responsibility and standards for the teaching profession.* Education Council, New Zealand. https://teachingcouncil.nz/professional-practice/our-code-our-standards/

Education Development Center. (2017). *Literacy, language, and learning initiative (L3).* Education Development Center. http://idd.edc.org/ projects/literacy-language-and-learning-l3-rwanda

Education Review Office. (2016). *School evaluation indicators.* New Zealand Government. https://ero.govt.nz/how-ero-reviews/schoolskura-english-medium/school-evaluation-indicators

Elipane, L. E. (2011). Incorporating lesson study in pre-service mathematics teacher education. In B. Ubuz, (Ed.), *Proceedings of the 35th conference of the international group for the psychology of mathematics education* (Vol. 2, pp 305-312). PME.

Elipane, L. E. (2012). Infrastructures within the student teaching practicum that nurture elements of lesson study. In T. Tso, (Ed.), *Proceedings of the 36th conference of the international group for the psychology of mathematics education* (Vol. 2, pp 219-226). PME.

Empson, S. B. (1995). Using sharing situations to help children learn fractions. *Teaching Children Mathematics*, *2*(2), 110–115. doi:10.5951/TCM.2.2.0110

Empson, S. B. (1999). Equal sharing and shared meaning: The development of fraction concepts in a first-grade classroom. *Cognition and Instruction*, *17*(3), 283–342. doi:10.1207/S1532690XCI1703_3

Empson, S. B., & Levi, L. A. (2011). *Extending children's mathematics: Fractions and decimals.* Heinemann.

Engelkamp, J., Zimmer, H. D., Mohr, G., & Sellen, O. (1994). Memory of self-performed tasks: Self-performing during recognition. *Memory & Cognition, 22*(1), 34–39. doi:10.3758/BF03202759 PMID:8035683

Engle, R. A., & Conant, F. R. (2002). Guiding principles for fostering productive disciplinary engagement: Explaining an emergent argument in a community of learners classroom. *Cognition and Instruction, 20*(4), 399–483. doi:10.1207/S1532690XCI2004_1

English, L. D., Ärlebäck, J. B., & Mousoulides, N. (2016). Reflections on progress in mathematical modelling research. In *The second handbook of research on the psychology of mathematics education* (pp. 383–413). Brill Sense. doi:10.1007/978-94-6300-561-6_11

Esen, B., & Saralar-Aras, I. (2021). Designing lesson plans for sustainable development focused education in the context of teaching polygons to middle school students. *Proceedings of the Second International Academician Studies Congress*, 1–8.

Evans, N., & Acquaye, E. (2019, March 25-29). *Problematizing the familiar in Ghana: Can an in-service model based on authentic problem-solving help teachers construct deeper understandings of new mathematics instructional models?* [Paper presentation]. Comparative and International Education Society 2018 Annual Conference, Mexico City, Mexico.

Evans, S., & Swan, M. (2014). Developing students' strategies for problem-solving in mathematics: The role of pre-designed "sample student work". *Educational Designer, 2*(7), 1–31. https://www.educationaldesigner.org/ed/volume2/issue7/article25/pdf/ed_2_7_evans_swan_14.pdf

Farnsworth, V., Kleanthous, I., & Wenger-Trayner, E. (2016). Communities of practice as a social theory of learning: A conversation with Etienne Wenger. *British Journal of Educational Studies, 64*(2), 139–160. doi:10.1080/00071005.2015.1133799

Fernández, C., Llinares, S., & Valls, J. (2011). Development of prospective mathematics teachers' professional noticing in a specific domain: Proportional reasoning. In *Proceedings of the 35th Conference of the International Group for the Psychology of Mathematics Education* (Vol. 2, pp. 329-336). PME.

Fives, H., Lacatena, N., & Gerard, L. (2015). Teachers' beliefs about teaching (And learning). In International Handbook of Research on Teachers' Beliefs (pp. 249-265). Taylor and Francis.

Fletcher, G. (n.d.). *3 act task file cabinet.* Retrieved from https://gfletchy.com/3-act-lessons/

Flores, M. M., Hinton, V. M., & Meyer, J. M. (2020). Teaching fraction concepts using the concrete-representational-abstract sequence. *Remedial and Special Education, 41*(3), 165–175. doi:10.1177/0741932518795477

Fosnot, C. T. (2007). *Field Trips and Fund-raisers: Introducing Fractions.* Heinemann.

Frank, K. (2021). The structure of the quadratic formula. *Mathematics Teacher: Learning and Teaching PK-12, 114*(5), 395-398.

Franke, M. L., Kazemi, E., & Battey, D. (2007). Mathematics teaching and classroom practice. In F. K. Lester (Ed.), *Second handbook of research on mathematics teaching and learning* (pp. 225–256). NCTM.

Fraser, B. J. (1998). Classroom environment instruments: Development, validity and applications. *Learning Environments Research, 1*(1), 7–33. doi:10.1023/A:1009932514731

Fredricks, J. A., Hofkens, T., Wang, M., Mortenson, E., & Scott, P. (2017). Supporting girls' and boys' engagement in maths and science learning: A mixed-methods study. *Journal of Research in Science Teaching, 55*(2), 271–298. doi:10.1002/tea.21419

Freeman, S., Eddy, S. L., McDonough, M., Smith, M. K., Okoroafor, N., Jordt, H., & Wenderoth, M. P. (2014). Active learning increases student performance in science, engineering, and mathematics. *Proceedings of the National Academy of Sciences of the United States of America, 111*(23), 8410–8415. doi:10.1073/pnas.1319030111 PMID:24821756

Freudenthal, H. (1971). Geometry between the devil and the deep sea. In H.-G. Steiner (Ed.), *The teaching of geometry at the pre-college level: Proceedings of the second CSMP international conference co-sponsored by Southern Illinois University and Central Midwestern Regional Educational Laboratory* (pp. 137–159). Springer Netherlands. 10.1007/978-94-017-5896-3_10

Freudenthal, H. (1973). *Mathematics as an educational task.* Reidel. doi:10.1007/BF00121085

Freudenthal, H. (1987). Mathematics starting and staying in reality. In I. Wirszup, & R. Street (Eds.), *Proceedings of the USCMP conference on mathematics education on development in school mathematics around the world.* NCTM.

Freudenthal, H. (1991). *Revisiting mathematics education: China lectures.* Kluwer Academic Publishers.

Frid, S. (2000). Constructivism and reflective practice in practice: Challenges and dilemmas of a mathematics teacher educator. *Mathematics Teacher Education and Development, 2*, 17–33.

Friedlander, A., & Tabach, M. (2001). Promoting multiple representations in algebra. In A. Cuoco (Ed.), *The roles of representation in school mathematics* (pp. 173–184). NCTM.

Friesen, M., & Kuntze, S. (2016). Teacher students analyse texts, comics and video-based classroom vignettes regarding the use of representations—Does format matter. In C. Csikos, A. Rausch, & J. Szitanyi (Eds.), *Proceedings of the 40th Conference of the International Group for the Psychology of Mathematics Education* (Vol. 2, pp. 259-266). PME.

Frykholm, J. A. (1999). The impact of reform: Challenges for mathematics teacher preparation. *Journal of Mathematics Teacher Education, 2*(1), 79–105. doi:10.1023/A:1009904604728

Fujita, T., Kondo, Y., Kumakura, H., & Kunimune, S. (2017). Students' geometric thinking with cube representations: Assessment framework and empirical evidence. *The Journal of Mathematical Behavior, 46*(1), 96–111. doi:10.1016/j.jmathb.2017.03.003

Fullan, M. (2009). Large-scale reform comes of age. *Journal of Educational Change, 10*(2-3), 101–113. doi:10.100710833-009-9108-z

Fullan, M., Rincón-Gallardo, S., & Hargreaves, A. (2015). Professional capital as accountability. *Education Policy Analysis Archives, 23*(15), 1–18. doi:10.14507/epaa.v23.1998

Furner, J., & Worrell, N. (2017). The importance of using manipulatives in teaching math today. *Transformations, 3*(1), 1–25. https://nsuworks.nova.edu/transformations/vol3/iss1/2

Fuson, K. C., & Briars, D. J. (1990). Using a base-ten blocks learning/teaching approach for first- and second-grade place-value and multidigit addition and subtraction. *Journal for Research in Mathematics Education, 21*(3), 180–206. doi:10.2307/749373

Gabriele, A. J., & Joram, E. (2007). Teachers' reflections on their reform-based teaching in mathematics: Implications for the development of teacher self-efficacy. *Action in Teacher Education, 29*(3), 60–74. doi:10.1080/01626620.2007.10463461

Gagatsis, A., & Monoyiou, A. (2011). Towards a comprehensive theoretical model of pre-service teachers' conceptual understanding of functions. In B. Ubuz (Ed.), *Proceedings of the 35th Conference of the International Group for the Psychology of Mathematics Education* (Vol. 2, pp. 369-376). PME.

Gainsburg, J. (2008). Real-world connections in secondary mathematics teaching. *Journal of Mathematics Teacher Education, 11*(3), 199–219. doi:10.100710857-007-9070-8

Gallego, F. A., Näslund-Hadley, E., & Alfonso, M. (2017). *Tailoring instruction to improve mathematics skills in preschools: A randomized evaluation.* https://www.povertyactionlab.org/sites/default/ files/publications/613_1026_Tailoring-Instructions-to- Improve-Mathematics-Skills-in-PreSchool_June2017. pdf

Gambari, I. A., Balogun, S. A., & Alfa, A. S. (2014). Efficacy of interactive whiteboard on psychomotor skills achievement of students in isometric and orthographic projection. *Contemporary Educational Technology, 5*(4), 316–330. doi:10.30935/cedtech/6133

Ganley, C. M., Schoen, R. C., LaVenia, M., & Tazaz, A. M. (2019). Construct validation of the math anxiety scale for teachers. *AERA Open, 5*(1), 1–16. doi:10.1177/2332858419839702

Garcia, N., Shaughnessy, M., & Pynes, D. (2021). Recording student thinking in a mathematics discussion. *Mathematics Teacher: Learning & Teaching PK-12, 114*(12), 926-932.

Garcia, M., Sanchez, V., & Escudero, I. (2007). Learning through reflection in mathematics teacher education. *Educational Studies in Mathematics, 64*(1), 1–17. doi:10.100710649-006-9021-9

Gee, J. P. (2017). Affinity Spaces and 21st Century Learning. *Educational Technology, 57*(2), 27–31. https://www.jstor.org/stable/44430520

Geertz, C. (1994). Thick description: Toward an interpretive theory of culture. *Readings in the philosophy of social science*, 213-231.

Gelfuso, A., & Dennis, D. V. (2014). Getting reflection off the page: The challenges of developing support structures for pre-service teacher reflection. *Teaching and Teacher Education, 38*, 1–11. doi:10.1016/j.tate.2013.10.012

Gemert, S. B. (2015). *All positive action starts with criticism: Hans Freudenthal and the didactics of mathematics.* Springer. doi:10.1007/978-94-017-9334-6

Genor, M., & Goodwin, A. L. (2005). Confronting ourselves: Using autobiographical analysis in teacher education. The *New Educator, 1*(4), 311–331. doi:10.1080/15476880500283021

Ghousseini, H., Lord, S., & Cardon, A. (2021). Supporting mathematics talk in kindergarten. *Mathematics Teacher: Learning and Teaching PK-12, 114*(5), 362-368.

Gilbertson, N. J. (2021). The opportunities of no-solution problems. *Mathematics Teacher: Learning and Teaching PK-12, 114*(3), 205-211.

Giles, R. M., Byrd, K. O., & Bendolph, A. (2016). An investigation of elementary preservice teachers' self-efficacy for teaching mathematics. *Cogent Education, 3*(1), 1160523.

Ginsburg, H. P., Kaplan, R., & Cannon, J. (2006) Helping early childhood educators to teach mathematics. In Critical issues in early childhood professional development (pp. 171-202). Paul H. Brookes.

Ginsburg, H. P., Hyson, M., & Woods, T. A. (2014). *Preparing early childhood educators to teach math: professional development that works.* Paul H. Brookes.

Ginsburg, H. P., Klein, A., & Starkey, P. (1998). The development of children's mathematical thinking: Connecting research with practice. In W. V. B. Damon (Ed.), Handbook of child psychology: Vol. 4. *Child psychology in practice* (5th ed., pp. 404–427). Wiley.

Goodall, J., Johnston-Wilder, S., & Russell, R. (2017). The emotions experienced while learning mathematics at home. In U. X. Eligio (Ed.), *Understanding 284 emotions in mathematical thinking and learning* (pp. 295–313). Academic Press. doi:10.1016/B978-0-12-802218-4.00011-X

Goodnough, K., & Hung, W. (2009). Enhancing pedagogical content knowledge in elementary science. *Teaching Education, 20*(3), 229–242. doi:10.1080/10476210802578921

Gough, N. (2013). Towards deconstructive nonalignment: A complexivist view of curriculum, teaching and learning. *South African Journal of Higher Education, 27*(5), 1213–1233. https://www.researchgate.net/publication/274375251_Towards_deconstructive_nonalignment_a_complexivist_view_of_curriculum_teaching_and_learning

Gravemeijer, K. P. E. (1994). *Developing realistic mathematics education.* CD-ß Press/Freudenthal Institute.

Gravemeijer, K. P. E. (1994). *Developing realistic mathematics education.* Utrecht, The Netherlands: CD-β Press.

Gravemeijer, K. P. E. (2001). Developmental research: Fostering a dialectic relation between theory and practice. In J. Anghileri (Ed.), *Principles and practice in arithmetic teaching* (pp. 147–161). Open University Press.

Gray, E. M., & Tall, D. O. (1994). Duality, Ambiguity, and Flexibility: A "Proceptual" View of Simple Arithmetic. *Journal for Research in Mathematics Education, 25*(2), 116–140. doi:10.2307/749505

Gresham, G. (2007). A study of mathematics anxiety in pre-service teachers. *Early Childhood Education Journal, 35*(2), 181–188. doi:10.1007/s10643-007-0174-7 doi:10.100710643-007-0174-7

Guajardo, J., Mabeti, F., & Kelly, M. (2013). *Numeracy Boost Malawi: Endline report July 2013.* Save the Children. http://resource-centre-uploads.s3.amazonaws.com/uploads/nb_malawi_zomba_endline_report_final.pdf

Guba, E. G., & Lincoln, Y. S. (1989). *Fourth generation evaluation.* Sage Publication, Inc.

Gunter, M. D., & Raymond, K. M. (2020). Communication and Metacognition in the Classroom. *Mathematics Teacher: Learning and Teaching PK-12, 113*(12), 1017-1022.

Gürbüz, M. Ç., & Agsu, M. (2017). Dialogic teaching model for ninth class students to conceptualize inequalities. *Journal of Education and Practice, 8*(1), 171–187. https://files.eric.ed.gov/fulltext/ED582086.pdf

Gutiérrez, R. (2012). Context matters: How should we conceptualize equity in mathematics education? In B. Herbel-Eisenmann, J. Choppin, D. Wagner, & D. Pimm (Eds.), *Equity in Discourse for Mathematics Education* (pp. 17–33). Springer. doi:10.1007/978-94-007-2813-4_2

Gutstein, E., Lipman, P., Hernandez, P., & de los Reyes, R. (1997). Culturally relevant teaching in a Mexican American context. *Journal for Research in Mathematics Education, 28*(6), 709–733. doi:10.2307/749639

Hallman-Thrasher, A., & Spangler, D. A. (2020). Purposeful questioning with high cognitive–demand tasks. *Mathematics Teacher: Learning and Teaching PK-12, 113*(6), 446-459.

Handal, B., & Herrington, A. (2003). Mathematics teachers' beliefs and curriculum reform. *Mathematics Education Research Journal, 15*(1), 59–69. doi:10.1007/BF03217369

Hardman, F., Abd-Kadir, J., Agg, C., Migwi, J., Ndambuku, J., & Smith, F. (2009). Changing pedagogical practice in Kenyan primary schools: The impact of school-based training. *Comparative Education, 45*(1), 65–86. doi:10.1080/03050060802661402

Harris, M. (2010). *Teacher efficacy beliefs: Understanding the relationship between efficacy and achievement in urban elementary schools.* UC Berkeley: Educational Leadership. https://escholarship.org/uc/item/44h29077

Harrow, C., & Merchant, N. (2020). Envelope curves unify sinusoidal graphing. *Mathematics Teacher: Learning and Teaching PK-12, 113*(4), 301-308.

Heck, D. J., Weiss, I. R., & Pasley, J. D. (2011). *A Priority research agenda for understanding the influence of the common core state standards for mathematics*. Horizon Research, Inc.

Hegarty, M., Stieff, M., & Dixon, B. L. (2013). Cognitive change in mental models with experience in the domain of organic chemistry. *Journal of Cognitive Psychology, 25*(2), 220–228. doi:10.1080/20445911.2012.725044

Heibert, J., & Stigler, J. W. (2004). A world of difference: Classrooms abroad provide lessons in teaching math and science. *Journal of Staff Development, 24*(4), 10–15.

Hembree, R. (1990). The nature, effects, and relief of mathematics anxiety. *Journal for Research in Mathematics Education, 21*(1), 33–46. doi:10.2307/749455

Hickey, D., Moore, A., & Pellegrino, J. (2001). The motivational and academic consequences of elementary environments: Do constructivist innovations and reforms make a difference? *American Educational Research Journal, 38*(3), 611–652. doi:10.3102/00028312038003611

Hicks, T., & Bostic, J. D. (2021). Formative assessment through think alouds. *Mathematics Teacher: Learning and Teaching PK-12, 114*(8), 598-606.

Hiebert, J., Gallimore, R., Garnier, H., Givvin, K. B., Hollingsworth, H., Jacobs, J., Chui, A. M.-Y., Wearne, D., Smith, M., Kersting, N., Manaster, A., Tseng, E., Etterbeek, W., Manaster, C., Gonzales, P., & Stigler, J. (2003). *Teaching mathematics in seven countries: Results from the TIMSS 1999 Video Study*. PsycEXTRA Dataset.

Hiebert, J., & Grouws, D. A. (2007). The effects of classroom mathematics teaching on students' learning. In F. K. Lester (Ed.), *Second handbook of research on mathematics teaching and learning* (pp. 371–404). National Council of Teachers of Mathematics.

Hiebert, J., Morris, A. K., Berk, D., & Jansen, A. (2007). Preparing teachers to learn from teaching. *Journal of Teacher Education, 58*(1), 47–61. doi:10.1177/0022487106295726

Hiebert, J., & Wearne, D. (1993). Instructional tasks, classroom discourse, and learners' learning in second-grade arithmetic. *American Educational Research Journal, 30*(2), 393–425. doi:10.3102/00028312030002393

Hill, H. C. (2021, February 4). *After 30 years of reforms to improve math instruction, reasons for hope and dismay*. Brookings Institute. https://www.brookings.edu/blog/brown-center-chalkboard/2021/02/04/after-30-years-of-reforms-to-improve-math-instruction-reasons-for-hope-and-dismay/

Hill, H. C., Sleep, L., Lewis, J. M., & Ball, D. L. (2007). Assessing teachers' mathematical knowledge: What knowledge matters and what evidence counts? In F. K. Lester (Ed.), *Second handbook of research on mathematics teaching and learning* (pp. 111–155). Information Age Publishing.

Hill, H., Litke, E., & Lynch, K. (2018). Learning lessons from instruction: Descriptive results from an observational study of urban elementary classrooms. *Teachers College Record, 120*(12), 120304. doi:10.1177/016146811812001207

Hincapie, S., & Hincapie, D. (2013). *La Fascia: Sistema de Unificación Estructural y Funcional del Cuerpo*. https://repository.ces.edu.co/bitstream/handle/10946/2630/Fascias.pdf?sequence=1&isAllowed=y

Hipp, K. (1996). *Teacher efficacy: Influence of the principal leadership, and student achievement.* Paper presented at the Annual Meeting of the American Educational Research Association, New York, NY.

Hodge, L. L., & Walther, A. (2017). Building a discourse community: Initial practices. *Mathematics Teaching in the Middle School, 22*(7), 430–437. doi:10.5951/mathteacmiddscho.22.7.0430

Hofmann, R., & Mercer, N. (2016). Teacher interventions in small group work in secondary mathematics and science lessons. *Language and Education, 30*(5), 400–416. doi:10.1080/09500782.2015.1125363

Honebein, P. C. (2009). Transmergent Learning and the Creation of Extraordinary Educational Experiences. *Educational Technology, 49*(1), 27–34. https://eric.ed.gov/?id=EJ829871

Horn, I. S. (2012). *Strength in numbers: Collaborative learning in secondary mathematics*. National Council of Teachers of Mathematics.

Horn, I. S., & Little, J. W. (2010). Attending to problems of practice: Routines and resources for professional learning in teachers' workplace interactions. *American Educational Research Journal, 47*(1), 181–217. doi:10.3102/0002831209345158

Ho, S. Y., & Lai, M. Y. (2012). Pre-service teachers' specialized content knowledge on multiplication of fractions. In T. Tso (Ed.), *Proceedings of the 36th Conference of the International Group for the Psychology of Mathematics Education* (Vol. 2, pp. 291-298). PME.

Hough, S., Solomon, Y., Dickinson, P., & Gough, S. (2017). Investigating the impact of a realistic mathematics education approach on achievement and attitudes in post-16 GCSE resit classes. Manchester Metropolitan University/The Nuffield Foundation.

Hoyles, C., & Noss, R. (2003). What can digital technologies take from and bring to research in mathematics education? In A. J. Bishop, M. A. Clements, C. Keitel, J. Kilpatrick, & F. K. S. Leung (Eds.), *Second international handbook of mathematics education* (pp. 323–349). Springer Netherlands. doi:10.1007/978-94-010-0273-8_11

Hoynes, S., Klemp, T., & Nilssen, V. (2018). Pre-service mathematics teachers' whole-class dialogs during field practice. In E. Bergqvist, M. Osterholm, C. Granberg, & L. Sumpter (Eds.), *Proceedings of the 42nd Conference of the International Group for the Psychology of Mathematics Education* (Vol. 3, pp. 123-130). PME.

Huffman, D., Thomas, K., & Lawrenz, F. (2008). Science and mathematics instruction in a reform-based teacher preparation program. *School Science and Mathematics, 108*(4), 137–145. doi:10.1111/j.1949-8594.2008.tb17820.x

Huinker, D., & Bill, V. (2017). *Taking action: Implementing effective mathematics teaching practices in grades K-Grade 5*. National Council of Teachers of Mathematics.

Hull, T. H., Miles, R. H., & Balka, D. S. (2012). *The common core mathematics standards: Transforming practice through team leadership*. NCTM and Corwin. doi:10.4135/9781544308692

Hume, A., & Coll, R. (2010). Authentic student inquiry: The mismatch between the intended curriculum and the student-experienced curriculum. *Research in Science & Technological Education, 28*(1), 43–62. doi:10.1080/02635140903513565

Humphreys, C., & Parker, R. (2015). *Making number talks matter: Developing mathematical practices and deepening understanding, Grades 4-10*. Stenhouse Publishers.

Hungi, N., & Ngware, M. (2018). The effects of preschool participation on mathematics achievement of Grade 6 pupils in Uganda. *Educational Research for Policy and Practice, 17*(2), 105-126. https://link.springer.com/article/10.1007/s10671-017-9218-y

Hunter, R. (2010). Changing roles and identities in the construction of a community of mathematical inquiry. *Journal of Mathematics Teacher Education, 13*(5), 397–409. doi:10.100710857-010-9152-x

Hunter, R., & Anthony, G. (2010). Developing mathematical inquiry and argumentation. In R. Averill & R. Harvey (Eds.), *Teaching primary school mathematics and statistics: Evidence-based practice* (pp. 197–206). NZCER Press.

Hynes, M. C. (1986). Selection Criteria. *The Arithmetic Teacher, 33*(6), 11–13. doi:10.5951/AT.33.6.0011

Illustrative Mathematics. (n.d.). *Half of a recipe.* http://tasks.illustrativemathematics.org/content-standards/5/NF/B/6/tasks/296

Inter-American Development Bank. (2015). *All children count.* Inter-American Development Bank. https://publications.iadb.org/en/publication/17038/all-children-count-early-mathematics-and-science-education-latin-america-and

Iskenderoglu, T. A. (2016). Multiplication and division problems posed by pre-service elementary mathematics teachers about fraction topic. In C. Csikos, A. Rausch, & J. Szitanyi (Eds.), *Proceedings of the 40th Conference of the International Group for the Psychology of Mathematics Education* (Vol. 2, pp. 35-42). PME.

Ivars, P., & Fernández, C. (2016). Narratives and the development of the skill of noticing. In C. Csikos, A. Rausch, & J. Szitanyi (Eds.), *Proceedings of the 40th Conference of the International Group for the Psychology of Mathematics Education* (Vol. 3, pp. 19-26). PME.

Ivars, P., Fernández, C., & Llinares, S. (2017). Pre-service teachers' uses of a learning trajectory to notice students' fractional reasoning. In *Proceedings of the 41st Conference of the International Group for the Psychology of Mathematics Education* (Vol. 3, pp. 25-32). PME.

Ivars, P., Fernández, C., & Llinares, S. (2019). Principles in the design of tasks to support preservice teachers' noticing enhancement. In M. Graven, H. Venkat, A. Essien, & P. Vale (Eds.), *Proceedings of the 43rd Conference of the International Group for the Psychology of Mathematics Education* (Vol. 2, pp. 408-415). PME.

Ivars, P., Fernández, C., Llinares, S., & Choy, B. H. (2018). Enhancing noticing: Using a hypothetical learning trajectory to improve pre-service primary teachers' professional discourse. *Eurasia Journal of Mathematics, Science and Technology Education, 14*(11), em1599. doi:10.29333/ejmste/93421

Ivy, J. T., Bush, S. B., & Dougherty, B. J. (2020). Stacking the deck: Reversibility and reasoning. *Mathematics Teacher: Learning and Teaching PK-12, 113*(1), 65-68.

Jacobs, V. R., & Empson, S. B. (2015). Responding to children's mathematical thinking in the moment: an emerging framework of teaching moves. *ZDM*, 1-13.

Jacobson, M. J., Levin, J. A., & Kapur, M. (2019). Education as a Complex System: Conceptual and Methodological Implications. *Educational Researcher, 48*(2), 112–119. doi:10.3102/0013189X19826958

Jacobs, V. R., Empson, S. B., Pynes, D., Hewitt, A., Jessup, N., & Krause, G. (2019). The responsive teaching in elementary mathematics project. In P. Sztajn & P. H. Wilson (Eds.), *Designing professional development for mathematics learning trajectories* (pp. 75–103). Teachers College Press.

Jacobs, V. R., & Philipp, R. A. (2010). Supporting children's problem solving. *Teaching Children Mathematics, 17*(2), 98–105. doi:10.5951/TCM.17.2.0098

Jacobs, V., Lamb, L., & Philipp, R. (2010). Professional noticing of children's mathematical thinking. *Journal for Research in Mathematics Education, 41*(2), 169–202. doi:10.5951/jresematheduc.41.2.0169

Jao, L. (2017). Shifting pre-service teachers' beliefs about mathematics teaching: The contextual situation of a mathematics methods course. *International Journal of Science and Mathematics Education, 15*(5), 895–914. doi:10.100710763-016-9719-9

Jarvis, D. (2016). Messy but meaningful: Exploring the transition to reform-oriented pedagogy with teachers of mathematics and coordinators in Ontario, Canada. *Teacher Development, 20*(1), 18–39. doi:10.1080/13664530.2015.1108928

Jaworski, B. (2003). Inquiry as a pervasive pedagogic process in mathematics education development. *Proceedings of the Third Conference of the European Society for Research in Mathematics Education.* http://erme.site/cerme-conferences/cerme3/cerme-3-proceedings/

Jaworski, B. (2008). Building and sustaining inquiry communities in mathematics teaching development. In K. Krainer & T. Wood (Eds.), *Participants in mathematics teacher education: Individuals, teams, communities and networks* (Vol. 3, pp. 309–330). Sense Publishers.

Jeschke, C., Lindmeier, A., & Heinze, A. (2017). What do Mathematics pre-service teachers lack for mastering instructional demands? In *Proceedings of the 41st Conference of the International Group for the Psychology of Mathematics Education* (Vol. 3, pp. 33-40). PME.

Johnson, B., & vanderSandt, S. (2011). "Math makes me sweat": The impact of pre-service courses on mathematics anxiety. *Issues in the Undergraduate Mathematics Preparation of School Teachers, 5*, 1–8.

Johnson, C. C. (2006). Effective professional development and change in practice: Barriers science teachers encounter and implications for reform. *School Science and Mathematics, 106*(3), 150–161. doi:10.1111/j.1949-8594.2006.tb18172.x

Johnson, D., & Johnson, F. (2016). *Joining together: Group theory and group skills.* Pearson.

Joldersma, C. W. (2002). Pedagogy of the Other: A Levinasian Approach to the Teacher-Student Relationship. *Philosophy of Education*, 181-188. https://www.researchgate.net/publication/314152336_Pedagogy_of_the_Other_A_Levinasian_Approach_to_the_Teacher-Student_Relationship

Jonason, C., Guajardo, J., Nath, B. D., & Hossain, M. (2014). *Literacy & Numeracy Boost: Bangladesh endline.* Save the Children. https://resourcecentre.savethechildren.net/pdf/literacy_boost_and_numeracy_boost-_bangladesh-_endline_report_april_2014.pdf/

Jones, K., Fujita, T., & Kunimune, S. (2012). Representations and reasoning in 3-D geometry in lower secondary school. In T. Y. Tso (Ed.), *Proceedings of the 36th conference of the international group for the psychology of mathematics education* (Vol. 2, pp. 229–346). The Psychology of Mathematics Education. https://eprints.soton.ac.uk/445627/1/Jones_etc_represent_reason3D_PME36_2012.pdf

Jong, C. (2016). Linking reform-oriented experiences to teacher identity: The case of an elementary mathematics teacher. *The Journal of Educational Research, 109*(3), 296–310. doi:10.1080/00220671.2014.947398

Joram, E., & Gabriele, A. J. (2012). How mathematics teachers can help curb childhood obesity. Iowa Council of Teachers of Mathematics Journal, 39, 3.

Jukes, M., & Sitabkhan, Y. (2021). *Understanding the social classroom: The basis of effective pedagogy?* RTI International SharEd. https://ierc-publicfiles.s3.amazonaws.com/public/resources/Jukes_Social_Classroom_Apr_24.pdf

Jung, H., & Brand, S. (2021). Student actions for social justice-oriented mathematical tasks. *Mathematics Teacher: Learning and Teaching PK-12, 114*(5), 378-385.

Kagan, S. (2001). *Kagan structures: Research and rationale.* Kagan Online Magazine.

Kalinec-Craig, C. A. (2017). The rights of the learner: A framework for promoting equity through formative assessment in mathematics education. *Democracy & Education, 25*(2), 1–11.

Kalyuga, S., Chandler, P., Tuovinen, J., & Sweller, J. (2001). When problem-solving is superior to studying worked examples. *Journal of Educational Psychology, 93*(3), 579–588. doi:10.1037/0022-0663.93.3.579

Kapur, M. (2014). Productive failure in learning math. *Cognitive Science, 38*(5), 1008–1022. doi:10.1111/cogs.12107 PMID:24628487

Kaput, J. J. (1992). Technology and mathematics education. In *Handbook of research on mathematics teaching and learning* (pp. 515–556). https://www.learntechlib.org/primary/p/5474/

Karp, A. (2010). Analyzing and attempting to overcome prospective teachers' difficulties during problem-solving instruction. *Journal of Mathematics Teacher Education, 13*(2), 121–139. doi:10.100710857-009-9127-y

Kar, T., Özdemir, E., Öçal, M. F., Güler, G., & Ipek, A. S. (2019). Indicators of prospective mathematics teachers' success in problem solving: The case of creativity in problem-posing. In M. Graven, H. Venkat, A. Eddien, & P. Vale (Eds.), *Proceedings of the 43rd Conference of the International Group for the Psychology of Mathematics Education* (Vol. 2, pp. 456-463). PME.

Kartal, O. (2015). *Enactment of common core state standards for mathematics: Relationship between teachers' choices of curriculum, teaching, and professional development* (Order No. 3733992). Available from Dissertations & Theses @ Illinois Institute of Technology; ProQuest Dissertations & Theses Global. (1732166593). http://216.47.136.160/docview/1732166593?accountid=28377

Katz, S., & Dack, L. A. (2014). Towards a culture of inquiry for data use in schools: Breaking down professional learning barriers through intentional interruption. *Studies in Educational Evaluation, 42*, 35–40. doi:10.1016/j.stueduc.2013.10.006

Kazak, S., Wegerif, R., & Fujita, T. (2015). The importance of dialogic processes to conceptual development in mathematics. *Educational Studies in Mathematics, 90*(2), 105–120. doi:10.100710649-015-9618-y

Kazemi, E., Franke, M., & Lampert, M. (2009). Developing pedagogies in teacher education to support novice teachers' ability to enact ambitious instruction. In R. Hunter, B. Bicknell, & T. Burgess (Eds.) *Crossing divides: Proceedings of the 32nd annual conference of the Mathematics Education Research Group of Australasia* (*Vol. 1*, pp. 12–30). Academic Press.

Kazemi, E., & Hintz, A. (2014). *Intentional talk: How to structure and lead productive mathematical discussions*. Stenhouse Publishers.

Kelley, T. R., Knowles, J. G., Holland, J. D., & Han, J. (2020). Increasing high school teachers self-efficacy for integrated STEM instruction through a collaborative community of practice. *International Journal of STEM Education, 7*(1), 1–13. doi:10.118640594-020-00211-w

Kelly, W. P., & Tomhave, W. K. (1985). A study of math anxiety/math avoidance in preservice elementary teachers. *The Arithmetic Teacher, 32*(5), 51–53. doi:10.5951/AT.32.5.0051

Kenyan Institute for Curriculum Design. (2017). *Curriculum designs-preprimary PP1-Grade4.* https://kicd.ac.ke/cbc-materials/curriculum-design/

Keune, F. J. (1998). Naar de knoppen [Inaugural lecture]. In *NRC/Handelsblad.* Catholic University. https://lib.ugent.be/en/catalog/rug01:000433729

Kieran, C. (2007). Learning and teaching of algebra at the middle school through college levels: Building meaning for symbols and their manipulation. In F. K. Lester (Ed.), *Second handbook of research on mathematics teaching and learning* (pp. 707–762). NCTM.

Kim, O., & Son, J. (2017). Preservice teachers' recognition of affordances and limitations of curriculum resources. In B. Kaur, W. K. Ho, T. L. Toh, & B. H. Choy (Eds.), *Proceedings of the 41st Conference of the International Group for the Psychology of Mathematics Education* (Vol. 3, pp. 57-64). PME.

King, G. A. (2015). *A multiple case study examining secondary mathematics teachers' perceptions of reform-oriented instruction and obstacles to instructional change* (Publication No. 3719749) [Doctoral dissertation, Middle Tennessee State University]. ProQuest Dissertations Publishing.

King, S., Korda, M., Nordstrum, L., & Edwards, S. (2015). *Liberia teacher training program: Endline assessment of the impact of early grade reading and mathematics interventions.* RTI International SharEd. https://shared.rti.org/content/liberia-teacher-training-program-endline-assessment-impact-early-grade-reading-and

Kinser-Traut, J. Y., & Turner, E. E. (2020). Shared authority in the mathematics classroom: Successes and challenges throughout one teacher's trajectory implementing ambitious practices. *Journal of Mathematics Teacher Education, 23*(1), 5–34. doi:10.100710857-018-9410-x

Kirkby, J., Walsh, L., & Keary, A. (2019). A case study of the generation and benefits of a community of practice and its impact on the professional identity of early childhood teachers. *Professional Development in Education, 45*(2), 264–275. doi:10.1080/19415257.2018.1449003

Kirschner, P. A., Sweller, J., & Clark, R. E. (2006). Why minimal guidance during instruction does not work: An analysis of the failure of constructivist, discovery, problem-based, experiential, and inquiry-based teaching. *Educational Psychologist, 41*(2), 75–86. doi:10.120715326985ep4102_1

Klančar, A., Starčič, A. I., Cotič, M., & Žakelj, A. (2021). Problem-based geometry in seventh grade: Examining the effect of path-based vs. conventional instruction on learning outcomes. *International Journal of Emerging Technologies in Learning, 16*(12), 16–35. doi:10.3991/ijet.v16i12.21349

Kleickmann, T., Tröbst, S., Heinze, A., Bernholt, A., Rink, R., & Kunter, M. (2017). *Teacher knowledge experiment: Conditions of the development of pedagogical content knowledge. Competence Assessment in Education.* Springer.

Klibanoff, R., Levine, S., Huttenlocher, J., Vasilyeva, M., & Hedges, L. (2006). Preschool children's mathematical knowledge: The effect of teacher math talk. *Developmental Psychology, 42*(1), 59–69. doi:10.1037/0012-1649.42.1.59 PMID:16420118

Knuth, E. J., Alibali, M. W., McNeil, N. M., Weinberg, A., & Stephens, A. C. (2005). Middle school students' understanding of core algebraic concepts: Equivalence & variable. *Zentralblattfür Didaktik der Mathematik, 37*(1), 68–76. doi:10.1007/BF02655899

Kogan, M., & Laursen, S. L. (2013). Assessing long-term effects of inquiry-based learning: A case study from college mathematics. *Innovative Higher Education, 39*(3), 183–199. doi:10.100710755-013-9269-9

Komatsu, H., Rappleye, J., & Silova, I. (2021). Student-centered learning and sustainability: Solution or problem? *Comparative Education Review, 65*(1), 6–33. doi:10.1086/711829

Konan, N., & Yilmaz, S. (2017). Teachers' views on teacher performance assessment: A mixed-method research. *Proceedings of the 26th International Educational Sciences Conference,* 1–3. https://dergipark.org.tr/download/journal-file/11372

Konan, N., & Yilmaz, S. (2018). Ogretmen iliskileri performans degerlendirmeye iliskin ogretmen gorusleri: Bir karma yontem arastirmasi. *Milli Egitim, 219*(1), 137–160. https://dergipark.org.tr/tr/pub/milliegitim/issue/39923/473428

Kong, S. F., & Mohd-Matore, M. E. E. (2022). Can a Science, Technology, Engineering, and Mathematics (STEM) approach enhance students' mathematics performance? *Sustainability, 14*(1), 379. doi:10.3390u14010379

Kormi-Nouri, R., Nyberg, L., & Nilsson, L. G. (1994). The effect of retrieval enactment on recall of subject-performed tasks and verbal tasks. *Memory & Cognition*, *22*(6), 723–728. doi:10.3758/BF03209257 PMID:7808281

Kosko, K., Rougee, A., & Herbst, P. (2014). What actions do teachers envision when asked to facilitate mathematical argumentation in the classroom? *Mathematics Education Research Journal*, *26*(3), 459–476. doi:10.100713394-013-0116-1

Krummheuer, G. (2007). Argumentation and participation in the primary mathematics classroom: Two episodes and related theoretical abductions. *The Journal of Mathematical Behavior*, *26*(1), 60–82. doi:10.1016/j.jmathb.2007.02.001

Kuntze, S., Friesen, M., Krummenauer, J., Skilling, K., Fernandez, C., Ivars, P., Llinares, S., Samkova, L., & Healy, L. (2021). Multi-criterion noticing: Pre-service teachers' difficulties in analyzing classroom vignettes. In M. Inprasitha, N. Changsri, & N. Boonsena (Eds.), *Proceedings of the 44th Conference of the International Group for the Psychology of Mathematics Education* (Vol. 3, pp. 173-183). PME.

Kusaka, S. (2019). Issue analysis of competency-based mathematics curriculum design in African countries: A case study of Mozambique's primary mathematics education. *Journal of Education and Learning*, *9*(1), 41–50. doi:10.5539/jel.v9n1p41

Lakatos, I. (1976). A renaissance of empiricism in the recent philosophy of mathematics. *The British Journal for the Philosophy of Science*, *27*(3), 201–223. doi:10.1093/bjps/27.3.201

Lammers, W. J., Gillaspy, J. A. Jr, & Hancock, F. (2017). Predicting academic success with early, middle, and late semester assessment of student–instructor rapport. *Teaching of Psychology*, *44*(2), 145–149. doi:10.1177/0098628317692618

Lamon, S. (1996). The development of unitizing: Its role in children's partitioning strategies. *Journal for Research in Mathematics Education*, *271*(2), 170–193. doi:10.2307/749599

Lamon, S. J. (2007). Rational numbers and proportional reasoning. In F. K. Lester Jr., (Ed.), *Second handbook of research on mathematics teaching and learning* (pp. 629–667). National Council of Teachers of Mathematics.

Lampert, M. (1990). When the problem is not the question and the solution is not the answer: Mathematical knowing and teaching. *American Educational Research Journal*, *27*(1), 29–63. doi:10.3102/00028312027001029

Lampert, M. (2001). *Teaching problems and the problems of teaching*. Yale University Press.

Lampert, M., & Cobb, P. (2003). Communication and language. In W. G. Kilpatrick & M. D. Schifter (Eds.), *A research companion to the principles and standards for school mathematics*. National Council of Teachers of Mathematics.

Lane, C., & Ríordáin, M. N. (2020). Out-of-field mathematics teachers' beliefs and practices: An examination of change and tensions using Zone Theory. *International Journal of Science and Mathematics Education*, *18*(2), 337–355. doi:10.100710763-019-09964-5

Langer, J., Gillette, P., & Arriaga, R. I. (2003). Toddlers' cognition of adding and subtracting objects in action and in perception. *Cognitive Development*, *18*(2), 233–246. doi:10.1016/S0885-2014(03)00022-4

Larkin, K., & Jorgensen, R. (2016). 'I hate maths: Why do we need to do maths?' Using iPad video diaries to investigate attitudes and emotions towards mathematics in year 3 and year 6 students. *International Journal of Science and Mathematics Education*, *14*(5), 925–944. doi:10.100710763-015-9621-x

Larrivee, B. (2000). Transforming teaching practice: Becoming the critically reflective teacher. *Reflective Practice*, *1*(3), 293–307. doi:10.1080/713693162

Larsen, S. P. (2013). A local instructional theory for the guided reinvention of the group and isomorphism concepts. *The Journal of Mathematical Behavior*, *32*(4), 712–725. doi:10.1016/j.jmathb.2013.04.006

Larsen, S., & Lockwood, E. (2013). A local instructional theory for the guided reinvention of the quotient group concept. *The Journal of Mathematical Behavior, 32*(4), 726–742. doi:10.1016/j.jmathb.2013.02.010

Laursen, S. L. (2013, November). From innovation to implementation: Multi-institution pedagogical reform in undergraduate mathematics. In D. King, B. Loch, & L. Rylands (Eds.), *Proceedings of Lighthouse Delta 2013: The 9th DELTA conference on the teaching and learning of undergraduate mathematics and statistics* (pp. 102-112). Retrieved from http://www.deltaconference.org/documents/program/Keynote-Monday-Laursen2013.pdf

Laursen, S. L., & Rasmussen, C. (2019). I on the prize: Inquiry approaches in undergraduate mathematics. *International Journal of Research in Undergraduate Mathematics Education, 5*(1), 129–146. doi:10.100740753-019-00085-6

Lavanderos Gallardo, L., & Oliva Figueroa, I. (2012). Desde la distinción a la configuración: complejidad, evaluación y aprendizaje desde la visión relacional. *Estudios Pedagógicos (Valdivia), 38*(1), 273–283. .273 doi:10.4067/s0718-07052012000100016

Lave, J. (1991). Situating learning in communities of practice. *Perspectives on Socially Shared Cognition, 2*, 63-82.

Lave, J., & Wenger, E. (1991). *Situated Learning: Legitimate Peripheral Participation.* Cambridge University Press. doi:10.1017/CBO9780511815355

Lavy, I., & Shriki, A. (2012). Engaging prospective teachers in the assessment of geometrical proofs. In T. Tso (Ed.), *Proceedings of the 36th Conference of the International Group for the Psychology of Mathematics Education* (Vol. 3, pp. 35-42). PME.

Leatham, K. R., Peterson, B. E., Stockero, S. L., & Van Zoest, L. R. (2015). Conceptualizing mathematically significant pedagogical opportunities to build on student thinking. *Journal for Research in Mathematics Education, 46*(1), 88–124. doi:10.5951/jresematheduc.46.1.0088

Lee, J., & Bednarz, R. (2009). Effect of GIS learning on spatial thinking. *Journal of Geography in Higher Education, 33*(2), 183–198. doi:10.1080/03098260802276714

Lee, M. Y. (2019). The development of elementary pre-service teachers' professional noticing of students' thinking through adapted lesson study. *Asia-Pacific Journal of Teacher Education, 47*(4), 383–398. doi:10.1080/1359866X.2019.1607253

Lee, Y., & Ertmer, P. A. (2006). Examining the effect of small group discussions and question prompts on vicarious learning outcomes. *Journal of Research on Technology in Education, 39*(1), 66–80. doi:10.1080/15391523.2006.10782473

LeFevre, J., Fast, L., & Sowinski, C. (2009) *Who's counting? Numeracy and literacy practices of early learning and child care practitioners.* Final report. Canadian Council on Learning. https://www.researchconnections.org/childcare/resources/19205

Leikin, R., & Zaslavsky, C. (1997). Facilitating student interactions in mathematics in a cooperative learning situation. *Journal for Research in Mathematics Education, 28*(3), 331–354. doi:10.2307/749784

Lemke, J. L., & Sabelli, N. H. (2008). Complex systems and educational change: Towards a new research agenda. *Educational Philosophy and Theory, 40*(1), 118-129. https://www.tandfonline.com/doi/abs/10.1111/j.1469-5812.2007.00401.x

Lepper, M. R. (1988). Motivational Considerations in the Study of Instruction. *Cognition and Instruction, 5*(4), 289–309. doi:10.12071532690xci0504_3

Levine, T. H. (2010). Tools for the study and design of collaborative teacher learning: The affordances of different conceptions of teacher community and activity theory. *Teacher Education Quarterly, 37*(1), 109–130.

Lewis, G. M. (2014). Implementing a reform-oriented pedagogy: Challenges for novice secondary mathematics teachers. *Mathematics Education Research Journal, 26*(2), 399–419. doi:10.100713394-013-0092-5

Liben, L. S. (2007). Education for spatial thinking. In Handbook of child psychology (Vol. 4). Wiley. doi:10.1002/9780470147658.chpsy0406

Lifshitz, C., Dolev, T., Margolin, T., & Agur, M. (2004). *The PACT Project for Ethiopian Israeli Immigrant Pre-schoolers and their Families Selected Findings from the Evaluation of the Second Year of Implementation in Netanya.* Academic Press.

Liljedahl, P. (2021). *Building thinking classrooms in mathematics.* Corwin.

Lim, K. H. (2020). Addressing the hammer-and-nail phenomenon. *Mathematics Teacher: Learning and Teaching PK-12, 113*(6), 487-495.

Lim, W., Jiang, B., DeVillar, R. A., & Lee, J. (2017). Patterns of language use in two modes of writing mathematical solutions. In B. Kaur, W. K. Ho, T. L. Toh, & B. H. Choy (Eds.), *Proceedings of the 41st conference of the international group for the psychology of mathematics education* (Vol. 3, pp. 169-176). PME.

Lim, W., Kim, H., Stallings, L., & Son, J. W. (2015). Celebrating diversity by sharing multiple solution methods. *Mathematics Teacher, 109*(4), 290–297. doi:10.5951/mathteacher.109.4.0290

Lin, P. (2018). Improving knowledge for teaching mathematical argumentation in primary classrooms. *Journal of Mathematics Education, 11*(1), 17–30. doi:10.26711/007577152790018

Lin, P. J., & Tsai, W. H. (2013). Enhancing pre-service teachers' knowledge of students' errors by using researched-based cases. In A. M. Lindmeier, & A. Heinze (Eds.), *Proceedings of the 37th Conference of the International Group for the Psychology of Mathematics Education* (Vol. 3, pp. 273-280). PME.

Lischka, A. E., & Stephens, D. C. (2020). The area model: Building mathematical connections. *Mathematics Teacher: Learning and Teaching PK-12, 113*(3), 186-195.

Lischka, A. E., Prince, K. M., & Reed, S. D. (2020). Implementing extended tasks: Developing perseverance. *Mathematics Teacher: Learning and Teaching PK-12, 113*(1), 47-52.

Little, J. W. (2003). Inside teacher community: Representations of classroom practice. *Teachers College Record, 105*(6), 913–945. doi:10.1177/016146810310500601

Livy, S., Downton, A., & Saito, E. (2017). Pre-service teachers' reflections of the summarise phase of a lesson study. In B. Kaur, W. K. Ho, T. L. Toh, & B. H. Choy (Eds.), *Proceedings of the 41st Conference of the International Group for the Psychology of Mathematics Education* (Vol. 3, pp. 185-192). PME.

London, R. (2016). The enneagram: A spiritual perspective for addressing significant problems through research. In J. Lin, T. Culham, & R. Oxford (Eds.), Toward a spiritual research paradigm: Exploring new ways of knowing, researching, and being (pp. 97 – 126). Information Age Publishing.

London, R. (2021). *A curriculum of nonroutine problems: A holistic approach to process-oriented mathematics education* [manuscript in preparation]. College of education, California State University, San Bernardino, CA.

London, R. (1976, April). *Process-oriented curriculum* [Paper presentation]. *American Educational Research Association Annual Meeting*, San Francisco, CA.

London, R. (1989). *Nonroutine problems: Doing mathematics.* Janson Publications.

London, R. (1993, April). *A curriculum of open-ended nonroutine problems* [Paper presentation]. *American Educational Research Association Annual Meeting*, New York, NY.

London, R. (2004). What is essential in mathematics education. *Encounter: Education for Meaning and Social Justice, 17*(3), 30–36.

London, R. (2007). What is essential in mathematics: A holistic view. *MSOR Connections, 7*(1), 30–34. doi:10.11120/msor.2007.07010030

Loucks-Horsley, S., Love, N., Stiles, K. E., Mundry, S., & Hewson, P. W. (2009). *Designing professional development for teachers of science and mathematics* (3rd ed.). Corwin Press.

Loucks-Horsley, S., Stiles, K., & Hewson, P. (1996). Principles of effective professional development for mathematics and science education: A synthesis of standards. *NISE Brief, 1*(1), n1.

Lucariello, J., Tine, M. T., & Ganley, C. M. (2014). A formative assessment of students' algebraic variable misconceptions. *The Journal of Mathematical Behavior, 33*, 30–41. doi:10.1016/j.jmathb.2013.09.001

Luft, J. A. (2001). Changing inquiry practices and beliefs: The impact of an inquiry-based professional development programme on beginning and experienced secondary science teachers. *International Journal of Science Education, 23*(5), 517–534. doi:10.1080/09500690121307

Lutfeali, S., Sitabkhan, Y., Ralaingita, W., & Piper, B. (2021). *Designing effective numeracy programs in low- and middle-income countries.* Science of Teaching. https://scienceofteaching.site/wp-content/uploads/2021/11/Practical-Guide-for-Numeracy_converted.pdf

Maasepp, B., & Bobis, J. (2015). Prospective primary teachers' beliefs about mathematics. *Mathematics Teacher Education and Development, 16*(2), 89–107.

MacDonald, B. L., Moss, D. L., & Hunt, J. H. (2020). Dominoes: Promoting units construction and coordination. *Mathematics Teacher: Learning and Teaching PK-12, 113*(7), 551-557.

Magiera, M. T. (2020). Prospective K-8 teachers' problem posing: Interpretations of tasks that promote mathematical argumentation. In A.I. Sacristán, J. C. Cortés-Zavala & P. M. Ruiz-Arias, (Eds.), *Mathematics education across cultures: Proceedings of the 42nd meeting of the North American Chapter of the International Group for the Psychology of Mathematics Education* (pp. 940-948). Cinvestav / AMIUTEM / PME-NA. 10.51272/pmena.42.2020-143

Magiera, M. T., & Zambak, V. S. (2020). Exploring prospective teachers' ability to generate and analyze evidence-based explanatory arguments. *International Journal of Research in Education and Science, 6*(2), 327–346. doi:10.46328/ijres.v6i2.765

Magiera, M. T., & Zambak, V. S. (2021). Prospective K-8 teachers' noticing of student justifications and generalizations in the context of analyzing written artifacts and video-records. *International Journal of STEM Education, 8*(1), 1–21. doi:10.118640594-020-00263-y

Makar, K. (2012). The pedagogy of mathematical inquiry. In R. Gillies (Ed.), *Pedagogy: New developments in the learning sciences* (pp. 371–397). Nova Science.

Ma, L. (2020). *Knowing and teaching elementary mathematics. Teacher understanding of fundamental mathematics in China and the United States*. Taylor & Francis: Routledge. doi:10.4324/9781003009443

Malinsky, M., Ross, A., Pannells, T., & McJunkin, M. (2006). Math anxiety in pre-service elementary school teachers. *Education, 127*(2), 274–279.

Manouchehri, A., & Goodman, T. (2000). Implementing mathematics reform: The challenge within. *Educational Studies in Mathematics, 42*(1), 1–34. doi:10.1023/A:1004011522216

Manousaridis, T. (2021). Promoting inquisitiveness. *Mathematics Teacher: Learning and Teaching PK-12, 114*(5), 369-377.

Marton, F., Booth, S., & Booth, S. (2013). *Learning and awareness*. Routledge., doi:10.4324/9780203053690

Marton, F., Tsui, A. B. M., Chik, P. P. M., Ko, P. Y., & Lo, M. L. (2004). *Classroom discourse and the space of learning*. Routledge., doi:10.4324/9781410609762

Marzolf, D. P., & DeLoache, J. S. (1994). Transfer in young children's understanding of spatial representations. *Child Development, 65*(1), 1–15. doi:10.2307/1131361 PMID:8131641

Mason, M. (Ed.). (2008). Complexity Theory and the Philosophy of Education. Wiley-Blackwell. .124 doi:10.1002/9781444307351

Mason, J. (2009). Teaching as disciplined enquiry. *Teachers and Teaching, 15*(2), 205–223. doi:10.1080/13540600902875308

Matusiak, B. S. (2017). Daylighting is more than an energy-saving issue. In Energy efficient buildings. doi:10.5772/65866

Mayer, R. E. (2004). Should there be a three-strikes rule against pure discovery learning? *The American Psychologist, 59*(1), 14–19. doi:10.1037/0003-066X.59.1.14 PMID:14736316

Mazzocchi, F. (2019). Scientific research across and beyond disciplines. *EMBO Reports, 20*(6), 1–6. doi:10.15252/embr.201947682 PMID:31040110

McCulloch, A. W., Marshall, P. L., DeCuir-Gunby, J. T., & Caldwell, T. S. (2013). Math autobiographies: A window into teachers' identities as mathematics learners. *School Science and Mathematics, 113*(8), 380–389. doi:10.1111/ssm.12041 doi:10.1111sm.12041

McEwan, P. J. (1998). The effectiveness of multigrade schools in Colombia. *International Journal of Educational Development, 18*(6), 435–452. doi:10.1016/S0738-0593(98)00023-6

McGee, J. R., Wang, C., & Polly, D. (2013). Guiding teachers in the use of a standards-based mathematics curriculum: Teacher perceptions and subsequent instructional practices after an intensive professional development program. *School Science and Mathematics, 113*(1), 16–28. doi:10.1111/j.1949-8594.2012.00172.x

McGinnis, J. R., Watanabe, T., & McDuffie, A. R. (2005). University mathematics and science faculty modeling their understanding of reform based instruction in a teacher preparation program: Voices of faculty and teacher candidates. *International Journal of Science and Mathematics Education, 3*(3), 407–428. doi:10.100710763-005-3591-3

McMillan, B. G., & Sagun, T. (2020). Extending choral counting. *Mathematics Teacher: Learning and Teaching PK-12, 113*(8), 618-627.

McNeil, N. M., & Alibali, M. W. (2005). Why won't you change your mind? Knowledge of operational patterns hinders learning and performance on equations. *Child Development, 76*(4), 883–899. doi:10.1111/j.1467-8624.2005.00884.x PMID:16026503

McPhan, G., Pegg, J., & Horarik, S. (2008). Feedback about professional growth for teachers of mathematics: A developmental perspective. In M. Goos, R. Brown, & K. Makar (Eds.), *Proceedings of the 31st Annual Conference of the Mathematics Education Research Group of Australasia* (pp. 345–352). Mathematics Education Research Group of Australasia Inc. https://www.merga.net.au/Public/Publications/Annual_Conference_Proceedings/2008_MERGA_CP.aspx

Mercer, N., & Sams, C. (2006). Teaching children how to use language to solve maths problems. *Language and Education, 20*(6), 507–528. doi:10.2167/le678.0

Merriam, S. B. (1998). *Qualitative research and case study applications in education. Revised and expanded from "Case study research in education"*. Jossey-Bass Publishers.

Metzler, J., & Woessmann, L. (2012). The impact of teacher subject knowledge on student achievement: Evidence from within-teacher within-student variation. *Journal of Development Economics*, *99*(2), 486–496. doi:10.1016/j.jdeveco.2012.06.002

Meyer, D. (n.d.). *Dan Meyer's three-act math tasks*. Google Docs. Retrieved from https://docs.google.com/spreadsheets/d/1jXSt_CoDzyDFeJimZxnhgwOVsWkTQEsfqouLWNNC6Z4/edit#gid=0

Michaels, S., O'Connor, M. C., Hall, M. W., & Resnick, L. B. (2010). Accountable talk® sourcebook. Pittsburg, PA: Institute for Learning University of Pittsburgh.

Middleton, J. A., & Spanias, P. (1999). Motivation for Achievement in Mathematics: Findings, Generalizations, and Criticisms of the Research. *Journal for Research in Mathematics Education*, *30*(1), 65–88. doi:10.2307/749630

Milinkovic, J. (2012). Pre-service teachers' representational preferences. In T. Tso (Ed.), *Proceedings of the 36th Conference of the International Group for the Psychology of Mathematics Education* (Vol. 3, pp 209-215). PME.

Mills, V. L. (2014). Mathematical goals: The alpha and omega of effective practice. *NCSM Summer Newsletter*, *44*(4), 2–3.

Ministry of Education. (2007). *The New Zealand Curriculum*. Learning Media.

Ministry of Education. (2008). *Book 7: Teaching fractions, decimals, and percentages*. Ministry of Education.

Mirvis, P. H., (1991). Flow: The Psychology of Optimal Experience. *Academy of Management Review*, *16*(3), 636–640. .3 doi:10.5465/amr.1991.4279513

Mishra, P., & Koehler, M. J. (2006). Technological pedagogical content knowledge: A framework for teacher knowledge. *Teachers College Record*, *108*(6), 1017–1054. doi:10.1177/016146810610800610

Moch, P. L. (2001). Manipulatives work! *The Educational Forum*, *66*(1), 81–87. doi:10.1080/00131720108984802

Monaghan, F. (2000). What difference does it make? Children's views of the differences between some quadrilaterals. *Educational Studies in Mathematics*, *42*(2), 179–196. doi:10.1023/A:1004175020394

MoNE. (2018). *Middle school mathematics program (Grades 5-8)*. MoNE. https://mufredat.meb.gov.tr/Dosyalar/2018813171044420-1-2018-80%20Matematik%20Uygulamalar%C4%B1%20Dersi%20(5-8.%20S%C4%B1n%C4%B1flar).pdf

Monson, D., Cramer, K., & Ahrendt, S. (2020). Using models to build fraction understanding. *Mathematics Teacher: Learning and Teaching PK-12*, *113*(2), 117-123.

Montessori, M. (1964). *The Montessori method* (A. E. George, Trans.). Robert Bentley.

Moore, W. P., & Esselman, M. E. (1992). *Teacher efficacy, empowerment, and a focused instructional climate: Does student achievement benefit?* Paper presented at the annual meeting of the American Educational Research Association, San Francisco, CA.

Moore, R. (2001). Basil Bernstein 1924-2000. *British Educational Research Journal*, *27*(3), 367–370. doi:10.1080/01411920120048359

Morin, E. (1999). *Seven complex lessons in education for the future*. UNESCO. https://unesdoc.unesco.org/ark:/48223/pf0000117740

Moschkovich, J. N. (1998). Resources for refining mathematical conceptions: Case studies in learning about linear functions. *Journal of the Learning Sciences*, *7*(2), 209–237. doi:10.120715327809jls0702_3

Moulds, P. (2004). Rich Tasks. *Educational Leadership*, *61*(4), 75–78.

Moyer-Packenham, P., & Bolyard, J. (2002). Exploring representation in the middle grades: Investigations in geometry with virtual manipulatives. *Australian Mathematics Teacher*, *58*(1), 19–25. https://digitalcommons.usu.edu/teal_facpub/52/

Muijs, D., & Reynolds, D. (2002). Teachers' beliefs and behaviors: What really matters? *Journal of Classroom Interaction*, *37*(2), 3–15. https://www.jstor.org/stable/23870407

Mulero, J., Segura, L., & Sepulcre, J. M. (2013). Is maths everywhere? Our students respond. *International Association of Technology, Education and Development (IATED)*, 1–10. http://rua.ua.es/dspace/handle/10045/27248

Murrey, D. L. (2008). Differentiating instruction in mathematics for the English language learner. *Mathematics Teaching in the Middle School*, *14*(3), 146–153. doi:10.5951/MTMS.14.3.0146

Naanou, P., & Rhodes, S. (2020). Cultural connections: Exploring the mathematics of qatayef. *Mathematics Teacher: Learning and Teaching PK-12*, *113*(12), 1034-1038.

Nabb, K. (2018a). Experiences using inquiry-oriented differential equations. *CODEE: Community of Ordinary Differential Equations Educators Journal*, *11*(1), 1-10. Available: https://scholarship.claremont.edu/codee/vol11/iss1/3

Nabb, K. (2018b). Reflecting on the fruits of active learning. *MathAMATYC Educator*, *9*(3), 9–11, 52.

Nabb, K. (2020). Active learning in undergraduate mathematics: The five practices and the five dimensions of TRU math. *MathAMATYC Educator*, *12*, 30–35, 68–69.

Nabb, K., Hofacker, E. B., Ernie, K. T., & Ahrendt, S. (2018). Using the 5 practices in mathematics teaching. *Mathematics Teacher*, *111*(5), 366–373. doi:10.5951/mathteacher.111.5.0366

Naidoo, K., & Naidoo, L. J. (2021). Designing teaching and reflection experiences to develop candidates' science teaching self-efficacy. *Research in Science & Technological Education*, •••, 1–21. doi:10.1080/02635143.2021.1895098

Nardi, E., & Steward, S. (2013). Is mathematics T.I.R.E.D? A profile of quiet disaffection in the secondary mathematics classroom. *British Educational Research Journal*, *29*(3), 345–366. doi:10.1080/01411920301852

Näslund-Hadley, E., Parker, S. W., & Hernandez-Agramonte, J. M. (2014). Fostering early math comprehension: Experimental evidence from Paraguay. *Global Education Review*, *1*(4).

National Council of Teachers of Mathematics (NCTM). (1989). *Curriculum and evaluation standards for school mathematics*. Reston, VA: NCTM.

National Council of Teachers of Mathematics (NCTM). (1991). *Professional standards for teaching mathematics*. Reston, VA: NCTM.

National Council of Teachers of Mathematics (NCTM). (1995). *Assessment standards for school mathematics*. Reston, VA: NCTM.

National Council of Teachers of Mathematics (NCTM). (2000). *Principles and standards for school mathematics*. Reston, VA: NCTM.

National Council of Teachers of Mathematics (NCTM). (2006). *Curriculum focal points for prekindergarten through grade 8 mathematics: A quest for coherence*. Reston, VA: NCTM.

National Council of Teachers of Mathematics (NCTM). (2014). *Principles to actions: Ensuring mathematical success for all*. Reston, VA: NCTM.

National Council of Teachers of Mathematics. (1989). *Curriculum and evaluation standards for school mathematics*. National Council of Teachers of Mathematics.

National Council of Teachers of Mathematics. (1991). *Professional standards for teaching mathematics*. National Council of Teachers of Mathematics.

National Council of Teachers of Mathematics. (1995). *Assessment Standards for School Mathematics*. National Council of Teachers of Mathematics.

National Council of Teachers of Mathematics. (2000). *Principles and Standards for School Mathematics*. National Council of Teachers of Mathematics.

National Council of Teachers of Mathematics. (2014). *Principles to actions*. National Council of Teachers of Mathematics.

National Council of Teachers of Mathematics. (2014). *Principles to Actions: Ensuring Mathematical Success for All*. Author.

National Council of Teachers of Mathematics. (2014). *Principles to actions: Ensuring mathematical success for all*. National Council of Teachers of Mathematics.

National Council of Teachers of Mathematics. (2016). *Illuminations playing fraction tracks*. Retrieved from https://illuminations.nctm.org/Activity.aspx?id=6637

National Council of Teachers of Mathematics. (2018). *Catalyzing change in high school mathematics: Initiating critical conversations*. National Council of Teachers of Mathematics.

National Governors Association Center for Best Practices & Council of Chief State School Officers. (2010). *Common core state standards for mathematics*. Author.

National Governors Association Center for Best Practices (NGA Center) & Council of Chief State School Officers (CCSSO). (2010). Common core state standards for mathematics. Washington, DC: NGA Center and CCSSO.

National Mathematics Advisory Panel. (2008). *National Mathematics Advisory Panel Foundations for success: The final report of the National Mathematics Advisory Panel*. https://www.ed.gov/about/bdscomm/list/mathpanel/report/final-report.pdf

National Research Council. (1996). *The preparation of mathematics teachers: Considerations and challenges*. National Academy Press.

National Research Council. (2000). *Inquiry and the national science education standards*. National Academy Press.

National Research Council. (2001). *Adding it up: Helping children learn mathematics*. National Academy Press.

National Science Foundation. (2002). Professional development that supports school mathematics reform. In *Foundations: A monograph for professionals in science, mathematics, and technology education* (Vol. 3, pp. 1-130). Arlington, VA: Author. Retrieved July 24, 2021, from https://www.nsf. gov/pubs/2002/nsf02084/start.htm

NCTM. (2000). *Principles and standards for school mathematics*. Author.

NCTM. (2014). *Principles to action: Ensuring mathematical success for all*. Author.

Neubrand, M., Seago, N., Agudelo-Valderrama, C., DeBlois, L., Leikin, R., & Wood, T. (2009) The balance of teacher knowledge: Mathematics and pedagogy. In RThe professional education and development of teachers of mathematics. New ICMI study series (Vol. 11, pp. 211-225). Springer. doi:10.1007/978-0-387-09601-8_21

Niess, M. (2008). Mathematics teachers developing technology, pedagogy and content knowledge (TPACK). *Society for Information Technology and Teacher Education Proceedings*, 5297–5304. https://www.learntechlib.org/primary/p/28121

Nieto, S. (2003). Teaching as autobiography. In S. Nieto (Ed.), *What keeps teachers going?* (pp. 22–36). Teachers College Press.

Nilsson, P. (2014). When teaching makes a difference: Developing science teachers' pedagogical content knowledge through learning study. *International Journal of Science Education, 36*(11), 1794–1814. doi:10.1080/09500693.2013.879621

Noormohammadi, S. (2014). Teacher reflection and its relation to teacher efficacy and autonomy. *Procedia: Social and Behavioral Sciences, 98*, 1380–1389. doi:10.1016/j.sbspro.2014.03.556

NRICH. (2018). *The Nrich Maths Project*. https://nrich.maths.org/

Nurlu, O. (2015). Investigation of teachers' mathematics teaching self-efficacy. *International Electronic Journal of Elementary Education, 8*(1), 21–40.

Nussbaum, E. M. (2008). Collaborative discourse, argumentation, and learning: Preface and literature review. *Contemporary Educational Psychology, 33*(3), 345–359. doi:10.1016/j.cedpsych.2008.06.001

NVivo qualitative data analysis software (Version 12). (2018). *QSR International*. Retrieved on October 1, 2021, from https://www.qsrinternational.com/nvivo-qualitative-data-analysis-software/home

O'Connor, K., Dearborne, E., & Casa, T. M. (2021). Inquiry + math workshop model = success! *Mathematics Teacher: Learning and Teaching PK-12, 114*(3), 187-195.

OECD. (2015). *Does math make you anxious? PISA in Focus, 48*. doi:10.1787/5js6b2579tnx-

Oh, Y. (2003). Applying theory of planned behavior model on studying teachers' change in mathematics instruction. In N. A. Payeman, B. J. Dougherty, & J. T. Zilliox (Eds.), *Proceedings of the 2003 Joint Meeting of PME and PMENA* (Vol. 4, pp. 405-412). PME.

Oldknow, A., & Tetlow, L. (2008). Using dynamic geometry software to encourage 3D visualisation and modelling. *The Electronic Journal of Mathematics & Technology, 2*(1), 1–8. https://link.gale.com/apps/doc/A178451802/AONE?u=anon~501ead4c&sid=googleScholar&xid=695b9a4a

Ontario Ministry of Education. (2005a). *The Ontario Curriculum, Grades 1–8: Mathematics, Revised*. Queen's Printer for Ontario.

Ontario Ministry of Education. (2005b). *The Ontario Curriculum, Grades 9 and 10: Mathematics, Revised*. Queen's Printer for Ontario.

Ontario Ministry of Education. (2007). *The Ontario Curriculum, Grades 11 & 12: Mathematics, Revised*. Queen's Printer for Ontario.

Ozmantar, M. F., Akkoc, H., Bingolbali, E., & Demir, S. (2011). Development of preservice teachers in linking multiple representations via technology. In B. Ubuz (Ed.), *Proceedings of the 35th Conference of the International Group for the Psychology of Mathematics Education* (Vol. 3, pp 313-320). PME.

Pang, J. (2011). What do prospective teachers analyze when they watch a mathematics lesson? In B. Ubuz (Ed.), *Proceedings of the 35th Conference of the International Group for the Psychology of Mathematics Education* (Vol. 3, pp 329-336). PME.

Paolucci, C. (2015). Changing perspectives: Examining the potential for advanced mathematical studies to influence preservice teachers' beliefs about mathematics. *Teaching and Teacher Education, 49*, 97–107. doi:10.1016/j.tate.2015.03.002

Pape, S. J., & Tchoshanov, M. A. (2001). The role of representation(s) in developing mathematical understanding. *Theory into Practice, 40*(2), 118–127. doi:10.120715430421tip4002_6

Park City Mathematics Institute. (2011). *Rubric for Implementing Standards for Mathematical Practice.* Retrieved March 15, 2016 from http://mathforum.org/pcmi/hstp/resources/rubric/rubric.pdf

Park Rogers, M., Cross, D. I., Sommerfield Gresalfi, M., Trauth-Nare, A., & Buck, G. (2011). First year implementation of a project-based learning approach: The need of addressing teachers orientations in the era of reform. *International Journal of Science and Mathematics Education, 9*(4), 893–917. doi:10.100710763-010-9248-x

Park, H., & Magiera, M. T. (2019). Pre-service teachers' conceptions of mathematical argumentation. In *Proceedings of the forty-first annual meeting of the North American Chapter of the International Group for the Psychology of Mathematics Education* (pp. 1264-1268). St Louis, MO: University of Missouri.

Park, H., & Magiera, M. T. (2020). Pre-service teachers' interpretations of mathematical reasoning. In *Mathematics Education Across Cultures: Proceedings of the 42nd Annual Meeting of the North American Chapter of the International Group for the Psychology of Mathematics Education* (pp. 1677-1681). Cinvestav/AMIUTEM/PME-NA. 10.51272/pmena.42.2020-265

Parrish, S. (2010). *Number talks: Helping children build mental math and computation strategies, grades K-5*. Math Solutions.

Parrish, S. (2010). *Number talks: Helping children build mental math and computation strategies, Grades K-5*. Math Solutions.

Parrish, S. D. (2011). Number talks build numerical reasoning. *Teaching Children Mathematics, 18*(3), 198–206. doi:10.5951/teacchilmath.18.3.0198

Patel, N., Franco, S., Miura, Y., & Boyd, B. (2012). Including curriculum focus in mathematics professional development for middle-school mathematics teachers. *School Science and Mathematics, 112*(5), 300–309. doi:10.1111/j.1949-8594.2012.00146.x

Peressini, D., & Knuth, E. (2000). The role of tasks in developing communities of mathematical inquiry. *Teaching Children Mathematics, 6*(6), 391–397. doi:10.5951/TCM.6.6.0391

Peretin, J. (2014). *The effectiveness of professional development using a checklist of common core state standards for mathematical practice for designing and delivering instruction on Algebra I achievement* (Doctoral dissertation). Robert Morris University.

Peterson, P. L., Fennema, E., Carpenter, T. P., & Loef, M. F. (1989). Teachers' pedagogical content beliefs in mathematics. *Cognition and Instruction, 6*(l), 1–40. doi:10.12071532690xci0601_1

Petit, M. M., Laird, R. E., Marsden, E. L., & Ebby, C. B. (2015). *A focus on fractions: Bringing research to the classroom* (2nd ed.). Routledge. doi:10.4324/9781315746098

Phan, H. P. (2014). Self-efficacy, reflection, and achievement: A short-term longitudinal examination. *The Journal of Educational Research, 107*(2), 90–102. doi:10.1080/00220671.2012.753860

Philipp, R., & Sowder, J. (2003). *Integrating mathematics and pedagogy (IMAP) web-based belief survey manual.* http://www.sci.sdsu.edu/CRMSE/IMAP/Beliefs-Survey_Manual.pdf

Philipp, R. A. (2007). Mathematics teachers' beliefs and affect. In F. Lester (Ed.), *Second handbook of research on mathematics teaching and learning* (Vol. 1, pp. 257–315). National Council of Teachers of Mathematics and Information Age Publishing.

Philipp, R. A., Ambrose, R., Lamb, L. L., Sowder, J. T., Schappelle, B. P., Sowder, L., & Chauvot, J. (2007). Effects of early field experiences on the mathematical content knowledge and beliefs of prospective elementary school teachers: An experimental study. *Journal for Research in Mathematics Education*, 438–476.

Piaget, J. (1962). *Play, dreams, and imitation in childhood* (C. Gattegno & F. M. Hodgson, Trans.). Norton.

Pierce, R., Stacey, K., Wander, R., & Ball, L. (2011). The design of lessons using mathematics analysis software to support multiple representations in secondary school mathematics. *Technology, Pedagogy and Education*, *20*(1), 95–112. doi:10.1080/1475939X.2010.534869

Piñero, J. L., Castro-Rodriguez, E., & Castro, E. (2018). Prospective primary teachers' conceptual understanding of mathematical problems and problem solving. In E. Bergqvist, M. Osterholm, C. Granberg, & L. Sumpter (Eds.), *Proceedings of the 42nd Conference of the International Group for the Psychology of Mathematics Education* (Vol. 3, pp. 465-472). PME.

Piper, B., & Mugenda, A. (2016). *USAID/Kenya Primary Math and Reading (PRIMR) Initiative: Endline impact evaluation* (revised edition). RTI International. https://learningportal.iiep.unesco.org/en/library/the-primary-math-and-reading-primr-initiative-endline-impact-evaluation

Piper, B., Ralaingita, W., Akach, L., & King, S. (2016). Improving mathematics outcomes in Kenya: The impact of the Primary Math and Reading (PRIMR) Initiative. *Journal of Development Effectiveness*. Advance online publication. doi:10.1080/19439342.2016.1149502

Pittalis, M., & Christou, C. (2010). Types of reasoning in 3D geometry thinking and their relation with spatial ability. *Educational Studies in Mathematics*, *75*(2), 191–212. doi:10.100710649-010-9251-8

Platas, L. M. (2014). Knowledge of Mathematical Development Survey: Testing the validity and reliability of the survey and interpreting its results. *NHSA Dialog*, *17*(1). https://journals.charlotte.edu/dialog/article/view/123

Platas, L. M. (2015). The Mathematical Development Beliefs Survey: Validity and reliability of a measure of preschool teachers' beliefs about the learning and teaching of early mathematics. *Journal of Early Childhood Research*, *13*(3), 295–310. doi:10.1177/1476718X14523746

Pollitt, R., Cohrssen, C., & Seah, W. T. (2020). Assessing spatial reasoning during play: Educator observations, assessment and curriculum planning. *Mathematics Education Research Journal*, *32*(2), 331–363. doi:10.100713394-020-00337-8

Polya, G. (1962). *Mathematical discovery* (Vol. 1). John Wiley & Sons.

Popovic, G. (2011). *Informal settings: Implications for secondary mathematics teachers' ability to recognize and apply mathematics* (Doctoral dissertation). Illinois Institute of Technology.

Pothier, Y., & Sawada, D. (1983). Partitioning: The emergence of rational number ideas in young children. *Journal for Research in Mathematics Education*, *14*(5), 307. doi:10.2307/748675

Prelli, G. E. (2016). How school leaders might promote higher levels of collective teacher efficacy at the level of school and team. *English Language Teaching*, *9*(3), 174. doi:10.5539/elt.v9n3p174

Pribyl, J. R., & Bodner, G. M. (2006). Spatial ability and its role in organic chemistry: A study of four organic courses. *Journal of Research in Science Teaching*, *24*(3), 229–240. doi:10.1002/tea.3660240304

Priestley, M. (2011). Schools, teachers, and curriculum change: A balancing act? *Journal of Educational Change*, *12*(1), 1–23. doi:10.100710833-010-9140-z

Pritchett, L., & Beatty, A. (2012). *The negative consequences of overambitious curricula in developing countries.* Center for Global Development Working Paper 293. https://www.cgdev.org/sites/default/files/1426129_file_Pritchett_Beatty_Overambitious_FINAL_0.pdf

Prothero, A. (2018). Charter Schools. *Education Week.* https://www.edweek.org/policy-politics/what-are-charter-schools/2018/08

Pryor, J., Akyeampong, K., Westbrook, J., & Lussier, K. (2012). Rethinking teacher preparation and professional development in Africa: An analysis of the curriculum of teacher education in the teaching of early reading and mathematics. *Curriculum Journal, 23*(4), 409–502. doi:10.1080/09585176.2012.747725

Puendetura, R. (2013). *SAMR: Moving from enhancement to transformation.* Retrieved on October 1, 2021, from http://www.hippasus.com/rrpweblog/archives/2013/05/29/SAMREnhancementToTransformation.pdf

Pyper, J. S., & MacGregor, S. (2018). Problem solving: How preservice teachers understand it during their preservice learning. In E. Bergqvist, M. Osterholm, C. Granberg, & L. Sumpter (Eds.), *Proceedings of the 42nd Conference of the International Group for the Psychology of Mathematics Education* (Vol. 4, pp. 19-26). PME.

Rakes, C. R., Kirvan, R. J., & Witkowski, A. (2020). Discovering radical simplification through geometry connections. *Mathematics Teacher: Learning and Teaching PK-12, 113*(9), 737-744.

Rasmussen, C., Keene, K. A., Dunmyre, J., & Fortune, N. (2018). *Inquiry oriented differential equations.* Course materials. Available at https://iode.wordpress.ncsu.edu

Rasmussen, C. L., & King, K. D. (2000). Locating starting points in differential equations: A realistic mathematics education approach. *International Journal of Mathematical Education in Science and Technology, 31*(2), 161–172. doi:10.1080/002073900287219

Rasmussen, C., & Marrongelle, K. (2006). Pedagogical content tools: Integrating student reasoning and mathematics in instruction. *Journal for Research in Mathematics Education, 37*, 388–420. doi:10.2307/30034860

Raymond, A. M. (1997). Inconsistency between a beginning elementary school teacher's mathematics beliefs and teaching practice. *Journal for Research in Mathematics Education, 28*(5), 550–576. doi:10.2307/749691

Read, N. (2017). Measures of learning and teaching material availability and use in Sub-Saharan Africa and other low-income countries. *European Journal of Education: Research, Development and Policy, 52*(4), 523–532. doi:10.1111/ejed.12242

Reid, A., Petocz, P., Smith, G., Wood, L., & Dortins, E. (2003). Mathematics students' conceptions of mathematics. *New Zealand Journal of Mathematics, 32*(Supplementary Issue), 163–172. http://www.sun.ac.za/english/learning-teaching/ctl/Documents/Reid%20et%20al%202003.pdf

Reinhold, S. (2015). Prospective elementary teachers' diagnostic proceeding in one-on-one diagnostic interviews: Facets of data collection and attention. In *Proceedings of the 39th Conference of the International Group for the Psychology on Mathematics Education* (Vol. 4, pp. 73-80). PME.

Reinholz, D. L. (2020). Five practices for supporting inquiry in analysis. *PRIMUS: Problems, Resources, and Issues in Mathematics Undergraduate Studies, 30*(1), 19–35. doi:10.1080/10511970.2018.1500955

Reisberg, D., & Heuer, F. (2005). Visuospatial images. In A. Miyake & P. Shah (Eds.), *The Cambridge handbook of visuospatial thinking* (pp. 35–80). Cambridge University Press. doi:10.1017/CBO9780511610448.003

Remillard, J. T., & Heck, D. J. (2014). Conceptualizing the curriculum enactment process in mathematics education. *ZDM - Mathematics Education, 46*(5), 705–718. .707 doi:10.1007/s11858-014-0600-4

Renkl, A. (2011). Instruction based on examples. In R. E. Mayer & P. A. Alexander (Eds.), *Handbook of research on learning and instruction* (pp. 272–295). Routledge. https://www.routledgehandbooks.com/doi/10.4324/9780203839089.ch3

Renkl, A. (2014). Toward an instructionally oriented theory of example-based learning. *Cognitive Science*, *38*(1), 1–37. doi:10.1111/cogs.12086 PMID:24070563

Renkl, A. (2017). Learning from worked-examples in mathematics: Students relate procedures to principles. *ZDM*, *49*(4), 571–584. doi:10.100711858-017-0859-3

Resnick, L. (1992). From protoquantities to operators: Building mathematical competence on a foundation of everyday knowledge. In G. Leinhardt, R. Putnam, & R. A. Hattrup (Eds.), *Analysis of arithmetic for mathematics teaching* (pp. 373–429). L. Erlbaum Associates.

Richardson, V. (2003). Preservice teachers' beliefs. In J. Raths (Ed.), *Teacher beliefs and classroom performance: The impact of teacher education* (pp. 1–22). Information Age Publishing.

Rimbey, K. A. (2013). *From the common core to the classroom: A professional development efficacy study for the common core state standards for mathematics* (Doctoral dissertation). Arizona State University.

Rittle-Johnson, B., & Koedinger, K. R. (2005). Designing knowledge scaffolds to support mathematical problem-solving. *Cognition and Instruction*, *23*(3), 313–349. doi:10.12071532690xci2303_1

Robertson, C., Nelson, T. A., Jelinski, D. E., Wulder, M. A., & Boots, B. (2009). Spatial-temporal analysis of species range expansion: The case of the mountain pine beetle, dendroctonus ponderosae. *Journal of Biogeography*, *36*(8), 1446–1458. doi:10.1111/j.1365-2699.2009.02100.x

Roberts, S. L. (2013). The "Chalk Talk" 2.0: Using google docs to improve the silent discussion in social studies. *Social Studies*, *104*(3), 130–136. doi:10.1080/00377996.2012.703972

Rockhurst University. (2018). *IBL calculus: Concepts and applications*. Retrieved from https://www.iblcalculus.com/

Rogers, K. C., & Kosko, K. W. (2019). How elementary and collegiate instructors envision tasks as supportive of mathematical argumentation: A comparison of instructors' task constructions. *The Journal of Mathematical Behavior*, *53*, 228–241. doi:10.1016/j.jmathb.2018.08.004

Romberg, T. A. (1989). *Curriculum and Evaluation Standards for School Mathematics*. The National Council of Teachers of Mathematics.

Ross, J. A., McDougall, D., & Hogaboam-Gray, A. (2002). Research on reform in mathematics education, 1993-2000. *The Alberta Journal of Educational Research*, *48*(2), 122–138.

Ross, S. H. (1989). Parts, wholes, and place value: A developmental view. *The Arithmetic Teacher*, *36*(6), 47–51. doi:10.5951/AT.36.6.0047

Roy, G. J., Allen, J. S., & Thacker, K. (2020). Linking factors and multiples to algebraic reasoning. *Mathematics Teacher: Learning and Teaching PK-12, 113*(3), 216-222.

Royal Society Te Apārangi. (2021). *Pāngarau mathematics and tuanga statistics in Aotearoa New Zealand: Advice on refreshing the English-medium Mathematics and Statistics learning area of the New Zealand Curriculum*. Royal Society Te Apārangi.

Royea, D., Nicol, C., & Osana, H. (2015). "Borrowing from the neighbour" – Preservice teacher' interpretations of student errors. In *Proceedings of the 39th Conference of the International Group for the Psychology on Mathematics Education* (*Vol. 4*, pp. 97-104). PME.

Rumack, A. M., & Huinker, D. (2019). Capturing mathematical curiosity with notice and wonder. *Mathematics Teaching in the Middle School, 24*(7), 394–399. doi:10.5951/mathteacmiddscho.24.7.0394

Russo, J., Russo, T., & Kalogeropoulos, P. (2020). Exploring exponential growth in elementary school. *Mathematics Teacher: Learning and Teaching PK-12, 113*(12), 989-994.

Ruthven, K., Mercer, N., Taber, K. S., Guardia, P., Hofmann, R., Ilie, S., Luthman, S., & Riga, F. (2017). A research-informed dialogic-teaching approach to early secondary school mathematics and science: The pedagogical design and field trial of the epiSTEMe intervention. *Research Papers in Education, 32*(1), 18–40. doi:10.1080/02671522.2015.1129642

Sakonidis, H., Tzekaki, M., & Kaldrimidou, M. (2001). Mathematics teaching practices in transition: Some meaning construction issues. In M. van den Heuvel-Panhuizen (Ed.), *Proceedings of the 25th Conference of the International Group for the Psychology of Mathematics Education* (Vol. 4, RR248). PME.

Saldaña, J. (2015). *The coding manual for qualitative researchers* (3rd ed.). Sage.

Salden, R. J. C. M., Aleven, V. A. W. M. M., Renkl, A., & Schwonke, R. (2009). Worked examples and tutored problem-solving. *Topics in Cognitive Science, 1*(1), 203–213. doi:10.1111/j.1756-8765.2008.01011.x PMID:25164806

Salomon, G., & Perkins, D. (2005). Do technologies make us smarter? Intellectual amplification with, of, and through technology. In R. J. Sternberg & D. D. Preiss (Eds.), *Intelligence and technology: The impact of tools on the nature and development of human abilities* (pp. 71–86). Erlbaum. doi:10.4324/9780203824252-11

Sánchez-Matamoros, G., Fernández, C., Valls, J., & Callejo de la Vega, M. L. (2018). Noticing students' mathematical thinking: Characterization, development and contexts. In E. Bergqvist, M. Osterholm, C. Granberg, & L. Sumpter (Eds.), *Proceedings of the 42nd Conference of the International Group for the Psychology of Mathematics Education* (Vol. 2, pp. 91-98). PME.

Sánchez, V., & Garcia, M. (2011). Socio-mathematical and mathematical norms in pre-service primary teachers' discourse. In B. Ubuz (Ed.), *Proceedings of the 35th Conference of the International Group for the Psychology of Mathematics Education* (Vol. 4, pp 105-112). PME.

Santagata, R., & Angelici, G. (2010). Studying the impact of the lesson analysis framework on preservice teachers' abilities to reflect on videos of classroom teaching. *Journal of Teacher Education, 61*(4), 339–349. doi:10.1177/0022487110369555

Santagata, R., & Yeh, C. (2014). Learning to teach mathematics and to analyze teaching effectiveness: Evidence from a video-and practice-based approach. *Journal of Mathematics Teacher Education, 17*(6), 491–514. doi:10.100710857-013-9263-2

Saralar, İ. (2020). *Designing lessons to help middle school students learn about orthogonal and isometric drawings of three-dimensional shapes* [Doctoral Thesis, The University of Nottingham]. Nottingham eThesis.

Saralar, İ., & Ainsworth, S. (2017). *An exploration of middle school mathematics teachers' beliefs and goals regarding GeoGebra: Four cases from the Turkish Republic* [Paper Presentation]. European Conference on Educational Research (ECER) 2017, Copenhagen, Denmark. https://eera-ecer.de/ecer-programmes/conference/22/contribution/39865/

Saralar-Aras, İ. (2022). An exploration of middle school mathematics teachers' beliefs and goals regarding a dynamic mathematics tool: GeoGebra. *Journal of Research in Science Mathematics and Technology Education, 5*(SI), 41–63. doi:10.31756/jrsmte.113SI

Sawyer, A. (2018). Factors influencing elementary mathematics teachers' beliefs in reform-based teaching. *The Mathematics Educator, 26*(2), 26–53.

Schank, R. C. (1994). Active learning through multimedia. *IEEE MultiMedia, 1*(1), 69–78. doi:10.1109/93.295270

Schmidt, W. H., Blömeke, S., Tatto, M. T., Hsieh, F. J., Cogan, L., Houang, R. T., & Schwille, J. (2011). *Teacher education matters: A study of middle school mathematics teacher preparation in six countries*. Teachers College Press, Columbia University.

Schmidt, W. H., Cogan, L., & Houang, R. (2011). The role of opportunity to learn in teacher preparation: An international context. *Journal of Teacher Education, 62*(2), 138–153. doi:10.1177/0022487110391987

Schmidt, W., Houang, R., & Cogan, L. (2002). A coherent curriculum. *American Educator, 26*(2), 1–18.

Schmoker, M. (2018). *Focus: Elevating the essentials to radically improve student learning*. Association of Supervisors and Curriculum Developers.

Schoenfeld, A. H. (1985). *Mathematical problem solving*. Academic Press.

Schoenfeld, A. H. (1985). *Mathematical problem-solving*. Academic Press.

Schoenfeld, A. H. (2006). Mathematics teaching and learning. In P. A. Alexander & P. H. Winne (Eds.), *Handbook of educational psychology* (pp. 479–510). Lawrence Erlbaum Associates Publishers.

Schoenfeld, A. H. (2013). Reflections on problem-solving theory and practice. *The Mathematics Enthusiast, 10*(1), 9–34. doi:10.54870/1551-3440.1258

Schoenfeld, A. H. (2014). What makes for powerful classrooms, and how can we support teachers in creating them? A story of research and practice, productively intertwined. *Educational Researcher, 43*(8), 404–412. doi:10.3102/0013189X14554450

Schoenfeld, A. H. (2016, April/May). What makes for powerful classrooms—And what can we do, now that we know? *MAA Focus, 36*(2), 11–14.

Schoenfeld, A. S. (1998). Toward a theory of teaching-in-context. *Issues in Education, 4*(1), 1–95. doi:10.1016/S1080-9724(99)80076-7

Schoen, H. L., Cebulla, K. J., Finn, K. F., & Fi, C. (2003). Teacher variables that relate to student achievement when using a standards-based curriculum. *Journal for Research in Mathematics Education, 34*(3), 228–259. doi:10.2307/30034779

Schön, D. A. (1983). *The reflective practitioner: How professionals think in action*. Temple Smith.

Schweisfurth, M. (2011). Learner-centred education in developing country contexts: From solution to problem? *International Journal of Educational Development, 31*(5), 425–432. doi:10.1016/j.ijedudev.2011.03.005

Schwille, J., & Dembélé, M. (2007). *Global perspectives on teacher learning: Improving policy and practice*. UNESCO International Institute for Educational Planning. https://files.eric.ed.gov/fulltext/ED496753.pdf

Seagal, S., & Horne, D. (1997). *Human Dynamics. A New Framework for Understanding People and Realizing the Potential in Our Organizations*. Pegasus Communications.

Seidman, I. (2005). Interviewing as Qualitative Research: A Guide for Researchers in Education and the Social Sciences. Teachers College Press.

Seung, E., Park, S., & Jung, J. (2014). Exploring preservice elementary teachers' understanding of the essential features of inquiry-based science teaching using evidence-based reflection. *Research in Science Education, 44*(4), 507–529. doi:10.100711165-013-9390-x

Sevimli, E., & Delice, A. (2011). Is what you prefer what you do? Representations in definite integral. In B. Ubuz (Ed.), *Proceedings of the 35th Conference of the International Group for the Psychology of Mathematics Education* (Vol. 4, pp 153-160). PME.

Sfard, A., & Prusak, A. (2005). Telling identities: In search of an analytic tool for investigating learning as a culturally shaped activity. *Educational Researcher, 34*(4), 14–22. doi:10.3102/0013189X034004014

Shaughnessy, M., Garcia, N. M., O'Neill, M. K., Selling, S. K., Willis, A. T., Wilkes, C. E. II, Salazar, S. B., & Ball, D. L. (2021). Formatively assessing prospective teachers' skills in leading mathematics discussions. *Educational Studies in Mathematics, 108*(3), 451–472. doi:10.100710649-021-10070-z PMID:34966189

Shaughnessy, M., Garcia, N., Selling, S. K., & Ball, D. L. (2016). What knowledge and skill do mathematics teacher educators need and (how) can we support its development? In M. B. Wood, E. E. Turner, M. Civil, & J. A. Eli (Eds.), *Proceedings of the 38th Annual Meeting of the North American Chapter of the International Group for the Psychology of Mathematics Education* (pp. 813-820). Tucson, AZ: The University of Arizona.

Sherin, M. G. (2002). When teaching becomes learning. *Cognition and Instruction, 20*(2), 119–150. doi:10.1207/S1532690XCI2002_1

Sherin, M. G., Jacobs, V. R., & Philipp, R. A. (Eds.). (2011). *Mathematics teacher noticing: Seeing through teachers' eyes*. Routledge. doi:10.4324/9780203832714

Shulman, L. (1986). Those who understand: Knowledge growth in teaching. *Educational Researcher, 15*(2), 4–14. doi:10.3102/0013189X015002004

Shulman, L. (1987). Knowledge and teaching: Foundations of the new reform. *Harvard Educational Review, 57*(1), 1–23. doi:10.17763/haer.57.1.j463w79r56455411

Shulman, L. S., & Shulman, J. H. (2004). How and what teachers learn: A shifting perspective. *Journal of Curriculum Studies, 36*(2), 257–271. doi:10.1080/0022027032000148298

Siegler, R., Carpenter, T., Fennell, F., Geary, D., Lewis, J., Okamoto, Y., & Wray, J. (2010). *Developing Effective Fractions Instruction for Kindergarten through 8th Grade. IES Practice Guide. NCEE 2010-4039*. What Works Clearinghouse.

Silver, E. A., Shapiro, L. J., & Deutsch, A. (1993). Sense making and the solution of division problems involving remainders: An examination of middle school students' solution processes and their interpretations of solutions. *Journal for Research in Mathematics Education, 24*(2), 117–135. doi:10.2307/749216

Simmons, P. E., Emory, A., Carter, T., Coker, T., Finnegan, B., Crockett, D., Richardson, L., Yager, R., Craven, J., Tillotson, J., Brunkhorst, H., Twiest, M., Hossain, K., Gallagher, J., Duggan-Haas, D., Parker, J., Cajas, F., Alshannag, Q., McGlamery, S., ... Labuda, K. (1999). Beginning teachers: Beliefs and classroom actions. *Journal of Research in Science Teaching, 36*(8), 930–954. doi:10.1002/(SICI)1098-2736(199910)36:8<930::AID-TEA3>3.0.CO;2-N

Simon, M. A. (1995). Reconstructing mathematics pedagogy from a constructivist perspective. *Journal for Research in Mathematics Education, 26*(2), 114–145. doi:10.2307/749205

Simpson, G., Hoyles, C., & Noss, R. (2006). Exploring the mathematics of motion through construction and collaboration. *Journal of Computer Assisted Learning, 22*(2), 114–136. doi:10.1111/j.1365-2729.2006.00164.x

Sinha, T., & Kapur, M. (2021). When problem-solving followed by instruction works: Evidence for productive failure. *Review of Educational Research, 91*(5), 761–798. doi:10.3102/00346543211019105

Sitabkhan, Y., & Platas, L. M. (2018). *Early Mathematics Counts: Promising Instructional Strategies from Low- and Middle-Income Countries*. RTI Press Publication No. OP-0055-1807. RTI Press. doi:10.3768/rtipress.2018.op.0055.1807

Sitabkhan, Y., Davis, J., Earnest, D., Evans, N., Ketterlin-Geller, L., Lutfeali, S., Ngware, M., Perry, L., Pinto, C., Platas, L., Ralaingita, W., Smith, K., & Srikantaiah, D. (2019). *Instructional strategies for mathematics in the early grades.* Mathematics Working Group Working Paper. USAID. https://shared.rti.org/content/instructional-strategies-mathematics-early-grades

Skattebol, J., & Arthur, L. M. (2014). Collaborative practitioner research: Opening a third space for local knowledge production. *Asia Pacific Journal of Education*, *34*(3), 351–365. doi:10.1080/02188791.2013.871690

Skemp, R. R. (2012). The psychology of learning mathematics (Expanded American Edition). Routledge. doi:10.4324/9780203396391

Skemp, R. R. (1987). *The psychology of teaching mathematics* (Revised American Edition). Lawrence Erlbaum. doi:10.4324/9781315621128

Skott, J. (2001). The emerging practices of a novice teacher: The roles of his school mathematics images. *Journal of Mathematics Teacher Education*, *4*(1), 3–28. doi:10.1023/A:1009978831627

Slick, S. K. (1998). The university supervisor: A disenfranchised outsider. *Teaching and Teacher Education*, *14*(8), 821–834. doi:10.1016/S0742-051X(98)00028-6

Smith, J. P. III. (1996). Efficacy and teaching mathematics by telling: A challenge for reform. *Journal for Research in Mathematics Education*, *27*(4), 387–402. doi:10.5951/jresematheduc.27.4.0387

Smith, M. S., Bill, V., & Hughes, E. K. (2008). Thinking through a lesson: Successfully implementing high-level tasks. *Mathematics Teaching in the Middle School*, *14*(3), 132–138. doi:10.5951/MTMS.14.3.0132

Smith, M. S., & Stein, M. K. (2011). *5 practices for orchestrating productive mathematics discussions.* National Council of Teachers of Mathematics.

Smith, M. S., & Stein, M. K. (2011). *Five practices for orchestrating productive mathematics discussions.* National Council of Teachers of Mathematics.

Smith, M., Bill, V., & Sherin, M. G. (2020). *The five practices in practice: Successfully orchestrating mathematics discussions in your elementary school classroom.* Corwin.

Smith, M., & Sherin, M. G. (2019). *The five practices in practice: Successfully orchestrating mathematics discussions in your middle school classroom.* Corwin.

Smith, M., Steele, M. D., & Sherin, M. G. (2020). *The five practices in practice: Successfully orchestrating mathematics discussions in your high school classroom.* Corwin.

Snider, R. B. (2021). Guiding questions for selecting mathematical examples. *Mathematics Teacher: Learning and Teaching PK-12, 114*(1), 63-68.

Son, J., & Lee, J. (2016). Pre-service teachers' understanding of fraction multiplication, representational knowledge, and computational skills. *Mathematics Teacher Education and Development*, *18*(2), 5–28.

Sowell, E. J. (1989). Effects of manipulative materials in mathematics instruction. *Journal for Research in Mathematics Education*, *20*(1), 498–505. doi:10.2307/749423

Spangler, D. A., & Wanko, J. J. (Eds.). (2017). Enhancing classroom practice with research behind principles to actions. NCTM, National Council of Teachers of Mathematics.

Stacey, K., & Vincent, J. (2009). Modes of reasoning in explanations in Australian eight-grade mathematics textbooks. *Educational Studies in Mathematics*, *72*(3), 271–288. doi:10.100710649-009-9193-1

Stake, R. E. (1995). *The art of case study research.* SAGE Publications Ltd.

Stake, R. E. (2005). Qualitative case studies. In N. K. Denzin & Y. S. Lincoln (Eds.), *The SAGE handbook of qualitative research* (3rd ed., pp. 443–466). Sage.

Staples, M. E., Bartlo, J., & Thanheiser, E. (2012). Justification as a teaching and learning practice: Its (potential) multifaceted role in middle grades mathematics classrooms. *The Journal of Mathematical Behavior, 31*(4), 447–462. doi:10.1016/j.jmathb.2012.07.001

Staples, M., & Newton, J. (2016). Teachers' contextualization of argumentation in the mathematics classroom. *Theory into Practice*, *55*(4), 294–301. doi:10.1080/00405841.2016.1208070

Star, J. R., Jeon, S., Comeford, R., Clark, P., Rittle-Johnson, B., & Durkin, K. (2021). Compare and discuss multiple strategies. *Mathematics Teacher: Learning and Teaching PK-12, 114*(11), 853-859.

Star, J. R., & Strickland, S. K. (2008). Learning to observe: Using video to improve preservice mathematics teachers' ability to notice. *Journal of Mathematics Teacher Education, 11*(2), 107–125. doi:10.100710857-007-9063-7

Steele, C. M., & Aronson, J. (1995). Stereotype threat and the intellectual test performance of African Americans. *Journal of Personality and Social Psychology, 69*(5), 797–811. doi:10.1037/0022-3514.69.5.797 PMID:7473032

Steffe, L. P. (2001). A new hypothesis concerning children's fractional knowledge. *The Journal of Mathematical Behavior, 20*(3), 267–307. doi:10.1016/S0732-3123(02)00075-5

Stein, M. K., & Bovalino, J. W. (2001). Manipulatives: One piece of the puzzle. *Mathematics Teaching in the Middle School, 6*(6), 356–359. doi:10.5951/MTMS.6.6.0356

Stein, M. K., Correnti, R., Moore, D., Russell, J. L., & Kelly, K. (2017). Using theory and measurement to sharpen conceptualizations of mathematics teaching in the common core era. *AERA Open, 3*(1), 1–20. doi:10.1177/2332858416680566

Stein, M. K., Engle, R. A., Smith, M. S., & Hughes, E. K. (2008). Orchestrating productive mathematical discussions: Five practices for helping teachers move beyond show and tell. *Mathematical Thinking and Learning, 10*(4), 313–340. doi:10.1080/10986060802229675

Stephan, M. L., Reinke, L. T., & Cline, J. K. (2020). Beyond hooks: Real-world contexts as anchors for instruction. *Mathematics Teacher: Learning and Teaching PK-12, 113*(10), 821-827.

Stockero, S. L. (2013). Student teacher noticing during mathematics instruction. In A. M. Lindmeier, & A. Heinze (Eds.), *Proceedings of the 37th Conference of the International Group for the Psychology of Mathematics Education* (Vol. 4, pp. 249-256). PME.

Stohlmann, M. S. (2020a). Planning questions to help surface understanding. *Mathematics Teacher: Learning and Teaching PK-12, 113*(4), 327-329.

Stohlmann, M. S. (2020b). Escape room math: Luna's lines. *Mathematics Teacher: Learning and Teaching PK-12, 113*(5), 383-389.

Sträßer, R. (2017). Cultural Framing of Teaching and Learning Mathematics. *Didactics of Mathematics as a Scientific Discipline*, 399-455. https://www.researchgate.net/publication/227113904_Cultural_Framing_of_Teaching_and_Learning_Mathematics

Straumanis, A., Bénéteau, C., Guadarrama, Z., Guerra, J. E., & Lenz, L. (2013). *Calculus I: A guided inquiry.* Wiley.

Strauss, A. L. (1987). *Qualitative analysis for social scientists.* Cambridge University Press. doi:10.1017/CBO9780511557842

Streefland, L. (1993). Fractions: A realistic approach. In T. P. Carpenter, E. Fennema, & T. A. Romberg (Eds.), *Rational numbers: An integration of research* (pp. 289–325). Routledge.

Su, F. (2017). Mathematics for human flourishing. *The American Mathematical Monthly, 124*(6), 483–493. doi:10.4169/amer.math.monthly.124.6.483

Sun, K. L., Baldinger, E. E., & Humphreys, C. (2018). Number talks: Gateway to sense making. *Mathematics Teacher, 112*(1), 48–54. doi:10.5951/mathteacher.112.1.0048

Supovitz, J. A., & Turner, H. M. (2000). The effects of professional development on science teaching practices and classroom culture. *Journal of Research in Science Teaching, 37*(9), 963–980. doi:10.1002/1098-2736(200011)37:9<963::AID-TEA6>3.0.CO;2-0

Suydam, M. N. (1984). Research report: Manipulative materials. *The Arithmetic Teacher, 31*(5), 27. doi:10.5951/AT.31.5.0027

Swan, P., & Marshall, L. (2010). Revisiting mathematics manipulative materials. *Australian Primary Mathematics Classroom, 15*(2), 13–19. https://files.eric.ed.gov/fulltext/EJ891801.pdf

Swars, S. L., Smith, S. Z., Smith, M. E., & Hart, L. C. (2009). A longitudinal study of effects of a developmental teacher preparation program on elementary prospective teachers' mathematics beliefs. *Journal of Mathematics Teacher Education, 12*(1), 47–66. doi:10.1007/s10857-008-9092-x doi:10.100710857-008-9092-x

Sztajn, P. (2003). Adapting reform ideas in different mathematics classrooms: Beliefs beyond mathematics. *Journal of Mathematics Teacher Education, 6*(1), 53–75. doi:10.1023/A:1022171531285

Tabulawa, R. (1997). Pedagogical classroom practice and the social context: The case of Botswana. *International Journal of Educational Development, 17*(2), 189–204. doi:10.1016/S0738-0593(96)00049-1

Tabulawa, R. (2003). International aid agencies, learner-centred pedagogy and political democratisation: A critique. *Comparative Education, 39*(1), 7–26. doi:10.1080/03050060302559

Tang, G., El Turkey, H., Cilli-Turner, E., Savic, M., Karakok, G., & Plaxco, D. (2017). Inquiry as an entry point to equity in the classroom. *International Journal of Mathematical Education in Science and Technology, 48*(sup1), S4-S15.

Tavil, Z. M. (2014). The effect of self reflections through electronic journals (e-journals) on the self efficacy of pre-service teachers. *South African Journal of Education, 34*(1), 1–20. doi:10.15700/201412120931

Taylor, C., & Lee, J. S. (2021). Ready, set, launch! -The engineering cycle for productive struggle. *Mathematics Teacher: Learning and Teaching PK-12, 114*(2), 117-124.

Taylor, M. P. (2002). Implementing the standards: Keys to establishing positive professional inertia in preservice mathematics teachers. *School Science and Mathematics, 102*(3), 137–142. doi:10.1111/j.1949-8594.2002.tb17907.x

Tekalmaz, G. (2019). Teacher reviews about reformed high school mathematics curriculum. *Kocaeli University Education Journal, 2*(1), 35–47. doi:10.33400/kuje.548562

Tekkumru-Kisa, M., Stein, M. K., & Doyle, W. (2020). Theory and research on tasks revisited. *Educational Researcher, 49*(8), 606–617. doi:10.3102/0013189X20932480

The Royal Society. (2001). *Teaching and learning geometry 11-19: Report of a Royal Society / Joint Mathematical Council working group*. The Royal Society. https://royalsociety.org/-/media/Royal_Society_Content/policy/publications/2001/9992.pdf

Thompson, D. M., & Gay, A. S. (2020). Fact strategy instruction + meaningful discourse = student understanding. *Mathematics Teacher: Learning and Teaching PK-12, 113*(7), 558-564.

Thompson, G., & Cook, I. (2014). Becoming-topologies of education: Deformations, networks and the database effect. In *Discourse: Studies in the Cultural Politics of Education.* Taylor & Francis.

Thomson, M. M., DiFrancesca, D., Carrier, S., & Lee, C. (2017). Teaching efficacy: Exploring relationships between mathematics and science self-efficacy beliefs, PCK and domain knowledge among preservice teachers from the United States. *Teacher Development, 21*(1), 1–20. doi:10.1080/13664530.2016.1204355

Thurtell, E., Forrester, T., & Chinnappan, M. (2019). Building conceptual knowledge of fraction operations among pre-service teachers: Effect of a representation-based approach within a teacher education program. *Mathematics Teacher Education and Development, 1*, 100–124.

Tindall-Ford, S., & Sweller, J. (2006). Altering the modality of instructions to facilitate imagination: Interactions between the modality and imagination effects. *Instructional Science, 34*(4), 343–365. doi:10.100711251-005-6075-5

Tobin, K. (2014). Transforming science education by expanding teacher and student collaboration. In A. L. Tan, C. L. Poon, & S. S. L. Lim (Eds.), *Inquiry into the Singapore science classroom* (pp. 47–66). Springer Singapore. doi:10.1007/978-981-4585-78-1_3

Toles, M., Colón-Emeric, C., Naylor, M. D., Barroso, J., & Anderson, R. A. (2016). Transitional care in skilled nursing facilities: A multiple case study. *BMC Health Services Research, 16*(1), 1–14. doi:10.118612913-016-1427-1 PMID:27184902

Toulmin, S. (1958). *The uses of argument.* Cambridge University Press.

Toulmin, S., Rieke, R., & Janik, A. (1984). *An introduction to reasoning* (2nd ed.). Macmillan.

Towers, J. (2012). Administrative supports and curricular challenges: New teachers enacting and sustaining inquiry in schools. *Canadian Journal of Education/Revue canadienne de l'éducation, 35*(1), 259-278.

Tschannen-Moran, M., & Hoy, A. W. (2001). Teacher efficacy: Capturing an elusive construct. *Teaching and Teacher Education, 17*(7), 783–805. doi:10.1016/S0742-051X(01)00036-1

Tschannen-Moran, M., Hoy, A. W., & Hoy, W. K. (1998). Teacher efficacy: Its meaning and measure. *Review of Educational Research, 68*(2), 202–248. doi:10.3102/00346543068002202

Tversky, B. (2005). Functional significance of visuospatial representations. In P. Shah & A. Miyake (Eds.), *The Cambridge handbook of visuospatial thinking* (pp. 1–34). Cambridge University Press. doi:10.1017/CBO9780511610448.002

UNESCO. (2012). *Challenges in basic mathematics education.* UNESCO. https://unesdoc.unesco.org/images/0019/001917/191776e.pdf

UNESCO. (2016). *School resources and learning environments in Africa: Key results from a regional survey on factors affecting quality of education.* UNESCO. http://uis.unesco.org/sites/default/files/school-resources-and-learning-environment-in-africa-2016-en/school-resources-and-learning-environment-in-africa-2016-en.pdf

Usiskin, Z. (1987). Resolving the continuing dilemmas in school geometry. In M. M. Lindquist & A. P. Shulte (Eds.), *Learning and Teaching Geometry, K12: 1987 Yearbook* (pp. 17–31). National Council of Teachers of Mathematics.

Uttal, D. H., DeLoache, J. S., & Scudder, K. V. (1997). Manipulatives as symbols: A new perspective on the use of concrete objects to teach mathematics. *Journal of Applied Developmental Psychology, 18*(1), 37–54. doi:10.1016/S0193-3973(97)90013-7

Uzun, M. S., Yüksel, N. S., & Dost, Ş. (2013). Investigation of pre-service teachers' mathematics teaching efficacy beliefs in terms of their reflective thinking tendencies. *Procedia: Social and Behavioral Sciences*, *106*, 1595–1602. doi:10.1016/j.sbspro.2013.12.180

Vaijayanti, K., Suma, M. N., & Mondal, A. (2016). *The impact of Akshara Ganitha—A longitudinal study 2012– 13 to 2014–15.* Akshara Foundation. https://akshara.org.in/en/resources-research/reports/

Vale, I., Pimentel, T., Cabrita, I., Barbosa, A., & Fonseca, L. (2012). Pattern problem solving tasks as a mean to foster creativity in mathematics. In T. Tso (Ed.), *Proceedings of the 36th Conference of the International Group for the Psychology of Mathematics Education* (Vol. 4, pp 171-178). PME.

Valentine, K. D., & Bolyard, J. (2019). Lived moments of shifts in prospective elementary teachers' mathematical learning. *Journal for Research in Mathematics Education*, *50*(4), 436–463. doi:10.5951/jresematheduc.50.4.0436

Van de Walle, J. A., Lovin, L. H., Karp, K. S., & Bay-Williams, J. M. (2014). Teaching student-centered mathematics: Developmentally appropriate instruction for grades pre-K–2 (Vol. 1). Teaching Student-Centered Mathematics Series. Pearson Higher Ed.

van de Walle, J. A., Karp, K. S., & Bay-Williams, J. M. (2010). Geometric thinking and geometric concepts. In *Elementary and middle school mathematics teaching developmentally* (7th ed., pp. 402–433). Pearson Education. http://xn--webducation-dbb.com/wp-content/uploads/2019/09/John-A.-Van-de-Walle-Karen-S.-Karp-Jennifer-M.-Bay-Williams-Elementary-and-Middle-School-Mathematics_-Teaching-Developmentally-Pearson-2012.pdf

Van de Walle, J. A., Karp, K. S., & Bay-Williams, J. M. (2019). *Elementary and middle school mathematics: Teaching developmentally* (10th ed.). Pearson Education, Inc.

van den Heuvel-Panhuizen, M. (2019). Didactics of mathematics in the Netherlands. In W. Blum, M. Artigue, M. A. Mariotti, R. Sträßer, & M. van den HeuvelPanhuizen (Eds.), European traditions in didactics of mathematics (pp. 57–94). Springer. doi:10.1007/978-3-030-05514-1_3

van den Heuvel-Panhuizen, M. (2003). The didactical use of models in realistic mathematics education: An example from a longitudinal trajectory on percentage. *Educational Studies in Mathematics*, *54*(1), 9–35. doi:10.1023/B:EDUC.0000005212.03219.dc

van den Heuvel-Panhuizen, M., & Drijvers, P. (2014). Realistic Mathematics Education. In S. Lerman (Ed.), *Encyclopedia of mathematics education* (pp. 521–525). Springer Netherlands. doi:10.1007/978-94-007-4978-8_170

Van den Heuvel-Panhuizen, M., & Fosnot, C. T. (2001). Assessment of mathematics achievements: Not only the answers count. In M. van den Heuvel-Panhuizen (Ed.), *Proceedings of the 25th Conference of the International Group for the Psychology of Mathematics Education* (Vol. 4, RR277). PME.

van Es, E. A., Cashen, M., Barnhart, T., & Auger, A. (2017). Learning to notice mathematics instruction: Using video to develop preservice teachers' vision of ambitious pedagogy. *Cognition and Instruction*, *35*(3), 165–187. doi:10.1080/07370008.2017.1317125

van Es, E. A., & Sherin, M. G. (2021). Expanding on prior conceptualizations of teacher noticing. *ZDM Mathematics Education*, *53*(1), 17–27. doi:10.100711858-020-01211-4

Vanoli, L., & Luebeck, J. (2021). Examining errors and framing feedback. *Mathematics Teacher: Learning and Teaching PK-12, 114*(8), 616-623.

Verstappen, P. (1994). Het dogma: Van de leefwereld naar de wiskundewereld [The dogma: From the daily life world to the world of mathematics]. *Tijdschrift Voor Didactiek der Bètawetenschappen, 12*(2), 104–129. http://www.fi.uu.nl/publicaties/literatuur/199402-verstappen.pdf

Vi-Nhuan, L., Stecher, B., Lockwood, J. R., Hamilton, L. S., Robyn, A., Williams, V., Ryan, G., Kerr, K. A., Mártinez, J., & Klein, S. (2006). *Improving mathematics and science education: A longitudinal investigation of the relationship between reform-oriented instruction and student achievement.* RAND.

von Suchodoletz, A., Jamil, F. M., Larsen, R. A., & Hamre, B. K. (2018). Personal and contextual factors associated with growth in preschool teachers' self-efficacy beliefs during a longitudinal professional development study. *Teaching and Teacher Education, 75*, 278–289. doi:10.1016/j.tate.2018.07.009

Vos, P., & Bos, K. (2001). Trends (1995-1999) in performances of Dutch grade 8 students in TIMSS against the background of the realistic mathematics curriculum. In M. van den Heuvel-Panhuizen (Ed.), *Proceedings of the 25th Conference of the International Group for the Psychology of Mathematics Education* (Vol. 4, RR045). PME.

Vula, E., Avdyli, R., Berisha, V., Saqipi, B., & Elezi, S. (2017). The impact of metacognitive strategies and self-regulating processes of solving math word problems. *International Electronic Journal of Elementary Education, 10*(1), 49–59. doi:10.26822/iejee.2017131886

Ward, C. (2001). Under construction: On becoming a constructivist in view of the Standards. *Mathematics Teacher, 94*(2), 94–96. doi:10.5951/MT.94.2.0094

Warshauer, H. K. (2015). Productive struggle in middle school mathematics classrooms. *Journal of Mathematics Teacher Education, 18*(4), 375–400. doi:10.100710857-014-9286-3

Warshauer, H. K. (2015). Strategies to support productive struggle. *Mathematics Teaching in the Middle School, 20*(7), 390–393. doi:10.5951/mathteacmiddscho.20.7.0390

Warwick, P., Vrikki, M., Vermunt, J. D., Mercer, N., & van Halem, N. (2016). Connecting observations of student and teacher learning: An examination of dialogic processes in Lesson Study discussions in mathematics. *ZDM, 48*(4), 555–569. doi:10.100711858-015-0750-z

Washor, E., & Mojkowski, C. (2006). Creating new steps: Innovating from the edge to the middle. *Phi Delta Kappan, 87*(10), 735–739. doi:10.1177/003172170608701006

Wasserman, N. H., Weber, K., Fukawa-Connelly, T., & Mejía-Ramos, J. P. (2020). Area-preserving transformations: Cavalieri in 2D. *Mathematics Teacher: Learning and Teaching PK-12, 113*(1), 53-60.

Watson, A., Mason, J., & Mason, J. (2006). *Mathematics as a constructive activity: Learners generating examples.* Routledge. doi:10.4324/9781410613714

Wawro, M., Rasmussen, C., Zandieh, M., Sweeney, G. F., & Larson, C. (2012). An inquiry-oriented approach to span and linear independence: The case of the magic carpet ride sequence. *PRIMUS: Problems, Resources, and Issues in Mathematics Undergraduate Studies, 22*(8), 577–599. doi:10.1080/10511970.2012.667516

Wearne, D., & Hiebert, J. (1988). A cognitive approach to meaningful mathematics instruction: Testing a local theory using decimal numbers. *Journal for Research in Mathematics Education, 19*(5), 371–384. doi:10.2307/749172

Wegerif, R., & Scrimshaw, P. (Eds.). (1997). *Computers and talk in the primary classroom.*

Weiss, I. R., & Pasley, J. D. (2004). What is high-quality instruction? *Educational Leadership, 61*(5), 24.

Wenger, E., McDermott, R., & Snyder, W. M. (2002b). Seven principles for cultivating communities of practice. *Cultivating Communities of Practice: A Guide to Managing Knowledge, 4.*

Wenger, E. (1998). *Communities of practice: Learning, meaning, and identity*. Cambridge University Press. doi:10.1017/CBO9780511803932

Wenger, E., McDermott, R. A., & Snyder, W. (2002a). *Cultivating communities of practice: A guide to managing knowledge*. Harvard Business Press.

Wheatley, G. H. (1992). The role of reflection in mathematics learning. *Educational Studies in Mathematics, 23*(5), 529–541. doi:10.1007/BF00571471

Whitenack, J., & Yackel, E. (2002). Making mathematical arguments in the primary grades: The importance of explaining and justifying ideas. *Teaching Children Mathematics, 8*(9), 524–528. doi:10.5951/TCM.8.9.0524

Widder, M., & Gorsky, P. (2013). How students solve problems in spatial geometry while using a software application for visualizing 3D geometric objects. *Journal of Computers in Mathematics and Science Teaching, 32*(1), 89–120. https://eric.ed.gov/?q=solve+AND+in&pg=11&id=EJ1006172

Wilder, S. (2015). Classroom challenge: A 3D snapshot of student learning in mathematics. *The Clearing House: A Journal of Educational Strategies, Issues and Ideas, 88*(3), 77–84. doi:10.1080/00098655.2015.1023244

Wilkerson, T., Kerschen, K., & Shelton, R. (2018). PreService teachers' critical connections to effective mathematical teaching practices: An instructional approach using vignettes. *Action in Teacher Education, 40*(4), 358–373. doi:10.1080/01626620.2018.1512430

Wilson, P. H., & Downs, H. A. (2014). Supporting mathematics teachers in the common core implementation. *AASA Journal of Scholarship and Practice, 11*(1), 38–47.

Wilson, P. S. (1986). Feature frequency and the use of negative instances in a geometric task. *Journal for Research in Mathematics Education, 17*(2), 130–139. doi:10.2307/749258

Wood, M. B., Sheldon, J., Felton-Koestler, M. D., Oslund, J., Parks, A. N., Crespo, S., & Featherstone, H. (2019). 8 teaching moves supporting equitable participation. *Teaching Children Mathematics, 25*(4), 218–223. doi:10.5951/teacchilmath.25.4.0218

Wood, T. (1999). Creating a context for argument in mathematics class. *Journal for Research in Mathematics Education, 30*(2), 171–191. doi:10.2307/749609

Wood, T., Cobb, P., & Yackel, E. (1991). Change in teaching mathematics: A case study. *American Educational Research Journal, 28*(3), 587–616. doi:10.3102/00028312028003587

Wyatt, M. (2016). "Are they becoming more reflective and/or efficacious?" A conceptual model mapping how teachers' self-efficacy beliefs might grow. *Educational Review, 68*(1), 114–137. doi:10.1080/00131911.2015.1058754

Xenofontos, C., & Kyriakou, A. (2016). Prospective elementary teachers' talk during collaborative problem solving. In C. Csikos, A. Rausch, & J. Szitanyi (Eds.), *Proceedings of the 40th Conference of the International Group for the Psychology of Mathematics Education* (Vol. 4, pp. 395-402). PME.

Yackel, E., & Cobb, P. (1996). Sociomathematical norms, argumentation, and autonomy in mathematics. *Journal for Research in Mathematics Education, 27*(4), 458–477. doi:10.5951/jresematheduc.27.4.0458

Yeager, D. S. (2012). *Productive persistence: A practical theory of community college student success.* Paper presented at the 2012 Annual Meeting for the American Educational Research Association, Vancouver, Canada.

Yee, S. P., Roy, G. J., & Graul, L. (2020). Conditional reasoning online with mastermind. *Mathematics Teacher: Learning and Teaching PK-12, 113*(5), 390-396.

Yeh, A., & Nason, R. A. (2004). Toward a semiotic framework for using technology in mathematics education: The case of learning 3d geometry. *ICCE2004 Proceedings,* 1–10.

Yeo, K. K. J., Goh, S., & Koh, E. Z. F. (2005). Exploring plan and elevation geometry with prodesktop. *Journal of Science and Mathematics Education, 28*(1), 125–137. https://repository.nie.edu.sg/bitstream/10497/15471/1/JSME-SEA-28-1-125.pdf

Yin, R. K. (2017). Case study research and applications: Design and methods. *Sage (Atlanta, Ga.).*

Yuan, F., & Evans, D. (2018) *The working conditions of teachers in low- and middle-income countries.* World Bank. https://riseprogramme.org/sites/default/files/inline-files/Yuan.pdf

Zambak, V. S., & Magiera, M. T. (2020). Supporting grades 1–8 pre-service teachers' argumentation skills: Constructing mathematical arguments in situations that facilitate analyzing cases. *International Journal of Mathematical Education in Science and Technology, 51*(8), 1196–1223. doi:10.1080/0020739X.2020.1762938

Zapatera, A., & Callejo, M. L. (2013). Pre-service primary teachers' noticing of students' generalization process. In A. M. Lindmeier, & A. Heinze (Eds.), *Proceedings of the 37th Conference of the International Group for the Psychology of Mathematics Education* (Vol. 4, pp. 425-432). PME.

Zawojewski, J. (2008). The banner problem [Class handout]. Illinois Institute of Technology., MSED 501.

Zee, M., & Koomen, H. M. Y. (2016). Teacher Self-Efficacy and Its Effects on Classroom Processes, Student Academic Adjustment, and Teacher Well-Being: A Synthesis of 40 Years of Research. *Review of Educational Research, 86*(4), 981–1015. doi:10.3102/0034654315626801

About the Contributors

Ozgul Kartal is an assistant professor at the University of Wisconsin-Whitewater. Her major area of research is the enhancement of teaching and learning in order to make mathematics accessible to a broader range of students by means of problem-solving and mathematical modeling, and implementation of effective mathematics teaching practices.

Gorjana Popovic Dr. Popovic is a Senior Lecturer in the Department of Applied Mathematics at the Illinois Institute of Technology. She received her PhD in Mathematics Education at the Illinois Institute of Technology. Her current interests include effective teaching practices, specifically planning for and implementing high quality tasks, and the role of language in teaching and learning of mathematics.

Susie Morrissey is an assistant professor at Mercer University. Her areas of interest center on effective teaching of mathematics and computer science. Research projects with colleagues across Mercer and at other universities are aimed at improving future opportunities for students through effective, equitable teacher education.

* * *

Tamar Apel Campo is an independent researcher, Head of the CimeH Institute in Israel. (Centre for investigation of mathematics with a humanistic approach). The Centre research scope, under her direction, includes the follow up of the implementation of the Teaching Mathematics with a Humanistic Approach program in Israel, Latin America and India. This includes the Let's Talk Math program designed for Special Students in Mumbai, which is offering the needed platform for successful mathematics learning. The doctoral thesis, The Relational Theory of Teaching and Learning Mathematics presents the Track Methodology, with a three dimensional model of research, includes dynamic interviews, analytical multi-leveled research and complex system analysis. The continuous follow up of the findings and their implementation are the essence of present activities in the educational field, leaving some time to grow vegetables in the organic garden in her house and enjoy the pleasure of family meetings.

Katherine Baker, Ed.D., is an Assistant Professor at Elon University where she supports prospective teachers in their mathematics education and instruction. She collaborates with prospective teachers, practicing teachers, and teacher educators to explore students' mathematical thinking and how to use that thinking in instructional decision making.

Rebecca Borowski is an assistant professor of mathematics education at Western Washington University. She earned her PhD in Mathematics Education from Indiana University in 2020. Her experience includes teaching kindergarten, first grade, and fifth grade, working as an instructional coach, and teaching undergraduate education majors. Her research interests include elementary teacher education and student thinking about quantities and measurement. She lives in Washington and enjoys spending time with her daughter.

Gamze Bulut graduated from Galatasaray University in Turkey in 2013. During her third year of college, she studied at the University of Sorbonne 6 in Paris, France. She received her Master's degree in Mathematics Education from Brooklyn College, the City University of New York, in 2017. After completing her Master's degree, Ms. Bulut started to work as a math teacher and is currently teaching for middle school in New York, NY.

Raewyn Eden has been a Lecturer in Mathematics Education at Massey University in Aotearoa New Zealand since 2016. She was awarded her PhD in 2019 by Te Herenga Waka Victoria University of Wellington and was made a Fellow of the Higher Education Authority in 2021. Drawing on her professional practice as a classroom teacher and tertiary educator, prevalent themes in her work include how collaboration can promote teacher learning, and issues of equitable educational opportunity. She is particularly interested in how teachers can be supported to engage in ambitious teaching that promotes rich learning for diverse (all) students, especially, but not only, in primary (elementary) mathematics classrooms. Raewyn works with pre-service teachers in Massey's Post Graduate Diploma of Learning and Teaching; facilitates professional learning and development in primary schools; and coordinates the Post Graduate Certificate in Tertiary Teaching and Learning.

Anthony Gabriele is a professor of Educational Psychology at the University of Northern Iowa. He received his Ph.D. from the University of Pittsburgh, and was awarded a James S. McDonnell Postdoctoral Fellowship in Cognitive Studies for Educational Practice which he completed at UCLA. His program of research has focused on the motivational and metacognitive processes involved in teacher and student learning, with a particular emphasis on learning within reform-oriented mathematics contexts. Over his career, Dr. Gabriele's research interests have led to several collaborations with math educators on topics including computational fluency, measurement estimation, and word problem-solving. Most recently, Dr. Gabriele's research has focused on studying preservice teachers' dispositions towards using research findings to inform and improve classroom practice.

Joan Gujarati, Ed.D., is an Assistant Professor in the Department of Curriculum and Instruction at the State University of New York (SUNY) at Oswego. Her prior positions have included Director of the Elementary Education Master of Arts in Teaching (MAT) Program at Brown University and Assistant Professor and Associate Dean for Accreditation and Technology at Manhattanville College. Dr. Gujarati received her Ed.D. in Curriculum and Teaching from Teachers College, Columbia University. Prior to receiving her doctorate, she was an elementary school teacher and a Math Teacher Leader in Massachusetts. Dr. Gujarati's research interests include early childhood and elementary mathematics education; teacher beliefs and identity; teacher quality, effectiveness, and retention; professional development; and curriculum development. She has published in peer-reviewed journals such as the Journal of Mathematical Behavior, Journal of Mathematics Education at Teachers College, Journal of Urban Learning, Teaching and Research, The Educational Forum, Kappa Delta Pi Record, New Teacher Advocate, i.e.: inquiry in education, and American Secondary Education.

Elana Joram is a professor of Educational Psychology at the University of Northern Iowa. She received her doctorate from the University of Toronto in Applied Cognitive Science. She has studied mathematical cognition and instruction, publishing articles on the numeracy demands of texts, measurement estimation, and the use of reference point benchmarks as a strategy to enhance estimation skills. In a second line of research, Dr. Joram has studied teachers' epistemic beliefs, particularly in relation to teachers' use of educational research for their practice. Her current research project investigates how teachers' beliefs about the certainty of knowledge predicts their receptivity to the use of educational research.

Sumeyra Karatas is a doctoral candidate in mathematics education at the University of Texas Rio Grande Valley. She received her Master`s degree from North American University in curriculum and instruction in math education in 2017. She has been working as a mathematics teacher since 2012. She has taught middle school, high school, and dual credit courses. She is recognized as the teacher of the year in her school district. Her research interest is effective mathematics teaching practices and meaningful mathematical learning.

Veysel Karatas is a doctoral candidate in mathematics education at the University of Texas Rio Grande Valley. He received his Master's degree in Master of education in curriculum and instruction in mathematics education at North American University in 2017. He is a certified mathematics teacher in grades (4-12). He worked in charter public schools as a mathematics teacher for nine years. He has been the district department head in mathematics last three years. He is working on his dissertation titled "Impact of using GeoGebra on students' perceptions and achievement in learning geometric transformations."

Brooke Krejci is a professor of Mathematics at the University of Wisconsin, River Falls. She received her doctorate in Curriculum and Instruction with an emphasis in Mathematics Education from the University of Northern Iowa. As a former middle school and high school mathematics teacher, her research focuses on improving instructional practices for preservice and practicing teachers. Specifically, Dr. Krejci has studied developing mathematics teaching efficacy in preservice elementary teachers. Her other research interests include developing high school geometry students' proof writing through peer critiquing and revising.

Robert (Bob) London is a professor emeritus at California State University, San Bernardino, and program coordinator for the MA in Holistic and Integrative Education. A major professional interest is writing and developing process-oriented mathematics curriculum, especially a curriculum of nonroutine problems that focuses on the process of problem solving in mathematics, and how that process can be applied to solving significant and meaningful problems in the student's life. In addition, I have written about transformative approaches to teacher education based on my work with the MA program in Holistic and Integrative Education.

Marta T. Magiera is an Associate Professor of Mathematics Education in the Department of Mathematical and Statistical Sciences at Marquette University. She is interested in the mathematical cognition of doing mathematics and conducts research in both theoretical and applied aspects of mathematics education and mathematics teacher education. Her primary work focuses on the nature of mathematical thinking, teaching, and learning and addresses practicing and pre-service mathematics teachers' development in the area of mathematical reasoning and proof, problem-solving, and mathematical modeling.

Keith Nabb is an Associate Professor of Mathematics at Piedmont Virginia Community College in Charlottesville, Virginia. He is also the Production Manager for the MathAMATYC Educator, the peer-reviewed journal of the American Mathematical Association of Two-Year Colleges. His current interests include active learning, mathematical knowledge for teaching, and social justice issues in mathematics teaching.

Tabitha Nduku is an Education Specialist and Researcher with over 10 years' experience in the education field. Tabitha provides technical support to projects in low- and middle-income countries in early mathematics, early literacy, play and Social Emotional learning. Her research interests focus on these areas also including Executive function and teacher professional development to support learning in low- and middle-income contexts. Tabitha has an Mphil in Critical Approaches to Children Literature from the University of Cambridge.

Hyejin Park is a Visiting Assistant Professor of Mathematics Education in the Department of Mathematics and Statistics at James Madison University. Her research focuses on mathematical argumentation and proof. More specifically, she is interested in teacher education, particularly in helping (prospective and practicing) teachers develop the knowledge and pedagogical skills that would enable them to strategically support students engaging in argumentation and proving activities in the mathematics classroom.

Linda M. Platas is the chair of the Child and Adolescent Development Department at SF State University. She has participated in the development of child assessment instruments including the Early Grades Math Assessment (EGMA) and the Measuring Early Learning Quality and Outcomes (MELQO) and served as an expert in mathematics and literacy development on many technical and policy groups. She is a member of the Development and Research in Early Math Education (DREME) Network.

Wendi Ralaingita is a specialist in curriculum and instruction with more than 25 years of experience in teaching, training, and teacher education and 15 years of experience in education program management. Dr. Ralaingita provides technical leadership and support to curriculum and materials development and teacher training interventions in mathematics and reading for the primary grades in Sub-Saharan Africa and South Asia.

Kathryn Rupe is an assistant professor of mathematics education at Western Washington University. Before working with preservice teachers, Katie taught middle school mathematics and worked as an elementary math instructional coach in Chicago Public Schools for ten years. Katie holds a BS in Elementary Education from Loyola University Chicago as well as an MS and PhD in Mathematics Education from the Illinois Institute of Technology. Her research interests include middle school students' algebraic reasoning and elementary mathematics teacher education.

Reyhan Safak is a Ph.D. candidate at The Ohio State University (OSU). She received her MA in Curriculum and Instruction from the American College of Education in 2015 while teaching at the

K-12 level. In 2018, she was hired as Graduate Assistant of the Office of Accreditation, Placement, and Licensure for OSU when she started the Ph.D. program in STEM. She has worked as a university supervisor at OSU since 2020. She writes and presents widely on issues of productive struggle, early algebraic reasoning, and clinical interviewing.

İpek Saralar-Aras is a national education expert at the Ministry of National Education in Turkey since July 2020. She completed her undergraduate education in B.Sc. Elementary Mathematics Education at the Middle East Technical University, with an M.Sc. degree in Elementary Science and Mathematics Education from the same university in Turkey. She is a qualified mathematics teacher, who has worked, volunteered and trained in numerous schools in Turkey and England. She has an MA in Learning, Technology and Education and a Ph.D. degree in Education from the University of Nottingham in England, the United Kingdom. She worked as a research assistant at the University of Nottingham's Centre for Research in Mathematics Education in 2019. Her current research mainly focuses on designing technology-enhanced lessons, improving middle school students' maths understanding, teacher education, geometry and educational technologies. She is a member of several educational hubs including EERA, EARLI, BSRLM, MCLS, and AERA.

Yasmin Sitabkhan is a Senior Mathematics Education Researcher at RTI International. She provides technical expertise to governments on early childhood and primary projects focusing on early mathematics learning in sub-Saharan Africa, South Asia, North Africa, and Central and South America. Her research interests center around uncovering a deeper understanding of teaching and learning mathematics in low- and middle-income contexts.

Vecihi Serbay Zambak is an Assistant Professor of Mathematics Education in the Department of Curriculum and Instruction at Monmouth University. In his research, he examines pre-service elementary and secondary teachers' mathematical knowledge for teaching argumentation and teacher noticing skills with the use of mathematics-specific technologies.

Index

F

G

H

I

J

K

L

M

N

O

P

Q

R

S

T

V

W

www.ingramcontent.com/pod-product-compliance
Lightning Source LLC
LaVergne TN
LVHW060405300926
849162LV00006B/1351
* 9 7 8 1 7 9 9 8 9 4 2 2 3 *